ᴬN

ONE WEEK LOAN

Structures

A revision of *Structures* by
P Bhatt and H M Nelson

P BHATT

University of Glasgow

LONGMAN

Addison Wesley Longman Limited
Edinburgh Gate
Harlow
Essex CM20 2JE
England

and Associated Companies throughout the World.

Typeset in 10/12.5pt Times by 35

First published 1999

ISBN 0-582-31222-1

British Library Cataloguing-in-Publication Data
A catalogue record for this book is available from the British Library

Library of Congress Cataloging-in-Publication Data
Bhatt, P.
    Structures / Prabhakara Bhatt. — [4th ed.]
        p.   cm.
    Rev. ed. of: Marshall & Nelson's structures / W. T. Marshall ; 3rd ed. rev. by P. Bhatt, H. M. Nelson, 1990.
    Includes bibliographical references and index.
    ISBN 0–582–31222–1
    I. Marshall, W. T. (William Thomas). Marshall & Nelson's structures.   II. Title.
TA645.B43   1999
624.1′71—dc21   98–36456
                                                            CIP

Printed in Malaysia, VVP

Contents

Preface

The object of this book is to produce a comprehensive modern textbook which covers the courses in Solid and Structural Mechanics taught at the first degree level in universities and polytechnics. Nowadays practical 'stress analysis' is done using widely available and highly sophisticated software. This demands that university courses emphasize more the fundamental theoretical basis of the computer-orientated methods used in practice and less the development of manual methods of calculation. The understanding of fundamental principles requires practice using hand calculation. To this end care is taken to use only those methods which, while suitable for hand calculation, at the same time clearly emphasize the fundamental concepts of equilibrium, compatibility, material laws. These methods are then generalized to result in modern computer-orientated methods of 'stress' analysis. This approach avoids the teaching of many different methods which give the appearance of the whole subject being a 'bag of tricks'. Great care is taken to explain the fundamental assumptions made in developing 'engineers' theory of stress analysis' and the implications and limitations of the method.

The book is organized into eleven chapters. The sole approach adopted is to derive the governing differential equation and use it to solve the problem for an element and then use this information to solve a variety of practical problems. Virtual work and energy principles are presented in a self-contained chapter and are not used elsewhere.

At the end of each chapter problems are given for practice, some of which are furnished with fairly detailed solutions while others are provided with answers only. It is hoped that students will find here a book which succeeds in giving a fundamental understanding of the fascinating subject of 'stress analysis'.

This book is partially based on an earlier edition co-authored with Mr H. M. Nelson. The present text incorporates a detailed discussion of statics which does not appear to be taught in schools any more. At the request of several readers, the chapter on Virtual Work and Energy Principles has been greatly expanded with many new examples including calculation of influence line coefficients for any statically indeterminate structure.

The author has endeavoured to make the book as free from theoretical and numerical errors as possible; but he hopes that readers will bring to his attention errors – numerical as well as theoretical and conceptual – and any areas of explanation lacking clarity.

The author would like to express his sincere thanks to his colleague Mr R. W. Watson for drawing the diagrams and to Mrs Tessa Bryden for typing the new material. A solutions manual which includes detailed solutions to the problems in the book has also been prepared.

Thanks are also due to Sheila, Arun and Ranjana for encouragement and moral support.

September 1998, Glasgow

Acknowledgements

The author would like to express his gratitude to the following individuals for discussions, criticisms, encouragement and other help rendered. Mr R. W. Watson, Lecturer in Civil Engineering at Glasgow University, who as ever with skill and patience draws all the diagrams appearing in his books. Dr D. V. Phillips, Senior Lecturer, Dr T. J. A. Agar, Lecturer, both at the Department of Civil Engineering, Glasgow University, for reading and commenting on certain chapters. The late Dr Tahiani, Associate Professor in Civil Engineering at Royal Military College, Kingston, Ontario, Canada, for many stimulating discussions which are reflected in the presentation of topics in many parts of this book. His teachers at National Institute of Engineering, Mysore, Karnataka, India who instilled in him a permanent fascination with the subject of stress analysis.

P. Bhatt, November 1988, Glasgow

Chapter one

Statics and analysis of statically determinate structures

A structure in the field of structural engineering can be defined as a body capable of resisting the applied forces without exceeding an acceptable limit of deformation of one part relative to the other. Structural designers are concerned with the behaviour of structures such as buildings, bridges, dams, aircraft and ships. The fundamental structural principles governing the design of these structures remain the same. The main difference is in the nature of forces which the structure has to resist. For example, a building structure has to resist gravity and wind loading while a bridge structure has to resist primarily vehicular loads, a dam has to resist the loads arising from water pressure and gravity, a ship structure has to resist hydrodynamic forces and an aircraft structure, on the other hand, has to resist aerodynamic forces and so on. Over the centuries different structural forms have evolved for different types of structures. For example, a building structure has essentially beams and columns to resist the applied loads. A bridge structure on the other hand can be constructed from beams, arches, cables and so on. All structures are an assembly of basic elements, some of which are shown in Figs 1.1 and 1.2.

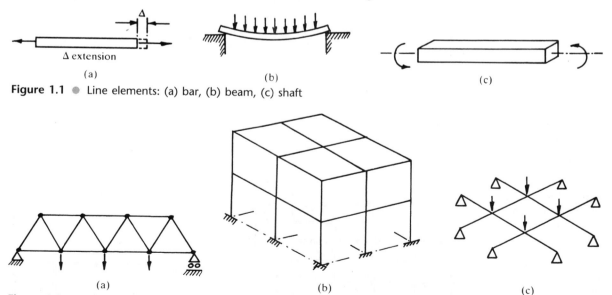

Figure 1.1 ● Line elements: (a) bar, (b) beam, (c) shaft

Figure 1.2 ● Different types of skeletal structures: (a) pin-jointed bridge or roof truss, (b) rigid-jointed building frame, (c) plane grid structure for a roof structure

1.1 ● Line elements

Figure 1.1(a) shows a bar which can resist a compressive or tensile force along its length. Under tensile force, the bar suffers an extension. Figure 1.1(b) shows a beam which is supported at its ends and can resist the loads acting normal to its axis. Under external loads, the beam bends. Similarly, Fig. 1.1(c) shows a shaft which can resist twisting moments acting on the member. Under the action of the twisting moment, the shaft twists. All these three elements – bar, beam and shaft – have their longitudinal dimension, i.e. length, very much larger than their other dimensions. That is why they are called one-dimensional or line elements. Figure 1.2 shows some structures assembled from line elements. The major portion of this book is devoted to the understanding of the behaviour of line elements. Structures assembled from line elements are often called skeletal structures because the load resisting part of the structure is the 'skeleton' on which the external 'skin' to weatherproof the building is fixed.

1.2 ● Continuum elements

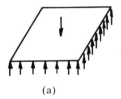

(a)

Figure 1.3(a) shows a slab or a plate which is used to resist loads applied normal to its plane, and which in many ways behaves like a beam. Just as a beam resists loads applied normal to it by bending, the plate also resists the loads by bending; but because of its two-dimensional nature, its behaviour is more complex. Figure 1.3(b) shows a wall which, like a bar, resists compressive and tensile loads applied to it and is a two-dimensional counterpart of a bar. The examples can be multiplied but the important point to note is that for these continuum elements, the length and breadth are very much larger than the thickness. These can therefore be classified as two-dimensional elements. Figure 1.4 shows a typical structure assembled from two-dimensional elements. It is useful to mention in passing that if all the dimensions of an element are approximately equal, then it is best described as a three-dimensional element.

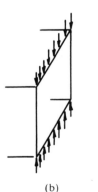

(b)

Figure 1.3 ● Two-dimensional elements: (a) slab or plate, (b) wall

Figure 1.4 ● A building assembled from slab and wall elements: (a) elevation, (b) plan

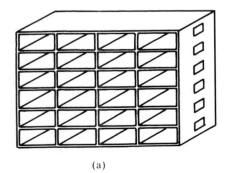

(a)

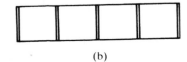

(b)

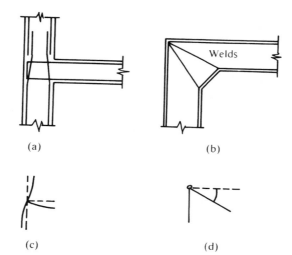

Figure 1.5 ● Joints at beam–column junction: (a) monolithic joint in a reinforced concrete structure, (b) welded joint in a steel structure, (c) joint rotation at a rigid-joint, (d) joint rotation at a pin-joint

1.3 ● Types of joints in a structure

Structures are assembled by joining elements at element intersections. In the case of steel structures, the elements are joined by welding or bolting; in reinforced concrete structures, the joint is made monolithic by ensuring proper disposition of the reinforcement; with timber structures glues, nails and various types of 'connectors' are used to join the members. Figure 1.5(a–b) shows some typical joints connecting beams and columns. Two typical types of joints, stiff or rigid-joint and pin-joint, are commonly used. The stiff or rigid-joint maintains the relative angle between members while the pin-joint allows a change in the angle between the members as shown in Fig. 1.5(c) and (d) respectively. Ideal pin-joints are hardly ever used but for the sake of simplifying force calculation, it is often assumed that the joint is a pin-joint.

Joints are used not only to connect other members but also to join the structure to its foundation. Various types of joints are used for this purpose. Ideal types of 'support joints' are classified as pinned or hinged support, roller support and fixed support. Since the type of support connection used determines the type of load that the joint can resist, it is useful to catalogue the support joints and the types of forces they can resist.

Roller support

Figure 1.6(a) is a schematic representation of a roller support. A roller support resists vertical movement only (in general normal to the roller path) but allows unrestricted horizontal and rotational movement. Therefore at a roller support the only force acting is a vertical force (in general a force normal to the roller path). A simple example of a beam on a roller support is shown in Fig. 1.6(b) where the end of the beam simply rests on a round bar.

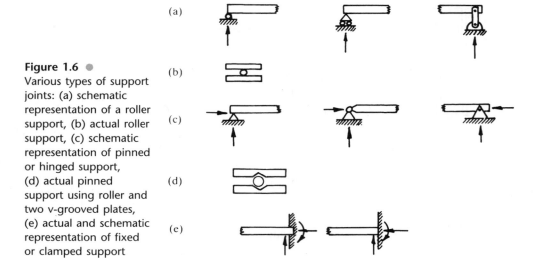

Figure 1.6 ●
Various types of support joints: (a) schematic representation of a roller support, (b) actual roller support, (c) schematic representation of pinned or hinged support, (d) actual pinned support using roller and two v-grooved plates, (e) actual and schematic representation of fixed or clamped support

Hinged or pinned support

Figure 1.6(c) is a schematic representation of a pinned support. A pinned support resists both horizontal and vertical movements but allows unrestricted rotational movement. Therefore at a hinged support the forces acting are a vertical force and a horizontal force. A simple example of a beam resting on a pinned support is shown in Fig. 1.6(d) where the beam rests on a round bar held between the two v-grooved plates.

Fixed support

Figure 1.6(e) is a schematic representation of a fixed support. In practice this is achieved by connecting the member to a heavy foundation. It resists horizontal, vertical and rotational movements. Therefore at a fixed support the forces acting are a vertical force, a horizontal force and a couple.

In practice, particularly in bridge structures, various types of proprietary 'support joints' are used to accommodate the necessary movements of structures.

1.4 ● Statics

A viable structure assembled from elements (one-, two- or three-dimensional as the case may be), and connected by joints (pin-jointed, rigid-jointed, etc.) and properly supported, must satisfy two important criteria. It must be able to resist the applied forces without the structure

* 'failing'
* suffering excessive deformation

In order to design structures to do this, the first step is to understand the laws governing forces. This field is called statics. Statics is defined as the study of bodies subjected to forces being at rest or moving in a straight line at constant velocity. Such a body is said to be in equilibrium. In practice statics is mainly concerned with bodies at rest. The 'bodies' that one is concerned with in structural engineering are buildings, bridges, dams, aircrafts, ships, space vehicles, etc. The science of statics is based on the famous Laws of Motion propounded by Sir Isaac Newton.

1.4.1 Newton's Laws

There are three important laws concerning the motion of bodies subjected to forces. These are:

Law 1 (Law of equilibrium): Every body in a state of rest or in motion in a straight line at uniform velocity continues to be in that state unless subjected to external forces.

Law 2 (Definition of force): Change of motion is proportional to the force impressed and takes place in the direction of the applied force. Mathematically it is expressed as:

$$F = Ma$$

where F = force, M = mass of the particle and a = acceleration of the particle caused by the force.

Law 3 (Action/reaction): Reaction is always equal and opposite to action. That is to say, the actions of two bodies upon each other are always equal and directly opposite.

1.4.2 Concept of force

The action on a body that tends to make it move, for example a push or a pull, is called a *force*. In practice force arises from several causes such as gravitational forces due to the pull of the earth on the body in question, magnetic forces, electrical forces, wave forces, wind and water pressure, and so on. In the field of structural engineering, one of the main forces to be considered is the weight of a material which arises from gravity. The main characteristics which describe a force are

- magnitude
- direction
- point of application

Graphically a force can be represented by a directed line segment AB as shown in Fig. 1.7(a). The length of the line represents, to some scale, the magnitude of the force and the direction of the arrow denotes the direction of the force. Quantities which need both magnitude and direction to describe them fully are called *vectors*. Force is a vector. The basic unit of force is called a Newton. One Newton is the force required to cause an acceleration of 1 metre per second per second on a mass

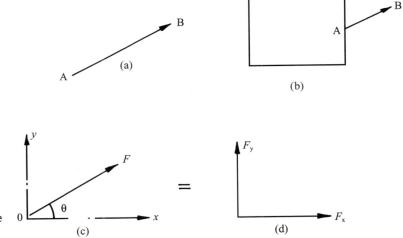

Figure 1.7 ● (a)–(d)
Representation of a force
and its components

of 1 kilogram. The weight of a body arises from the gravitational pull of the earth on the body. The value of acceleration due to gravity is usually denoted by g. Thus a body of mass M has a weight W and the two are related by g, the acceleration due to gravity. This is expressed mathematically as:

$$W = Mg$$

In SI units, $g = 9.897$ metres per second per second (m s^{-2}) on earth. For example if $M = 575$ kg, then $W = 575 \times 9.897 = 5639$ Newtons $= 5.64$ kN.

Clearly the weight of a body is dependent on its mass and the acceleration due to gravity. The acceleration due to gravity varies not only between different planets (for example the earth and the moon) but also between different places on earth. The mass of a body, however, is invariant. It is the variation in the value of acceleration due to gravity which makes the same mass possess different weights.

In structural engineering, the magnitude of a force is normally expressed in terms of kiloNewtons or kN for short.

Figure 1.7(b) shows a force acting on a body. The diagram fully describes the line of action of the force being along AB from A to B; the point of application of the force is at A.

1.4.3   Resolution of a force into its components

Consider a force F whose direction is along a line inclined at an angle θ to the x-axis as shown in Fig. 1.7(c). The positive direction of θ is the smallest angle measured in the anticlockwise direction from the positive x-axis to the line of action of the force. Because of the fact that force is a vector, the effect of the force F on the body is same as the application of two forces F_x and F_y acting along the x- and y-axes respectively, as shown in Fig. 1.7(d). The magnitudes of F_x and F_y are given by

$$F_x = F \cos \theta \text{ and } F_y = F \sin \theta$$

The forces F_x and F_y are called the components of F along the x- and y-axes respectively. Note that

$$F_x^2 + F_y^2 = F^2 \cos^2 \theta + F^2 \sin^2 \theta = F^2 (\cos^2 \theta + \sin^2 \theta)$$

but

$$\cos^2 \theta + \sin^2 \theta = 1$$

therefore

$$F_x^2 + F_y^2 = F^2$$

$$\tan \theta = F_x/F_y$$

$$F = \sqrt{(F_x^2 + F_y^2)} \text{ and } \theta = \tan^{-1}(F_y/F_x)$$

Example 1

Figure 1.8 shows two forces $F_1 = 50$ kN and $F_2 = 90$ kN acting on a body. Calculate the x and y components of the two forces.

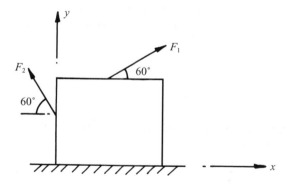

Figure 1.8 ● Forces acting on a structure

Solution
Force F_1: magnitude $= 50$ kN, $\theta = 30°$
$F_x = 50 \cos (30) = 43.30$ kN and $F_y = 50 \sin (30) = 25$ kN
Force F_2: magnitude $= 90$ kN
Note that the direction of the force is inclined at an angle of $(180 - 60) = 120°$ to the direction of the positive x-axis. Therefore $\theta = (180 - 60) = 120°$
$F_x = 90 \cos (120) = -45.0$ kN and $F_y = 90 \sin (120) = 77.94$ kN
The negative sign for F_x indicates that it acts in the negative direction of the x-axis, i.e. from right to left.

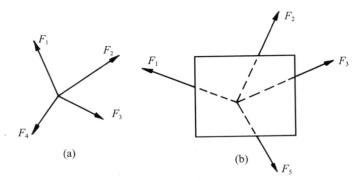

Figure 1.9 ● (a), (b)
Concurrent force system

1.4.4 Concurrent force system

A collection of forces treated as a single group is called a force system. A system of forces is said to be concurrent if the lines of action of all the forces meet at a point. Figure 1.9(a) is an example of a concurrent system of forces where all the forces of the system act at the same point. It is important to appreciate that all the forces *need not act at the same point*, only their lines of action must meet at a common point. Figure 1.9(b) is an example of a concurrent force system where, although the forces act at different points on the body, their lines of action, *when extended*, meet at a single point. Irrespective of whether the forces act at the same point or not, concurrent forces can be considered as equivalent to a single force acting at the common point of intersection. This force is called the resultant of the concurrent force system. The components in the *x*- and *y*-directions of the resultant force are the algebraic sum of the corresponding components due to individual forces.

Example 2

Calculate the magnitude and direction of the resultant of the concurrent force system shown in Fig. 1.10(a).

Solution

F	θ	$F_x = F \cos\theta$	$F_y = F \sin\theta$
90	70	30.78	84.57
50	30	43.30	25.00
40	−50 or 310	25.71	−30.64
		$\Sigma F_x = 99.79$	$\Sigma F_y = 78.93$

The resultant force R therefore has *x*-component = 99.79 kN and *y*-component = 78.93 kN. Therefore

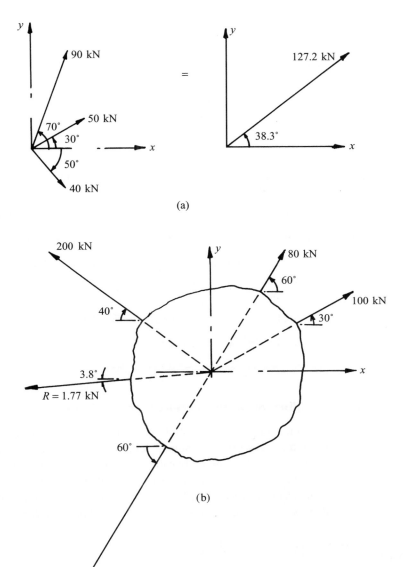

Figure 1.10 ●
(a),(b) Resultant of a
concurrent force system

$$R = \sqrt{(F_x^2 + F_y^2)} = \sqrt{(99.79^2 + 78.93^2)} = 127.23$$

$$\theta = \tan^{-1}(F_y/F_x) = \tan^{-1}(78.93/99.79) = 38.34°$$

Example 3

Calculate the magnitude and direction of the resultant of the concurrent force system shown in Fig. 1.10(b).

Solution

F	θ	$F_x = F \cos \theta$	$F_y = F \sin \theta$
100	30	86.60	50.00
80	60	40.00	69.28
200	$(180 - 40) = 140$	−153.21	128.56
300	$(180 + 60) = 240$	−150.00	−259.81
		$\Sigma F_x = -176.61$	$\Sigma F_y = -11.97$

The resultant force R therefore has x-component $= -176.61$ kN and y-component $= -11.97$ kN. Therefore

$$R = \sqrt{(F_x^2 + F_y^2)} = \sqrt{((-176.61)^2 + (-11.97)^2)} = 177.02 \text{ kN}$$

$$\theta = \tan^{-1}(F_y/F_x) = \tan^{-1}((-11.97)/(-176.61)) = 3.88°$$

Note that since both the x- and y-components of the resultant are negative, the angle of inclination is $\theta = 180 + 3.88 = 183.88$. As a check

$$R \cos \theta = 177.02 \cos (183.88) = -176.61 \text{ kN}$$

$$R \sin \theta = 177.02 \sin (183.88) = -11.97 \text{ kN}$$

1.4.5 Equilibrium of concurrent force systems

Newton's first law of motion states that a particle is in equilibrium (generally meaning at rest), if no resultant force acts on it. Clearly therefore if the resultant force is zero, then the body cannot accelerate and the body is in equilibrium. This important fact can be used to solve very many interesting problems in the field of concurrent force systems. It is worth repeating that for a resultant force to be zero, then *both the x- as well as the y-component must be zero*.

Example 4

A particle is acted on by forces as shown in Fig. 1.11. Determine the magnitude and direction of the unknown force R required to maintain equilibrium.

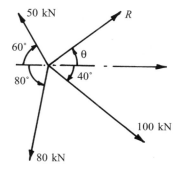

Figure 1.11 ● Forces acting on a particle

Solution Let the magnitude of the unknown force be R and its inclination to the x-axis be θ.

F	θ	$F_x = F \cos \theta$	$F_y = F \sin \theta$
50	$180 - 60 = 120$	-25.00	43.30
80	$80 + 180 = 260$	-13.89	-78.79
100	$-40 + 360 = 320$	76.60	-64.28
R	θ	$R \cos \theta$	$R \sin \theta$
		$\Sigma F_x = 37.71 + R \cos \theta$	$\Sigma F_y = -99.77 + R \sin \theta$

For equilibrium $\Sigma F_x = 37.71 + R \cos \theta = 0$ and $\Sigma F_y = -99.77 + R \sin \theta = 0$. Therefore

$$-37.71 = R \cos \theta \quad \text{and} \quad 99.77 = R \sin \theta$$

Squaring both sides and adding, and remembering that $\cos^2 \theta + \sin^2 \theta = 1$,

$$R^2 = (-37.71)^2 + (99.77)^2 \text{ or } R = 106.66 \text{ kN}$$

$$\theta = \tan^{-1} \{(99.77)/(-37.71)\} = -69.30°$$

Because the y-component is positive and the x-component is negative, the inclination from the x-axis is $180 - 69.30 = 110.71°$

1.4.6 Concept of a free body

A very powerful concept useful in the solution of problems in structural engineering is the concept of a free body. The important idea is that if a body is in equilibrium, then any part of it isolated from the rest of the body is also in equilibrium as long as all the forces acting on that portion *including any interaction forces between that portion and the rest of the body are* included. Consider the example of a particle of weight W held in equilibrium by two strings as shown in Fig. 1.12(a).

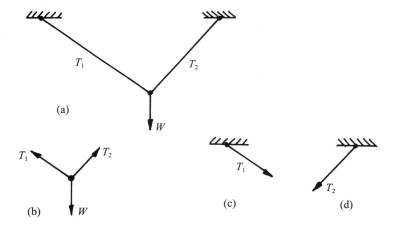

Figure 1.12 ● (a)–(d) Free-body diagrams

Let the tension in the strings be T_1 and T_2. Imagine the particle isolated from the supports but with all the forces acting on it as shown in Fig. 1.12(b). Clearly the forces acting on the particle are its weight W and the tensions in the two strings T_1 and T_2 These three forces must maintain the particle in equilibrium. Note that the tensions in the strings 'pull it up'. However, if the free body for the support regions are considered (Fig. 1.12(c)), then the tensions in the strings 'pull the support down'. As can be clearly seen, because of Newton's third law of motion, 'action and reaction are equal and opposite'.

Example 5

A particle of weight 100 kN is kept in equilibrium by tensions in two strings as shown in Fig. 1.13(a). If the magnitude and direction of the tension in the left string is 80 kN at 45° to the horizontal and the right string is inclined at θ to the horizontal, determine the tension T and the inclination θ of the second string.

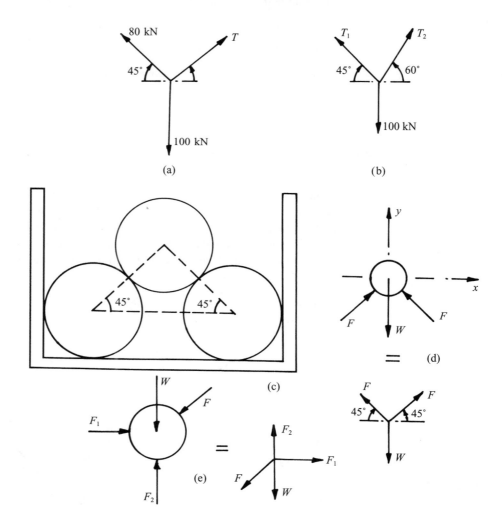

Figure 1.13 ●
(a), (b) A particle held in equilibrium by two strings, (c)–(e) three discs in a box

Solution

F	θ	$F_x = F \cos \theta$	$F_y = F \sin \theta$
100	$180 + 90 = 270$	0.0	-100
80	$180 - 45 = 135$	-56.57	56.57
T	θ	$T \cos \theta$	$T \sin \theta$
		$\Sigma F_x = -56.57 + T \cos \theta$	$\Sigma F_y = -43.33 + T \sin \theta$

For equilibrium $\Sigma F_x = -56.57 + T \cos \theta = 0$ and $\Sigma F_y = -43.33 + T \sin \theta = 0$. Therefore

$$56.57 = T \cos \theta \quad \text{and} \quad 43.33 = T \sin \theta$$

Squaring both sides and adding, and remembering that $\cos^2 \theta + \sin^2 \theta = 1$, $T^2 = (56.57)^2 + (43.33)^2$ or $T = 71.34$ kN and $\theta = \tan^{-1} \{(43.33)/(56.57)\} = 37.45°$

Example 6

A particle of weight 100 kN is kept in equilibrium by tensions in two strings as shown in Fig. 1.13(b). If the magnitude and direction of the tension in the left string is T_1 kN at 45° to the horizontal and the magnitude and direction of the tension in the right string is T_2 kN at 60° to the horizontal, determine the tensions in the strings.

Solution

F	θ	$F_x = F \cos \theta$	$F_y = F \sin \theta$
100	-90	0.0	-100.0
T_1	135	$-0.707T_1$	$0.707T_1$
T_2	60	$0.5T_2$	$0.866T_2$
		$\Sigma F_x = -0.707T_1 + 0.5T_2$	$\Sigma F_y = -100.0 + 0.707T_1 + 0.866T_2$

For equilibrium $\Sigma F_x = 0$ and $\Sigma F_y = 0$. Therefore

$$-0.707T_1 + 0.5T_2 = 0, \quad -100.0 + 0.707T_1 + 0.866T_2 = 0$$

Solving the simultaneous equations, $T_1 = 51.77$ kN, $T_2 = 73.21$ kN

Example 7

Three smooth cylinders of identical diameters are stacked in a box as shown in Fig. 1.13(c). Each cylinder has a weight of 2.5 kN. Calculate the forces exerted by the side cylinders on the middle cylinder. Also calculate the forces exerted on the box by the end cylinders.

Solution Consider the free body consisting of the middle cylinder. The contact between the cylinders is smooth, therefore the direction of the interactive forces at the contact must be directed along the normal to the interface, which in this case is along the radius at the contact surface. Therefore the two equal supporting forces F from the end cylinders is directed along a line connecting the centres of the cylinders as shown in Fig. 1.13(d). Clearly the weight of the cylinder and the supporting forces from the lower cylinders form a system of concurrent forces. The system is in equilibrium. Therefore the resultant of the force system must be zero.

Force	θ	$F_x = F \cos \theta$	$F_y = F \sin \theta$
2.5	−90	0	−2.5
F on the left	45	$0.707F$	$0.707F$
F on the right	$180 - 45 = 135$	$-0.707F$	$0.707F$
		$\Sigma F_x = 0.0$	$\Sigma F_y = -2.5 + 1.414F$

Equating $\Sigma F_y = 0$, $F = 1.768$ kN

In order to calculate the supporting forces from the box, consider the free body diagram of the left end cylinder as shown in Fig. 1.13(e). The cylinder is in equilibrium under the action of four concurrent forces. The forces are the weight of the cylinder equal to 2.5 kN which acts vertically down, the supporting forces from the side of the box labelled F_1 acting horizontally and force F_2 due the support from the base acting vertically upwards and finally the action from the middle cylinder calculated above and equal to 1.768 kN. Note that when considering the middle cylinder, the force of 1.768 kN was 'supporting' the middle cylinder (Fig. 1.13(d)) but when considering the end cylinder, the same force 'presses down' on the end cylinder (Fig. 1.13(e)). This is another instance of the application of the principle, 'action and reaction are equal and opposite'.

Force	θ	F_x	F_y
weight 2.5	−90	0	−2.5
1.768	$180 + 45 = 225^*$	−1.25	−1.25
F_1	0	F_1	0
F_2	90	0	F_2
		$\Sigma F_x = F_1 - 1.25$	$\Sigma F_y = F_2 - 3.75$

* Note: The direction of the force is 225° not 45°.

Equating $\Sigma F_x = 0$, $\Sigma F_y = 0$
$F_1 - 1.25 = 0$, $F_2 - 3.75 = 0$
Therefore $F_1 = 1.25$, $F_2 = 3.75$ kN
The end cylinder pushes the box out with a force of 1.25 kN and pushes down on the base with a force of 3.75 kN.

Example 8

Two smooth discs of identical diameters are stacked in a cavity as shown in Fig. 1.14(a). Each disc has a weight of 1.5 kN. Calculate the forces exerted by the discs on each other as well as on the bottom and side supports.

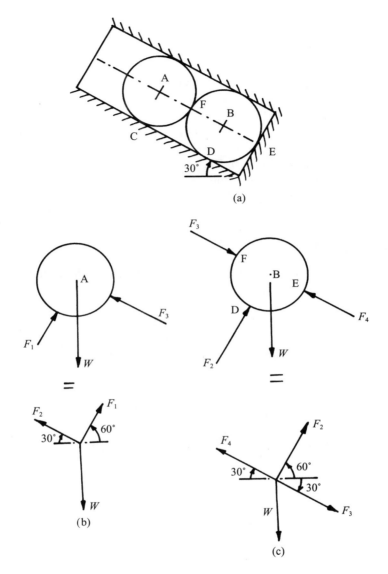

Figure 1.14 ● (a)–(c)
Equilibrium of a set of
discs in a cavity

Solution Let the unknown forces be as shown in Fig. 1.14(b). Considering the free-body
diagrams for each cylinder in turn, as shown in Figs 1.14(b) and 1.14(c), we have

Upper disc

Force	θ	F_x	F_y
Weight 1.5	−90	0	−1.5
F_1	60	$0.5F_1$	$0.866F_1$
F_3	$180 − 30 = 150$	$−0.866F_2$	$0.5F_2$

Equating $\Sigma F_x = 0$, $\Sigma F_y = 0$
$0.5F_1 − 0.866F_3 = 0$, $−1.5 + 0.866F_1 + 0.5F_3 = 0$
Therefore $F_1 = 1.3$ kN, $F_3 = 0.75$ kN

Lower disc

Force	θ	F_x	F_y
Weight 1.5	−90	0	−1.5
F_2	60	$0.5F_2$	$0.866F_2$
$F_3 = 0.75$	−30	0.65	−0.375
F_4	$180 - 30 = 150$	$-0.866F_4$	$0.5F_4$

Equating $\Sigma F_x = 0$, $\Sigma F_y = 0$
$0.65 + 0.5F_2 - 0.866F_4 = 0$, $-1.875 + 0.866F_2 + 0.5F_4 = 0$
Therefore $F_2 = 1.3$ kN, $F_4 = 1.5$ kN

Example 9

For the pin-jointed truss shown in Fig. 1.15, draw the free-body diagrams for each of the joints A to E.

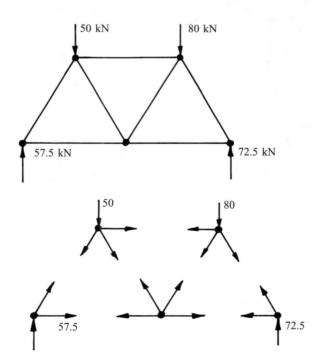

Figure 1.15 ● Free bodies for the joints of a pin-jointed truss

Solution Because the truss is pin-jointed (i.e. each member is pinned at its two ends), the only force that exists is a tensile or compressive force in each member. Assuming that the force in each member is tensile, the free-body diagram for each joint is as shown in Fig. 1.15. Note that because the force in each member is assumed to be tensile, each member exerts a pull on the two joints which it connects.

1.4.7   General technique for the solution of problems

It is clear from the examples in the previous section that the technique for solving problems essentially consists of drawing free-body diagrams and equating ΣF_x and ΣF_y of the forces involved to zero. Because there are only two equations involved, it is important to make sure that there are no more than two unknown forces involved.

1.4.8   Concept of moment

Consider a force F acting on a particle and on a wheel whose axis is at A (Fig. 1.16(a) and (b)). The action of the force on the particle is to make it displace in the direction of the force. However, the action of the force on the wheel is to make it rotate about the axis. Clearly if the force acts at the axis itself, then the force cannot cause any rotation. This simple example shows that the ability of a force to cause rotation depends not only on the magnitude of the force, but also on the distance that the force acts from the axis about which rotation is being considered. As shown in Fig. 1.16(c), there are clearly an infinite number of distances from the axis of rotation to the force. The important question is which distance should be used in assessing the potential of the force to cause rotation. The question can be answered

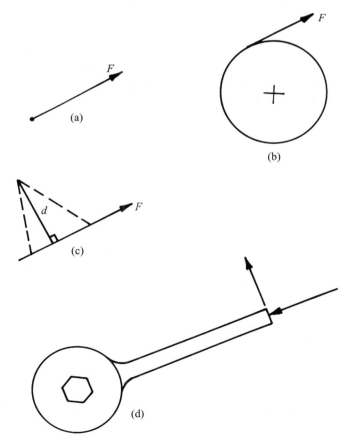

Figure 1.16 ● (a)–(d) Force acting on different positions on a wheel and lever arm of a force

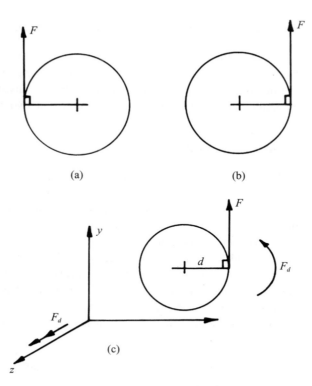

Figure 1.17 ● (a)–(c)
Magnitude and direction
of a moment

by considering the case of a spanner tightening a bolt as shown in Fig. 1.16(d). Clearly if the force is applied along the length of the spanner, then no rotation will be caused. For maximum tightening effect, the force should be applied in a direction such that the component of the force which acts along the length of the spanner which is ineffective is zero. In other words, the force should act perpendicular to the length of the spanner. This simple example demonstrates the fact that the distance to be used is the *perpendicular distance* from the axis of rotation to the line of action of the force.

The moment M of a force F, which is a measure of its potential to cause rotation, is defined as $M = Fd$, the product of the force F and the *perpendicular distance d* from the axis of rotation to the line of action of the force.

The distance d is normally known as *lever arm* or *moment arm*. The magnitude of the moment is given by Fd. Because the moment is a product of a force (normally given in kN) and a distance (normally given in metres), moment is expressed in units of kiloNewton metres or kN m for short. However, it has magnitude as well as direction, as can be seen by considering two cases of a wheel where the same force F is applied to the rim at a distance d as shown in Fig. 1.17(a) and (b). In the first case the wheel rotates in the clockwise direction while in the second case it rotates in the anticlockwise direction. In other words the moment M not only has a magnitude of Fd but also has a direction. Moment is, like force, a vector.

Clearly the direction associated with a moment is the direction of the potential rotation that the moment can cause. Considering the situation shown in Fig. 1.17(c), the positive direction is defined as follows. If the rotation is taking place about the *z*-axis, then using the fingers of the right hand, if the fingers curl in the direction of the potential rotation, and the thumb points to the positive *z*-axis, then that moment is positive. In other words anticlockwise rotation is positive.

In order to distinguish a moment vector from a force vector, it is usual to show a moment vector with a double-headed arrow as opposed to a force which is shown with a single-headed arrow. It is worth repeating the fact that the direction of the moment is not determined by the direction of the force, rather it is determined by the direction of the *rotation* that the force is likely to cause.

Example 10

Calculate the moment of the force F of 20 kN about A as shown in Fig. 1.18(a).

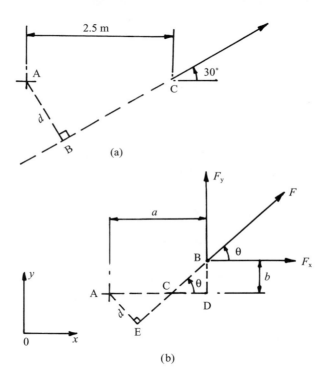

Figure 1.18 ● (a), (b)
Lever arm of a force

Solution The perpendicular distance from A to the line of force is

$$d = 2.5 \sin 30 = 1.25$$

$$M = Fd = 20 \times 1.25 = 25 \text{ kN m}$$

The moment causes anticlockwise rotation and is therefore positive.

Example 11

Show that the moment of a force is equal to the algebraic sum of the moments due to the components of the force.

Solution Consider the force F as shown in Fig. 1.18(b). The line of action of the force is inclined to the x-axis at an angle θ and the point of application of the force has coordinates (a, b) *from the axis of rotation*. From geometry

$$CD = b \cot \theta$$

$$AC = AD - CD = a - b \cot \theta$$

$$d = AC \sin \theta = (a - b \cot \theta) \sin \theta = a \sin \theta - b \cos \theta$$

$$M = Fd = F(a \sin \theta - b \cos \theta)$$

$$= aF \sin \theta - bF \cos \theta = aF_y - bF_x$$

Because the lever arms for F_x and F_y are respectively b and a, and in addition F_x causes clockwise rotation and is thus negative and F_y causes anticlockwise rotation and is therefore positive, the moment of force F is thus equal to the algebraic sum of the moments due to its components. Note that the dimensions (a, b) are measured from the *axis of rotation*.

Example 12

Calculate the resultant moment about A due to the two forces shown in Fig. 1.19. A force of 30 kN at 60° to the horizontal acts at a point with co-ordinates (12, 10) and another force of 50 kN at 30° to the horizontal acts at a point with co-ordinates (5, 15). The point A about which moments are required has co-ordinates (3, 4).

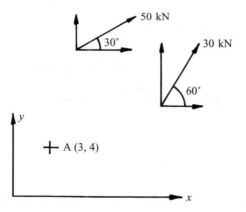

Figure 1.19 ●
Resultant force and
moment of a set of
non-concurrent forces

Solution 30 kN force: acts at 30° to the horizontal with $a = (12 - 3) = 9$, $b = (10 - 4) = 6$
50 kN force: acts at 60° to the horizontal with $a = (5 - 3) = 2$, $b = (15 - 4) = 11$

F	θ	$F_x = F \cos \theta$	$F_y = F \sin \theta$	a	b	$-bF_x$	aF_y
30	60	15.0	25.981	9	6	−90.0	233.83
50	30	43.30	25.0	2	11	−476.3	50.0
						$\Sigma = -566.3$	$\Sigma = 283.83$

$$M = \Sigma(aF_y - bF_x) = -566.3 + 283.83 = -282.47 \text{ kN m}$$

The two forces produce a net moment of 282.47 kN m causing clockwise rotation.

1.4.9 Couples

In the previous examples we considered force systems which gave rise not only to a resultant force but also to a moment about a specified point. Couples are special cases of force system which have zero resultant force but a non-zero moment. A couple is defined as a pair of equal and opposite parallel forces at a distance of d. Figure 1.20 shows such a system. An important property of a couple is that apart from the fact that the resultant force is zero, the magnitude of the moment of the couple about any axis is a constant and is equal to Fd. This can be easily proved. Considering the couple shown in Fig. 1.20, taking moments of the forces about the axis passing through point A,

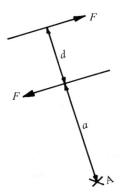

Figure 1.20 ●
Example of a couple

$$M = Fa - F(a + d) = -Fd$$

In other words, the magnitude of the couple is Fd irrespective of the value of a, defining the position of the axis about which the moment of the couple is required.

1.4.10 Plane parallel force systems

In the previous sections attention has been focused on determining the conditions for equilibrium of a concurrent force system. It was shown that for equilibrium of a concurrent system, the resultant force must be zero. It was also shown that a measure of the potential of a force to cause rotation about an axis is given by the moment of that force about the axis.

In this section we will deal with a force system consisting of a set of parallel forces. Consider the set of parallel forces shown in Fig. 1.21(a). First of all let us determine the resultant of the force system. Because all the forces act in the vertical y-direction, ΣF_x is automatically zero and ΣF_y is simply the algebraic sum of the forces. In this case $\Sigma F_y = -10 - 30 + 20 - 60 = -80$ kN.

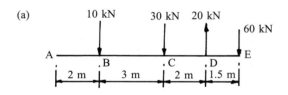

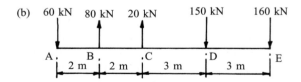

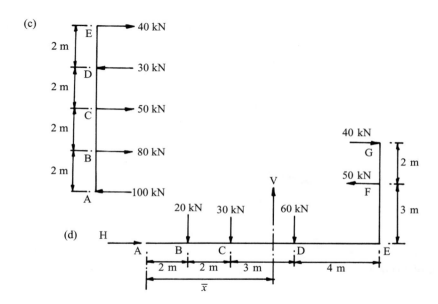

Figure 1.21 ● (a)–(d)
Parallel force systems

Let us also calculate the moment of the force system about A:

$$M = -(10 \times 2) - (30 \times 5) + (20 \times 7) - (60 \times 8.5) = -540 \text{ kN m}$$

If this system is to be in equilibrium, then clearly two conditions have to be satisfied:

- the resultant force must be zero
- the moment of the force system must also be zero about any axis

Clearly the above system is not in equilibrium. In order to maintain equilibrium, then we need to apply an unknown force R at an unknown distance $\bar{x}$ from A.
In this case $\Sigma F_y = -10 - 30 + 20 - 60 + R = 0$, $R = 80$ kN

$$\Sigma M = -(10 \times 2) - (30 \times 5) + (20 \times 7) - (60 \times 8.5) + (R\bar{x}) = 0$$

$$R\bar{x} = 540 \text{ kN m}, \; R = 80 \text{ kN}, \; \bar{x} = 540/80 = 6.75 \text{ m}$$

It should be noted that for equilibrium, it is only necessary to check that the resultant R is zero and that the resultant moment about any axis in the plane is also zero. The position of the axis is dictated by convenience in carrying out arithmetic/algebraic calculations.

As an illustration of the above point, if the moments are calculated about B in Fig. 1.21(a), then

$$\Sigma M = (10 \times 0) - (30 \times 3) + (20 \times 5) - (60 \times 6.5) + (R \times a) = 0$$

$R = 80$ kN as before and $a = 4.75$ m

The resultant R acts at 4.75 m to the right from B or at $(4.75 + 2.0) = 6.75$ m from A. This is the same value as calculated before by taking moments about A.

Example 13

Calculate the magnitude and position of the force R needed to maintain equilibrium of the system shown in Fig. 1.21(b). Check your calculations by taking moments about A and also about C.

Solution In this case $\Sigma F_y = -60 + 80 + 20 - 150 - 160 + R = 0$, $R = 270$ kN.
ΣM about A $= (60 \times 0) + (80 \times 2) + (20 \times 4) - (150 \times 7) - (160 \times 10) + (R \bar{x}) = 0$.
$R \bar{x} = 2410$ kN m, $R = 270$ kN, $\bar{x} = 2410/270 = 8.926$ m to the right of A.
ΣM about C $= (60 \times 4) - (80 \times 2) + (20 \times 0) - (150 \times 3) - (160 \times 6) + (R \bar{x}) = 0$.
$R \bar{x} = 1330$ kN m, $R = 270$ kN, $\bar{x} = 1330/270 = 4.926$ m to the right of C.
Note that it is very important to ensure that the *correct signs* are used when taking moments.

Example 14

Calculate the magnitude and position of the force R needed to maintain equilibrium of the system shown in Fig. 1.21(c). Check your calculations by taking moments about A and also about C.

Solution In this case $\Sigma F_y = 0$. $\Sigma F_x = -100 + 80 + 50 - 30 + 40 + R = 0$, $R = -40$ kN. Negative sign indicates that the resultant force acts from right to left.
ΣM about A $= (100 \times 0) - (80 \times 2) - (50 \times 4) + (30 \times 6) - (40 \times 8) + \{(R = 40) \bar{y}\}$
$= 0$. $R \bar{y} = 500$ kN m, $R = 40$ kN, $\bar{y} = 12.5$ m above A.
ΣM about C $= -(100 \times 4) + (80 \times 2) + (50 \times 0) + (30 \times 2) - (40 \times 4) + \{(R = 40) \bar{y}\} = 0$. $R \bar{y} = 340$ kN m, $R = 40$ kN, $\bar{y} = 8.5$ m above C.

1.4.11 Non-parallel non-concurrent force systems

Concurrent force systems and parallel force systems are special cases of general force systems which occur in practice. A simple example of such a general force system is shown in Fig. 1.21(d).

Example 15

Calculate the magnitude and position of the force V and the magnitude of H in order to maintain equilibrium of the system shown in Fig. 1.21(d).

Solution $\Sigma F_x = 40 - 50 + H = 0$, $H = 10$ kN
$\Sigma F_y = -20 - 30 - 60 + V = 0$, $V = 110$ kN
ΣM about A $= -(20 \times 2) - (30 \times 4) - (60 \times 7) + (V\bar{x}) + (50 \times 3) - (40 \times 5) = 0$
$V\bar{x} = 630$, $\bar{x} = 5.73$ m

1.4.12 Distributed loads

In the previous sections, the loads considered acted at a single point. They are called concentrated loads. In practice we frequently have to deal with distributed loads as well as concentrated loads. For example, the self-weight of a beam is not concentrated at a point but is distributed along its length. If the total weight of the beam is 100 kN and its length is 20 m, then we say that the self-weight of the beam is a uniformly distributed load equal to $100/20 = 5$ kN per metre (or kN m^{-1} for short) of its length.

Another example of distributed load often encountered is hydrostatic pressure due to fluids such as water as shown in Fig. 1.22.

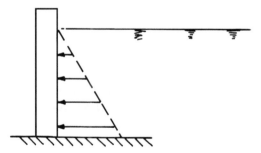

Figure 1.22 ●
Distributed force due to hydrostatic pressure

In dealing with uniformly distributed loads, the idea is to treat them as concentrated loads with the magnitude equal to the intensity of the load times its length and the load acting at the middle of its length.

Example 16

Determine the resultant force and its position from A for the loading shown in Fig. 1.23(a).

Solution   First of all convert the uniformly distributed loads into equivalent concentrated loads.

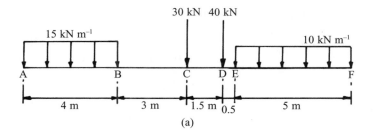

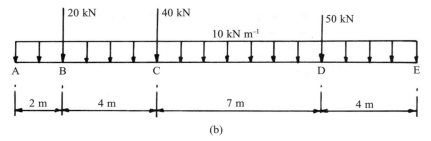

Figure 1.23 ● (a), (b) A system of uniformly distributed and concentrated parallel forces

(a) Load in region AB: Total load = intensity × length = −15 × 4 = −60 kN
It acts at the middle of its length. Therefore the equivalent concentrated load is a load of −60 kN at a distance of 4/2 = 2.0 m from A.

(b) Load in region EF: Total load = intensity × length = −10 × 5 = −50 kN
It acts at the middle of its length. Therefore the equivalent concentrated load is a load of −50 kN at a distance of 5/2 = 2.5 m from E or at a distance of (4 + 3 + 1.5 + 0.5 + 5/2) = 11.5 m from A.

ΣF_y = −(60 due to load on AB) − (30 at C) − (40 at D) − (50 due to load on EF)
        = −180 kN

Resultant load = −180 kN

ΣM about A = −(60 × 2.0) due to load on AB − (30 × 7.0) load at C
            − (40 × 8.5) load at D − (50 × 11.5) due to load on EF
        = −1245.0 kN m

Example 17

Determine the resultant force and its position from A for the loading shown in Fig. 1.23(b).

Solution   First of all convert the uniformly distributed loads into equivalent concentrated loads.

(a) Load in region AE: Total load = intensity × length = −10 × 17 = −170 kN
It acts at the middle of its length. Therefore the equivalent concentrated load is a load of −170 kN at a distance of 17/2 = 8.5 m from A.

ΣF_y = −(170 due to load on AE) − (20 at B) − (40 at C) − (50 at D)

        = −280 kN

Resultant load = −280 kN

ΣM about A = −(170 × 8.5) due to load on AB − (20 × 2.0) load at B
− (40 × 6.0) load at C − (50 × 13.0) load at D

= −2375.0 kN m

1.4.13 Summary of requirements for equilibrium of bodies in two dimensions

It was stated that in the case of particles, they are capable of translating along x- and y-directions only. In order for a particle to be in equilibrium, it is therefore necessary for the resultant force on the particle to be zero. This was expressed by stating that, taking into account all the forces acting on the particle,

$$\Sigma F_x = 0 \quad \text{and} \quad \Sigma F_y = 0$$

In the case of two-dimensional bodies lying in the x–y plane, the body cannot only translate along the x- and y-directions as a particle can but in addition it can also rotate about an axis perpendicular to the x–y plane, namely an axis parallel to the z-axis. For equilibrium, the conditions are

$$\Sigma F_x = 0, \quad \Sigma F_y = 0 \quad \text{and} \quad \Sigma M_z = 0$$

These are the general conditions to be satisfied by the force system acting on a two-dimensional body in the x–y plane.

Example 18

A beam AF loaded by concentrated loads, as shown in Fig. 1.24(a), is held in equilibrium by vertical forces V_A and V_F. Determine the values of these forces.

Solution Because the beam is in equilibrium, the resultant force and the resultant moment of the forces about any axis must be zero. As only vertical forces are involved, the conditions to be satisfied are $\Sigma F_y = 0$, $80 - 50 - 60 - 90 + V_A + V_F = 0$, $V_A + V_F = 120$ kN.
Taking moments about the z-axis passing through A,

$$\Sigma M \text{ about A} = V_A \times 0 + 80 \times 3 - 50 \times 4 - 60 \times 8 - 90 \times 11 + V_F \times 14 = 0$$

$$V_F = 107.86 \text{ kN}, \quad V_A = 12.14 \text{ kN}$$

As a check, taking moments about the z-axis passing through F,

$$-V_A \times 10 - 80 \times 12 + 50 \times 10 + 60 \times 6 + 90 \times 3 + V_F \times 0 = 0$$

$$V_A = 12.14 \text{ kN}$$

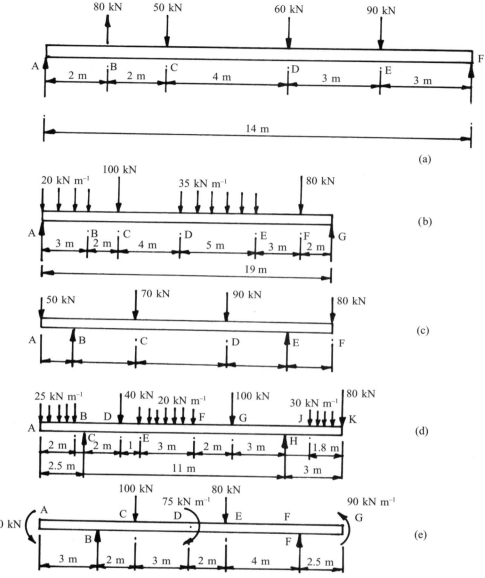

Figure 1.24 ● (a)–(e) A parallel force system acting on a beam

Example 19

A beam AG loaded by concentrated loads and uniformly distributed loads, as shown in Fig. 1.24(b), is held in equilibrium by vertical forces V_A and V_G. Determine the values of these forces.

Solution   Because the beam is in equilibrium, the resultant force and the resultant moment of the forces about any axis must be zero. As only vertical forces are involved, the conditions to be satisfied are $\Sigma F_y = 0$, $-20 \times 3 - 100 - 35 \times 5 - 80 + V_A + V_G = 0$, $V_A + V_G = 415$ kN.

Taking moments about the z-axis passing through A,

$$\Sigma M \text{ about A} = V_A \times 0 - 20 \times 3 \times (3/2) - 100 \times 5 - 35 \times 5 \times (9 + 5/2)$$
$$- 80 \times 17 + V_G \times 19 = 0$$

$V_G = 208.55$ kN, $V_A = 206.45$ kN

As a check, taking moments about the z-axis passing through G,

$$-V_A \times 19 + 20 \times 3 \times (19 - 3/2) + 100 \times 14 + 35 \times 5 \times (5 + 5/2)$$
$$+ 80 \times 2 + V_G \times 0 = 0$$

$V_A = 206.45$ kN

Example 20

A beam AF loaded by concentrated loads, as shown in Fig. 1.24(c), is held in equilibrium by vertical forces V_B and V_E. Determine the values of these forces.

Solution   As the beam is in equilibrium, the resultant force and the resultant moment of the forces about any axis must be zero. Because only vertical forces are involved, the conditions to be satisfied are $\Sigma F_y = 0$, $-50 - 70 - 90 - 80 + V_B + V_E = 0$, $V_B + V_E = 290$ kN.

Taking moments about the z-axis passing through B,

$$\Sigma M \text{ about B} = V_B \times 0 + 50 \times 2 - 70 \times 4 - 90 \times 10 - 80 \times 17 + V_E \times 14 = 0$$

Note that the moment of force at A about B is positive as it causes an anticlockwise rotation.

$V_E = 174.29$ kN, $V_B = 115.71$ kN

As a check, taking moments about the z-axis passing through E,

$$50 \times 16 - V_B \times 14 + 70 \times 10 + 90 \times 4 - 80 \times 3 = 0$$

$V_B = 115.71$ kN

Example 21

A beam AK loaded by concentrated and distributed loads, as shown in Fig. 1.24(d), is held in equilibrium by vertical forces V_C and V_H. Determine the value of these forces.

Solution   As the beam is in equilibrium, the resultant force and the resultant moment of the forces about any axis must be zero. Because only vertical forces are involved, the conditions to be satisfied are $\Sigma F_y = -(25 \times 2)$ distributed load over AB $- 40 -$

(20×3) distributed load over EF $- 100 - 80 - (30 \times 1.8)$ distributed load over JK $+ V_C + V_H = 0$, $V_C + V_H = 384$ kN.

Taking moments about the z-axis passing through C,

$$V_C \times 0 + 25 \times 2 \times (2.5 - 2/2) - 40 \times 2 - 20 \times 3 \times (2 + 1 + 3/2) - 100 \times 8$$
$$- 30 \times 1.8 \times (11 + 3 - 1.8/2) - 80 \times 14 + V_H \times 11 = 0$$

$$V_H = 263.86 \text{ kN}$$

$$V_C + V_H = 384$$

$$V_C = 384 - V_H = 120.15 \text{ kN}$$

As a check, taking moments about the z-axis passing through H,

$$-V_C \times 11 + 25 \times 2 \times (11 + 2.5 - 2.0/2) + 40 \times 9 + 20 \times 3 \times (5 + 3.0/2)$$
$$+ 100 \times 3 - 80 \times 3 - 30 \times 1.8 \times (3 - 1.8/2) + V_H \times 0 = 0$$

$$V_C = 120.15 \text{ kN}$$

Example 22

A beam AG loaded by concentrated loads and couples, as shown in Fig. 1.24(e), is held in equilibrium by vertical forces V_B and V_F. Determine the value of these forces.

Solution As the beam is in equilibrium, the resultant force and the resultant moment of the forces about any axis must be zero. Because only vertical forces are involved, the conditions to be satisfied are $\Sigma F_y = 0$, $-100 - 80 + V_B + V_F = 0$, $V_B + V_F = 180$ kN. Note that the couples do not enter into the above equation of equilibrium.

Taking moments about the z-axis passing through B,

$$V_B \times 0 + 50 - 75 + 90 - 100 \times 2 - 80 \times 7 + V_F \times 11 = 0$$

$$V_F = 63.18 \text{ kN}, \quad V_B = 116.82 \text{ kN}$$

Note that the couples appear as themselves without any lever arm being associated with them.

As a check, taking moments about the z-axis passing through F,

$$-V_B \times 11 + 50 - 75 + 90 + 100 \times 9 + 80 \times 4 + V_F \times 0 = 0$$

$$V_B = 116.82 \text{ kN}$$

1.4.14   Problems in three dimensions

In the previous sections only two-dimensional problems were considered. In three-dimensional problems, a body is capable of translations along the three Cartesian directions x, y and z. In addition, it can also rotate about the three axes. It is therefore necessary for equilibrium that all possible accelerations associated with three translations and three rotations are prevented. The equations of equilibrium are therefore:

$$\Sigma F_x = 0 \quad \text{and} \quad \Sigma F_y = 0$$

$$\Sigma F_z = 0 \quad \text{and} \quad \Sigma M_x = 0$$

$$\Sigma M_y = 0 \quad \text{and} \quad \Sigma M_z = 0$$

Example 23

Forces act on the plate as shown in Fig. 1.25(a). The plate is held in equilibrium by three vertical forces at the corners A, B and C. Calculate the values of these forces. AD = 4 m, AB = 2 m.

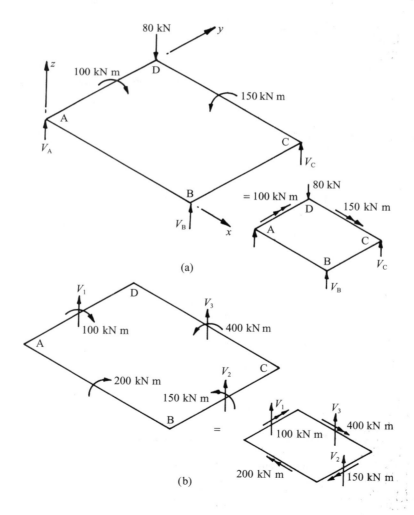

Figure 1.25 ● (a), (b)
A parallel force
system acting on
a 2-D structure

Solution   The equations of equilibrium that have to be satisfied are
$\Sigma F_x = 0$, $\Sigma F_y = 0$, $\Sigma F_z = 0$ in terms of the forces and
$\Sigma M_x = 0$, $\Sigma M_y = 0$, $\Sigma M_z = 0$ in terms of the moments about the x-, y- and z-axes respectively.

Let the vertical forces at the corners A, B and C be V_A, V_B and V_C. Because no forces are applied in the x- and y-directions, $\Sigma F_x = 0$ and $\Sigma F_y = 0$ are automatically satisfied and $\Sigma F_z = 0$ leads to

$$V_A + V_B + V_C - 80 = 0$$

Considering $\Sigma M_x = 0$, taking moments about AB and parallel to the x-axis,

$$150 - 80 \times 4 + V_C\,4 = 0, \; V_C = 42.5 \text{ kN}$$

Considering $\Sigma M_y = 0$, taking moments about BC and parallel to the y-axis,

$$100 - 80 \times 2 + V_A\,2 = 0, \; V_A = 30 \text{ kN}$$

Because no forces or couples contribute to ΣM_z it is automatically equal to zero.

$$V_B = -V_A - V_C + 80$$
$$V_B = 7.5 \text{ kN}$$

Example 24

Couples and a force of 50 kN at the centre act on the plate as shown in Fig. 1.25(b). Calculate the value of three unknown vertical forces V_1, V_2 and V_3 needed to maintain equilibrium. AD = 3 m, AB = 4 m.

Solution   Because no forces act in the x- and y-directions, it is only necessary to consider the following three equations of equilibrium.
$\Sigma F_z = 0$ in terms of the forces and
$\Sigma M_x = 0$, $\Sigma M_y = 0$ in terms of the moments about the x- and y-axes.
$\Sigma F_z = 0$ leads to $V_1 + V_2 + V_3 - 50 = 0$
$\Sigma M_x = 0$ about AB leads to $-200 + 400 - 50 \times (3/2) + V_3 \times 3 + V_1\,(3/2) + V_2(3/2) = 0$

$\Sigma M_y = 0$ about BC leads to $100 - 150 + 50 \times (4/2) + V_1 \times 4 + V_3\,(4/2) = 0$
Solving $V_1 = 104.2$, $V_2 = 79.2$, $V_3 = -133.4$ kN

1.5 ● Statically determinate structures

In many types of structures, it is possible to determine the forces in the members of the structure using only the equations of statics. Such structures are known as statically determinate structures. Structures in which the forces in the members cannot be determined using only equations of statics are known as statically indeterminate or hyperstatic structures. In practice both statically determinate and indeterminate structures are extensively used. Further discussion on these matters is postponed until the reader has gained sufficient understanding of the determination of forces in structural members.

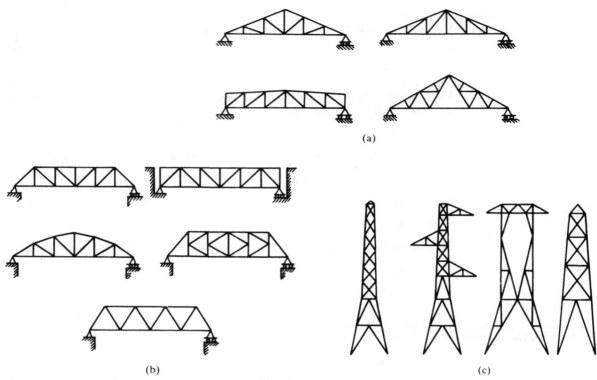

Figure 1.26 ● Typical pin-jointed structures: (a) roof trusses, (b) bridge trusses, (c) towers

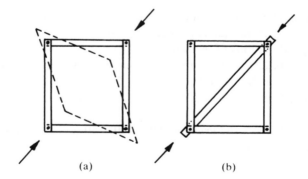

Figure 1.27 ●
Pin-jointed structures:
(a) quadrilateral,
(b) triangulated

(a)   (b)

1.6 ● Simple pin-jointed trusses

Trusses are a common form of structure generally assembled so as to create a series of triangles. Typical trusses used for roof, bridge and tower structures are shown in Fig. 1.26. It is *assumed* that the connection between the members is pinned and the only force in the members is an axial force, either tensile or compressive. With this simple assumption it is now easy to see why the structure is assembled so as to create in most cases a series of triangles.

Consider four bars pinned together so as to form a quadrilateral as shown in Fig. 1.27(a). If an external load is applied, then the quadrilateral offers little resistance and suffers very large distortions as shown by the dotted lines. Remembering that a structure has to resist external forces without undergoing large relative displacements, it is obvious that the quadrilateral with pin-joints at the corners is not a viable structure. If, on the other hand, an additional member is introduced as shown in Fig. 1.27(b), then the structure is well able to resist the external loads acting on the structure. The reason for this is that a triangle with pin-joints at the corners is a 'stiff' structure in the sense that it is able to resist forces without undergoing large deformations. Therefore any structure assembled from a series of triangles will also be able to satisfy the requirement of not undergoing large displacements under external loads.

1.6.1 Classification of statically determinate pin-jointed trusses

Trusses in practice can be statically determinate or indeterminate. Statically determinate trusses can be quite quickly analysed by manual methods. For manual analysis purposes, it is useful to classify statically determinate trusses as follows.

Simple truss

Figure 1.28(a) shows how starting from a simple triangle 1–2–3, and using two members at a time to create an additional joint, quite an elaborate truss can be formed. Similarly, Fig. 1.28(b) shows that, instead of starting from a basic triangle, starting from a rigid base such as a wall and proceeding as in Fig. 1.28(a) a truss can be formed. Such trusses are known as simple trusses.

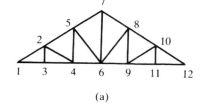
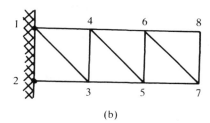

Figure 1.28 ● (a), (b) Simple pin-jointed trusses

(a)

(b)

Compound truss

Compound trusses use simple trusses as building blocks. Figure 1.29 shows a compound truss where two simple trusses, ABC and DEC, are connected at C and by a bar BD so as to create an additional triangle BCD. In fact, as shown in Fig. 1.30(c), two 'stiff structures' such as simple trusses can be connected together

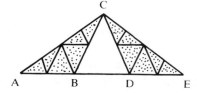

Figure 1.29 ●
A compound truss

Figure 1.30 ● Stable and unstable compound trusses: (a) unstable structure as free sway of ABCD with respect to EFGH, (b) unstable structure as ABCD can rotate freely about I, (c) stable structure

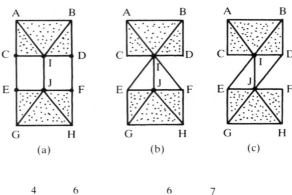

Figure 1.31 ●
(a) Simple truss: transformed into a complex truss (b)

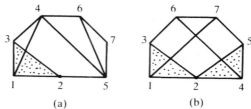

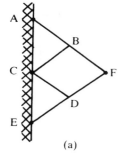

Figure 1.32 ●
Pin-jointed structures with quadrilaterals:
(a) stable, (b) unstable configurations

by three bars which are neither parallel nor concurrent to produce a compound truss. If the three bars are parallel as in Fig. 1.30(a), then clearly large relative displacement between ABCD and EFGH can occur, thus ruling out the resulting compound truss as a viable structure. Similarly, if the three bars are concurrent, as in Fig. 1.30(b), then clearly ABCD can rotate about I and large deformations are possible so that the resulting compound truss is not an acceptable structure. A structure which permits large relative deformations is said to be an unstable structure or mechanism.

Complex trusses

This category includes all those statically determinate trusses which are neither simple nor compound. Figure 1.31(a) is a simple truss. The joint numbering indicates the order in which the joints were formed. Figure 1.31(b) shows a complex truss. It is obtained from the simple truss shown in Fig. 1.31(a) by a simple re-arrangement of bars to produce a symmetrical truss.

So far it has been emphasized that a pin-jointed quadrilateral is not an acceptable structure and that for creating a 'stiff' pin-jointed structure, it is necessary to ensure that the structure can be divided into a series of triangles. This is true, but the presence of a quadrilateral in a structural configuration does not necessarily rule it out from being an acceptable structure. The important criterion is whether the quadrilateral allows large deformation to take place under external loading. To emphasize this point, consider the structures shown in Fig. 1.32(a) and (b).

1. Figure 1.32(a) shows a structure with a quadrilateral in it. The structure was assembled as follows.

   (a) Starting from the rigid base ACE, joint B was created using members AB and BC. Triangle ABC is a 'stiff' structure and joint B is held firmly in place.

(b) Similarly, joint D was created using members CD and DE. Triangle CDE is a stiff structure and joint D is held firmly in place.

(c) Starting from joints B and D, which are held firmly in place, using members BF and DF, joint F is created. Because joints B and D are held in place, joint F can be created without any danger of creating an unacceptable structure in spite of the fact that in the final structure a quadrilateral CBFD is present.

2. Figure 1.32(b) also shows a pin-jointed structure with a quadrilateral CEFD. The structure was assembled as follows.

(a) Starting from the rigid base AB, joint C is created using members AC and BC. Joint C is thus firmly held in place.

(b) Starting from the joints B and C, both of which are firmly held in place, joint D is created using members BD and CD. Joint D is thus firmly held in place.

(c) Triangle EFG is assembled.

(d) Using parallel members CE and DF, triangle EFG is connected to joints C and D.

It is evident from the above description that joints E and F are never firmly held in place. Therefore the quadrilateral CEFD can undergo large deformations under external load. Thus the presence of the quadrilateral CEFD rules out this structure from being an acceptable structure.

This example shows clearly that it is necessary to examine in a systematic manner whether each joint of the structure is firmly held in place before pronouncing a judgement on the acceptability or otherwise of a given structure. Whenever a 'suspect' truss is encountered, one should consider carefully how the truss was constructed and whether at any stage there is a possibility that it could become unstable leading to an unacceptable structure or mechanism. This could be particularly important during the erection stages of complex structures.

1.7 ● Analysis of simple pin-jointed trusses

The analysis of simple trusses can be carried out using only the equations of statics. In other words, they are statically determinate. In order to demonstrate the procedure adopted, a simple example will be solved. However, before doing so an important concept called free-body diagram will be explained.

1.7.1 Free-body diagram

Because a structure as a whole body is in equilibrium, any part of it is also in equilibrium. A part of the structure can therefore be isolated (in imagination) from the rest of the structure and it will be in equilibrium under the action of the external forces acting on it *and* the internal forces in the members which are cut.

Consider the truss shown in Fig. 1.33(a). If the joint C is isolated from the rest of the structure as shown in Fig. 1.33(b), then joint C is in equilibrium under the action of the load applied to joint C and the forces in the members meeting at joint

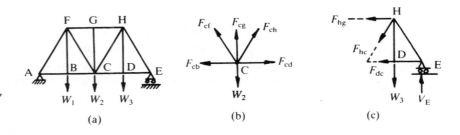

Figure 1.33 ●
(a) Pin-jointed truss,
(b) free body of a joint,
(c) free body of a part
of a truss

C. Similarly, if the portion of the structure shown in Fig. 1.33(c) is isolated from the structure, then it is in equilibrium under the action of external loads at joints D, E and H and the internal forces in the members which are cut, i.e. DC, HC and HG. When considering the equilibrium of the free body it is *convenient* to *assume* that the forces in the members of the truss are *tensile*.

1.7.2 Method of joints

One of the most popular methods for the determination of the axial forces in the members of a simple pin-jointed truss is the method of joints. In this method, using the free-body diagram for each joint of the truss, equilibrium of the joints of the truss is considered one at a time (for simplicity). At each joint it is necessary to satisfy the equations of statics, namely

$$\sum \text{Forces in } x\text{-direction} = 0 \quad \text{and} \quad \sum \text{Forces in } y\text{-direction} = 0$$

$\Sigma M_z = 0$ is automatically satisfied because all the forces at the joint pass through the joint and therefore the lever arm for every force is zero.

Since there are only the above two equations to be satisfied at each joint, only two unknown forces can be determined. It is therefore necessary to proceed in a systematic fashion from joint to joint ensuring that a maximum of only two unknown forces exist when the equilibrium of a particular joint is being considered. For convenience, it is assumed that the forces in the members are tensile forces. The members therefore exert a pull on the joints they connect. The method is illustrated by a simple example.

Example 1

Calculate the forces in the members of the pin-jointed truss shown in Fig. 1.34.

Solution   The given truss is a simple truss. The solution steps are as follows.

1. Determine the reactions at the supports. The support A is a pinned support and support B is on rollers. There are at support A a vertical (V_A) and horizontal (H_A) reaction but at B only a vertical (V_B) reaction. Assuming that all external forces acting to the right or upwards are positive and the moment about the z-axis is positive if clockwise, equilibrium requires that

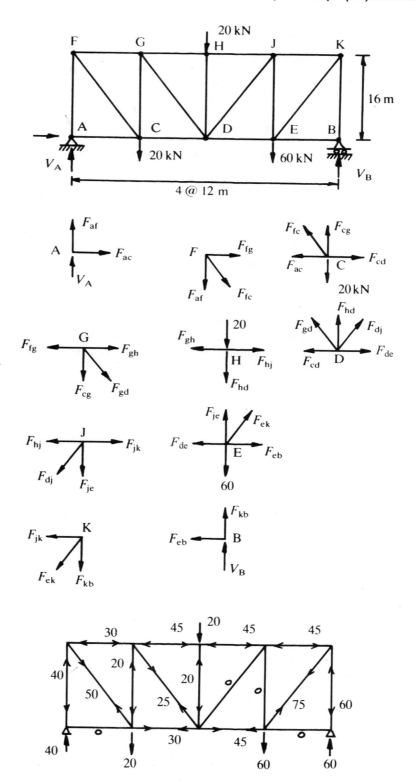

Figure 1.34 ● Analysis of pin-jointed truss by the method of joints

$$H_A + 0 = 0$$

Therefore $H_A = 0$

$$V_A + V_B - 20 \text{ (at C)} - 20 \text{ (at H)} - 60 \text{ (at E)} = 0$$

Therefore

$$V_A + V_B = 100$$

Taking moments at A about an axis parallel to the z-axis, we have

$$H_A \times 0 + V_A \times 0 - V_B \times 48$$
$$+ 20 \times 12 \text{ at C} + 20 \times 24 \text{ at H} + 60 \times 36 \text{ at E} = 0$$

Therefore

$$V_B = 60 \text{ kN}$$

$$V_A = 100 - V_B = 40 \text{ kN}$$

Note: Because there were only three reactions, it was possible to determine them using only the three equations of statics. If there were more than three reactions like, for example, when joint D is also on a roller support, then there will be four reactions, H_A, V_A, V_D and V_B. In this case the reactions cannot be determined from the equations of statics only. The structure is a statically indeterminate or hyperstatic structure. Solution of such problems will be discussed in Chapters 6 and 10.

2. Determination of axial forces in the members of the truss.

Joint A. Referring to the free-body diagram for joint A, the forces acting are $V_A = 40$, $H_A = 0$, forces F_{af} and F_{ac} in members AF and AC respectively. Using the equations of statics

$$V_A + F_{af} = 0, \ H_A + F_{ac} = 0$$

Therefore

$$F_{af} = -V_A = -40 \text{ kN}$$

The negative sign indicates that the force in member AF is compressive: $F_{ac} = -H_A = 0$.

Note that F_{ac} or F_{ca} indicate the same axial force in the member connecting joints C and A. This applies to all other members.

Joint F. The members meeting at F are FA (or AF), FG and FC. However, the force F_{af} in member AF is already known. Therefore there are only two unknown forces, F_{fg} and F_{fc}, and they can therefore be determined from the equations of statics. Referring to the free-body diagram for joint F,

$$F_{fg} + F_{fc} \cos (CFG) = 0, \ F_{af} + F_{fc} \sin (CFG) = 0$$

$\cos (CFG) = 3/5$, $\sin (CFG) = 4/5$ and $F_{af} = -40$ kN. Therefore

$$F_{fc} = -F_{af}/\sin (CFG) = 50 \text{ kN}$$

and

$$F_{fg} = -F_{fc} \cos (CFG) = -(50)(3/5) = -30 \text{ kN}$$

Similarly, considering the rest of the joints in turn we have:

Joint C. Unknown forces F_{cg} and F_{cd}.

$$F_{cd} - F_{ac} - F_{fc} \cos (ACF) = 0$$

$$F_{cg} + F_{fc} \sin (ACF) - 20 = 0$$

$$F_{cg} = 20 - F_{fc} \sin (ACF) = 20 - 50(4/5) = -20 \text{ kN}$$

$$F_{cd} = F_{ac} + F_{fc} \cos (ACF) = 0 + 50(3/5) = 30 \text{ kN}$$

Joint G. Unknown forces F_{gh} and F_{gd}.

$$F_{gh} - F_{fg} + F_{gd} \cos (HGD) = 0$$

$$-F_{cg} - F_{gd} \sin (HGD) = 0$$

$$F_{gd} = -F_{cg}/\sin (HGD) = -(-20)/(4/5) = 25 \text{ kN}$$

$$F_{gh} = F_{fg} - F_{gd} \cos (HGD) = -30 - 25(3/5) = -45 \text{ kN}$$

Joint H. Unknown forces F_{hj} and F_{hd}.

$$F_{hj} - F_{gh} = 0, \quad -F_{hd} - 20 = 0$$

Therefore

$$F_{hj} = F_{gh} = -45 \text{ kN}, \quad F_{hd} = -20 \text{ kN}$$

Joint D. Unknown forces F_{dj} and F_{de}.

$$F_{de} + F_{dj} \cos (EDJ) - F_{gd} \cos (CDG) - F_{cd} = 0$$

$$F_{hd} + F_{dj} \sin (EDJ) + F_{gd} \sin (CDG) = 0$$

$$F_{dj} = \{-F_{hd} - F_{gd} \sin (CDG)\}/\sin (EDJ)$$

$$= \{20 - 25(4/5)\}/(4/5) = 0$$

$$F_{de} = -F_{dj} \cos (EDJ) + F_{gd} \cos (CDG) + F_{cd}$$

$$= 0 + 25(3/5) + 30 = 45$$

Joint J. Unknown forces F_{jk} and F_{je}.

$$F_{jk} - F_{dj} \cos (DJH) - F_{hj} = 0, \quad -F_{je} - F_{dj} \sin (DJH) = 0$$

$$F_{jk} = F_{dj} \cos (DJH) + F_{hj} = 0 + (-45) = -45 \text{ kN}$$

$$F_{je} = -F_{dj} \sin (DJH) = 0$$

Joint E. Unknown forces F_{ek} and F_{eb}.

$$F_{eb} + F_{ek} \cos (BEK) - F_{de} = 0$$

$$F_{je} - 60 + F_{ek} \sin (BEK) = 0$$

$$F_{ek} = \{-F_{je} + 60\}/\sin (BEK) = \{0 + 60\}/(4/5) = 75 \text{ kN}$$

$$F_{eb} = -F_{ek} \cos (BEK) + F_{de} = -75(3/5) + 45 = 0$$

Joint K. Unknown force F_{kb} only.

$$-F_{jk} - F_{ek} \cos(EKJ) = 0.$$

This equation serves as a check on the previous calculations, $-(-45) - 75(3/5) = 0.$

$$-F_{kb} - F_{ek} \sin(EKJ) = 0$$

$$F_{kb} = -F_{ek} \sin(EKJ) = -75(4/5) = -60 \text{ kN}$$

Joint B. No unknown forces exist but equilibrium will serve as a check on the previous calculations.

$$-F_{eb} = 0, \quad V_B + F_{kb} = 0$$

$F_{eb} = 0$ checks with the previous calculation at joint E and $F_{kb} = -V_B = -60$ checks with the previous calculation at joint K.

Figure 1.34 shows the forces in the members. The notation used is if the force is tensile then the force in the member exerts a pull on the joints connected by the member, therefore the tensile force in a member in indicated by →←. Similarly, a compressive force is indicated by ←→, indicating that the member exerts a push on the joints it connects.

Example 2

Determine the forces in the members of the pin-jointed truss shown in Fig. 1.35.

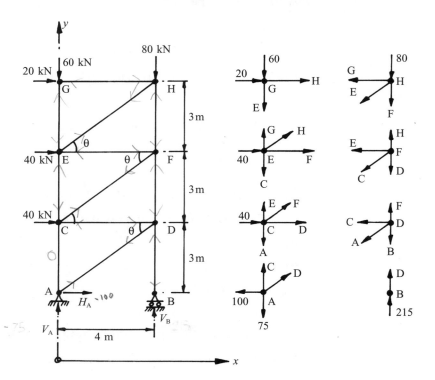

Figure 1.35 ●
A pin-jointed truss

Solution Calculate the reactions at the supports. Because the joint A is pinned, there are two possible reactions, horizontal and vertical. At support B, there is only a vertical reaction because it is on rollers. From the equilibrium condition, $\Sigma F_x = 0$, $H_A + 20 + 40 + 40 = 0$, $H_A = -100$, i.e. $H_A = 100$ kN from right to left.

From the equilibrium condition, $\Sigma F_y = 0$, $V_A + V_B - 60 - 80 = 0$

Taking moments about A, $40 \times 3 + 40 \times 6 + 20 \times 9 - 60 \times 0 + 80 \times 4 - V_B \times 4 = 0$. $V_B = 215$ kN and $V_A = -75$ kN, i.e. $V_A = 75$ kN acting vertically down.

Consider the free body of each joint in turn and making sure that there are only two unknown joint forces, determine the forces in the members of the truss. Initially assume that the axial force in all the members is tensile.

Note that $\tan \theta = 3/4$, $\sin \theta = 0.6$ and $\cos \theta = 0.8$

Remember that $F_{ad} = F_{da}$, $F_{bd} = F_{db}$ and so on.

Joint A. $\Sigma F_x = 0$, $-100 + F_{ad} \cos \theta = 0$, $F_{ad} = 125$ kN

$\Sigma F_y = 0$, $F_{ac} + F_{ad} \sin \theta - 75 = 0$, $F_{ac} = 0$

Joint B. $\Sigma F_y = 0$, $F_{bd} + 215 = 0$, $F_{bd} = -215$ (compression)

Joint D. $\Sigma F_x = 0$, $-F_{da} \cos \theta - F_{dc} = 0$, $F_{dc} = -100$ kN (compression)

$\Sigma F_y = 0$, $F_{df} - F_{da} \sin \theta - F_{db} = 0$, $F_{df} = -140$ kN (compression)

Joint C. $\Sigma F_x = 0$, $40 + F_{cd} + F_{cf} \cos \theta = 0$, $F_{cf} = 75$ kN

$\Sigma F_y = 0$, $F_{ce} + F_{cf} \sin \theta - F_{ca} = 0$, $F_{ce} = -45$ kN (compressive)

Joint F. $\Sigma F_x = 0$, $-F_{ef} - F_{cf} \cos \theta = 0$, $F_{ef} = -60$ kN (compressive)

$\Sigma F_y = 0$, $F_{fh} - F_{fd} - F_{fc} \sin \theta = 0$, $F_{fh} = -95$ kN (compressive)

Joint E. $\Sigma F_x = 0$, $40 + F_{ef} + F_{eh} \cos \theta = 0$, $F_{eh} = 25$ kN

$\Sigma F_y = 0$, $F_{eg} + F_{eh} \sin \theta - F_{ec} = 0$, $F_{eg} = -60$ kN (compressive)

Joint H. $\Sigma F_x = 0$, $-F_{hg} - F_{he} \cos \theta = 0$, $F_{hg} = -20$ kN (compressive)

$\Sigma F_y = 0$, $-80 - F_{hf} - F_{he} \sin \theta = 0$, $-80 - (-95) - (25)0.6 = 0$, check on previous calculations.

Joint G. $\Sigma F_x = 0$, $20 + F_{gh} = 0$, $F_{gh} = -20$ kN, check on previous calculations.

$\Sigma F_y = 0$, $-60 - F_{ge} = 0$, $F_{ge} = -60$ kN (compressive), check on previous calculations.

1.7.3 Analysis of simple pin-jointed trusses – method of sections

In the method of joints, success depended on having a maximum of only two unknowns to be determined at any one joint. It will fail in the case of a compound truss because at some of the joints more than two unknowns need to be determined. To illustrate this point, consider the compound truss shown in Fig. 1.36(a). Assume that the reactions at the supports are known from the equilibrium consideration of the entire truss. Starting from joint A, forces F_{ab} and F_{ac} can be determined. Then moving successively to joints C and B forces F_{cd}, F_{cb}, F_{bd} and F_{be} can be determined. Similarly, starting from joint K and then moving on to joints M and L, forces F_{kl}, F_{kj}, F_{mp}, F_{ml}, F_{lq} and F_{lp} can be determined. No further progress can be made because, at other joints, there are three unknown forces. What needs to be done in order for the method of joints to be successful is that all the reactions on each of the simple trusses from which the compound truss is assembled must be determined. The compound truss consists of two simple trusses AJE and KJQ as shown in Fig. 1.36(b). The free-body diagrams for the two simple trusses indicate

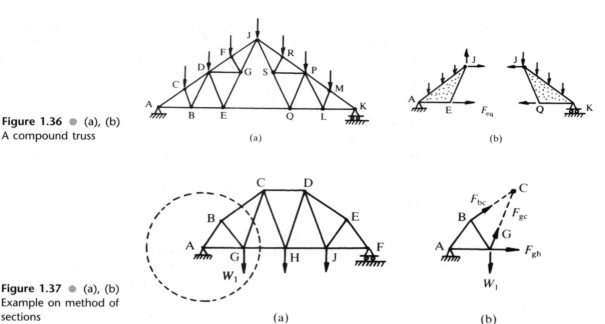

Figure 1.36 ● (a), (b)
A compound truss

Figure 1.37 ● (a), (b)
Example on method of sections

that even without knowing the reactive forces at J, by taking moments about J, the force F_{eq} in member EQ can be determined. Once the force F_{eq} is known, then the method of joints can be used to determine the forces in the other members of the simple trusses. Of course there is no reason why the method of sections cannot be applied to simple trusses. For example, for the truss shown in Fig. 1.37(a), the free body in Fig. 1.37(b) can be used to determine the forces F_{gh} and F_{bc} by taking moments about C and G respectively. However, in general it is not an efficient procedure unless for some reason forces in only a few selected members are required. Some examples are given in Chapter 2, Section 2.6.

1.8 ● Relationship between the number of joints and members in statically determinate trusses

Before proceeding to discuss further the analysis of statically determinate trusses, it is useful to establish a simple relationship that exists between the number of joints and number of members in a statically determinate truss.

Simple truss

In a simple truss, one starts with a basic triangle consisting of three members and three joints. *Each additional joint requires two additional members to create it.* Therefore if the number of members in the structure is m and the number of joints is j, then the total number of *additional* joints is equal to $(j - 3)$ and of *additional* members is equal to $(m - 3)$. The relationship between the additional members and additional joints is therefore

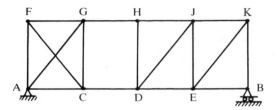

Figure 1.38 ●
An unstable pin-jointed
structure

$$(m - 3) = 2(j - 3) \text{ or } m = 2j - 3$$

In counting the number of members m, if the structure is started from two joints on a rigid base as shown in Fig. 1.28(b), then the *rigid base should also be counted as a member*.

As a simple example consider the truss shown in Fig. 1.28 for which $j = 10$, $m = 17$. This satisfies the condition $m = (2j - 3)$ because $2 \times 10 - 3 = 17$. It should be emphasized that the equation deals with a simple count of the overall number of joints and members, with the result that a wrong arrangement of members for a given number of joints can produce an unstable structure in the sense that they allow large relative displacements between the parts of the structure. The fact that the structure is unstable will not be detected by the formula. To illustrate the danger of relying on a simple count of the number of joints and members in deciding whether a simple truss is statically determinate or not, consider the truss shown in Fig. 1.38. It is the same truss as in Fig. 1.34, except that member GD is replaced by GA. This does not affect the overall count of m and j, but quite clearly the resulting structure is unstable because of the presence of the pin-jointed quadrilateral CGHD. This should be a good warning to the reader not to use formulae blindly. Rather, one should have an understanding of the basic concept behind the formulae before they are used.

Compound truss

Compound trusses use simple trusses as building blocks. Therefore, for each simple truss 'i' from which the compound truss is synthesized, the formula $m_i = (2j_i - 3)$ holds good. As already explained, a compound truss can be formed from, say, two simple trusses by connecting them together by three bars which are neither parallel nor concurrent or any equivalent type of connection to prevent large relative deformation between the simple trusses. Figure 1.39 shows two such trusses.

Consider the truss shown in Fig. 1.39(a). If m_1 and j_1 refer to the total number of members and joints respectively of the first simple truss ABC and similarly m_2 and j_2 refer to the second simple truss DEF, then $m_1 = 2j_1 - 3$ and $m_2 = 2j_2 - 3$. The two simple trusses are connected by three bars AE, FB and CD without any increase in the number of joints over the sum of the joints of the two simple trusses. Therefore for the compound truss $j = j_1 + j_2$. The number of members is increased by three giving $m = m_1 + m_2 + 3$. Therefore for the compound truss

$$m = m_1 + m_2 + 3 = (2j_1 - 3) + (2j_2 - 3) + 3$$

$$= 2(j_1 + j_2) - 3 = 2j - 3$$

Figure 1.39 ●
Compound trusses:
(a) two triangles
connected by three bars,
(b) two triangles
connected at a common
joint and a bar

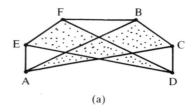
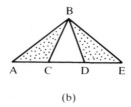

(a)   (b)

In this case therefore the rule which was applicable to simple trusses also holds good for the compound truss.

Now consider the truss shown in Fig. 1.39(b). Clearly the compound truss is formed by connecting two simple trusses ABC and BDE at a common joint B and adding an additional member CD to complete the connection. In this case $m = m_1 + m_2 + 1$ and $j = j_1 + j_2 - 1$ because the two trusses have a common joint. Therefore

$$m = m_1 + m_2 + 1 = (2j_1 - 3) + (2j_2 - 3) + 1$$
$$= 2(j_1 + j_2 - 1) - 3 = 2j - 3$$

Therefore this compound truss also obeys the $m = 2j - 3$ rule.

Complex trusses

Consider the two trusses shown in Fig. 1.31. Figure 1.31(a) is a simple truss. The joints have been numbered to indicate the order in which the joints were formed from the basic triangle 1–2–3. Evidently for this truss $m = 2j - 3$ holds good. A careful inspection of the truss shown in Fig. 1.31(b) indicates that it is obtained from the simple truss shown in Fig. 1.31(a) by a simple rearrangement of the bars. In other words, for this truss also $m = 2j - 3$. This truss is however neither simple nor compound. It is an example of a complex truss. It is interesting to note that although the truss is statically determinate, the method of joints fails because there are at least three unknowns at every joint. Similarly, the method of sections also fails because it is impossible to find convenient sections to isolate parts of the truss.

Although many specialized procedures have been developed for the manual solution of such trusses, they are best analysed by the general computer-orientated procedure given in Chapter 6.

1.9 ● Some aspects of the behaviour of statically determinate trusses

Statically determinate structures possess some attractive features in terms of response to support settlement, tolerance problems during construction and temperature changes. These aspects are discussed below in some detail.

Effect of settlement of supports

Consider the two trusses shown in Fig. 1.40. Both are simple trusses. Figure 1.40(a) is on two supports only, requiring the determination of two vertical and one hor-

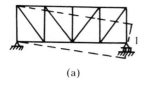

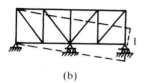

Figure 1.40 ● Support settlement of pin-jointed structures: (a) statically determinate structure, (b) statically indeterminate structure

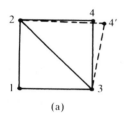

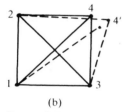

Figure 1.41 ● Thermal deformation in pin-jointed structures: (a) statically determinate structure, (b) statically indeterminate structure

izontal reaction which can be determined from statics. On the other hand, the truss shown in Fig. 1.40(b) is on three supports, requiring the determination of four reactions consisting of three vertical reactions and one horizontal reaction which cannot be determined from the considerations of statics only. It is therefore a statically indeterminate structure. Let support 1 settle vertically downwards by a small amount. In the case of the truss shown in Fig. 1.40(a), the truss undergoes a rigid body rotation about the left-hand support as shown by the dotted lines. Assuming that the deflection at the support is small, the rigid body rotation does not result in any change in the forces because both the loads and the statically determinate reactions remain unaltered. In the case of the statically indeterminate truss shown in Fig. 1.40(b), a simple rigid body rotation about the right-hand support will violate the zero vertical deflection condition at the central vertical support. Thus the vertical reaction at the middle support changes. This requires that all other vertical reactions also change. This new set of reactions will result in a new set of forces in the members of the truss. This simple example leads to the following general rule.

In a statically determinate structure, small settlement of supports does not affect the forces in the members of the truss. The member forces are governed only by the external loads acting on the truss. This statement is generally not true in the case of statically indeterminate structures. This insensitivity to support settlement is useful if the truss is used for, say, a bridge structure, where there is a possibility of the supports settling.

Response to thermal changes

Consider the two trusses shown in Fig. 1.41. Figure 1.41(a) shows a statically determinate truss ($m = 5$, $j = 4$ and hence $m = 2j - 3$), the joint numbers indicating the order in which the joints were formed. The truss in Fig. 1.41(b) is formed from the truss in Fig. 1.41(a) by the addition of an extra member 1–4. For this truss $m = 5$, $j = 4$ and $m = (2j - 3) + 1$. The structure has one more member than is required for a statically determinate configuration. The structure is said to be one degree statically indeterminate. If the member 2–4 is subjected to a temperature change which alters its length by Δ, then in the case of the statically determinate structure shown in Fig. 1.41(a), members 2–4 and 3–4 rotate to meet at a new position 4′. The configuration is altered but because there are no external loads at the joints, the member forces are zero. Therefore, in the case of the statically determinate truss, changes in temperature cause displacements of the joints but do not result in member forces. In the statically indeterminate structure shown in Fig. 1.41(b), the position of joint 4 is fixed by the unaltered lengths of members 1–4 and 3–4. Therefore member 2–4 has to be 'forced' to fit in to joint 4. Thus there exists a force in member 2–4. If there is a force in member 2–4, then, for equilibrium, forces exist in other members. Thus, in the case of a statically indeterminate structure, changes in lengths of the members due to temperature change (or any other cause such as incorrect length during fabrication) result in displacements of the joints *and* forces in the members of the structure.

Incorrect lengths of members

The discussion with respect to thermal changes also shows that if the change in length Δ is due to incorrect length of member 2–4 due to an error during manufacture, then in the case of a statically determinate structure it affects slightly the final geometry of the structure but does not induce any forces in the members. However, in the case of a statically indeterminate structure, incorrect lengths of members induce forces in the members of the truss.

1.10 ● Analysis of statically determinate beams

In the previous sections, the analysis of pin-jointed structures was discussed. In the case of pin-jointed structures, the force in a member consists only of an axial force and generally it is constant throughout the member. In the case of a beam subjected to loads normal to its axis, there are two 'forces' at a cross-section and these forces vary along the length of the beam. As an example, consider the cantilever beam shown in Fig. 1.42. It is fixed at one end to a wall and loaded at the other end by a load W normal to the axis of the beam. Loads acting normal to the axis of the beam are often called lateral loads. Consider the equilibrium of the free body on which the external load acts. It is clear that for equilibrium of the free body, at the right-hand end, a vertical upward force Q equal to W, and a clockwise couple $M = Wx$, must act at the cut face. According to Newton's third law of motion, action and reaction are always equal and opposite, and so the forces at the cut section of the free body fixed to the support must be a vertical downward force Q equal to W, and an anticlockwise couple $M = Wx$. Thus there are two internal forces at the cut section. The first is a force Q which acts normal to the axis of the beam and is known as shear force. The second is a couple M and is known as the bending moment. The bending moment and shear force can be defined as follows.

Shear force: the algebraic sum of all forces normal to the beam axis acting *either to the left or the right of the section.*

Bending moment: the algebraic sum of the moments about an axis parallel to the z-axis and passing through the cut section of all the forces acting *either to the left or the right of the section.*

Because both shear force Q and bending moment M vary along the length of the beam, it is useful to show their distribution graphically in diagrams known as a

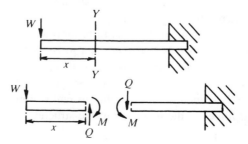

Figure 1.42 ●
A cantilever beam

Shear Force Diagram (SFD for short) and a Bending Moment Diagram (BMD) respectively.

1.10.1 Sign convention for bending moment and shear force

The sign convention adopted for bending moment and shear force are as follows:

(a) Bending moment M is positive if it causes the beam to bend so as to cause tension on the bottom face of a beam. Another way of expressing this is to say that if the bending moment causes the beam to sag then it is positive and if it makes the beam hog, then it is negative.

(b) Shear force Q which produces an anticlockwise distortion of the beam is positive. The above convention is shown in Fig. 1.43.

Figure 1.43 ● Sign convention for shear force and bending moment: (a) forces on a beam segment, (b) positive shear force and the corresponding deformation pattern, (c) positive bending moment and the corresponding deformation pattern

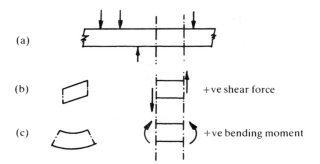

1.11 ● Examples of bending moment and shear force diagrams

Some simple cases of loading on beams and cantilevers commonly met in practice will be used to draw the bending moment diagram (BMD) and the shear force diagram (SFD). The bending moment diagram will be drawn on the tension side of the beam. The shear force diagram will be drawn so that positive shear force is above and negative shear force is below the beam.

1.11.1 Cantilevers

A cantilever is a beam which is free (unsupported) at one end and embedded in a wall or a 'heavy' foundation at the other end. The distribution of forces in the embedded section is complex, as shown in Fig. 1.44. Because the forces in the

Figure 1.44 ●
A cantilever beam:
(a) beam with loads,
(b) forces at the fixed support, (c) resultant forces at the wall due to the forces over the fixed length

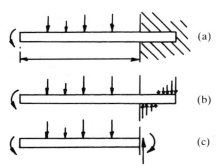

embedded section are generally not known, the SFD and BMD are constructed only up to the section where the cantilever meets the wall.

Example 1

Cantilever with an end couple M_1 as shown in Fig. 1.45. Considering the free body, from equilibrium we have

$$Q = 0$$

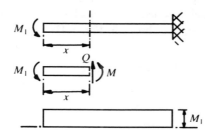

Figure 1.45 ●
Bending moment and shear force distribution in a cantilever beam with an end couple

Taking moments about the cut section,

$$M + M_1 = 0$$

Therefore

$$Q = 0, \ M = -M_1$$

Q is zero and M is a constant equal to $-M_1$. The corresponding BMD is shown in Fig. 1.45. The moment M causes tension on the top face of the beam.

Example 2

Cantilever with an end load W as shown in Fig. 1.46. Considering the free body, from equilibrium we have

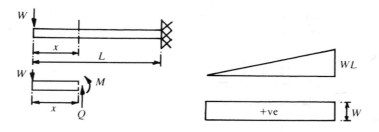

Figure 1.46 ●
Bending moment and shear force distribution in a cantilever beam with an end load

$$Q - W = 0, \text{ and } M + Wx = 0$$

Therefore

$$Q = W, \ M = -Wx$$

where Q is constant while M is a linear function of x. The corresponding SFD and BMD are shown in Fig. 1.46; $M_{max} = -WL$ at the support causing tension at the top face.

Example 3

A cantilever with a uniformly distributed load of q per unit length as shown in Fig. 1.47.

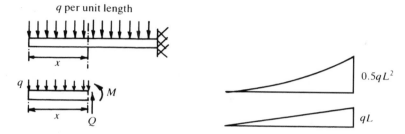

Figure 1.47 ●
Bending moment and shear force distribution in a cantilever beam with uniformly distributed load q

Considering the free body, for equilibrium

$$Q - qx = 0, \quad M + (qx)x/2 = 0$$

Therefore

$$Q = qx, \quad M = -qx^2/2$$

Note that qx is the total load on the beam of length x and its resultant is at $0.5x$ from the cut end. The SFD and BMD are shown in Fig. 1.47; $M_{max} = -qL^2/2$ causing tension at the top face and $Q_{max} = qL$ both occurring at the support.

Example 4

A cantilever with a linearly varying load as shown in Fig. 1.48.

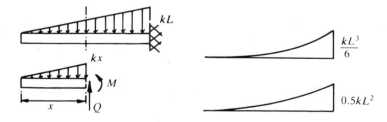

Figure 1.48 ●
Bending moment and shear force distribution in a cantilever beam with linearly varying load

This type of loading commonly occurs when designing retaining walls, water tanks, etc., where the load due to 'fluid pressure' increases linearly with 'depth'. The lateral load q is given by $q = kx$, where k is a constant. Considering the free body shown in Fig. 1.48, for equilibrium

$$Q - 0.5(kx)x = 0, \quad M + (0.5kx^2)x/3 = 0$$
$$Q = 0.5kx^2, \quad M = -kx^3/6$$

Note that kx is the height of the load triangle at a distance x from the free end. The total load on the segment of length x is the area of the load triangle $= 0.5(kx)x = 0.5kx^2$ and its resultant is at $x/3$ from the cut end. The SFD and BMD are shown in Fig. 1.48; $M_{max} = -kL^3/6$ causing tension at the top and $Q_{max} = 0.5kL^2$ both occur at the support.

In the previous examples, single expressions for bending moment and shear force were valid for the entire cantilever. This is not always the case. In the next two examples the loading is variable along the span. In such cases, different expressions for BM and SF are needed to take account of the influence of loading in different parts of the structure.

Example 5

Figure 1.49 shows a cantilever with two concentrated loads and a distributed load covering part of the span. Draw the BMD and the SFD.

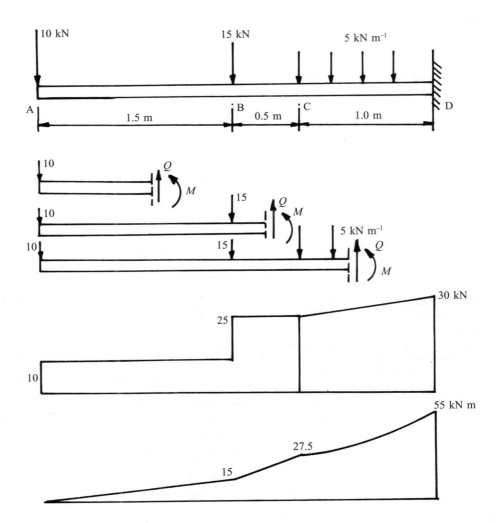

Figure 1.49 ● Forces acting on a cantilever

Solution   Divide the span into segments AB, BC and CD and draw a free-body diagram with a cut section in each these segments.

(a) Segment AB: $0 < x \leqslant 1.5$
$Q - 10.0 = 0$, $Q = 10$, constant
$M + 10x = 0$, $M = -10x$, linear variation
$x = 0$, $Q = 10$, $M = 0.0$ and $x = 1.5$, $Q = 10$ and $M = -15.0$

(b) Segment BC: $1.5 < x \leqslant 2.0$
$Q - 10.0 - 15 = 0$, $Q = 25$, constant
$M + 10x + 15(x - 1.5) = 0$, $M = -25x + 22.5$, linear variation
$x = 1.5$, $Q = 25$, $M = -15.0$ and $x = 2.0$, $Q = 25$ and $M = -27.5$

(c) Segment CD: $2.0 < x \leqslant 3.0$
$Q - 10 - 15 - 5(x - 2.0) = 0$, $Q = 15 + 5x$, linear variation
$M + 10x + 15(x - 1.5) + 5(x - 2.0)^2/2 = 0$, $M = -25x + 22.5 - 5(x - 2.0)^2/2$, parabolic variation
$x = 2.0$, $Q = 25$, $M = -27.5$ and $x = 3.0$, $Q = 30$ and $M = -55.0$

Example 6

Figure 1.50 shows a cantilever with two concentrated loads and two distributed loads covering parts of the span. Draw the BMD and the SFD.

Solution   Divide the span into segments AB, BC, CD and DE and draw a free-body diagram with a cut section in each of these segments.

(a) Segment AB: $0 < x \leqslant 1.0$
$Q - 10.0 - 2x = 0$, $Q = 10 + 2x$, linear variation.
$M + 10x + 2x^2/2 = 0$, $M = -10x - x^2$, parabolic variation
$x = 0$, $Q = 10$, $M = 0.0$ and $x = 1.0$, $Q = 12$ and $M = -11.0$

(b) Segment BC: $1.0 < x \leqslant 1.5$
$Q - 10.0 - 2(1) = 0$, $Q = 12$, constant
$M + 10x + 2(1)(x - 1.0/2) = 0$, $M = -12x + 1.0$, linear variation
$x = 1.0$, $Q = 12$, $M = -11.0$ and $x = 1.5$, $Q = 12$ and $M = -17.0$
Note how in the segment AB, the uniformly distributed load component causes parabolic variation of the bending moment but the same load only causes linear variation in the segment BC.

(c) Segment CD: $1.5 < x \leqslant 2.2$
$Q - 10.0 - 2(1) - 20 = 0$, $Q = 32$, constant
$M + 10x + 2(1)(x - 1.0/2) + 20(x - 1.5) = 0$, $M = -32x + 31.0$, linear variation
$x = 1.5$, $Q = 32$, $M = -17.0$ and $x = 2.2$, $Q = 32$ and $M = -39.4$

(d) Segment DE: $2.2 < x \leqslant 3.2$
$Q - 10.0 - 2(1) - 20 - 4(x - 2.2) = 0$, $Q = 23.2 + 4x$, linear variation
$M + 10x + 2(1)(x - 1.0/2) + 20(x - 1.5) + 4(x - 2.2)^2/2 = 0$
$M = -32x + 31.0 - 2(x - 2.2)^2$, parabolic variation
$x = 2.2$, $Q = 32$, $M = -39.4$ and $x = 3.2$, $Q = 36$ and $M = -73.4$

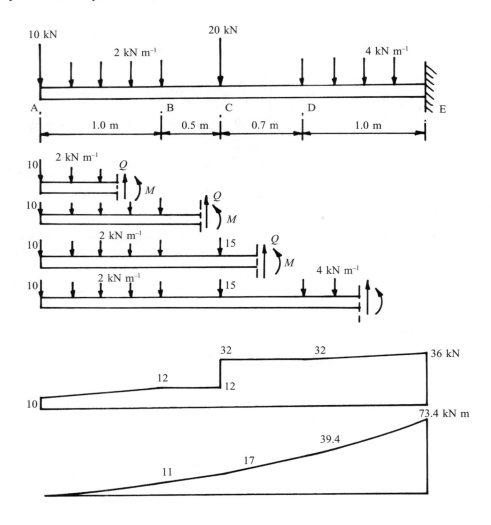

Figure 1.50 ● Forces acting on a cantilever

Example 7

Figure 1.51 shows a cantilever with an inclined load. Calculate the BM and the SF distribution in the beam.

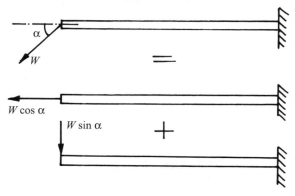

Figure 1.51 ● A cantilever with an inclined load

Solution　The inclined load W can be resolved into horizontal and vertical components $W \cos \alpha$ and $W \sin \alpha$ respectively. Clearly the horizontal component $W \cos \alpha$ induces only axial tension while the vertical component $W \sin \alpha$ induces BM and SF as in Example 1.

This example shows that only the components of load acting *perpendicular to the axis of the beam* induce BM and SF.

General observation. In a cantilever, the maximum bending moment and shear force occur at the support. It is interesting to note that when expressed as a polynomial, M is always one degree higher than Q. For example, if M is linear in x then Q is constant. Similarly, if M is quadratic then Q is linear in x and so on. This observation is general and is applicable to all beams. It will be proved in Section 1.12.

1.11.2　Simply supported beams

In a simply supported beam, one end is supported on a pin-support or hinge which prevents the beam from moving either horizontally or vertically. The other end is supported on a roller support which prevents vertical movement. At both the supports, the beam rotates freely. The roller support also allows for free expansion due to thermal changes, etc.

Example 1

A simply supported beam with a central concentrated load W as shown in Fig. 1.52.

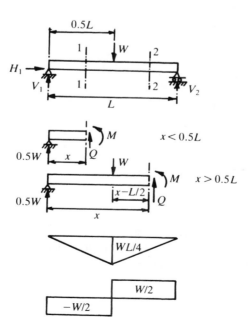

Figure 1.52 ●
Bending moment and shear force distribution in a simply supported beam with a central concentrated load

From symmetry, the vertical reactions V_1 and V_2 are $0.5W$. Because there are no horizontal loads, the horizontal reaction at the left support is also zero. Considering the free-body diagrams, for equilibrium in vertical direction

$$Q + V_1 = 0 \text{ if } x < 0.5L \text{ and } Q + V_1 - W = 0 \text{ if } x > 0.5L$$

As $V_1 = 0.5W$

$$Q = -0.5W \text{ for } x < 0.5L \text{ and } Q = 0.5W \text{ if } x > 0.5L$$

This indicates that there is a discontinuity in the distribution of SF at $x = 0.5L$

$$M - V_1x = 0 \text{ for } x \leqslant 0.5L \text{ and } M - V_1x + W(x - 0.5L) = 0 \text{ for } x \geqslant 0.5L$$

As $V_1 = 0.5W$

$$M = 0.5Wx \text{ for } x \leqslant 0.5L, \ M = 0.5W(L - x) \text{ for } x \geqslant 0.5L$$

The SFD and BMD are shown in Fig. 1.52; $M_{max} = WL/4$ at the centre and $Q = \pm 0.5W$ over the whole beam with an abrupt change in shear force at the midspan. The abrupt change in SF is equal to the load acting at the section.

Example 2

A simply supported beam with a uniformly distributed load q per unit length as shown in Fig. 1.53.

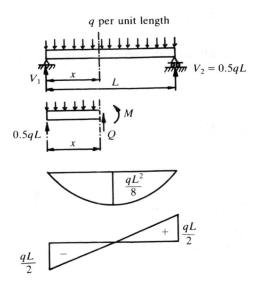

Figure 1.53 ● BM and SF distribution in a simply supported beam with u.d.l. q

From symmetry, the vertical reactions V_1 and V_2 are $0.5qL$. Because there are no horizontal loads, the horizontal reaction at the left support is also zero. Considering the free-body diagram, we have

$$Q + V_1 - qx = 0$$

As $V_1 = 0.5qL$

$$Q = -q(0.5L - x)$$

Similarly

$$M - V_1x + (qx)(x/2) = 0, \text{ therefore } M = 0.5qx(L - x)$$

The SFD and BMD are shown in Fig. 1.53. Note that the BMD is a parabola. $M_{max} = qL^2/8$ at the centre and $Q_{max} = \pm0.5qL$ at the ends. The total load W on the beam is given by $W = qL$. Therefore $M_{max} = qL^2/8 = WL/8$ which is only half the corresponding maximum bending moment if the load W was a concentrated load at the centre as in Example 1. The maximum shear force $Q_{max} = \pm0.5W$. An important lesson to be learnt from this is that *spreading the load decreases the intensity of maximum bending moment in the beam*. The maximum value of the shear force is not greatly influenced by the spread of loading.

Example 3

A simply supported beam with a clockwise couple M_1 at the left-hand end as shown in Fig. 1.54.

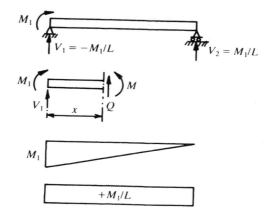

Figure 1.54 ● BM and SF distribution in a simply supported beam with an end couple

Taking moments about the left-hand support, $M_1 - V_2L = 0$, $V_2 = M_1/L$. Because $V_1 + V_2 = 0$, $V_1 = -M_1/L$, i.e. V_1 acts vertically downwards. Considering the free-body diagram

$$Q + V_1 = 0 \text{ or } Q = -V_1 = M_1/L$$

$$M - M_1 - V_1x = 0, M = M_1 - M_1x/L = M_1(1 - x/L)$$

The SFD and BMD are shown in Fig. 1.54; $M_{max} = M_1$ causing tension at the bottom face and occurs at the left-hand support section. $Q = M_1/L$ is constant over the whole beam.

Example 4

A simply supported beam with a concentrated clockwise couple M_c at a distance 'a' from the left support as shown in Fig. 1.55.

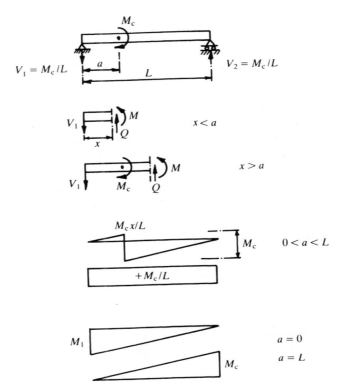

Figure 1.55 ● BM and SF distribution in a simply supported beam with an external couple in the span

Taking moments about the left-hand support, $M_c - V_2L = 0$. Therefore $V_2 = M_c/L$. Because $V_1 + V_2 = 0$, $V_1 = -M_c/L$, i.e. V_1 acts vertically downwards. Considering the free-body diagram

$$Q + V_1 = 0 \text{ giving } Q = -V_1 = M_c/L$$

$$M + V_1x = 0 \text{ if } x < a \text{ and } M - M_c + V_1x = 0 \text{ if } x > a,$$

$$M = -V_1x = -M_cx/L \text{ if } x < a \text{ and } M = M_c - M_cx/L = M_c(1 - x/L) \text{ if } x > a$$

The SFD and BMD are shown in Fig. 1.55. It is interesting to note that similar to the discontinuity in the case of the SFD with a central concentrated load (see Example 1, Section 1.11.1), in this case the BMD is discontinuous. The abrupt change in the bending moment at the section is equal to the concentrated couple acting at the section.

M_{max} = maximum of either $M_c(a/L)$ or $M_c(1 - a/L)$ depending on whether the ratio a/L is greater than or less than 0.5. $Q = M_c/L$ and remains constant over the whole beam. The BMD for the case $a = 0$ and $a = L$ is also shown in Fig. 1.55. By superposition, the BMD in the case when the couples at the supports are either both equal or equal and opposite can be obtained and are shown in Fig. 1.56.

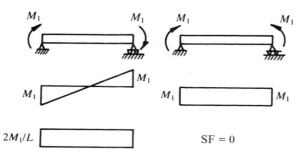

Figure 1.56 ●
BM and SF distribution
in a simply supported
beam due to symmetric
and antisymmetric
system of end couples

Example 5

A simply supported beam with a triangularly distributed load $q(x) = kx$ where k is a constant as shown in Fig. 1.57.

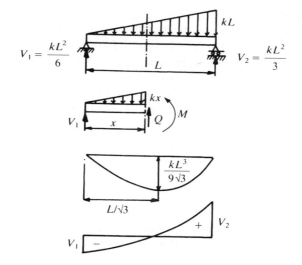

Figure 1.57 ●
Bending moment and
shear force distribution
in a simply supported
beam under triangularly
distributed load

The total load on the beam is $0.5(kL)L = 0.5kL^2$. The resultant load acts at a distance of $2L/3$ from the left-hand support. Taking moments at the left-hand support, $(0.5kL^2)(2L/3) - V_2L = 0$ or $V_2 = kL^2/3$. Since $V_1 + V_2 = $ total load on the beam $= 0.5kL^2$,

$$V_1 = 0.5kL^2 - V_2 = kL^2/6$$

Considering the free-body diagram

$$Q + V_1 - 0.5(kx)x = 0, \ Q = 0.5kx^2 - V_1$$

Therefore

$$Q = \{kL^2/6\}[3(x/L)^2 - 1]$$

$$M - V_1x + 0.5(kx)x(x/3) = 0$$

Table 1.1

Loading	M_{max}	Q_{max}
Midspan concentrated load W	$0.25WL$	$0.5W$
Uniformly distributed load	$0.125WL$	$0.5W$
Triangular load	$0.128WL$	$0.67W$

Therefore

$$M = \{kL^2x/6\}[1 - (x/L)^2]$$

The SFD and BMD are shown in Fig. 1.57. The maximum bending moment occurs when

$$dM/dx = 0 = \{kL^2/6\}[1 - 3x^2/L^2]$$

This occurs when

$$\{1 - 3(x/L)^2\} = 0 \text{ or } (x/L) = 1/\sqrt{3} = 0.577$$

Evidently in this case the maximum bending moment occurs off centre. Substituting the value of $x = L/\sqrt{3}$, $M_{max} = kL^3/(9\sqrt{3})$. The total load W on the beam is $0.5(kL)L = W$. Therefore $M_{max} = 2WL/(9\sqrt{3}) = 0.128WL$. As can be seen by comparison with the maximum bending moment of $WL/8 = 0.125WL$ when the total load W is uniformly distributed, the maximum bending moment in the case of a triangular load is approximately equal to the value of the corresponding bending moment in the uniformly distributed load (u.d.l.) case. $Q_{max} = +kL^2/3 = 0.667W$ at the right-hand support.

The results from Examples 1, 2 and 5 are shown in Table 1.1 when a total load W is distributed in different ways.

Example 6

Figure 1.58 shows a simply supported beam with distributed and concentrated loads and a couple acting on it. Draw the BMD and the SFD.

Solution First of all determine the reactions. As there are only vertical loads acting, the reactions at A and J will also be vertical.
For vertical equilibrium,
$V_A + V_J = 20 + 12 \times 1.5 + 50 + 15 \times 1.8 = 115$ kN
Taking moments about A,
$20 \times 1.0 + 12 \times 1.5 \times (1.5 + 1.5/2) + 50 \times 3.7 + 40$
$+ 15 \times 1.8(5.4 + 1.8/2) - V_J \times 9.2 = 0$
Note that the couple appears as 40, the value of the couple without any associated lever arm, because the moment of a couple about any point is the same as the value of the couple itself.
Solving, $V_J = 49.52$ kN and $V_A = 65.48$ kN
As a check, taking moments about J,
$-15 \times 1.8(2.0 + 1.8/2) - 40 - 50 \times 5.5 - 12 \times 1.5(6.2 + 1.5/2)$
$- 20 \times 8.2 + V_A \times 9.2 = 0$
From which $V_A = 65.48$ kN as before.

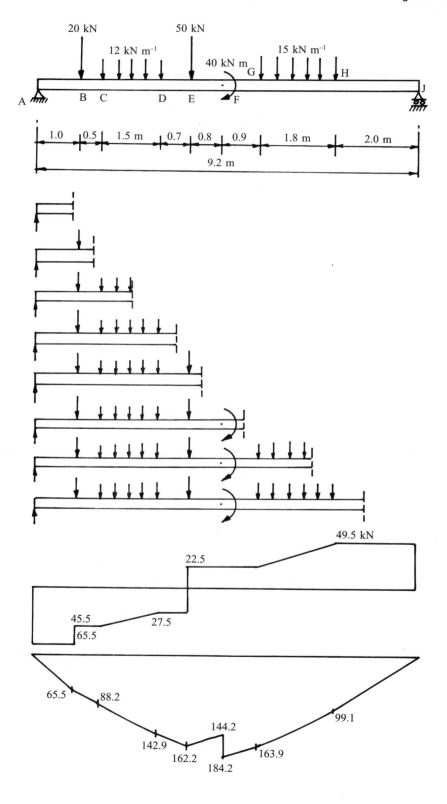

Figure 1.58 ●
Forces acting on a
simply supported beam

Using the free-body diagrams for different sections, write down the expressions for BM and SF.

Using Macaulay brackets, the expressions for SF and BM are given by

$$Q = -65.48 + 20\langle x - 1\rangle^0 + 12\{\langle x - 1.5\rangle - \langle x - 3\rangle\} + 50\langle x - 3.7\rangle^0$$
$$+ 15\{\langle x - 5.4\rangle - \langle x - 7.2\rangle\} - 49.52\langle x - 9.2\rangle^0$$
$$M = -65.48x + 20\langle x - 1\rangle^1 + 12\{\langle x - 1.5\rangle^2 - \langle x - 3\rangle^2\}/2 + 50\langle x - 3.7\rangle^1$$
$$+ 40\langle x - 4.5\rangle^0 + 15\{\langle x - 5.4\rangle^2 - \langle x - 7.2\rangle^2\}/2 - 49.52\langle x - 9.2\rangle^1$$

Or taking section by section:

(a) Segment AB: $0 < x \leqslant 1.0$

$Q + (V_A = 65.48) = 0$, $Q = -65.48$, constant

$M = V_A x$, $M = 65.48x$, linear variation

$x = 0$, $Q = -65.48$, $M = 0.0$ and $x = 1.0$, $Q = -65.48$ and $M = 65.48$

(b) Segment BC: $1.0 < x \leqslant 1.5$

$Q + V_A - 20.0 = 0$, $Q = -45.48$, constant

$M = V_A x - 20(x - 1.0)$, $M = 45.48x + 20.0$, linear variation

$x = 1.0$, $Q = -45.48$, $M = 65.48$ and $x = 1.5$, $Q = -45.48$ and $M = 88.22$

(c) Segment CD: $1.5 < x \leqslant 3.0$

$Q + V_A - 20.0 - 12(x - 1.5) = 0$, $Q = -63.48 + 12x$, linear variation

$M = V_A x - 20(x - 1.0) - 12(x - 1.5)^2/2 = 0$

$M = 45.48x + 20.0 - 6(x - 1.5)^2$, parabolic variation

$x = 1.5$, $Q = -45.48$, $M = 88.22$ and $x = 3.0$, $Q = -27.48$ and $M = 142.94$

(d) Segment DE: $3.0 < x \leqslant 3.7$

$Q + V_A - 20.0 - 12(1.5) = 0$, $Q = -27.48$, constant

$M = V_A x - 20(x - 1.0) - 12(1.5)(x - 1.5 - 1.5/2) = 0$

$M = 27.48x + 60.5$, linear variation

$x = 3.0$, $Q = -27.48$, $M = 142.94$ and $x = 3.7$, $Q = -27.48$ and $M = 162.18$

(e) Segment EF: $3.7 < x \leqslant 4.5$

$Q + V_A - 20.0 - 12(1.5) - 50.0 = 0$, $Q = 22.52$, constant

$M = V_A x - 20(x - 1.0) - 12(1.5)(x - 1.5 - 1.5/2) - 50(x - 3.7) = 0$

$M = -22.52x + 245.5$, linear variation

$x = 3.7$, $Q = 22.52$ and $M = 162.18$ and $x = 4.5$, $Q = 22.52$ and $M = 144.16$

(f) Segment FG: $4.5 < x \leqslant 5.4$

$Q + V_A - 20.0 - 12(1.5) - 50.0 = 0$, $Q = 22.52$, constant

$M = V_A x - 20(x - 1.0) - 12(1.5)(x - 1.5 - 1.5/2) - 50(x - 3.7) + 40 = 0$

$M = -22.52x + 285.5$, linear variation

$x = 4.5$, $Q = 22.52$ and $M = 184.16$ and $x = 5.4$, $Q = 22.52$ and $M = 163.89$

(g) Segment GH: $5.4 < x \leqslant 7.2$

$Q + V_A - 20.0 - 12(1.5) - 50.0 - 15(x - 5.4) = 0$, $Q = -58.48 + 15x$, linear variation

$M = V_A x - 20(x - 1.0) - 12(1.5)(x - 1.5 - 1.5/2) - 50(x - 3.7) + 40$
$- 15(x - 5.4)^2/2 = 0$

$M = -22.52x + 285.5 - 7.5(x - 5.4)^2$, parabolic variation

$x = 5.4$, $Q = 22.52$ and $M = 163.89$ and $x = 7.2$, $Q = 49.52$ and $M = 99.06$

(h) Segment HJ: $7.2 < x \leqslant 9.2$

$Q + V_A - 20.0 - 12(1.5) - 50.0 - 15(1.8) = 0$, $Q = 49.52$, constant

$M = V_A x - 20(x - 1.0) - 12(1.5)(x - 1.5 - 1.5/2) - 50(x - 3.7) + 40$
$- 15(1.8)(x - 6.3) = 0$

$M = -49.52x + 455.6$, linear variation

$x = 7.2$, $Q = 49.52$ and $M = 99.06$ and $x = 9.2$, $Q = 49.52$ and $M = 0$

Example 7

Figure 1.59 shows a simply supported beam with uniformly and also triangularly distributed loads and a concentrated load acting on it. Draw the BMD and the SFD.

Solution First of all determine the reactions. As there are only vertical loads acting, the reactions at A and F will also be vertical.

$V_A + V_F = 20 \times 3 + 100 + \{0.5 \times 3 \times 40\} = 220$ kN

Note that for a triangularly distributed load, the total load is equal to (length of load × maximum intensity)/2

Taking moments about A,

$-20 \times 3 \times (3/2) - 100 \times 5 - (3 \times 40/2)\{6 + (2/3)3\} + V_F \times 10 = 0$

Note that for a triangularly distributed load, the resultant of the load acts from the start of the load at (2/3) of the length of the load.

$V_F = 107$ kN and $V_A = 113$ kN

As a check, taking moments about F,

$20 \times 3 \times (10 - 3/2) + 100 \times 5 + (3 \times 40/2)\{1 + (1/3)3\} - V_A \times 10 = 0$

$V_A = 113$ kN as before.

Using Macaulay brackets, the expressions for SF and BM become

$Q = -113 + 20\{x - \langle x - 3 \rangle\} + 100\langle x - 5 \rangle^0 + 40\{0.5\langle x - 6 \rangle^2 - \langle x - 9 \rangle - 0.5\langle x - 9 \rangle^2\}$
$- 107\langle x - 10 \rangle^0$

$M = -113x + 20\{x^2 - \langle x - 3 \rangle^2\}/2 + 100\langle x - 5 \rangle^1$
$+ 40\{0.5\langle x - 6 \rangle^3/3 - \langle x - 9 \rangle^2/2 - 0.5\langle x - 9 \rangle^3/3\} - 107\langle x - 10 \rangle^1$

(a) Segment AB: $0 < x \leqslant 3.0$

$Q + (V_A = 65.48) - 20x = 0$, $Q = -113 + 20x$, linear variation

$M = V_A x - 20x(x/2)$, $M = 113x - 10x^2$, parabolic variation

$x = 0$, $Q = -113.0$, $M = 0.0$ and $x = 3.0$, $Q = -53.0$ and $M = 249.0$

(b) Segment BC: $3.0 < x \leqslant 5.0$

$Q + (V_A = 65.48) - 60 = 0$, $Q = -53$, constant

$M = V_A x - 20(3)(x - 3/2)$, $M = 53x + 90.0$, linear variation

$x = 3.0$, $Q = -53.0$, $M = 249.0$ and $x = 5.0$, $Q = -53.0$ and $M = 355.0$

(c) Segment CD: $5.0 < x \leqslant 6.0$

$Q + (V_A = 65.48) - 20(3) - 100 = 0$, $Q = 47.0$, constant

$M = V_A x - 20(3)(x - 3/2) - 100(x - 5.0)$, $M = -47x + 590.0$, linear variation

$x = 5.0$, $Q = 47.0$, $M = 355.0$ and $x = 6.0$, $Q = 47.0$ and $M = 308.0$

(d) Segment DE: $6.0 < x \leqslant 9.0$

$Q + (V_A = 65.48) - 20(3) - 100 - (x - 6)\{(40/3)(x - 6)\}/2 = 0$

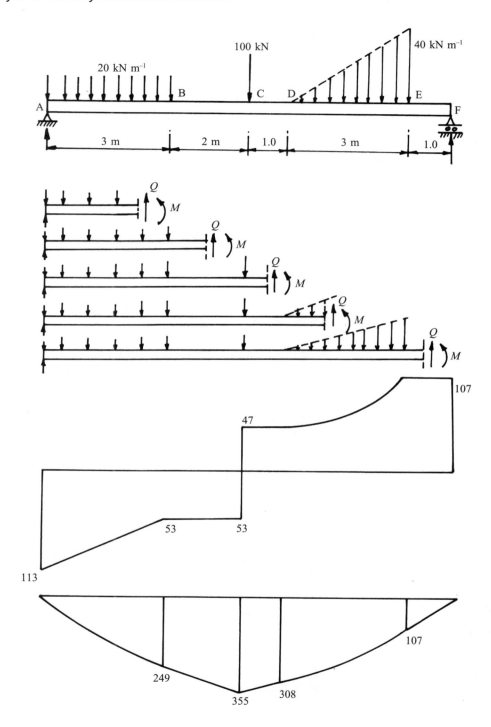

Figure 1.59 ● Forces acting on a simply supported beam

Note that if the length of the triangular load involved is $(x - 6)$, the intensity of the load at the end of 3 m is 40, then the intensity of the load at the end of $(x - 6)$ is $\{(40/3)(x - 6)\}$.

$Q = 47.0 + (20/3)\{x - 6\}^2$, parabolic variation

$M = V_{A}x - 20(3)(x - 3/2) - 100(x - 5.0) - [0.5(x - 6)(40/3)(x - 6)]\{(x - 6)/3\}$

Note that the terms inside square brackets represent the triangular part of the load in the region $(x - 6)$ and the terms inside curly brackets represent the distance to the triangular part of the load involved from the section under consideration.

$M = -47x + 590.0 - (20/9)(x - 6)^3$, cubic variation

$x = 6.0$, $Q = 47.0$, $M = 308.0$ and $x = 9.0$, $Q = 107.0$ and $M = 107.0$

(e) Segment EF: $9.0 < x \leqslant 10.0$

$Q + (V_{A} = 65.48) - 20(3) - 100 - 0.5(3)(40) = 0$, $Q = 107.0$, constant

$M = V_{A}x - 20(3)(x - 3/2) - 100(x - 5.0) - (0.5)(3)(40)\{x - 6 - (2/3)3\}$

$M = -107x + 1070.0$, linear variation

$x = 9.0$, $Q = 107.0$, $M = 107.0$ and $x = 10.0$, $Q = 107.0$ and $M = 0.0$

Example 8

A simply supported beam of 12 m span carrying a distributed load of 2 kN m^{-1} over the whole span and two concentrated loads of 15 kN and 25 kN acting at 4 m and 7 m from the left-hand end as shown in Fig. 1.60.

Taking moments about the left-hand support, $V_{2}12 - (2 \times 12)6$ due to u.d.l. $-15 \times 4 - 25 \times 7 = 0$, thus $V_{2} = 31.58$ kN. Since $V_{1} + V_{2} = $ total load on the beam $= 2 \times 12$ due to u.d.l. $+ 15 + 25 = 64$ kN, then $V_{1} = 64 - V_{2} = 32.42$ kN. Proceeding in the usual way, by taking Sections between AB, BC and CD

$Q + V_{1} - 2x = 0$ for $0 \leqslant x < 4$

$Q + V_{1} - 2x - 15 = 0$ for $4 \leqslant x < 7$

$Q + V_{1} - 2x - 15 - 25 = 0$ for $7 \leqslant x < 12$

In fact, the above three cases can be written as a single equation:

$Q + V_{1} - 2x - 15\langle x - 4 \rangle^0 - 25\langle x - 7 \rangle^0 = 0$

where the convention adopted is that

$\langle x - a \rangle = 0$ if $x \leqslant a$

$\qquad = (x - a)$ if $x > a$

This convention is called Macaulay brackets or Singularity functions,

$Q = -V_{1} + 2x + 15\langle x - 4 \rangle^0 + 25\langle x - 7 \rangle^0$

$M = V_{1}x + (2x)(x/2) + 15\langle x - 4 \rangle + 25\langle x - 7 \rangle$, $0 \leqslant x \leqslant 12$

The variation of shear force Q is clearly linear and at specific points it is given by:

At $x = 0$, $Q = -V_{1} = -32.42$ kN

To the left of B, $Q = -V_{1} + 2(x = 4) = -24.42$ kN

To the right of B, $Q = -V_{1} + 2(x = 4) + 15 = -9.42$ kN

To the left of C, $Q = -V_{1} + 2(x = 7) + 15 = -3.42$ kN

To the right of C, $Q = -V_{1} + 2(x = 7) + 15 + 25 = 21.58$ kN

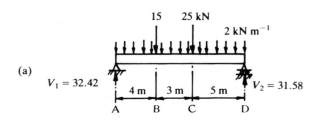

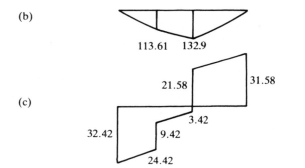

Figure 1.60 ● (a)–(c) Simply supported beam under uniformly distributed and concentrated loads

To the left of D, $Q = -V_1 + 2(x = 12) + 15 + 25 = 31.58$ kN
To the right of D, $Q = -V_1 + 2(x = 12) + 15 + 25 - V_2 = 0$ kN
The SFD can now be drawn as shown in Fig. 1.60(c). It is important to note that at the positions where concentrated loads or reactions act, there is a discontinuity in the SFD equal to the value of concentrated load or reaction acting at that point. Similarly
$M - V_1x + (2x)(x/2) = 0$, for $0 \leqslant x \leqslant 4$
$M - V_1x + (2x)(x/2) + 15(x - 4) = 0$, for $4 \leqslant x \leqslant 7$
$M - V_1x + (2x)(x/2) + 15(x - 4) + 25(x - 7) = 0$, for $7 \leqslant x \leqslant 12$
    Adopting the Macaulay convention, $M - V_1x + (2x)(x/2)$ due to u.d.l. $+ 15\langle x - 4 \rangle + 25\langle x - 7 \rangle = 0$.

$$M = V_1x - x^2 - 15\langle x - 4 \rangle - 25\langle x - 7 \rangle$$

The variation of bending moment M is clearly parabolic and at specific points it is given by:
At A, $x = 0$, $M = V_1x = 0$
At B, $x = 4$, $M = V_1x - (2x)(x/2) = 113.61$ kN m
At C, $x = 7$, $M = V_1x + (2x)(x/2) + 15(x - 4) = 132.9$ kN m
At D, $x = 12$, $M = V_1x + (2x)(x/2) + 15(x - 4) + 25(x - 7) = 0$
The BMD can now be drawn. It is shown in Fig. 1.60(b).

1.11.3 General expressions for BM and SF using Macaulay brackets

It is useful to have general expressions for bending moment M and shear force Q at a section x from the origin due to discontinuous applied loads.

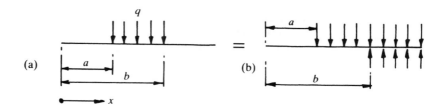

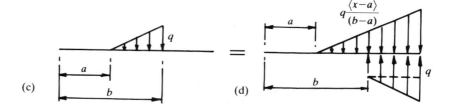

Figure 1.61 ● (a)–(d)
General uniformly and
triangularly distributed
loading

1. Set of N concentrated loads with load W_i at a distance a_i from the origin:

$$Q = Q_0 + \sum W_i \langle x - a_i \rangle^0$$

$$M = M_0 - \sum W_i \langle x - a_i \rangle^1$$

where Q_0 = shear force at the section due to forces like reactions, etc. at the origin, M_0 = bending moment at the section due to forces like reactions, etc. at the origin, and Σ is the summation over the loads 1 to N.

2. Set of N concentrated couples with M_{ci} at a distance a_i from the origin:

$$Q = Q_0$$

$$M = M_0 - \sum M_{ci} \langle x - a_i \rangle^0$$

3. Set of N uniformly distributed loads with q_i extending from a_i to b_i from the origin (see Fig. 1.61(a)): it should be noted that when using the Macaulay bracket convention, there is no procedure for cancelling the effect of a load except by applying a reverse load. Thus a load q_i extending from a_i to b_i can be thought of as a load q_i extending from a_i to infinity plus a load $-q_i$ extending from b_i to infinity (see Fig. 1.61(b)).

$$Q = Q_0 + \sum q_i \left\{ \langle x - a_i \rangle - \langle x - b_i \rangle \right\}$$

$$M = M_0 - \sum 0.5 q_i \left\{ \langle x - a_i \rangle^2 - \langle x - b_i \rangle^2 \right\}$$

4. Set of N triangularly distributed loads with an apex height of q_i at $x = b_i$ and extending from a_i to b_i from the origin (see Fig. 1.61(c)): in this case the given load can be treated as the sum of three loads, i.e.

(a) given load starting from $x = a_i$ and extending to infinity. The height of the load at any distance x is $q_i\langle x - a_i\rangle/(b_i - a_i)$. The total value W_i of the load which is triangularly distributed and extending from a_i to x is given by

$$W_i = 0.5\langle x - a_i\rangle \text{ height}$$
$$= 0.5q_i\langle x - a_i\rangle^2/(b_i - a_i)$$

Similarly, the bending moment caused by this load whose centroid is at a distance $\langle x - a_i\rangle/3$ from the section is given by

$$M_i = -W_i \langle x - a_i\rangle/3 = 0.167q_i\langle x - a_i\rangle^3/(b_i - a_i)$$

(b) a uniformly distributed load q_i starting from b_i and extending to infinity which can be treated as in (3) above.

(c) the given load starting from b_i and extending to infinity which can be treated as in (a) above.

The above three loads are shown in Fig. 1.61(d). The final expressions for M and Q are as follows:

$$Q = Q_0 + \sum q_i \left\{ -\langle x - b_i\rangle + \frac{0.5\langle x - a_i\rangle^2}{(b_i - a_i)} - \frac{0.5\langle x - b_i\rangle^2}{(b_i - a_i)} \right\}$$

$$M = M_0 - \sum q_i \left\{ -0.5\langle x - a_i\rangle^2 + \frac{0.167\langle x - a_i\rangle^3}{(b_i - a_i)} - \frac{0.167\langle x - b_i\rangle^3}{(b_i - a_i)} \right\}$$

Note: The observation previously made in the case of cantilever beams, namely that the polynomial expression for BM is always one degree higher than that for SF, is also true in the case of simply supported beams.

1.12 ● Relationship between bending moment and shear force

It was observed in connection with the examples on BM and SF calculation for simple loading on cantilevers and simply supported beams, that the polynomial expression for BM was always one degree higher than that for SF. It will now be proved that this is a general property linking BM and SF. Consider an infinitesimal length dx of the beam shown in Fig. 1.62. Since the BM and SF vary from section to section, let the bending moment on the left and right faces of the infinitesimal beam segment be M and $\{M + (dM/dx)dx\}$ respectively. dM/dx is the rate of increase of M with x and $\{(dM/dx)dx\}$ is the increase in M over a length dx. Similarly, let the shear force on the left and right faces of the infinitesimal segment be Q and $\{Q + (dQ/dx)dx\}$ respectively. Let the applied lateral load on the beam be $q(x)$, indicating that it is a function of x. Let $q(x)$ be positive in the positive y-direction. Considering the vertical and rotational equilibrium of the element, we have

$$\{Q + (dQ/dx)dx\} - Q + q(x) = 0$$

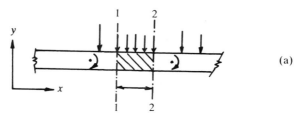

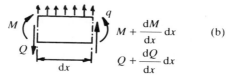

Figure 1.62 ●
Relationship between
BM and SF at a section

i.e. $dQ/dx + q(x) = 0$

Taking moments about an axis parallel to the z-axis through the right-hand end

$$\{M + (dM/dx)\, dx\} + Q\, dx - M - (q\, dx)\, dx/2 = 0$$

Ignoring terms of second order we have

$$dM/dx + Q = 0$$

Differentiating with respect to x

$$d^2M/dx^2 + dQ/dx = 0$$

Eliminating dQ/dx, we have

$$d^2M/dx^2 = q(x)$$

The above equation is often stated as the fundamental equation of equilibrium for beam elements.

As shown in Fig. 1.62(c), if a concentrated external load W acts at a section, then, for equilibrium, the shear force on sections immediately to either side of the load should differ by the load W. Similarly, if a concentrated external couple M_c acts at a section as shown in Fig. 1.62(d), then, for equilibrium, the bending moment on sections immediately to either side of the couple should differ by the applied couple M_c.

Table 1.2

q	Q	M
0	Constant	Linear
Constant	Linear	Parabolic
Linear	Parabolic	Cubic

Consider first the equation $dQ/dx = -q(x)$. This indicates that the rate of change of shear force is directly proportional to the lateral load on the beam. If q is zero, then dQ/dx is zero, indicating that Q is constant. Similarly, if q is constant, then dQ/dx is constant and therefore Q is linear and if q is a linear function of x (e.g. $q = kx$), then Q is quadratic and so on.

Next consider the equation $dM/dx = -Q$. The relationship that has been observed with respect to q and Q applies respectively to Q and M. For simple loading cases commonly occurring in practice, the information in Table 1.2 will be useful for checking the shape of the SFD and BMD.

It can therefore be concluded that in the case of a beam loaded by a series of concentrated loads, where $q = 0$ the SFD consists of a series of constant value segments. At the positions where the concentrated loads act, there is an abrupt change in the value of the shear force equal to the concentrated load at the section. The BMD consists of a series of linear segments.

Another useful result is that symmetrical loading on symmetrical structures leads to symmetrical distribution of bending moment but asymmetrical distribution of shear force.

The reader should check the SFD and BMD for the examples in previous sections to verify that the above rules relating q, Q and M have been satisfied.

Since for maximum (or minimum) values of the bending moment M, $dM/dx = 0$ and since dM/dx = shear force Q, then for maximum (minimum) M, $Q = 0$. This is a convenient way of determining the position where the maximum (or minimum) bending moment occurs. As a simple example, it can be seen that in the case of symmetrically loaded simply supported beams the maximum bending moment occurs at the centre where the shear force is zero because of symmetry.

1.12.1 Shear force near concentrated loads and bending moment near concentrated couples

It was noticed in connection with SFD that at the position of concentrated forces (loads or reactions), there is discontinuity equal to the concentrated force acting at that position. In practice concentrated knife edge loads do not exist. They are meant to represent heavy loads acting over a small area. For students, discontinuities are very confusing. For example, in the case of a simply supported beam with a concentrated load (Fig. 1.63(a)), students often have trouble in visualizing what exactly is the shear force at the position of the concentrated load. In order to avoid uncertainty about the exact shear force at a section, it is suggested that the given concentrated force is imagined to be distributed over a small width and the corresponding SFD sketched. This is shown in Fig. 1.63(b) by a 'full line' for the case of a simply supported beam with a central concentrated load. Comparing the 'full line' SFD

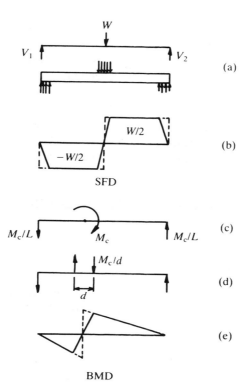

Figure 1.63 ● (a)–(e) SF near a concentrated load and BM near a concentrated couple

in Fig. 1.63(b) with the corresponding SFD in Fig. 1.52 (shown by a 'dotted line' in Fig. 1.63(b)), it is clear that the SFD in Fig. 1.63(b) avoids concentrated discontinuities and thus avoids any ambiguity about the value of the shear force at a section. For example, in the above it is quite clear that the shear force at midspan is zero and the shear force at a small distance from the midspan is almost equal to $0.5W$. The technique for avoiding uncertainty in calculating SF can also be applied in the case of the BMD. As was shown in Fig. 1.55, at the position of a concentrated couple, there is a discontinuity equal to the value of the couple acting at that section. A couple M_c can always be replaced by a set of equal and opposite parallel forces F at a distance d apart as shown in Fig. 1.63(d), since for equilibrium $Fd = M_c$, $F = M_c/d$. Keeping d small, the corresponding BMD can be drawn as shown in Fig. 1.63(e). Comparing the 'full line' BMD in Fig. 1.63(e) with the BMD shown in Fig. 1.55 (shown by the 'dotted line' in Fig. 1.63(e)), it is clear that Fig. 1.63(e) avoids concentrated discontinuity in the BMD and that the BM is zero at the section where the concentrated couple acts.

1.13 ● Statically determinate 2-D rigid-jointed frames in the vertical plane

In Section 1.11, cantilevers and simply supported beams were considered. In this section statically determinate rigid-jointed structures assembled from several beam elements will be discussed. It is assumed that both the structure and the loads lie in the vertical x–y plane.

For the purposes of constructing the SFD and BMD it is convenient to separate the given frame into a series of cantilever and simply supported elements, ensuring the equilibrium of each. In this connection, the designation 'cantilever element' simply means an element supported at one end only. It is not necessary that translation and rotation at the supported end be zero. Similarly, the designation 'simply supported element' is used to mean an element connected at both ends to other elements. These ideas are illustrated by simple examples.

1.14 ● Examples of statically determinate 2-D rigid-jointed frames

Example 1

Figure 1.64 shows a beam on two simple supports with 'overhangs' over each support and subjected to a uniformly distributed load q per unit length.

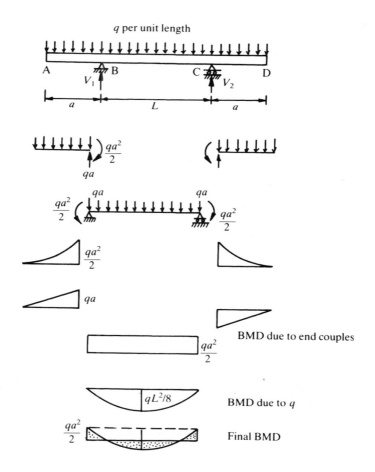

Figure 1.64 ●
BM and SF distribution
in a simply supported
beam with overhangs

Solution The given structure can be divided into two cantilever segments (AB and CD) and a simply supported beam segment (BC). Considering the equilibrium of the cantilever segments AB and DC, it is clear that at the supports B and C respectively an upward vertical force (qa) and a couple $0.5qa^2$ (clockwise at B and anticlockwise at C) are required. These reactions are supplied by the simply supported beam BC. According to Newton's third law of motion, action and reaction are equal and opposite, and so a downward vertical force qa and a couple $qa^2/2$ (anticlockwise at B and clockwise at C) act at the ends B and C of the beam.

Cantilever segments AB and DC. Using the information from Example 3, Section 1.11.1, the BMD and SFD can be drawn. The maximum bending moment and shear force are respectively $0.5qa^2$ and qa and both occur at the supports B and C.

Simply supported segment BC. The beam is loaded by the uniformly distributed load (u.d.l.) q over its length and at its ends by couples $qa^2/2$ and vertical forces qa as actions from the cantilever segments AB and DC. From Example 2 of Section 1.11.2 the BMD for q is parabolic with a central ordinate of $qL^2/8$. Similarly, from Example 4 of Section 1.11.2 the BMD due to couples equal to $qa^2/2$ is a constant value of $qa^2/2$. The two bending moment diagrams are shown separately. The final BMD is obtained by the superposition of the two separate BMDs. The final SFD is due to the u.d.l. only because the equal and opposite couples $0.5qa^2$ do not induce any shear force.

The maximum bending moment M_{span} at the centre of the simply supported span BC is equal to $M_{span} = (qL^2/8 - qa^2/2)$. The maximum bending moment at the supports $M_{support}$ at B and C is equal to $M_{support} = qa^2/2$. Very often in practical design it is useful to make the two values equal. If $M_{span} = M_{support}$, then $(qL^2/2 - qa^2/2) = qa^2/2$ or $L = \sqrt{2}a$. This is a simple instance where designers may proportion their structure to optimize the use of material.

Example 2

A right-angle rigid-jointed frame loaded by a uniformly distributed load q and a concentrated load at the tip equal to qL as shown in Fig. 1.65.

The frame can be divided into two cantilever segments BC and AB. Since all the forces acting on the cantilever BC except the 'support reactions' at B are known, it is convenient to start the analysis from segment BC.

1. Cantilever BC. The reactions at the support B for equilibrium are a vertical force to support a total force of (qL due to concentrated load + qL due to u.d.l.) = $2qL$ and a couple {($qL)L$ due to concentrated load + $0.5qL^2$ due to u.d.l.} = $1.5qL^2$. The BMD and SFD can be drawn easily by superimposing the case of tip load $W = qL$ and u.d.l. q respectively from the information in Examples 2 and 3 of Section 1.11.1.

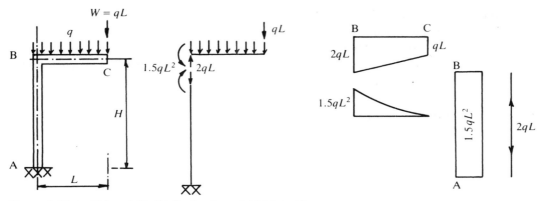

Figure 1.65 ● BM and SF distribution in a rigid-jointed frame

2. Cantilever AB. This is loaded at B by the reactions from the cantilever BC. The actions consist of a vertical force $2qL$ which acts in the axial direction of the beam AB and is therefore an axial force. This force does not produce any bending moment in the cantilever AB. The second action is a clockwise couple $1.5qL^2$. Note that there cannot be a horizontal force at B because no horizontal forces act on the cantilever BC. The reactions at A consist of a vertical reaction equal to $2qL$ to balance the axial force applied at B and an anticlockwise couple $1.5qL^2$ to balance the clockwise couple $1.5qL^2$ applied at B. The cantilever is thus subjected to equal and opposite couples of $1.5qL^2$ at the ends which induces constant bending moment in the beam as in Fig. 1.45(b) and to an axial force of $2qL$ which induces an axial compression of $2qL$ in the beam. The BMD can be easily drawn.

This example shows that in rigid-jointed structures carrying loads in their own plane, the forces at a section in a member in general consist of an axial force, a shear force and a bending moment.

Example 3

Draw the BMD and SFD for the rigid-jointed frame in Fig. 1.66.

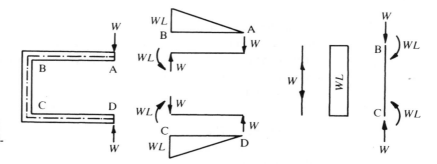

Figure 1.66 ● BM and SF distribution in a rigid-jointed frame

Solution The structure can be separated into two cantilever segments BA and CD and a simply supported segment CB. The cantilevers require reactions consisting of a vertical force W and a couple WL at B (and C). These act as actions on the simply supported beam BC. The forces on BC consist of axial force W and a couple WL at each end. The corresponding BMD and SFD can be easily drawn. The segment BC is subjected to an axial compressive force W and a constant bending moment of WL.

Example 4

Draw the BMD and the SFD for the rigid-jointed triangular frame resting on pinned and roller supports as shown in Fig. 1.67.

Solution   Because the applied load is vertical and one of the supports is on rollers, the reactions are vertical and by symmetry 10 kN each. Considering member AB, this is a cantilever supported at B and subjected to an inclined load of 10 kN at A. The component normal to AB is $10 \cos 30 = 8.66$ kN and the component along AB is $10 \sin 30 = 5$ kN. The length of AB is $1.5 \sec 30 = 1.732$ m. The member is subjected to an axial load of 5 kN compression, a constant shear force of 8.66 kN and

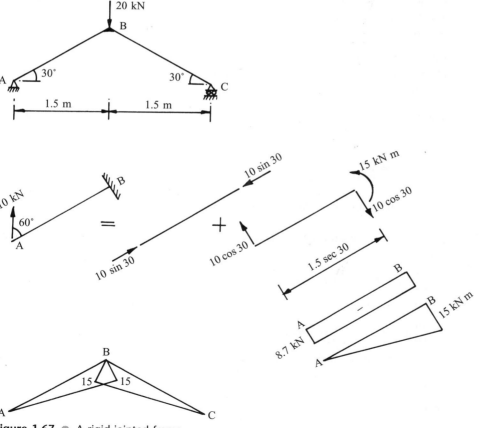

Figure 1.67 ● A rigid-jointed frame

a linearly varying BM of zero at A and $(8.66 \times 1.732) = 15$ kN m at B. It causes tension on the inside of the frame.

1.15 ● Structures on multiple supports made statically determinate by internal hinges

Consider the two beams shown in Fig. 1.68(a) and (b). Both beams have four external supports but the beam shown in Fig. 1.68(b) has two internal hinges. The beam in Fig. 1.68(a) is clearly statically indeterminate because (assuming that only vertical forces are present) there are only two equations of statics, i.e. $\Sigma F_y = 0$ and $\Sigma M_z = 0$ which can be used to solve for the four unknown vertical reactions. The beam in Fig. 1.68(b) has two additional internal hinges incorporated at C and D. Naturally, at a hinge the bending moment is zero. Thus the presence of two internal hinges provides two additional independent conditions which, together with the equations of statics $\Sigma F_y = 0$ and $\Sigma M_z = 0$, enable the determination of all the vertical reactions of the beam.

An internal hinge is a device used in practice to convert a basic statically indeterminate structure into a statically determinate structure. The reasons for preferring a statically determinate structure to a statically indeterminate one as discussed in Section 1.9 are

(a) statically determinate structures are simple to analyse;
(b) they do not have stresses induced by support settlement, temperature changes, etc.

The analysis of statically determinate structures with internal hinges is illustrated by simple examples.

Example 1

Analyse the statically determinate structure shown in Fig. 1.68(b).

Solution Let the vertical reactions at A, B, E and F respectively be V_1 to V_4. The equations of statics are

$$\sum F_y = V_1 + V_2 + V_3 + V_4 - q(4L) = 0 \tag{1.1}$$

Taking moments about an axis parallel to the z-axis and passing through A,

$$\sum M_z = (4qL)(2L) - V_2L - V_3 3L - V_4 4L = 0 \tag{1.2}$$

Considering free bodies ABC and DEF as in Fig. 1.68(c) and (d), we have

Moment at C = 0,

therefore

$$V_1(1.5L) + V_2(0.5L) - (q1.5L)(0.75L) = 0 \tag{1.3}$$

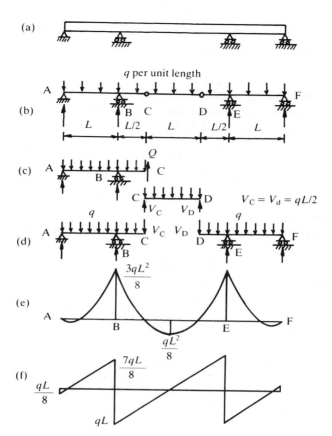

Figure 1.68 ● (a)–(f) BM and SF distribution in a balanced cantilever structure

Moment at D = 0,

therefore

$$-V_4(1.5L) - V_3(0.5L) + (q1.5L)(0.75L) = 0 \qquad [1.4]$$

The four equations can now be solved. However, in this example, the work is simplified because by symmetry, $V_1 = V_4$ and $V_2 = V_3$. Therefore,

$$\sum F_y = 0 = 2V_1 + 2V_2 = 4qL, \text{ i.e. } V_1 + V_2 = 2qL$$

Moment at C = 0,

therefore

$$V_1(1.5L) + V_2(0.5L) - (1.5qL)(0.75L) = 0$$

$$\text{i.e. } 1.5V_1 + 0.5V_2 = 1.125qL$$

Solving the equations

$$V_1 + V_2 = 2qL, \quad 1.5V_1 + 0.5V_2 = 1.125qL$$

$$V_1 = V_4 = qL/8, \quad V_2 = V_3 = 15qL/8$$

Note: No independent additional equations are obtained by taking moments about C or D of the *entire* structure, because this does not impose the condition that the bending moment at the hinge is zero. The bending moment at the hinge can be made equal to zero *only by considering the moment of all forces either to the left or the right of the hinge.*

Calculation of BM and SF values. Having calculated the values of the reactions, the bending moment and shear force can be easily calculated using the expressions developed in Section 1.11.3.

$$Q = -V_1 + qx - V_2\langle x - L\rangle^0 - V_3\langle x - 3L\rangle^0 - V_4\langle x - 4L\rangle^0$$

$$M = V_1x - qx^2/2 + V_2\langle x - L\rangle + V_3\langle x - 3L\rangle + V_4\langle x - 4L\rangle$$

A better understanding of the load distribution in the structure can be obtained by separating the element CD between the hinges from the rest of the structure and studying the behaviour of each segment ABC, CD and DEF separately.

Element CD. The moments at C and D are zero and only a vertical force exists at each hinge. From symmetry, the force at the hinge must be $0.5qL$ to keep the total load qL on the element CD in equilibrium. The vertical reactions are provided by the beams ABC and DEF. The element CD therefore behaves like a simply supported beam of span L. The maximum bending moment is $qL^2/8$ at the centre.

Element ABC or DEF. The elements are supported on two vertical supports, A (or F) and B (or E), and are loaded by a uniformly distributed load q and actions from the beam CD which are $0.5qL$ acting vertically downwards at C and D. Note that as the simply supported span CD increases, the vertically downward force at C (or D) also increases proportionately. This tends to increase the bending moment, causing tension on the top surface at B (or E) but decreases the bending moment causing tension at the bottom face in the segment AB (or EF). Thus by choosing suitable values for the sections AB (or FE), BC (or ED) and CD, a very 'efficient' bending moment distribution, i.e. maximum values at several sections being approximately equal can be obtained.

The structure analysed above is called a balanced cantilever and is quite a common type of bridge structure constructed in both steel and reinforced concrete. Figure 1.68(g) shows a steel balanced cantilever bridge formed from trusses.

Figure 1.68 ● (g) A balanced cantilever steel truss bridge

(g)

Example 2

Draw the BMD and SFD for the rigid-jointed structure with pinned supports A and D and an additional internal hinge at E and loaded as shown in Fig. 1.69. Also calculate the distribution of axial force (AF) in the members of the frame.

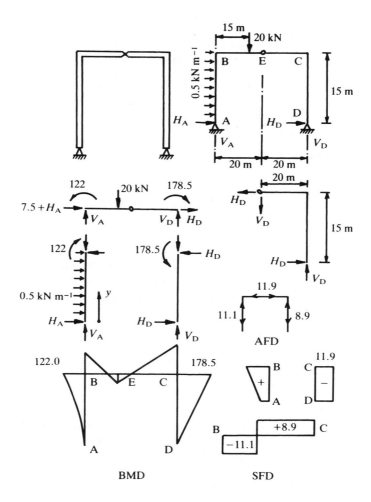

Figure 1.69 ● BM, SF and AF distribution in a three-pinned portal frame

Solution As there are only two reactions at each pinned support, namely a horizontal and a vertical reaction, the total number of reactions is equal to four. The equations of statics are

(i) $\sum F_x = 0$, $H_A + H_D + 0.5(15) = 0$

$$H_A + H_D = -7.5$$

(ii) $\sum F_y = 0$, $V_A + V_D - 20 = 0$

$$V_A + V_D = 20.0$$

(iii) Taking moments about A,

$$\sum M_z = 0, -V_D 40 + \{(0.5)(15)\}(15/2) + (20)15 = 0.$$

Therefore $V_D(40) = 356.25$, $V_D = 8.91$.

Since $V_D = 8.91$ and $V_A = 20 - V_D = 11.09$, the vertical reactions can be determined from statics only. However, without an additional independent equation H_A

and H_D cannot be determined and the problem becomes statically indeterminate. However, the presence of an internal hinge at E provides an additional equation. Isolating the free body ECD and taking moments about E we have

$$-H_D(15) - V_D(20) = 0, \text{ i.e. } H_D = -1.33V_D$$

where

$$V_D = 8.91, H_D = -11.88,$$

therefore

$$H_A = -7.5 - H_D = -7.5 + 11.88 = 4.38$$

Thus all the reactions are known and the bending moment and shear force values at various sections can be determined.

Segment AB.

(i) Shear force $Q = H_A + 0.5y$, where the origin for y is at A. The horizontal reaction at $B = H_A + (0.5)15 = 11.88$ to the left.
(ii) Bending moment $M = -(H_A)y - (0.5y)(y/2)$. The couple at B is a clockwise couple

$$M_B = H_A(15) + \{(0.5)(15)\}(15/2) = 122.$$

(iii) Axial force $F = V_A$ compressive.

Segment CD.

(i) Shear force $Q = H_D$
(ii) Bending moment $M = -(H_D)y$. The horizontal reaction at $C = H_D$ to the left $= 11.88$. The couple at C is a clockwise couple $M_c = H_D15 = -178.2$.
(iii) Axial force $F = V_D$ compressive.

Segment BEC. The actions at B are a vertical force V_A, a horizontal force to the right of 11.88 and an anticlockwise couple 122. Similarly, the reactions at C are a vertical force V_D, a horizontal force to the left of 11.88 and an anticlockwise couple of 178.2. Therefore,

(i) Shear force $Q = -V_A + 20\langle x - 15 \rangle^0$.
(ii) Bending moment $M = -122 + V_Ax - 20\langle x - 15 \rangle$.
 Note: $M_E = -122 + V_A20 - 20\langle 20 - 15 \rangle = 0$ as it should be, providing a check on the calculation.
(iii) Axial compressive force $F = 11.88$.
 The BMD, SFD and AFD (axial force diagram) can be drawn as shown in Fig. 1.69.

1.16 ● Structures curved in elevation – arches

Portal frames are assembled from linear elements, namely columns and beams. A structure need not necessarily consist of linear elements only. For aesthetic and construction reasons, a structure could be curved in elevation resulting in an arch. Various types of arches are shown in Fig. 1.70. Depending on the number of hinges present, they are classified as:

(a) Three-hinged arch where the hinges are present at supports and in the span. This is a curved element counterpart of the frame in Fig. 1.69 and is therefore a statically determinate structure.

(b) Two-hinged arch where the hinges are present at supports only. This is a statically indeterminate structure with four reactions, i.e. with one reaction more than can be determined from statics; this is statically indeterminate to the first degree.

(c) Hingeless or fixed arches have six reactions because at each support there exist three reactions – a horizontal, a vertical and a couple. There are three reactions more than can be determined from statics and this is therefore statically indeterminate to the third degree.

Figure 1.70 ●
Three-pin, two-pin and fixed arches

Common shapes of arches are semicircular (as in Roman bridges), segments of an arc of a circle and parabolic. One important aspect of the effect of the shape of the arch on the bending moment due to vertical loads must be appreciated. Consider the horizontally unrestrained arch and the corresponding beam shown in Fig. 1.71. When vertical loads are applied, the arch 'spreads' out. There is a 'pull in' in the case of the beam but this is generally very small. The vertical reactions are the same in both the beam and the arch and the distribution of bending moment in the arch and in the beam is therefore the same and it produces tension on the bottom face. If in the case of the arch the horizontal movement at the roller support is prevented, then a horizontal reaction, H, develops at the supports.

Figure 1.71 ●
Behaviour of a simply supported beam and a two-pinned arch

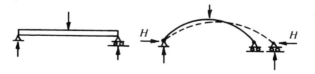

The bending moment due to H at any section is then Hy, where y equals the height above the supports of the arch at the section, producing tension at the top face and thus reducing the net bending moment due to vertical loads. Thus an arch can be economically used for spans much larger than the corresponding beam. Arches are a common form of structure used especially in the construction of both reinforced concrete and steel bridges. Since the curved form of the arch by itself does not allow the construction of a horizontal deck, depending on the level of the roadway with respect to the supports of the arch, loads are transmitted from the deck or roadway to the arch through a series of 'columns' and/or 'hangers' connected to the arch and the horizontal deck as shown in Fig. 1.72.

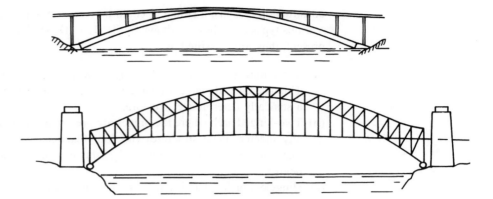

Figure 1.72 ●
Concrete and steel
bridges

1.16.1 Internal forces at a section in an arch segment

Example 1

Analyse the three-pinned parabolic arch with a uniformly distributed load q as shown in Fig. 1.73. The arch has a span of 120 m and a rise of 20 m. The equation for the parabola is given by

$$y = \{4r/L^2\}x(L - x)$$

where L = span, r = rise and the origin for x and y is at the left-hand support.

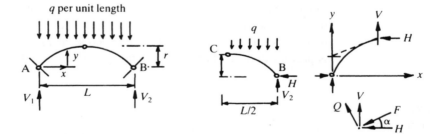

Figure 1.73 ● Three-pinned parabolic arch

Solution The vertical reactions are, by symmetry, $V_1 = V_2 = 0.5qL$. Considering the free body of CB, for moment at C to be zero,

$$-V_2(0.5L) + Hr + q(0.5L)(0.25L) = 0$$

$$H = 0.125qL^2/r$$

Note that the smaller the value of the rise, r, the larger the value of the horizontal thrust, H. Thus arches which are 'flat' need to be well restrained horizontally.

The bending moment is given by

$$M = V_1x - (qx)(x/2) - Hy$$

$$= 0.5qL^2\{(x/L) - (x/L)^2 - 0.25(y/r)\}$$

Substituting for y,

$$M = 0.5qL^2(x/L)\{1 - x/L - (1 - x/L)\} = 0$$

Thus the arch rib has zero bending moment at all sections.

Considering the free body of the arch rib, the vertical and horizontal forces at a section are (a) a vertical upward force $V = qx - V_1 = q(x - 0.5L)$; (b) a horizontal force to the left equal to H. The above forces are not along the normal and tangent to the section. Therefore they are *not* the shear force Q and the axial force F at the section. The shear force is the normal component and the axial force is the tangential component of the resultant due to H and V forces to the axis of the arch. If at the section the tangent is inclined at an angle of α to the horizontal, then resolving V and H in Q and F directions, we have

Shear force $Q = H \sin \alpha + V \cos \alpha$

$$= (0.125qL^2/r) \sin \alpha + q(x - 0.5L) \cos \alpha$$

$$Q = q \cos \alpha \{(0.125L^2/r) \tan \alpha + (x - 0.5L)\}$$

Since $\tan \alpha = dy/dx = (4r/L^2)(L - 2x)$, Q can be expressed as $Q = q \cos \alpha \{0.5(L - 2x) + (x - 0.5L)\} = 0$

Axial compressive force $F = H \cos \alpha - V \sin \alpha$

$$= q \cos \alpha \{0.125L^2/r - (x - 0.5L)(4r/L^2)(L - 2x)\}$$

$$= \frac{(qL^2 \cos \alpha)}{\gamma}\left\{0.125 + \frac{2r^2}{L^2}\left(\frac{2x}{L} - 1\right)^2\right\}$$

The arch rib is therefore free from bending moment or shear force. It is under pure axial compression. This example shows that in the case of arches, by a suitable choice of the geometry of the arch, it is possible to minimize greatly the internal forces in an arch rib.

Example 2

Analyse the three-pinned semicircular arch under uniformly distributed load q as shown in Fig. 1.74.

Solution Proceeding as in Example 1, the vertical reactions are by symmetry $V_1 = V_2 = qR$. Considering the free body CB, for moment at C to be zero,

$$-V_2R + HR + qR(R/2) = 0$$

$$H = 0.5qR$$

The bending moment M at a section at an angle α from the horizontal is given by

$$M = V_1R(1 - \cos \alpha) - q\{R(1 - \cos \alpha)\}^2/2 - HR \sin \alpha$$

Substituting for $V_1 = qR$ and $H = 0.5qR$ and simplifying

$$M = 0.5qR^2\{1 - \sin \alpha - \cos^2 \alpha\}$$

Substituting $\cos^2 \alpha = 1 - \sin^2 \alpha$, $M = 0.5qR^2 \sin \alpha (\sin \alpha - 1)$

M is a maximum, when $\sin \alpha = 0.5$, $\alpha = 30°$

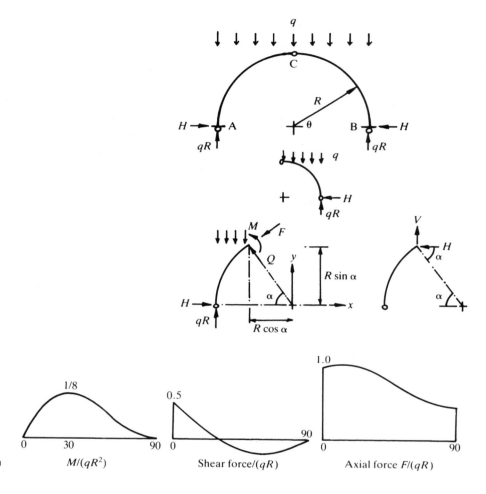

Figure 1.74 ● Three-pinned semicircular arch

$M/(qR^2)$ Shear force/(qR) Axial force $F/(qR)$

Quite clearly $M = 0$ when $\alpha = 0$ or π (i.e. at the supports) and at $\alpha = \pi/2$ (the central hinge). However, unlike the parabolic arch, the bending moment is not equal to zero everywhere.

Considering the free body of the arch rib, the vertical (upward) and horizontal forces at a section are respectively:

(a) $V = \{qR(1 - \cos \alpha) - V_1\} = -qR \cos \alpha$;
(b) $H = 0.5qR$.

Shear force $Q = V \sin \alpha + H \cos \alpha = 0.5qR \cos \alpha \{1 - 2\sin \alpha\}$, and axial compressive force

$$F = -V \cos \alpha + H \sin \alpha = 0.5qR\{\sin \alpha + 2\cos^2 \alpha\}$$

Figure 1.74 shows the BMD, SFD and AFD diagrams.

The two examples clearly show the important influence the shape of the arch has on the distribution of internal forces in the arch rib.

1.16.2 Forces at a section in portal frames and arches

It should be appreciated that in all the 2-D structures considered so far – beams, portal frames and arches and so on – the loads were applied in the same vertical x–y plane as that in which the structure was positioned. In such a structure, the forces acting at a cross-section consist in general of an axial force, a shear force and a bending moment. This is valid irrespective of how complex the structure is and whether it is statically determinate or indeterminate.

1.17 ● 3-D rigid-jointed structures

In the previous sections of this chapter, 2-D statically determinate beams and rigid-jointed structures where the structure and the loads lie in the vertical x–y plane were considered. In a member of such structures, the internal forces present at a section are a bending moment, a shear force and an axial force. In the case of 3-D structures in general, at a cross-section in a member three forces and three couples can exist. Taking the x-axis to coincide with the axis of the member, and considering a section normal to the axis of the member as shown in Fig. 1.75, we have

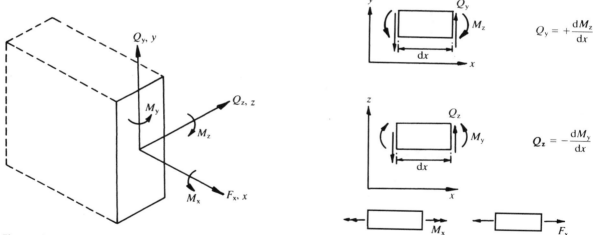

Figure 1.75 ● Forces at a cross-section in a member of a 3-D structure

(a) Axial force = force in the x-direction = F_x.

(b) Shear forces: there exist two components of shear force, namely Q_y in the y-direction, associated with bending in the vertical x–y plane, as in 2-D structures considered previously, and Q_z in the z-direction associated with bending in the horizontal x–z plane.

(c) Bending moments: a couple M_z about the z-axis which causes bending in the vertical x–y plane. Because of equilibrium considerations with respect to rotation about the z-axis (as discussed in Section 1.12), M_z and Q_y are related by $Q_y = dM_z/dx$. Similarly, a couple M_y about the y-axis which causes bending in the horizontal x–z plane. As before for equilibrium with respect to rotation about the y-axis, M_y and Q_z are related by $Q_z = \mathrm{d}M_y/\mathrm{d}x$ as shown in Fig. 1.75.

(d) Torsional or twisting moment: a couple M_x about the x-axis.

1.17.1 Representation of a couple as a double-headed arrow

Before considering some examples, it is useful to introduce a simple vector notation for couples which will assist in the representation of couples in 3-D systems. Using the right hand, if the couple acts in the direction in which the fingers curl, then the couple is represented by a double-headed arrow (to distinguish from a force which is represented by a single-headed arrow) pointed in the direction of the thumb. This is shown in Fig. 1.76. Thus a double-headed arrow in the x-direction represents a torsional moment. Similarly, a double-headed arrow in the y- and z-directions represents bending moments M_y and M_z respectively. The reader should become familiar with this notation as it will be of considerable assistance in the study of the later sections of the book.

Figure 1.76 ●
Representation of couples as double-headed arrows

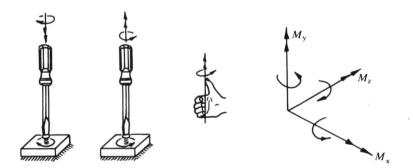

1.18 ● Floor grids and bow girders

One special class of 3-D rigid-jointed structures are those structures which lie in the horizontal plane but in which the loads are applied normal to the plane. Such structures are called plane grids. The structure shown in Fig. 1.77 is a simple example of a plane grid. More complex grids of beams are used to carry floor loads as shown in Fig. 1.78(a). Plane grids using members curved in plan are quite common in curved bridge structures as shown in Fig. 1.78(b). When only a single curved member in plan is used, it is known as a bow girder. But however complex the grid work, at a cross-section, the elements of the grid are subjected to bending moment, twisting moment and shear force.

1.18.1 Example of plane grids

Example 1

A right-angle cantilever bent loaded normal to its plane as shown in Fig. 1.77.

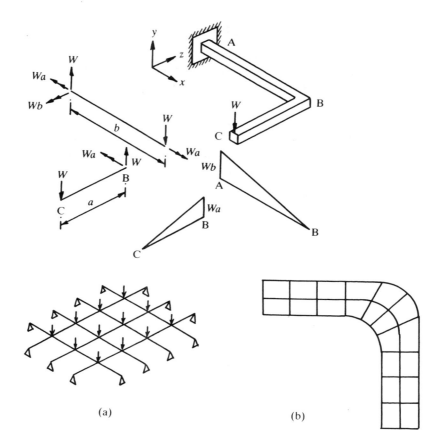

Figure 1.77 ● Bending and twisting moments and shear force distribution in a plane grid

Figure 1.78 ● Plane grids: (a) floor grid for a roof, (b) plan of a curved plane grid for a bridge structure

(a)　　　　　　(b)

Solution　The structure can be divided into two cantilever segments, AB and BC.

Segment BC. For equilibrium, the forces at B are a vertical upward force W in the y-direction and a couple Wa in the negative x-direction. The segment therefore is subjected to a constant shear force and a linearly varying bending moment.

Segment AB. The actions at B from segment BC consist of a vertical downward force W and a couple Wa about the positive x-axis. The reaction at A consists of (a) a force W in the y-direction to balance the load W at B; (b) a couple Wa in the negative x-direction to balance the couple Wa at B; (c) a couple Wb about the z-axis to balance the couple about the z-axis caused by the loads W at A and B. The couple about the x-axis is the twisting moment acting on the element. The vertical force W at B causes the bending of the beam in the x–y plane. Thus the segment is subjected to a twisting moment Wa, a constant shear force W and a linearly varying bending moment about the z-axis with a maximum value at A of Wb.

The reader should note that at joint B, the internal couple Wa about the x-axis is a bending moment with respect to the beam BC but is a torsional moment with respect to beam BA. Thus a couple about an axis can have different effects depending on the member's orientation to the direction of the couple.

1.18.2 A simple bow girder

A bow girder is the equivalent of an arch when the loads act normal to the plane of the structure. The calculation of forces at a cross-section in a bow girder is best illustrated by a simple example.

Example 1

A quarter circle bow girder of radius R carrying a uniformly distributed load q as shown in Fig. 1.79.

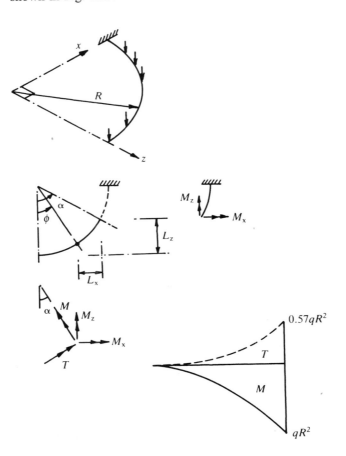

Figure 1.79 ●
Quarter circle bow girder under uniformly distributed load

Solution Looking in plan, at any section at an angle α from the z-axis, as shown in Fig. 1.79, a vertical shear force Q and couples M_x and M_z act about the x- and z-axes respectively. The curved length $\mathrm{d}s$ of an infinitesimal segment subtending an angle $\mathrm{d}\phi$ is given by

$$\mathrm{d}s = R\,\mathrm{d}\phi$$

The vertical load $\mathrm{d}W$ acting on length $\mathrm{d}s$ is

$$\mathrm{d}W = q\,\mathrm{d}s = q(R\,\mathrm{d}\phi)$$

The load dW acts at distances L_x and L_z parallel to the x- and z-axes respectively from the section considered. L_x and L_z are given by

$$L_z = R(\cos \phi - \cos \alpha), \; L_x = R(\sin \alpha - \sin \phi)$$

For equilibrium of the free body,

$$-Q - \int dW = 0, \; -M_x + \int dW L_z = 0$$

and

$$-M_z - \int dW L_x = 0$$

Substituting for $dW = qR \, d\phi$, the above equations are integrated with respect to ϕ varying from 0 to α:

$$Q = -\int qR \, d\phi = -qR \, \alpha$$

$$M_z = -\int dW L_x = -\int qR \, d\phi \{R(\sin \alpha - \sin \phi)\}$$

$$= -qR^2 \int (\sin \alpha - \sin \phi) \, d\phi = -qR^2 \{\alpha \sin \alpha + (\cos \alpha - 1)\}$$

$$M_x = \int dW L_z = \int qR \, d\phi \{R(\cos \phi - \cos \alpha)\}$$

$$= qR^2 \int (\cos \phi - \cos \alpha) \, d\phi = qR^2 \{\sin \alpha - \alpha \cos \alpha\}$$

Q is obviously the shear force but the couples M_x and M_z are neither the bending moment M nor the torsional moment T because they do not act about the normal and tangent to the section. However, because couples are also vectors, they can be resolved into normal and tangential components as follows:

$$M = -M_x \sin \alpha + M_z \cos \alpha, \; T = M_x \cos \alpha + M_z \sin \alpha$$

Substituting for M_x and M_z and noting that $\sin^2 \alpha + \cos^2 \alpha = 1$,

$$T = qR^2\{\alpha - \sin \alpha\}$$

$$M = qR^2\{\cos \alpha - 1\}$$

$$Q = -qR\alpha$$

Note: In the above expressions, α should be in radians.

Thus all the force actions at a section are determined. Figure 1.79 shows the BMD and TMD (twisting moment diagram).

1.19 ● Example of a 3-D rigid-jointed structure

This chapter will be concluded with an example of a simple statically determinate 3-D structure.

Example 1

Analyse the structure shown in Fig. 1.80.

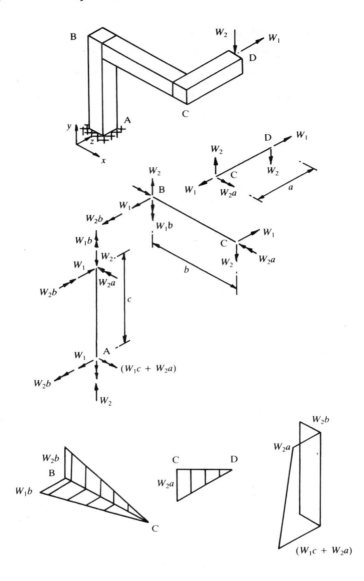

Figure 1.80 ● Force distribution in a 3-D structure

Solution Separate the structure into three cantilever segments and consider the equilibrium of each including external forces and actions from members framing into it.

Segment CD. For equilibrium, the forces required at C are: (a) force W_1 in the z-direction which is an axial force; (b) force W_2 in the y-direction which is a shear force; (c) a couple W_2a about the x-axis which is a bending moment. Thus the element is subjected to an axial tensile force of W_1, a constant shear force of $-W_2$ and a linearly varying bending moment with a maximum value of Wa at C. The bending moment causes tension on the top face. The segment bends in the y–z plane.

Segment BC. The reactions from the segment CD at C act as actions on segment BC at C. The forces required at B for equilibrium are:

(a) force W_2 in the y-direction which is a shear force;
(b) a couple W_2b about the z-axis to maintain equilibrium with respect to W_2 forces in the y-direction acting at C and B;
(c) force W_1 in the z-direction which is also a shear force;
(d) a couple W_1b about the y-axis to maintain equilibrium with respect to W_1 forces in the z-direction acting at C and B;
(e) a couple W_2a about the x-axis to maintain equilibrium with respect to the couple W_2a acting at C about the x-axis. This is a torsional moment.

Thus this element is subjected to twisting moment W_2a and constant shear forces $-W_2$ in the y-direction and W_1 in the z-direction. Associated with the shear forces W_2 and W_1 are respectively the maximum bending moments of W_2b causing tension at the top face and the maximum bending moment of W_1b causing tension on the negative z-face. The beam is thus subjected to bending moments about both the y- and z-axes. This is a case of biaxial bending.

Segment AB. The actions from segment BC at B consist of:

(a) forces W_2 and W_1 in the y- and z-directions which are respectively axial force and shear force along the z-axis;
(b) a couple W_1b about the y-axis which is a twisting moment;
(c) a couple $-W_2a$ about the x-axis and a couple $-W_2b$ about the z-axis.

Both of these couples are bending moments. The element is subjected to biaxial bending.
   To keep the element in equilibrium, the forces required at A are:

(a) an axial force W_2 in the z-direction;
(b) a couple $-W_1b$ about the y-axis which is a twisting moment;
(c) a force W_1 about the z-axis which is a shear force;
(d) associated with the shear force about the z-axis are the bending moment about the x-axis equal to $\{W_2a$ applied at $B + W_1c$ to balance the force W_1 applied at B};
(e) a couple W_2b about the z-axis which balances the couple $-W_2b$ applied at B.

This element is thus subjected to:

(a) constant axial compressive force of W_2;
(b) a constant twisting moment of W_1b;
(c) a constant bending moment W_2b causing tension on the negative x-face;
(d) a constant shear force W_1 in the z-direction;
(e) a linearly varying bending moment causing tension on the negative z-face.

1.20 ● Behaviour of statically indeterminate structures

It was stated in Section 1.9 that compared with statically indeterminate structures, statically determinate structures possess the following useful behaviour:

(a) the structure is not stressed when subjected to 'movement' of supports;
(b) changes in temperature of members cause the structures to deform but do not stress the structure;

(c) incorrect lengths of members cause only deformation but do not stress the members.

These are important aspects of behaviour which are of significance to the structural designer. In addition to these behavioural aspects, there is also the additional advantage that the structure is amenable to manual calculation, although this is not an important consideration in the present age of electronic computers.

In spite of these advantages, apart from simply supported beams and cantilevers which are statically determinate, the majority of rigid-jointed structures, such as frames, grids and 3-D structures, are in practice more likely to be statically indeterminate than determinate. The reasons for this are not hard to find. Some of these reasons are to do with manufacturing/construction aspects while others are related to the beneficial aspects of statically indeterminate structures.

Construction aspects. In reinforced concrete structures, it is simpler to produce rigid-joints than pin-joints. Similarly, in the case of steel structures, rigid-joints produced by welding are simpler, 'neater' and easier to maintain than bolted joints.

For example, consider the statically indeterminate continuous beam shown in Fig. 1.81(a) and the corresponding statically determinate structure consisting of three simply supported beams shown in Fig. 1.81(b). In the case of reinforced concrete construction cast *in situ*, it is much simpler to reinforce and cast the continuous beam than to interrupt the reinforcement in order to produce 'breaks' between the beams.

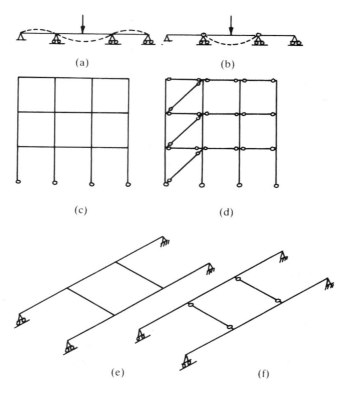

Figure 1.81 ● (a)–(f) Statically determinate and indeterminate structures

Also, considering the two multistorey structures shown in Fig. 1.81(c) and (d), if the structures are made from reinforced concrete, then the statically indeterminate structure is simpler to construct than the corresponding statically determinate structure. Note that the structure shown in Fig. 1.81(d) requires diagonal members to prevent the 'deck of cards' type of collapse under horizontal loads.

Behaviour aspects. Under this heading three aspects should be considered. They are:

(a) Ability to transfer loads from one part to another. Consider the beams shown in Fig. 1.81(a) and (b). It is obvious that if a support is removed in the simply supported beams shown in Fig. 1.81(b), then the structure collapses. The same thing does not happen in the case of the continuous beam in Fig. 1.81(a). This highlights one very important aspect of statically indeterminate structures, namely the ability to 'transfer' loads from one part of the structure to another.

This can also be seen in the two grid structures shown in Fig. 1.81(e) and (f). In the statically determinate structure in Fig. 1.81(f), load on a longitudinal beam is not transferred to another longitudinal beam because the transverse beams are hinged at their ends. On the other hand, if the transverse beams are rigidly connected to the longitudinal beams, then load transfer from the loaded to the unloaded longitudinal beams can take place.

(b) Stiffness of members. If a load is applied to the central span of the beams shown in Fig. 1.81(a) and (b), then one can intuitively sense that the deflection under the load in the statically determinate beam shown in Fig. 1.81(a) will be larger than that in the case of the statically indeterminate beam shown in Fig. 1.81(b). This is again due to the ability of statically indeterminate structures to transfer load from the loaded to the unloaded parts of the structure.

(c) Stress control. The effect observed in paragraph (b) above means that by suitable choice of the dimensions of members in a statically indeterminate structure, it is possible to control the forces acting at a section by 'channelling' more force to be transferred to one part than to the other.

These are some of the very important advantages of the statically indeterminate structure. It is likely that at this stage the reader might not appreciate all the aspects discussed in this section. It is therefore strongly recommended that this section be studied again after reading Chapter 6. The main disadvantage of the statically indeterminate structure is that it is complex to analyse manually. However, now that computers are commonly available, complexity of analysis is not a serious obstacle.

1.21 ● Degree of statical indeterminacy of 2-D rigid-jointed structures

In Section 1.8, a simple rule was developed relating the number of joints j to the number of members m of a statically determinate 2-D pin-jointed structure, $m = 2j - 3$. Therefore the degree of statical indeterminacy of a 2-D pin-jointed structure is given by $\{m - (2j - 3)\}$. Similar but more complex rules have been developed for

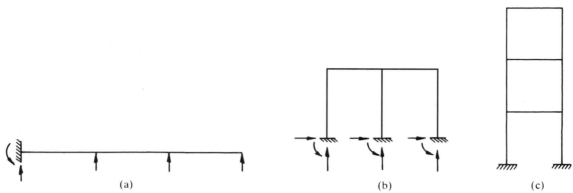

Figure 1.82 ● (a)–(c) Statically indeterminate structures

calculating the degree of statical indeterminacy of rigid-jointed structures. Even now, when structures are commonly analysed by computers, fairly large statically determinate pin-jointed structures are amenable to reasonably quick manual calculations. Therefore there is a need to know whether a particular structure is statically determinate or not. However, rigid-jointed structures, except the most elementary, are not really suitable for manual methods of analysis. The modern method of structural analysis to be presented in Chapter 6 does not require any knowledge of the degree of statical indeterminacy of the structure. Even so, purely as an 'intellectual exercise' some simple rules will be given for calculating the degree of statical indeterminacy of the typical 2-D rigid-jointed structures shown in Fig. 1.82.

Continuous beams. Beams are generally only subjected to vertical loads. Cantilevers are fixed at one end and free at the other. They have at their support a vertical reaction and a moment reaction. Similarly, simply supported beams have a vertical reaction at each support. The two reactions can be determined from the two equations of statics, i.e. $\Sigma F_y = 0$ and $\Sigma M_z = 0$. Therefore, if a beam has more than two reactions, then it becomes a statically indeterminate structure. For example, the continuous beam shown in Fig. 1.82(a) has four vertical reactions and one moment reaction at the fixed end. Thus the structure has a total of five reactions. Since only two reactions can be determined from statics, because the equation $\Sigma F_x = 0$ is automatically satisfied, the degree of statical indeterminacy of this structure is $5 - 2 = 3$. Thus, for continuous beams, the rule is given by

Degree of statical indeterminacy = No. of reactions − 2

Multistorey rigid-jointed frames. A single-bay single-storey frame with 'fixed base' has six reactions because at each support there is a horizontal, a vertical and a moment reaction. Since only three equations of statics are available, the degree of statical indeterminacy for this structure is three. If a further bay is added, it results in three additional reactions. Thus for a single-storey multibay rigid-jointed frame with a 'fixed base' the degree of statical indeterminacy is three times the number of bays.

If, on the other hand, the frame has a 'pinned base', then at each support there is no moment reaction and the column at its junction with the foundation is allowed to rotate freely, i.e. is released from any constraint to rotation. This is called a 're-lease' and results in a reduction in the number of reactions by one. Thus a general rule for single-storey multibay rigid-jointed structures is

Degree of statical indeterminacy = 3 × No. of bays − releases

Consider a single-bay multistorey rigid-jointed structure. Each storey uses the storey below it as the 'foundation'. Because a single-storey frame with 'fixed base' is three times statically indeterminate, the general rule for single-bay multistorey structures is

Degree of statical indeterminacy = 3 × No. of storeys − releases

Combining the above two rules, the general rule for multistorey, multibay 2-D rigid-jointed frames is

Degree of statical indeterminacy = 3 × No. of (bays + storeys) − releases

1.22 ● References

General texts

Buckle I (ed.) 1977 *Morgan's Elements of Structures* Longman (A delightful book.)
Francis A J 1980 *Introducing Structures* Pergamon
Gordon J E 1978 *Structures* Penguin
Gordon J E 1979 *The New Science of Strong Materials* (2nd edn) Pitman
Salvadori M, Heller R 1975 *Structure in Architecture* (2nd edn) Prentice-Hall (An excellent book.)

Jackson J H, Wirtz G 1983 *Statics and Strength of Materials* McGraw-Hill (A large number of worked examples.)
Pilkey W D, Pilkey O H 1974 *Mechanics of Solids* Quantum (A good book.)
Simon A L, Ross D A 1983 *Principles of Statics and Strength of Material* William Brown & Co. Iowa (Good on statics, truss analysis, BMD, SFD, etc.)
Wayne Brown G 1985 *Basic Statics and Stress Analysis* McGraw-Hill, Ryerston (Extensive treatment of statics. Good on truss analysis, bending moment and shear force diagrams.)

A selection of texts on specific subject matter

Hsieh Yuan-Yu 1982 *Elementary Theory of Structures* (2nd edn) Prentice-Hall (Good on truss analysis.)

1.23 ● Problems

1. Are the pin-jointed structures shown in Fig. E1.1 stable? If not, suggest a suitable modification to render them stable. Ensure the stability of the structure by checking if each joint is properly 'braced', i.e. 'rigidly' held in place.

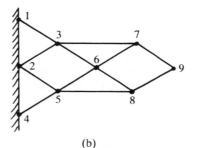

(a)　　　　　　　　　(b)

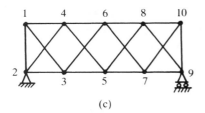

(c)

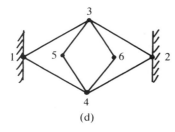

(d)

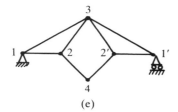

(e)

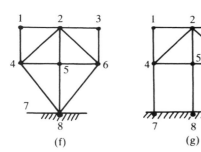

(f)   (g)

Figure E1.1 ● (a)–(g)

Answers:

1.1(a): Unstable. Add member 6–4' or 5–3' to make it stable.

1.1(b): Stable. The numbering of joints shows the order in which the structure can be built.

1.1(c): Stable. Note that there are no joints at the point where the diagonals intersect.

1.1(d): Stable. The numbering of joints shows the order in which the structure can be built.

1.1(e): Unstable. Remove 2–4 and 2'–4 and replace by 2–2' or just add 2–2'.

1.1(f): Unstable, as all the three supports, 4–8, 5–8 and 6–8, are concurrent at 8 and the

structure can rotate as a rigid body about joint 8 (see Section 1.6.1). Make it stable by replacing 5–8 by 7–5.

1.1(g): Unstable, as all the three supports, 7–4, 8–5 and 9–6, are parallel and the structure can sway as a rigid body (see Section 1.6.1). Make it stable by replacing 6–9 by 7–5.

2. Determine the forces in the members of the trusses shown in Fig. E1.2(a)–(c). Use the method of joints.

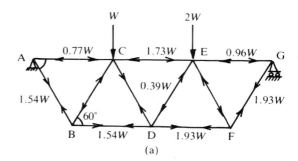

(a)

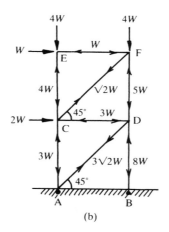

(b)

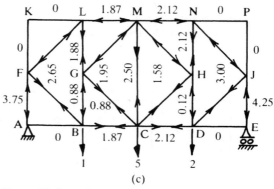

(c)

Figure E1.2 ● (a)–(c)

Answers:

1.2(a): $V_A = 1.33W$, $V_G = 1.67W$. Use $\sin 60 = \sqrt{3}/2$
and $\cos 60 = 0.5$.

1.2(b): $V_B = 8W$, $V_A = 0$, $H_A = 3W$. Use
$\sin 45 = \cos 45 = 1/\sqrt{2}$.

1.2(c): $V_A = 3.75$, $V_E = 4.25$. All diagonal members
are inclined at 45° to the horizontal.

3. Using the method of sections, show that for the truss
shown in Fig. E1.3, the tensile force in member
HJ is equal to $8WL/H$. (L = horizontal distance
between the loads and H is the rise of the truss.)

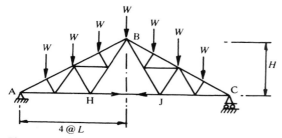

Figure E1.3

4. Show that the force in any internal member of the
truss shown in Fig. E1.4 is zero.

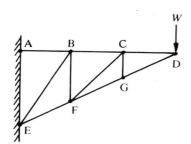

Figure E1.4

5. Using the method of sections, show that the
forces in members CE and BD of the truss shown
in Fig. E1.2(a) are $-1.73W$ and $1.54W$ respectively.

6. Draw the BMD and SFD for the structures shown
in Figs E1.5 to E1.7.

Answer:

(a) Figure E1.5: $V_A + V_D = 1 \times 6 + 8 = 14$

$V_D 20 = (1 \times 6)\{4 + 6/2\} + 8 \times 10$

$V_D = 6.1$, $V_A = 7.9$

$Q = -7.9 + 1\{\langle x - 4 \rangle - \langle x - 10 \rangle\}$
$+ 8\langle x - 10 \rangle^0 - 6.1\langle x - 20 \rangle^0$

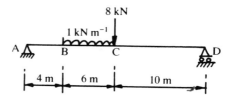

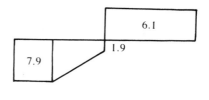

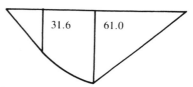

Figure E1.5

$M = 7.9x - (1/2)\{\langle x - 4 \rangle^2 - \langle x - 10 \rangle^2\}$
$- 8\langle x - 10 \rangle + 6.1\langle x - 20 \rangle$

$Q = 0$ at $x = 14$, $M_{max} = 61$

(b) Figure E1.6: $V_A + V_C = 8 \times 15 + 5 = 125$

$V_C 20 = (8 \times 15)(15/2) + 5 \times 23$

$V_A = 74.25$, $V_C = 50.75$

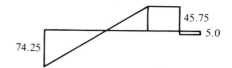

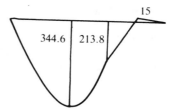

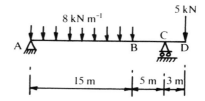

Figure E1.6

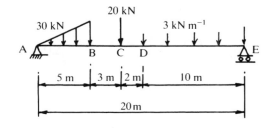

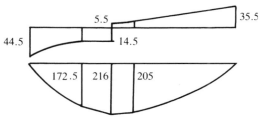

Figure E1.7

$Q = -74.25 + 8\{x - \langle x - 15 \rangle\}$
$\quad - 50.75\langle x - 20 \rangle^0 + 5\langle x - 23 \rangle^0$

$M = 74.25x - (8/2)\{x^2 - \langle x - 15 \rangle^2\}$
$\quad + 50.75\langle x - 20 \rangle - 5\langle x - 23 \rangle$

$Q = 0$ at $x = 9.25$. $M_{max} = 344.56$

$Q = 0$ at $x = 20$, $M_{max} = -15$

(c) Figure E1.7: $V_E + V_A = 30 + 20 + 3 \times 10 = 80$

$V_E 20 = \{30\}\{(2/3\}5) + 20(8)$
$\quad + (3 \times 10)\{10 + 5\}$

$V_E = 35.5$, $V_A = 44.5$

Note that at B the intensity of the load was 12 kN m^{-1}.
Considering each of the differently loaded sections,
$Q = -44.5 + 0.5x \, (12/5)x^2$

$\quad = -44.5 + 1.2x^2 \qquad\qquad 0 \le x \le 5$
$\quad = -44.5 + 30 = -14.5 \qquad\quad 5 < x \le 8$
$\quad = -44.5 + 30 + 20 = 5.5 \qquad 8 < x \le 10$
$\quad = -44.5 + 30 + 20 + 3(x - 10) \quad 10 < x < 20$
$\quad = -44.5 + 30 + 20 + 3(10) = 35.5 \qquad x = 20$

$M = 44.5x - 0.5x(12/5)x(x/3)$
$\quad = 44.5x - 0.4x^3 \qquad\qquad 0 \le x \le 5$
$\quad = 44.5x - 0.5(12)(5)\{x - 5(2/3)\}$
$\quad = 14.5x + 100 \qquad\qquad 5 \le x < 8$
$\quad = 44.5x - 0.5(12)(5)\{x - 5(2/3)\}$
$\quad\quad - 20(x - 8) \qquad\qquad 8 \le x \le 10$

Simplifying

$\quad = -5.5x + 260 \qquad\qquad 8 \le x \le 10$
$\quad = 44.5x - 0.5(12)(5)\{x - 5(2/3)\}$
$\quad\quad - 20(x - 8) - (3/2)(x - 10)^2 \quad 10 \le x \le 20$
Simplifying
$M = -1.5x^2 + 24.5x + 110 \qquad 10 \le x \le 20$
OR using the formulae developed in Section 1.11.2, Example 6:

$Q = -44.5 + 20\langle x - 10 \rangle^0$ due to concentrated load
$\quad + 3\{\langle x - 10 \rangle - \langle x - 20 \rangle\}$ due to u.d.l.
$\quad + 12\{0.5x^2/5 - \langle x - 5 \rangle - 0.5\langle x - 5 \rangle^2/5\}$
$\quad$ due to triangularly distributed load.

$M = 44.5x - 20\langle x - 10 \rangle - (3/2)\{\langle x - 10 \rangle^2$
$\quad - \langle x - 20 \rangle^2\} - 12\{0.167x^3/5$
$\quad - 0.5\langle x - 5 \rangle^2 - 0.167\langle x - 5 \rangle^3\}$

$Q = 0$ when $x = 8$. $M_{max} = 216$.

7. Draw the BMD and SFD for the beam with an internal hinge which is fixed at one end, simply supported at the other end, as shown in Fig. E1.8.

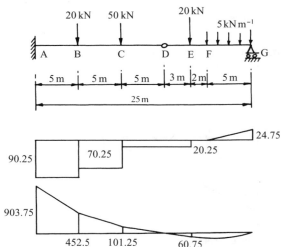

Figure E1.8

Answer: $V_A + V_G = 20 + 50 + 20 + 5 \times 5 = 115$
Taking moments about the hinge of the free body DEFG,

$\quad 20 \times 3 + 5 \times 5 \times (5 + 5/2) = V_G 10$, $V_G = 24.75$

and

$\quad V_A = 90.25$

For overall equilibrium, taking moments about A assuming M_A is anticlockwise,

$\quad -M_A + 20 \times 5 + 50 \times 10 + 20 \times 18$
$\quad + 5 \times 5 \times (25 - 5/2) - V_G 25 = 0$,

$M_A = 903.75$

$Q = -90.25 + 20\langle x - 5\rangle^0 + 50\langle x - 10\rangle^0$
$+ 20\langle x - 18\rangle^0 + 5\langle x - 20\rangle^1 - 24.75\langle x - 25\rangle^0$

$M = -M_A + 90.25x - 20\langle x - 5\rangle^1 - 50\langle x - 10\rangle^1$
$- 20\langle x - 18\rangle^1 - (5/2)\langle x - 20\rangle^2$
$+ 24.75\langle x - 25\rangle^1$

$Q = 0$ at A to the left of the support,
$M_{max} = -903.75$
$Q = 0$ at 4.95 from G. $M_{max} = 61.25$

8. Draw the BMD and SFD for the beam with two internal hinges which is fixed at both ends, as shown in Fig. E1.9.

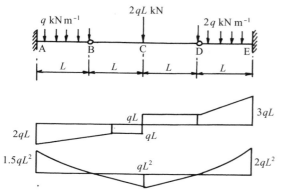

Figure E1.9

Answer: Considering the free body BCD, from symmetry vertical reactions at B and D are qL. Considering these to act as *downward* loads at B on free body AB and at D on free body DE, we have

$V_A = q(L)$ on AB $+ qL$ at B $= 2qL$,
$-M_A + qL(L/2)$ on AB $+ qL(L)$ at C $= 0$.

Therefore, $M_A = 1.5qL^2$ anticlockwise.

$V_E = 2q(L)$ on DE $+ qL$ at D $= 3qL$

and

$M_E - 2qL(L/2)$ on DE $- qL(L)$ at D $= 0$.

Therefore, $M_E = 2qL^2$ clockwise.

$Q = -V_A + q\{x - \langle x - L\rangle^1\} + 2qL\langle x - 2L\rangle^0$
$+ 2q\{\langle x - 3L\rangle^1 - \langle x - 4L\rangle^1\} - V_E\langle x - 4L\rangle^0$

$M = -M_A + V_Ax - 0.5q\{x^2 - \langle x - L\rangle^2\}$
$- qL\langle x - 2L\rangle^1 - q\{\langle x - 3L\rangle^2 - \langle x - 4L\rangle^2\}$
$- V_E\langle x - 4L\rangle^1 + M_E\langle x - 4L\rangle^0$

9. Draw the BMD, SFD and AFD (axial force diagram, also called thrust diagram) for the portal frame with pinned supports and an internal hinge as shown in Fig. E1.10.

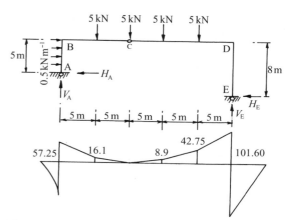

Figure E1.10

Answer: $H_A + H_E = 5(0.5) = 2.5$, $V_A + V_E = 5 + 5 + 5 + 5 = 20$
Taking moments about A for the whole structure
$0.5 \times 5 \times (5/2) + 5 \times 5 + 5 \times 10 + 5 \times 15$
$+ 5 \times 20 + H_E(8 - 5) = V_E25$
Therefore

$V_E25 - H_E3 = 256.25$

Taking moments about C for the free body CDE only

$5 \times 5 + 5 \times 10 + H_E8 = V_E15$

Therefore

$V_E30 - H_E8 = 150$

Solving $V_E = 11.77$, $H_E = 12.70$. Therefore $H_A = -10.20$, $V_A = 8.23$.

Section AB: Axial force $F = -V_A = -8.23$,
$Q = -H_A + 0.5y = 10.20 + 0.5y$,
$M = H_Ay - 0.5y^2/2 = -10.20y - 0.25y^2$ where the origin for y is at A. At B, $y = 5$. $M_B = 57.25$, $Q_B = 12.70$.

Section ED: $F = -V_E = -11.77$, $Q = -H_E = -12.70$, $M = H_Ey = 12.70y$, where the origin for y is at E. At D, $y = 8$. $M_D = 101.60$, $Q_D = -12.70$

Section BCD: Consider the free body BCD with the actions from AB at B and from ED at D.

$F = -H_E = -12.70$

$Q = -V_A + 5\langle x - 5\rangle^0 + 5\langle x - 10\rangle^0 + 5\langle x - 15\rangle^0$
$+ 5\langle x - 20\rangle^0 - V_E\langle x - 25\rangle^0$

$M = -57.25$ from moment at A due to reaction from AB at A $+ V_Ax - 5\langle x - 5\rangle^1 - 5\langle x - 10\rangle^1$
$- 5\langle x - 15\rangle^1 - 5\langle x - 20\rangle^1 + V_E\langle x - 25\rangle^1$
$+ 101.60\langle x - 25\rangle^0$ from moment at D due to reaction from ED at D.

10. Draw the AFD for the bars loaded as shown in Fig. E1.11(a) and (b)

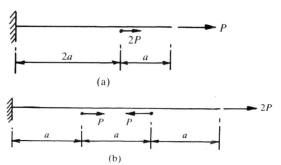

(a)

(b)

Figure E1.11 ● (a), (b)

Answer:
(a) $F = P + 2P\langle x - a\rangle^0 - 3P\langle x - 3a\rangle^0$
(b) $F = 2P - P\langle x - a\rangle^0 + P\langle x - 2a\rangle^0 - 2P\langle x - 3a\rangle^0$

11. Draw the BMD, SFD and AFD for the three-pinned parabolic arch shown in Fig. E1.12. It is given that $y = (4h/L^2)x(L - x)$, $h = 0.2L$

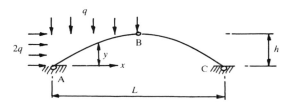

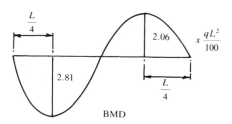

BMD

Figure E1.12

Answer: Assuming that H_A and H_C act to the left,

$$V_A + V_C = q(0.5L), \quad H_A + H_C = 2qh = 0.4qL$$

$V_CL = 2qh(h/2) + q(0.5L)(0.25L)$, $V_C = 0.165qL$,
$V_A = 0.335qL$. Taking moments about B for the free body BC, $H_Ch = V_C(0.5L)$, $H_C = 0.4125qL$,
$H_A = -0.0125qL$.

For AB:
$$M = V_Ax - q(x^2/2) + H_Ay - 2q(y^2/2), 0 \leqslant x \leqslant 0.5L$$

$$= qxL\{-0.64(x/L)^3 + 1.28(x/L)^2 - 1.13(x/L) + 0.325\}$$

$$V = -V_A + qx, \quad H = H_A + qy$$

where V and H are the vertically upward and horizontal to the left forces at the section on the positive face.

$$Q = V \cos \alpha + H \sin \alpha, \quad F = -V \sin \alpha + H \cos \alpha,$$

where $\tan \alpha = dy/dx = (4h/L^2)(L - 2x)$
$$= 0.8\{1 - 2(x/L)\}$$
For BC: $M = V_C(L - x) - H_Cy$, $0.5L \leqslant x \leqslant L$
$$= 0.165qL(L - x)\{1 - 2(x/L)\}$$

$$V = V_C, \quad H = H_C$$

12. Draw the BMD, SFD and TMD (twisting moment diagram) for the plane grid structure shown in Fig. E1.13.

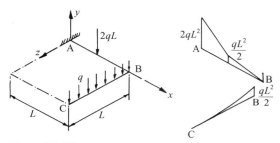

Figure E1.13

Answer: CB is a cantilever beam segment under u.d.l. of q and bends in the y–z plane. The bending moment of $0.5qL^2$ and shear force of qL at B on BC act respectively as twisting moment and shear force on AB at B. AB is another cantilever beam segment and bends in the y–x plane and twists about the x-axis.

BC: $Q = -qL + qz = -q(L - z)$

M about x-axis $= qLz - 0.5qL^2 - 0.5qz^2$
$$= -0.5q(L - z)^2$$

AB: $Q_A = qL$ at B + $2qL$ at D = $3qL$,

M_z at A = $qL(L)$ at B + $2qL(0.5L)$ at D = $2qL^2$,

M_x at A = $-0.5qL^2$

$Q = -V_A + 2qL\langle x - 0.5L\rangle^0$
$$= -3qL + 2qL\langle x - 0.5L\rangle^0$$

M about z-axis $= -2qL^2$ due to the reaction couple at
A + $V_Ax - 2qL\langle x - 0.5L\rangle^1$
$M = -2qL^2 + 3qLx - 2qL\langle x - 0.5L\rangle^1$

T about x-axis $= 0.5qL^2$

13. Draw the BMD, SFD and TMD (twisting moment diagram) for the plane grid structure shown in Fig. E1.14.

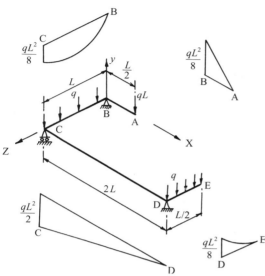

Figure E1.14

Answer: Determine the vertical reactions at B, C and D.

$$V_B + V_C + V_D = qL + q(L) + q(0.5L) = 2.5qL$$

Taking moments about CD: $V_B L - qL(L)$ at $A - q(L)(0.5L)$ on BC $- q(0.5L)(0.25L)$ on DE = 0

$$V_B = 1.625qL$$

Taking moments about CB: $V_D(2L) - qL(0.5L)$ at A $- q(0.5L)(2L)$ on DE = 0

$$V_D = 0.75qL$$

Therefore

$$V_C = 0.125qL$$

AB: Is a cantilever segment with an end load of qL at B and bends in the x–y plane. To maintain equilibrium, the reactions at B on AB are a vertical force of qL and a bending moment of $0.5qL^2$ about the z-axis. These act as actions on the element CB at B. The couple of $0.5qL^2$ at B acts as a twisting moment on CB at B.

CB: This member bends in the y–z plane and twists about the z-axis. The bending moment acts about the x-axis. At B the loads are $(V_B - qL)$ in the vertical direction and a twisting moment about the z-axis of $-0.5qL^2$ from the member AB.

To maintain equilibrium, at C the reactions are vertical force $= q(L) - \{V_B - qL\} = 0.375qL$, a clockwise couple $= (V_B - qL)L - q(L)\{0.5L\} = 0.125qL^2$

about the x-axis which acts as bending moment at C on CB and acts on CD at C as a twisting moment and a twisting moment of $0.5qL^2$ about the z-axis which acts as a bending moment at C on CD.

CD: At C the loads are a vertical force $= V_C - 0.375qL = -0.25qL$ (acts downward) moment of $0.5qL^2$ about the z-axis and twisting moment of $0.125qL^2$ about the x-axis. To maintain equilibrium, at D the forces are a vertically upward force of $0.25qL$, a twisting moment of $0.125qL^2$ about the x-axis and zero bending moment about the z-axis.

DE: This is a cantilever subjected to u.d.l. of q and bending in the y–z plane.

BA: $M = q(x - 0.5L)$, $Q = -qL$, $T = 0$

CB: $M = 0.625qLz - 0.5qz^2$, $Q = 0.625qL - qz$, $T = 0.5qL^2$

CD: $M = 0.5qL^2 - 0.25qLx$, $Q = 0.25qL$, $T = -0.125qL^2$

ED: $M = 0.5qx^2$, $Q = -qx$, $T = 0$

14. Draw the BMD, SFD, AFD (axial force diagram) and TMD (twisting moment diagram) for the 3-D structure shown in Fig. E1.15.

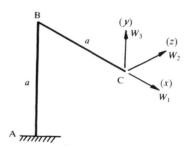

Figure E1.15

Answer: Member BC: This is subjected to an axial tensile force of W_1. It acts as a cantilever subjected to end loads W_3 and W_2 and bending in the x–y and x–z planes respectively. The reactions at B are forces W_1, W_3 and W_2 in the x-, y- and z-directions respectively and couples W_2a and W_3a about the z- and y-axes respectively.

Member AB: This is subjected to known reactions at B from AB. It is under an axial tensile force of W_3, a twisting moment about the y-axis of W_2a. For bending in the x–y plane, it is subjected at B to a load W_1 in the x-axis and a couple W_3a about the z-axis. For bending in the y–z plane, it is subjected at B to a load of W_2 in the z-direction.

1.24 ● Examples for practice

1. Determine the resultant of the following coplanar, concurrent forces.

(a) 120 kN at 45° to the horizontal and 80 kN at 30° to the horizontal.

Answer: 198.35 kN @ 39° to the horizontal.

(b) 200 kN at −60° to the horizontal, and 80 kN at 45° to the horizontal, 40 kN at 120° to the horizontal and 100 kN at 240° to the horizontal.

Answer: 189.52 kN @ 297.2° to the horizontal.

2. (a) A system of parallel vertical concentrated forces act on a beam at horizontal positions indicated from the left-hand support. Positive sign indicates a downward force. Calculate the resultant force and its position from the left-hand support.
100 kN @ 5 m, −80 kN @ 6 m, 200 kN @ 12 m and −150 kN @ 14 m.

Answer: 70 kN @ 4.57 m.

(b) A system of parallel concentrated vertical forces act on a beam at horizontal positions indicated from the left-hand support. Positive sign indicates a downward force. Calculate the resultant force and its position from the left-hand support.
Note: Negative sign for position indicates that the load acts to the left of the support.
200 kN @ −5 m, 100 kN @ 8 m, 150 kN @ 12 m and −90 kN @ 15 m.

Answer: 360 kN @ 0.694 m.

3. A system of parallel concentrated and distributed vertical forces act on a beam at horizontal positions indicated from the left-hand support. Positive sign indicates a downward force. Calculate the resultant force and its position from the left-hand support.
Note: Negative sign for position indicates that the load acts to the left of the support.
100 kN @ 5 m, 20 kN m^{-1} from 3 m to 12 m, 200 kN @ 10 m, −30 kN m^{-1} from −3 to 10 m.

Answer: 90 kN @ 27.61 m.

4. A system of parallel concentrated horizontal forces act on a beam at *vertical* positions indicated from the bottom support. Positive sign indicates a horizontal force acting from left to right. Calculate the resultant force and its position from the left-hand support.
Note: Negative sign for position indicates that the load acts below the support.
150 kN @ 3 m, −80 kN @ 4 m, −60 kN @ 6 m and 40 kN @ 8 m.

Answer: 50 kN @ 2.25 m.

5. A system of parallel concentrated and distributed horizontal forces act on a beam at *vertical* positions indicated from the bottom support. Positive sign indicates a horizontal force acting from left to right. Calculate the resultant force and its position from the left-hand support.
Note: Negative sign for position indicates that the load acts below the support.
30 kN m^{-1} from 2 m to 8 m, 80 kN @ 6 m, 20 kN m^{-1} from −4 m to 6 m and 100 kN @ 12 m.

Answer: 560 kN @ 4.96 m.

6. The following forces act on a body in the direction and at coordinates indicated. Calculate the resultant force and moment about the point indicated.
(a) Point about which moment is required is (4, 12).
100 kN at 60° to the horizontal @ (8, 6).
200 kN at 30° to the horizontal @ (12, −8).
80 kN at 135° to the horizontal @ (−6, 14).

Answer: 294.79 kN at 55.58° to the horizontal and moment = −4458.0 kN m.

(b) Point about which moment is required is (8, 6).
100 kN at 61.35° to the horizontal @ (−8, −12).
200 kN at 0° to the horizontal @ (8, −6).
50 kN at −30° to the horizontal @ (−12, 8).

Answer: 520.83 kN at 24.85° to the horizontal and moment = −3421.7 kN m.

7. A system of concentrated horizontal and vertical forces act on a body at positions indicated. Calculate the value of the resultant force and moment about the origin.
Vertical forces: positive down.
20 kN @ (4, 3), −50 kN @ (−8, 12) and 100 kN @ (6, 10).
Horizontal forces: positive left to right.
−80 kN @ (3, 8), 100 kN @ (10, 10) and 40 kN @ (6, 12).

Answer: 92.2 kN @ 40.6° to the horizontal.
Moment = 1920 kN m.

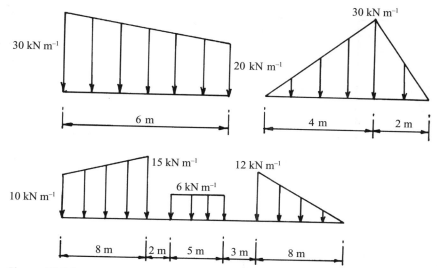

Figure E1.16

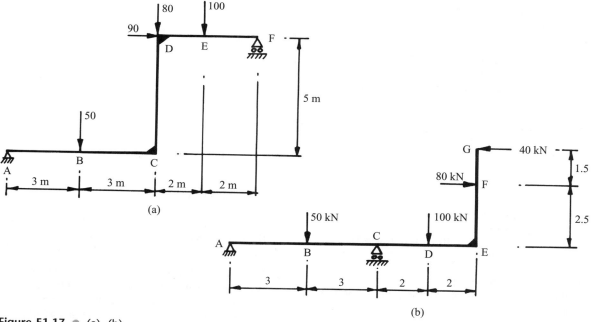

Figure E1.17 ● (a), (b)

8. Figure E1.16 shows distributed loads acting on a beam. Calculate the value of the resultant and its position from the left-hand end.

Answer:

(a) Trapezoidal distributed load of intensity 30 kN m^{-1} and 20 kN m^{-1} at the left- and right-hand ends respectively.
Answer: 150 kN @ 2.8 m

(b) Triangular load of 30 kN m^{-1} at the apex.
Answer: 90 kN @ 3.33 m

(c) Combined trapezoidal, uniformly distributed and triangular loads.
Answer: 178 kN @ 10.08 m

9. Figure E1.17 shows forces acting on a structure. Calculate the reactions at the supports.

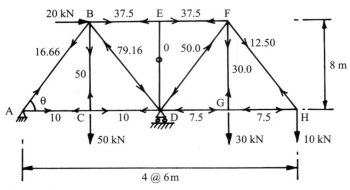

Figure E1.18

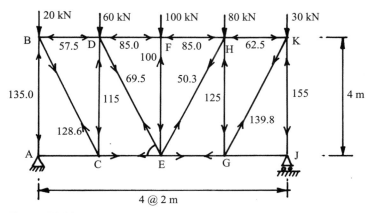

Figure E1.19

Answer:

(a) $V_A = 42$ kN, $H_A = 90$ kN (right to left),
   $V_F = 188$ kN
(b) $V_A = -15$ kN, $H_A = 40$ kN (right to left),
   $V_C = 165$ kN

10. Figure E1.18 shows a pin-jointed truss. Calculate the forces in the members, using the method of joints.

Answer:

(a) $V_A = -13.33$ kN (down), $H_A = -20$ kN (right to left), $V_D = 103.33$ kN
   AB = 16.66, AC = CD = 10.0, BC = 50.0,
   BE = EF = 37.50, BD = -79.16, DE = 0,
   DF = -50.0, GD = GH = -7.50, GF = 30.0,
   FH = 12.50

11. Figure E1.19 shows a pin-jointed truss. Calculate the forces in the members, using the method of sections.

$V_A = 135$ kN, $V_J = 155$ kN
AC = GJ = 0, AB = -135, BD = -57.5,
BC = 128.58, CE = 57.5, DC = -115, DE = 69.496,
DF = FH = -85, FE = -100, EH = 50.32, EG = 62.5,
HK = -62.5, HG = -125, GK = 139.76, KJ = -155.

12. Figure E1.20 shows a simply supported beam with an overhang, subjected to concentrated and uniformly distributed vertical loads and concentrated couples. Calculate the SF and BM at all critical sections.

Answer: $V_A = 197.92$ kN, $V_G = 372.08$ kN
SF:
$Q = -197.92 + \{x - \langle x - 4\rangle\} + 50\langle x - 2\rangle^0 + 100\langle x - 6\rangle^0 + 40\{\langle x - 8\rangle - \langle x - 12\rangle\} + 150\langle x - 10\rangle^0 - 372.08\langle x - 12\rangle^0 + 10\{\langle x - 12\rangle - \langle x - 13\rangle\} + 20\langle x - 13\rangle^0$
A = -197.92, B = -157.92 and -107.92, C = -67.92, D = -67.92 and 32.08, E = 32.08,
F = 112.08 and 262.08, G = 342.08 and -30.0,
H = -20.0

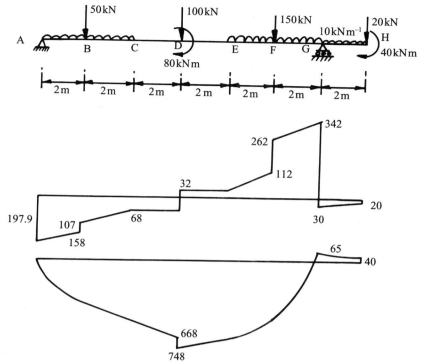

Figure E1.20

BM:
$M = -197.92x + \{x^2 - \langle x - 4\rangle^2\}/2 + 50\langle x - 2\rangle$
$+ 100\langle x - 6\rangle - 80\langle x - 6\rangle^0 + 40\{\langle x - 8\rangle^2 - \langle x - 12\rangle^2\}/$
$2 + 150\langle x - 10\rangle - 372.08\langle x - 12\rangle + 10\{\langle x - 12\rangle^2\}/2$
$+ 20\langle x - 13\rangle - 40\langle x - 13\rangle^0$
Negative value causes tension on the bottom face.

A = 0, B = 355.84, C = 531.68, D = 667.52 and
747.52, E = 683.76, F = 539.16, G = −65.0,
H = −40.0
Note: Positive moment causes tension at the
bottom face.

Chapter two

Influence lines for statically determinate structures

In the previous chapter, diagrams have been drawn which show internal forces such as bending moment at any section on a beam when the loads occupy a given position. However, many structures, such as bridge structures, crane girders, etc., have to support loads which move. In such a case it is important from the design point of view to calculate the maximum stress resultants at a given section in a member arising from all possible load positions on the structure. A very useful concept often employed for this purpose is the influence line diagram. In this chapter the concept of the influence line diagram and its evaluation for statically determinate structures will be explained.

2.1 ● Moving loads and influence line diagrams

Consider the simply supported beam shown in Fig. 2.1(a). Figure 2.1(b–e) shows the bending moment diagrams when a unit load is acting at different positions on

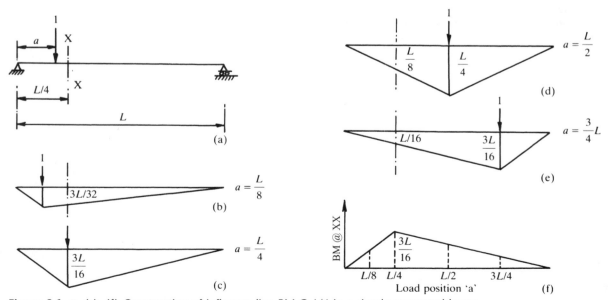

Figure 2.1 ● (a)–(f) Construction of influence line BM @ L/4 in a simply supported beam

the span. From these bending moment diagrams, it is possible to draw a diagram (Fig. 2.1(f)) showing on the y-axis the value of the bending moment at a *given section* such as, say, the *quarter span* section and the corresponding position of the unit load on the x-axis. Such a diagram is called an influence line diagram for the bending moment at the given section, which in this case is the quarter span of a simply supported beam. It shows the variation of bending moment at quarter span section as a unit load travels from one support to the other. In a similar way consider the pin-jointed truss shown in Fig. 2.2 which is used as a bridge girder. By positioning a unit load at joints in the bottom chord, axial force in the members of the truss can be determined. From this information it is possible to construct an influence line for the axial force in any particular member such as members 3–5 and 4–5.

From the above examples, it can be seen that an influence line can be defined as a diagram which shows the variation of a function such as bending moment, axial force, etc. *at a given* position in a structure as a unit load travels from one support to the other. The italicized words are important, for the influence line must be drawn for a given stress resultant at a given position in the structure. The reader should appreciate the difference between, for example, a bending moment diagram which is drawn for the whole structure and an influence line for bending moment at a particular section. A bending moment diagram shows the distribution of bending moment at all sections in the span with the loads in a given position. An influence line for bending moment at a given position is, however, a diagram which shows how the bending moment *at that* section varies as a unit load is placed at different positions in the span.

2.2 ● Influence lines for bending moment and shearing force in a simply supported beam

Before the influence diagrams can be drawn, the section for which the influence line needs to be drawn must be selected. Take the section X in Fig. 2.3(a). The problem then is to draw a diagram showing how the bending moment at X varies as a concentrated unit load travels across the span. This diagram will be the influence line for bending moment at X.

At any given instant let the load be at a distance a from A. Then considering the forces to the right of X gives the bending moment M_x at X as

$$M_x = V_B(L - x) - 1\langle a - x \rangle$$

Note the use of Macaulay's brackets where

$$\langle a - x \rangle = 0 \text{ if } a \leqslant x \quad \text{and} \quad \langle a - x \rangle = (a - x) \text{ if } a > x$$

By taking moments about A, $V_B = a/L$.

$$M_x = \frac{a}{L}(L - x) - 1\langle a - x \rangle$$

Therefore

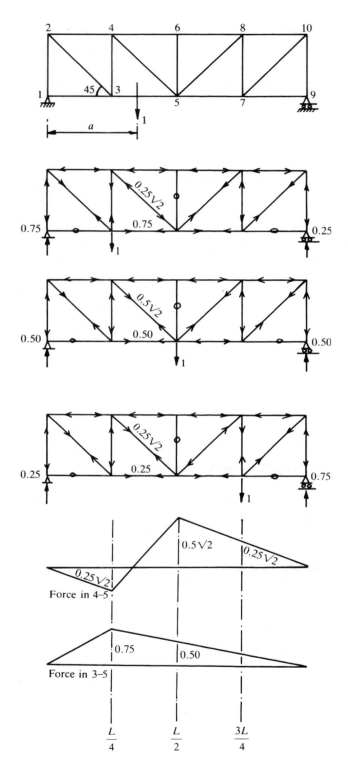

Figure 2.2 ● Force analysis and influence line construction for axial forces in members 3–5 and 4–5

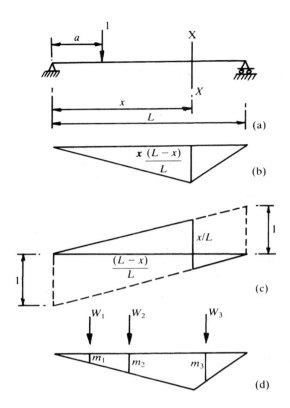

Figure 2.3 ● (a)–(d) Influence lines for BM and SF in a simply supported beam

If $a \leqslant x$, $M_x = a(1 - x/L)$

If $a > x$, $M_x = x(1 - a/L)$

Thus M_x is a linear function in a and reaches its maximum value when $a = x$. $M_{max} = x(1 - x/L)$. When $a = 0$ and $a = L$ the values of M_x are zero and the influence line for bending moment is shown in Fig. 2.3(b). The shearing Q_x at X is given by

$$Q_x = V_B - 1\langle a - x \rangle^0$$
$$= a/L - 1\langle a - x \rangle^0$$

Therefore if $a \leqslant x$, then $Q_x = a/L$ and if $a > x$, then $Q_x = (a/L - 1)$. These are linear relationships in a and the diagram consists of two parallel lines as shown in Fig. 2.3(c).

Influence lines are drawn for unit loads because the effect of any given load W is obtained simply by multiplying the effect of a unit load by W. If, as shown in Fig. 2.3(d), at any given instant loads W_1, W_2 and W_3 occupy positions on the span such that the corresponding ordinates to the bending moment influence line are m_1, m_2 and m_3 respectively, then the moment at X when the loads are in this position is given by

$$M_x = W_1 m_1 + W_2 m_2 + W_3 m_3$$

If W_1, W_2 and W_3 are a given set of loads which can cross the beam, then the bending moment at X has its maximum value when the function $W_1 m_1 + W_2 m_2 + W_3 m_3$ has its maximum value.

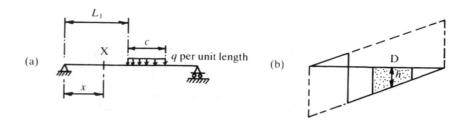

Figure 2.4 ● (a), (b)
Calculation of shear
force due to uniformly
distributed load

It should be noted that although an influence line is constructed for a point load moving across the span, it can also be used for a distributed load situation as follows.

Consider the beam of Fig. 2.4(a), where the influence line for shear at X is as shown in Fig. 2.4(b). The shearing force at this section due to a uniformly distributed load of intensity q starting from L_1 from the left-hand support and extending over the length c is obtained as follows.

Consider an element of load at D of length dx. The equivalent concentrated load due to q on this element is q dx and if the corresponding ordinate to the influence line is h, then the shear at X due to the element of load is qh dx. Therefore

Total shear at X due to load $cq = q \int_{L_1}^{c+L_1} h \, dx$

But the term under the integral sign is the area under the influence line bounded by the loaded length c.

Thus with a uniformly distributed load, the area of the influence line diagram under the uniformly distributed load is multiplied by the intensity of loading to give the net effect of the uniformly distributed load on the value of the function.

The maximum value of the bending moment at X due to a given load system will occur when the function ΣWm has its maximum value. This will depend on the load system. The following moving load systems will be examined in turn:

(a) Single concentrated load W.
(b) Uniformly distributed load of intensity q and length $\geqslant L$.
(c) Any system of concentrated loads.

2.3 ● Maximum bending moment and shearing force due to a single concentrated moving load

As shown in Fig. 2.3(b), the influence line for the bending moment at X has its maximum ordinate when the load is at X and the magnitude of moment is $x(L - x)/L$. Thus the maximum bending moment at X due to a single moving load W is $Wx(L - x)/L$. The diagram showing how this function varies with x is known as a *maximum bending moment diagram* and for a single moving load is the parabola shown in Fig. 2.5(a). The maximum value of this function occurs when $x = \frac{1}{2}L$, in which case M_x has the value $\frac{1}{4}WL$. It should be appreciated that the maximum bending moment diagram describes the maximum bending moment that occurs <u>at any section</u> in the beam for all positions of a moving concentrated load W on the beam. It is not a bending moment diagram for a stationary position of unit load.

Figure 2.5 ● (a), (b)
Maximum BM and SF
distribution in a simply
supported beam due to
a moving concentrated
load *W*

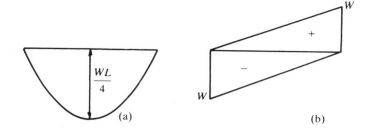

$$\frac{WL}{4}$$

(a)

W W

(b)

The influence line for shearing force given in Fig. 2.3(c) shows that the maximum values of both positive and negative shear occur when the load is immediately to the left or right of X, their values being x/L and $(L - x)/L$ respectively. Thus the maximum positive shear at X due to a single load W is Wx/L and the maximum negative shear at X is $W(L - x)/L$. The maximum shearing force diagram is given in Fig. 2.5(b). An important fact which should be noted from this diagram is that there are both positive and negative values for the maximum shearing force at any section. This may be of importance in the design of reinforced concrete beams where the shear resistance is provided by the use of 'bent up' bars.

2.4 ● Maximum bending moment and shearing force due to a uniformly distributed load longer than the span

In dealing with a uniformly distributed load it is the area under that part of the influence line covered by the load which is critical. Examination of the influence line for bending moment at X in Fig. 2.3(b) shows that this area has its maximum value when the whole span is covered by the load, in which case

$$M_x = \tfrac{1}{2}qL\,\frac{x(L - x)}{L} = \tfrac{1}{2}qx(L - x)$$

This function, shown in Fig. 2.6(a), is a parabola, giving a maximum bending moment diagram for a uniformly distributed load longer than the span similar to Fig. 2.5(a), but with the maximum value of $qL^2/8$ when $x = \tfrac{1}{2}L$. Inspection of the influence line for shearing force at X in Fig. 2.3(c) shows that the maximum positive area corresponds to the case where the load covers the portion AX of the span, and the maximum negative area when the length BX is covered. The values of the maximum shearing forces are given as $qx^2/(2L)$ for the positive case and $q(L - x)^2/(2L)$ for the negative case. Each is a parabolic relationship and the diagrams for maximum shearing force for a uniformly distributed load longer than the span are given in Fig. 2.6(b). It should be noted that the maximum shear force value is obtained by covering the beam either to the left or right of the section. It is *not* obtained by covering the whole span as in the case of maximum bending moment.

The maximum bending moment and shear force diagrams due to a single concentrated load and uniform load shown in Figs 2.5 and 2.6 respectively are particularly useful in the design of simply supported bridge girders. This is because notional traffic loads on bridges are generally specified in terms of a uniformly distributed load and an additional concentrated load.

Figure 2.6 ● (a), (b)
Maximum BM and SF
distribution in a simply
supported beam due to
a moving uniformly
distributed load longer
than the span

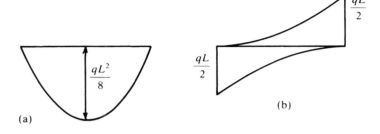

(a)

(b)

Figure 2.7 ● A simply
supported beam with
two moving
concentrated loads

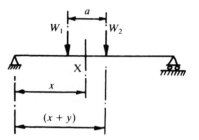

2.5 ● Maximum bending moment and shearing force due to a system of concentrated loads

Consider in the first instance the two loads W_1 and W_2 separated by a distance a as shown in Fig. 2.7. When the load W_2 is at X, the moment at X is M_{x2} and is given by

$$M_{x2} = V_B(L - x), \text{ where } V_B = \{W_1(x - a) + W_2 x\}/L, \ x > a$$

Therefore

$$M_{x2} = \frac{W_2 x(L - x)}{L} + \frac{W_1(x - a)(L - x)}{L}$$

Similarly, when the load W_1 is at X, the moment at X is M_{x1} and is given by

$$M_{x1} = V_A x, \text{ where } V_A = \{W_2(L - a - x) + W_1(L - x)\}/L$$

Therefore

$$M_{x1} = \frac{W_2 x(L - x - a)}{L} + \frac{W_1 x(L - x)}{L}$$

When neither load is at X but the load W_2 is at a distance y to the right of X (Fig. 2.7), the moment at x is M_{xY} and is given by

$$M_{xY} = V_A x - W_1(a - y)$$

where

$$V_A = \{W_1(L - x - y + a) + W_2(L - x - y)\}/L$$

$$M_{xY} = V_A x - W_1(a - y)$$

$$= \{W_1(L - x - y + a) + W_2(L - x - y)\}x/L - W_1(a - y)$$

Since if $(y - a) = 0$, then $M_{xY} = M_{x1}$ and if $y = 0$, then $M_{xY} = M_{x2}$, M_{xY} can be expressed in two different ways as follows:

1. M_{xY} in terms of M_{x1}: express M_{xY} in terms of M_{x1} and additional terms multiplied by $(a - y)$.

$$M_{xY} = \{W_1(L - x) + W_2(L - x - a)\}x/L + (a - y)\{W_1(L - x) + W_2x\}/L$$
$$= M_{x1} + (a - y)\{W_2x - W_1(L - x)\}/L$$

2. M_{xY} in terms of M_{x2}: express M_{xY} in terms of M_{x2} and additional terms multiplied by y.

$$M_{xY} = [\{W_1(L - x)x - a(L - x)\} + W_2(L - x)x]/L + y\{W_1(L - x) - W_2x\}/L$$
$$= M_{x2} - y\{W_2x - W_1(L - x)\}/L$$

It is seen that the term in the curly brackets, namely $\{W_2x - W_1(L - x)\}$ is common to both expressions for M_{xY}. If it is positive, i.e. $W_2x > W_1(L - x)$, then $M_{x2} > M_{xY} > M_{x1}$. Therefore the maximum bending moment occurs when the load W_2 is on the section. On the other hand, if it is negative, i.e. $W_2x < W_1(L - x)$, then $M_{x1} > M_{xY} > M_{x2}$. In this case the maximum bending moment occurs when the load W_1 is on the section. Thus the maximum bending moment at a section occurs when a load is at that section.

In the case when many point loads move across the span, similar arguments apply except that it is more complex to determine which specific load is to be placed at the section. The problem can be approached as follows. Let the train of loads (Fig. 2.8a) have a resultant ΣW which at any given position is distance y from the left-hand support. Let ΣW_L be the resultant of the loads to the left of X and ΣW_R be the resultant of loads to the right of X (Fig. 2.8b). Note that $\Sigma W = \Sigma W_L + \Sigma W_R$. Let the distance between ΣW and ΣW_L be a. The position is then as shown in Fig. 2.8(b) and the moment M_X at X is obtained by taking moments of the forces to the left and is given by

$$M_x = V_A x - \sum W_L(a - y + x), \text{ where } V_A = \sum W(L - y)/L$$

$$M_x = \sum W(L - y)x\frac{1}{L} - \sum W_L(x - y + a)$$

The variable is y, and for maximum or minimum M_x,

$$\frac{dM_x}{dy} = \frac{-\Sigma Wx}{L} + \sum W_L$$
$$= -\left(\sum W_R + \sum W_L\right)x/L + \sum W_L = 0$$

Therefore

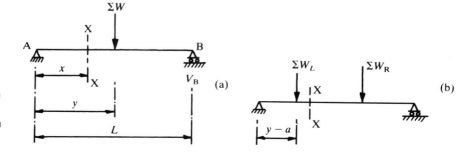

Figure 2.8 ● (a), (b) Simply supported beam under a set of concentrated loads with a resultant ΣW

$$\left\{ -\sum W_R x + \sum W_L (L - x) \right\}/L = 0$$

Since the values of $\sum W_R$ and $\sum W_L$ are discontinuous, we cannot find the value of x for which $dM_x/dx = 0$. What we can find is the load positions for which dM_x/dx *changes sign*.

For a maximum value of M_x, dM_x/dy passes from positive through zero to a negative value. dM_x/dx is positive if

$$\left\{ -\sum W_R x + \sum W_L (L - x) \right\} > 0$$

i.e.

$$\sum W_L/x > \sum W_R/(L - x)$$

i.e. *average* load on the left of section is > *average* load on the right of section.

Similarly, dM_x/dy is negative when the *average* load on the left of section is less than the *average* load on the right of section. Hence the critical load is the one which causes the value of {average load on left of section – average load on right of section} to *change sign*.

A problem which is similar to that of finding the load position to give the maximum bending moment at a given section is that of finding the load position to give the maximum bending moment under a given load. The object of this is to determine the maximum bending moment anywhere in the span of the beam. This is dealt with in the following way. Let W_i be the given load in a system and it is required to find the load position such that the maximum bending moment occurs under this load. Let $\sum W$ be the resultant of all the loads in the system and $\sum W_L$ be the resultant of the loads to the left of W_i as shown in Fig. 2.9. Let W_i be at a distance x from the left-hand support when the maximum moment occurs under it. Let $\sum W$ and $\sum W_L$ be distance a_1 and a_2 respectively, from W_i.

Then the moment M_i under W_i is given by

$$M_i = V_A x - \sum W_L a_2$$

where $\quad V_A = \dfrac{\sum W}{L}(L - x - a_1)$

$$\therefore M_i = +\frac{\sum W}{L} x(L - x - a_1) - \sum W_L a_2$$

For this to be a maximum

$$\frac{dM_i}{dx} = 0, \text{ or } L - 2x - a_1 = 0$$

$$x = \tfrac{1}{2}L - \tfrac{1}{2}a_1$$

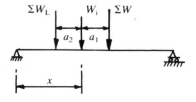

Figure 2.9 ● Load position for maximum bending moment at x

Thus the maximum bending moment under a given load occurs when the centre of the span bisects the distance between that load and the resultant of the system.

It is not possible to give any definite rule for the load position for maximum shearing force as this depends on the load system. From the influence line for shear force at a section it appears that because the maximum ordinate, whether positive or negative, is at the section itself, if all the loads in the system are approximately of the same value, then the maximum positive shear occurs when the extreme right-hand load of the system is at the section and all the loads wholly to the left, and the maximum negative shear when the extreme left-hand load is at the section and all the loads wholly to the right. In some cases, however, where the end loads of a system are small compared with the other loads, this rule will not apply. Trial and error methods must be used, but in practice very few trials are necessary to establish the most critical load position.

Example 1

The span of a girder is 10 m. The live load system shown in Fig. 2.10(a) may cross the span in either direction. Determine the maximum bending moment in the girder

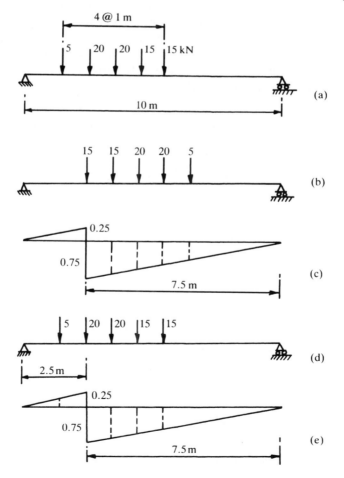

Figure 2.10 ●
(a)–(e) Load positions
for maximum positive
and negative shear force
at 2.5 m from left

and obtain the equivalent uniformly distributed loading to cause the same value of maximum moment. Draw the influence line for shear at the left-hand quarter point and calculate the maximum value.

Solution To determine the maximum bending moment in the girder, it is necessary to find the position of the centre of gravity of the load system. The resultant load

$$\sum W = (5 + 20 + 20 + 15 + 15) = 75 \text{ kN}$$

Let this be at a distance $\bar{x}$ from the 5 kN load at the left-hand end. Taking moments about the line of action of this load gives

$$75\bar{x} = 5 \times 0 + 20 \times 1 + 20 \times 2 + 15 \times 3 + 15 \times 4$$

leading to

$$\bar{x} = 2.2 \text{ m}$$

Since the loads in the system, except for the 5 kN, are all approximately equal, the maximum bending moment anywhere in the beam will occur under either load 3 or 4 when these loads are near the middle of the beam. The distance between load 3 and the resultant is equal to $(2.2 - 2) = 0.2$ m. Therefore the maximum bending moment occurs under load 3 when it is at $0.2/2 = 0.1$ m to the left of the centre and the resultant load of 75 kN is at 0.1 m to the right of mid span. Similarly, the distance between load 4 and the resultant is equal to $(3.0 - 2.2) = 0.8$ m. Therefore the maximum bending moment under load 4 occurs when it is at $0.8/2 = 0.4$ m from the centre and to the right.

(a) Load 3 at 0.1 m to the left of centre. With the load in this position, the resultant load of 75 kN is at 4.9 m from the right-hand support. Therefore the reaction at the left-hand end is $(4.9/10) \times 75$ kN $= 36.75$ and the moment M_3 under the load 3 is given by

$$M_3 = 36.75 \times 4.9 - 5 \times 2 \text{ due to load } 1 - 20 \times 1$$
$$\text{due to load } 2 = 150 \text{ kN m}$$

(b) Load 4 at 0.4 m to the right of centre. With the load in this position the resultant load of 75 kN is at 4.6 m from the left-hand support. Therefore the reaction at the right-hand end is $(4.6/10) \times 75$ kN $= 34.50$ and the moment M_4 under the load 4 is given by

$$M_4 = 34.50 \times 4.6 - 15 \times 1 \text{ due to load } 5 = 143.7 \text{ kN m}$$

The maximum bending moment anywhere in the beam is therefore 150.0 kN m. The equivalent uniformly distributed load q_e to give the same maximum bending moment anywhere in the girder is given by

$$q_e L^2/8 = 150$$

Therefore

$$q_e = 12 \text{ kN m}^{-1}$$

The influence line for shearing force at the left-hand quarter-point is shown in Fig. 2.10(c). One possible load position for the maximum value under the given load system is when load 5 is at quarter span and the load system (reversed from the order shown in Fig. 2.10(a)) is placed wholly to the right as shown in Fig. 2.10(b). The ordinates at the load positions can be calculated using the similar triangles shown in Fig. 2.10(c). Then, the shearing force Q_5 is given by

$$Q_5 = \frac{3}{4}\left(15 + \frac{6.5}{7.5} \times 15 + \frac{5.5}{7.5} \times 20 + \frac{4.5}{7.5} \times 20 + \frac{3.5}{7.5} \times 5\right)$$

$$= -44.6 \text{ kN}$$

Another possibility is for load 1 to be to the left, load 2 at the quarter-point and the remaining loads to the right as shown in Fig. 2.10(d). The ordinates at the load positions can again be determined from similar triangles shown in Fig. 2.10(e). In this case the shearing force Q_2 is given by

$$Q_2 = -\frac{3}{4}\left(20 + \frac{6.5}{7.5} \times 20 + \frac{5.5}{7.5} \times 15 + \frac{4.5}{7.5} \times 15\right) + \frac{1}{4} \times 5 \times \frac{1.5}{2.5}$$

$$= -42.4 \text{ kN}$$

This is less than Q_5, hence the maximum shearing force is 44.6 kN.

2.6 ● Influence lines for axial force in members of pin-jointed structures

A pin-jointed truss is frequently used for bridge structures subject to moving vehicular loads. In a beam the important stress resultants at a section are the bending moment and shear force. Therefore the influence lines are drawn for these stress resultants. In a pin-jointed structure, an influence line is drawn for the axial force in a specific member. As shown in Section 2.1, if the influence lines are required for all the members of a truss, then the simplest procedure is to place a unit load successively at all the joints where loads will act and determine the axial forces in all the members of the truss. From this information, the influence lines for any member can be easily constructed. However if the influence lines are required for only a few selected members, as often happens at preliminary stages of design of the structure, then the following method can be employed. The method is called the method of sections. The technique is to study the equilibrium of a portion of the truss to determine forces in specific members.

Let the problem be to draw the influence lines for the axial forces in the members EF which is a typical chord member and DE which is a typical 'web' member of the parallel chord truss of span L and depth d shown in Fig. 2.11(a).

(a) Influence line for the axial force in the 'web' member DE. There are three possibilities for unit load to be at x where:

$$x < L_1, \ x > L_2 \text{ and } L_2 \geqslant x \geqslant L_1.$$

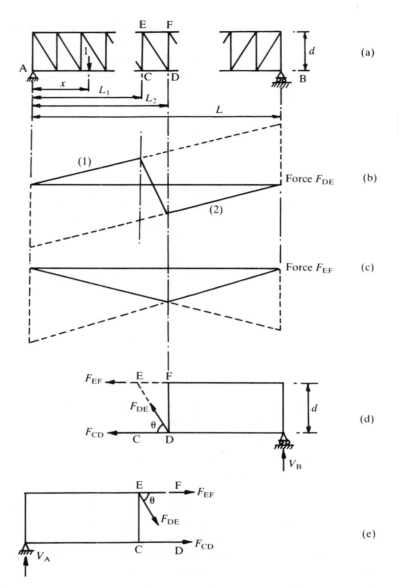

Figure 2.11 ●
(a)–(e) Construction of influence lines for forces in members DE and EF

(i) $x < L_1$. In this case, $V_B = x/L$ and if a section is cut through the panel, then assuming that the force in DE is tensile, for vertical equilibrium, as shown in Fig. 2.11(d),

$$F_{DE} \sin \theta + V_B = 0, \text{ where } V_B = x/L$$

Therefore

$$F_{DE} = -\frac{x}{L} \operatorname{cosec} \theta$$

the negative sign denoting compression and θ = angle CDE.

(ii) $x > L_2$. In this case, $V_A = (L - x)/L$ and if a section is cut through the panel to the right of CE, assuming that the force in DE is tensile, for vertical equilibrium, as shown in Fig. 2.11(e),

$$F_{DE} \sin \theta - V_A = 0$$

Therefore

$$F_{DE} = \left(\frac{L - x}{L}\right) \text{cosec } \theta$$

(iii) $L_2 > x > L_1$. In this case the load is on the panel CD. Member CD acts as a simply-supported beam transferring part of the load to each of the joints C and D. The span of the beam is $(L_2 - L_1)$ and the load acts at a distance $(x - L_1)$ from C. Therefore the reaction at D is

$$\frac{x - L_1}{L_2 - L_1}$$

and since $V_B = x/L$ for vertical equilibrium of the free body in Fig. 2.11(d), then

$$F_{DE} = -\left\{\frac{x}{L} - \frac{(x - L_1)}{(L_2 - L_1)}\right\} \text{cosec } \theta$$

The three equations are each linear in x and the form of the influence line is shown by the full lines in Fig. 2.11(b). The dotted lines will form a part of the influence line for any member sloping in the same direction as DE; the change of sign from line 1 to line 2 occurs at the panel of which the actual member forms a part.

(b) Influence line for the axial force in the chord member EF. Consider the section in Fig. 2.11(d), and take moments of the forces about D. The lines of action of the forces in the members DE and CD pass through D and will have no moment about this point.

(i) $x < L_2$. In this case using the free body shown in Fig. 2.11(d), and taking moments about D, for equilibrium

$$F_{EF}d + V_B(L - L_2) = 0, \text{ where } V_B = x/L$$

Therefore

$$F_{EF} = -\frac{V_B(L - L_2)}{d} = -\frac{x(L - L_2)}{Ld}$$

(ii) $x > L_2$. In this case using the free body shown in Fig. 2.11(e) and taking moments about D, for equilibrium

$$V_A L_2 + F_{EF}d = 0, \text{ where } V_A = (L - x)/L$$

Therefore

$$F_{EF} = -\frac{L_2}{d}(L - x)$$

The minus sign denotes compression. The two equations are straight lines which meet vertically below D and give the influence line for the force in EF as the full lines in Fig. 2.11(c).

Member EF is a typical top or bottom chord member and one feature of the influence line is that it is of the same sign for all values of x and it is similar to the influence line for the bending moment in a simply-supported beam. On the other hand, the influence line for the force in DE, which has a change of sign as the load crosses the panel, is typical of an internal diagonal member and is similar to the influence line for shear at a given section in a simply-supported beam.

Example 1

A non-parallel boom symmetrical pin-jointed bridge truss which is simply supported across a span of 60 m and has a maximum depth of 12 m at the centre of the span is as shown in Fig. 2.12(a). The form of the top boom is parabolic, the depths

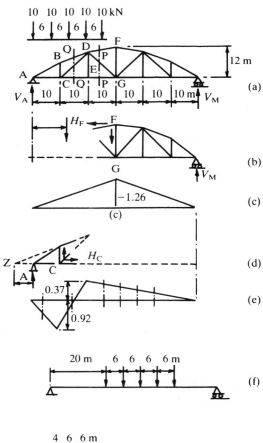

Figure 2.12 ●
(a)–(g) Construction of influence lines for axial forces in a truss with inclined top chord

at the panel points being 6.67, 10.67, 12.0, 10.67 and 6.67 m, but all members are straight between panel points. The load system shown is supported at the bottom chord joints. Draw the influence line diagrams for the forces in the members DF and CD, and determine the maximum possible forces in these members as the given load system crosses the span.

Solution The technique used is the method of sections. Note that because the top chord is inclined to the horizontal, the forces in the 'diagonal' members cannot be determined from vertical equilibrium only. The 'trick' involved is to resolve the force in a given member into its horizontal and vertical components and take moments of the forces acting on a free body such that only a single component of the force is involved.

(a) Force in member DF. To find the force in member DF take a section P–P through the panel and study the equilibrium of the free body to the right of P–P when the unit load is to the left of P–P. Resolve F_{DF} into two components H_F and V_F at F.

However,

$$H_F = F_{DF} \cos \theta$$

where

$$\tan \theta = (GF - DE)/10 = (12.0 - 10.67)/12 = 0.133$$

Therefore

$$\cos \theta = 0.9912 \text{ and } H_F = 0.9912 F_{DF}$$

Taking moments about G (Fig. 2.12(b)), i.e.

$$12 H_F + 30 V_M = 0, \text{ where } V_M = x/60$$

Therefore

$$H_F = -x/24$$

and

$$F_{DF} = -(x/24)/0.9912 = -x/23.79$$

This is linear in x and when $x = 30$, $F_{DF} = -30/(23.79) = -1.26$.

Similarly, for the unit load to the right of P–P and using the left-hand free body, F_{DF} is again seen to be a linear function of x, as shown in Fig. 2.12(c).

The maximum value for the force in this member occurs when the central 10 kN load is at the centre of the span. The ordinates to the influence line for the outer and inner loads are then 0.75 and 1.01 respectively, and

$$F_{DF \text{ max}} = 10(1.26 + 2 \times 1.01 + 2 \times 0.75) = 47.8 \text{ kN}$$

(b) Force in member CD. To obtain the influence line diagram for the force in F_{CD}, draw a section Q–Q and examine the equilibrium of the free body to the left of Q–Q (Fig. 2.12 (d)) when the unit load is to the right of Q–Q. Resolve the force in F_{CD}

(assumed tensile) into its horizontal and vertical components V_C and H_C at C, then

$$V_C = F_{CD} \sin \theta$$

$$\tan \theta = DE/CE = 10.67/10 = 1.067$$

$$\sin \theta = 0.7296$$

Therefore

$$V_C = F_{CD}\ 0.7296$$

Since the cut Q–Q involves members BD and CD, the force in member BD can be eliminated if moments are taken about a point Z where BD meets the horizontal line AM. Inclination of BD to the horizontal is

$$\tan \theta = (DE - BC)/10 = (10.67 - 6.67)/10 = 0.4$$

Therefore

$$DE/ZE = 0.4 \text{ and so } ZE = 10.67/0.4 = 26.675$$

$$ZC = 26.67 - 10 = 16.67$$

$$ZA = ZC - 10 = 6.67$$

Taking moments about Z

$$-V_C ZC - V_A ZA = 0, \text{ where } V_A = (60 - x)/60$$

Therefore

$$V_C = -(1 - x/60)0.4$$

$$F_{CD} = V_C/0.7296 = -(1 - x/60)\ 0.5483$$

When $x = 20$, $F_{CD} = -0.366$

In a similar way, when the unit load is to the left of the panel, the equilibrium of the free body to the right of Q–Q, again taking moments about Z, gives

$$V_C ZC - V_M ZM = 0$$

$$16.67 V_C - 66.67 V_M = 0, \text{ therefore } V_C = 4 V_M$$

$$V_C = F_{CD}\ 0.7296 \text{ as before and } V_M = x/60$$

Therefore

$$F_{CD} = (x/60)4/0.7296 = 0.0915x. \text{ When } x = 10, F_{CD} = 0.915$$

The influence line diagram is given in Fig. 2.12(e).

For the maximum compression in F_{CD} place the wheels as shown in Fig. 2.12(f), which gives

$$F_{CD} = 10(0.37 + 0.31 + 0.25 + 0.20 + 0.15) = 12.8 \text{ kN}$$

For the maximum tension in F_{CD} the wheels are placed as shown in Fig. 2.12(g), which gives

$$F_{CD} = 10(0.37 + 0.92 + 0.15) = 14.4 \text{ kN}$$

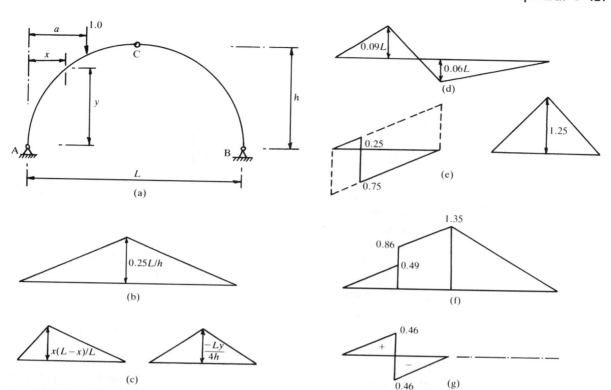

Figure 2.13 ● (a)–(g) Construction of influence lines for a three-pinned arch

2.7 ● Influence lines for forces in a three-pin arch

The three-pin arch, being statically determinate, is a useful structure with which to demonstrate a more general derivation of influence line diagrams. The arch shown in Fig. 2.13(a) is traversed by a unit load. The span is L and the rise at the centre is h. The position of a unit load from the left-hand support is a. Taking moments about A:

$$V_B = a/L \text{ and } V_A = (L - a)/L$$

(a) Influence line for horizontal reaction H.

    (i) When $a \leqslant 0.5L$, then taking moment about C of the portion CB of the arch, for equilibrium

$$Hh = 0.5LV_B = 0.5a$$

    (ii) When $a \geqslant 0.5L$, then taking moment about C of the portion AC of the arch, for equilibrium

$$Hh = 0.5LV_A = 0.5(L - a)$$

The above two linear relationships in a give the influence line for H as shown in Fig. 2.13(b).

(b) Influence line for bending moment at a section x from left-hand support.

    (i) $a \leqslant x$. Taking moments about X of the forces on the right-hand portion XB, we have

$$M = V_B(L - x) - Hy, \text{ where } y = \text{ordinate to the arch at } x$$
$$= (a/L)(L - x) - Hy$$

    (ii) $a \geqslant x$. Taking moments about X of the forces on the left-hand portion AX, we have

$$M = V_A x - Hy, \text{ where } y = \text{ordinate to the arch at } x$$
$$= (1 - a/L)x - Hy$$

The moment due to the vertical reactions is identical to that in a simply supported beam of span L. Therefore the equation for the influence line for the bending moment at a section in the three-pin arch can be expressed as {Influence line for bending moment at the corresponding section in a simply supported beam $- y$. Influence line for horizontal thrust H}

    The two diagrams are shown in Fig. 2.13(c).

    As an example, if the arch is parabolic, then

$$y = (4h/L^2)x(L - x), h = 0.2L$$

If the influence line is required for say $x = 0.25L$, then $y = 0.75h$. The net influence line is shown in Fig. 2.13(d).

(c) Influence line diagrams for shear force and axial force at a section.

    (i) $a < x$: the vertical V and horizontal H forces at the cut section are an upward force of V_B and a horizontal force of H to the left. Thus $V = V_B = a/L$, $H = H$.

    (ii) $a > x$: the vertical V and horizontal H forces at the cut section are a downward force of V_A and a horizontal force of H to the left. Thus $V = -V_A = -(1 - a/L)$, $H = H$.

    If the tangent to the section is inclined at an angle α to the horizontal, then the vertical and horizontal forces can be resolved along the tangent and the normal to the section to give the axial and shear force at the section as explained in Section 1.16.1.

    Axial force (Compression) $= H \cos \alpha - V \sin \alpha$

    Shear force $= V \cos \alpha + H \sin \alpha$

The influence lines for V and H are shown in Fig. 2.13(e). The influence line for V is the same as the influence line for shear force in a corresponding simply supported beam. The influence lines for axial thrust and shear force at a section can be expressed as the superposition of V and H

    Influence line for axial thrust
    $= \cos \alpha$. {Influence line for H}
      $- \sin \alpha$. {Influence line for shear force in a corresponding simply supported beam}

    Influence line for shear force
    $= \sin \alpha$. {Influence line for H}
      $+ \cos \alpha$. {Influence line for shear force in a corresponding simply supported beam}

If the section chosen is $x = 0.25L$, then $y = 0.75h$ and

$$\tan \alpha = dy/dx = (4h/L^2)(L - 2x) = 2h/L.$$

Assuming that $h = 0.2L$, then $\tan \alpha = 0.4$, $\sin \alpha = 0.3714$ and $\cos \alpha = 0.9285$. The influence lines for the shear and axial forces at $x = 0.25L$ are shown in Fig. 2.13(f) and (g) respectively.

2.8 ● Influence lines for forces in a balanced cantilever structure

Figure 2.14(a) shows a balanced cantilever structure. In this structure the connection between portions ABC and DEF is made through the simply supported beam CD. The interaction between the segments takes place through the vertical reactions V_C and V_D at the hinges as shown in Fig. 2.14(b). When drawing influence lines for any stress resultant in the portion ABC, it is useful to remember that

(a) When the load is in the region AB, the whole structure behaves as a simply supported beam AB.

(b) When the load is in the region BC, then the beam ABC behaves like a beam simply supported at A and B with an overhang BC.

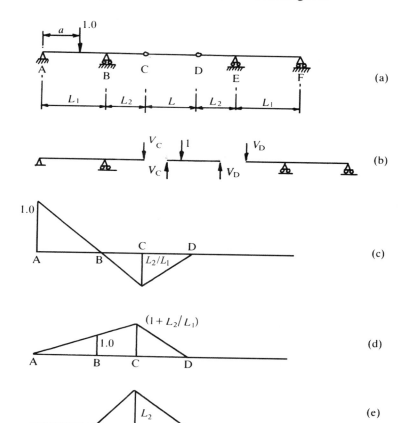

Figure 2.14 ● (a)–(e) Construction of influence lines for forces in a balanced cantilever structure

(c) When the load is moving from C to D, V_C changes from 1 to zero. The portion ABC behaves as though it is loaded by a vertically downward load V_C at C.

(d) Once the unit load is in the region DEF, the forces in the section ABC will be zero because V_C will be zero.

1. INFLUENCE LINE FOR REACTION AT A.

Let the load be at a distance of a from the left-hand support A.

(a) Load in the region AB, i.e. $L_1 \geqslant a \geqslant 0$: $V_A = 1 - a/L_1$.

(b) Load in the region BC, i.e. $L_1 + L_2 \geqslant a \geqslant L_1$: taking moments about B, $V_A = -(a - L_1)/L_1$. When $a = L_1 + L_2$, then $V_A = -L_2/L_1$.

(c) Load in the region CD: taking moments about B, $V_A = -V_C L_2/L_1$. V_C varies linearly from 0 to 1 as the load moves form C to D. Thus the influence line for V_A will be as shown in Fig. 2.14(c).

2. INFLUENCE LINE FOR REACTION AT B.

(a) Load in the region AB and BC: $V_B = 1 - V_A$.

(b) Load in the region CD: $V_B + V_A = V_C$. $V_C = 1$ at C and 0 at D. The influence line for V_B is shown in Fig. 2.14(d).

3. INFLUENCE LINE FOR BENDING MOMENT AT B.

(a) Load in the region AB: $M_B = 0$ because AB behaves as a simply supported beam.

(b) Load in the region BC and CD: $M_B = V_A L_1$. Thus the influence line for the bending moment at B is shown as in Fig. 2.14(e).

2.9 ● Summary

Previous sections have shown that the influence lines for stress resultants for statically determinate structures consist of a series of straight line segments. This is not true if the structure is statically indeterminate. Influence lines for statically indeterminate structures are considered in Chapter 10, Section 10.9.

In the case of a simply supported beam and pin-jointed parallel chord truss, it is useful to remember that

(a) The influence line for the bending moment at a section x in a simply supported beam is a triangle with the height of the triangle equal to $x(1 - x/L)$ as shown in Fig. 2.3(b). Because the top and bottom chords in a parallel chord truss resist the bending moment at a section, the influence lines for axial forces in these members also resemble the influence line bending moment as shown in Fig. 2.11(c).

(b) The influence line for shear force at a section x in a simply supported beam consists of two triangles which form part of two triangles with unit height as shown in Fig. 2.3(c). Because in a parallel chord pin-jointed truss the diagonal members alone resist the shear force at a section, the influence line for axial force in a diagonal member also resembles the influence line for shear force at a section in a simply supported beam, as shown in Fig. 2.11(b).

2.10 ● References

Hsieh Yuan-Yu 1982 *Elementary Theory of Structures* (2nd edn) Prentice-Hall

2.11 ● Problems

1. Construct the influence lines for the axial force in members BC and BG of the Warren truss shown in Fig. E2.1. All members are 20 m long. The loads travel on the top chord. Calculate the maximum compressive force in BC if a u.d.l. of 2 kN m^{-1} and 20 m long crosses the span.

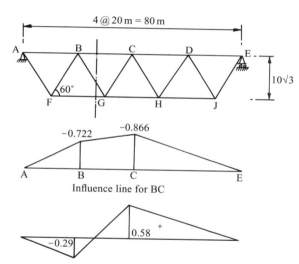

Figure E2.1

Answer: $V_A = (1 - x/80)$, $V_E = x/80$
(a) F_{BC}: take a vertical section through G.
   (i) If the unit load is to the left of G, i.e. $0 \leqslant x \leqslant 30$, then taking moment about G of the free body to the right of G,

$$-F_{BC}10\sqrt{3} - V_E 50 = 0.$$

Therefore

$$F_{BC} = -x/16\sqrt{3},$$

if the load is at B, then $x = 20$,

$$F_{BC} = -0.722.$$

  (ii) If the unit load is to the right of G, i.e. $30 \leqslant x \leqslant 80$ then taking moment about G of the free body to the left of G,

$$F_{BC}10\sqrt{3} + V_A 30 = 0.$$

Therefore

$$F_{BC} = -(1 - x/80)\sqrt{3}$$

if the load is at C, then $x = 40$,

$$F_{BC} = -0.866.$$

(iii) When the load is between B and C, the load is shared linearly between the 'reactions' at B and C. The influence line is a straight line connecting values of -0.722 at B and -0.866 at C.
Load position for maximum force in BC:
(a) If the load is between B and C, then

$$F_{BC} = 2(\text{area under load})$$
$$= 2\{20(0.5)(-0.722 - 0.866)\}$$
$$= -31.76 \text{ kN}$$

(b) If the load is at C and beyond, then

$$F_{BC} = 2[20(0.5)\{-0.866 - 0.866(20/40)\}]$$
$$= -25.98 \text{ kN}$$

(c) If the load is partly in BC and CD, then let the load be at a distance x from B. Then because the load is 20 m long, it is also at a distance of x from D. The areas A_1 in portion BC and A_2 in CD are given by

$$A_1 = 0.5(20 - x)\{-0.722 + (-0.866 + 0.722)x/20 - 0.866\}$$
$$= (20 - x)(-0.794 - 0.0072x)$$

$$A_2 = 0.5x\{-0.866 - 0.866(40 - x)/40\}$$
$$= -0.433x(2 - x/40)$$

For maximum force, $(A_1 + A_2)$ must be a maximum. Differentiating $(A_1 + A_2)$ w.r.t. x, and equating to zero, $x = 5.0$. $A_1 = -12.18$, $A_2 = -4.06$, and

$$F_{BC} = 2(A_1 + A_2) = -32.48 \text{ kN}.$$

(b) Influence line for BG.
   (i) Load to the left of B, i.e. $0 \leqslant x \leqslant 20$, considering the free body to the right of G, is

$$F_{BG} \sin 60 + V_E = 0;$$

therefore

$$F_{BG} = -(x/80)(2/\sqrt{3}) = -x/40\sqrt{3}$$

and when $x = 20$,

$$F_{BG} = -0.29 \text{ compression.}$$

   (ii) Load to the right of C, i.e. $40 \leqslant x \leqslant 80$, considering the free body to the left of G, is

$$F_{BG} \sin 60 = V_A;$$

therefore

$$F_{BG} = (1 - x/80)(2/\sqrt{3})$$

and when $x = 40$,

$$F_{BG} = 0.577 \text{ tension.}$$

2. Construct the influence lines for the axial force in members AB and AD of the truss shown in Fig. E2.2. Member BD = 5 m, AE = 10 m.

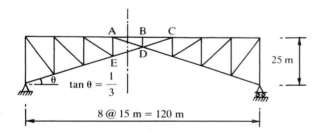

Influence line for AB

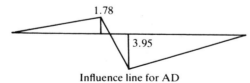

Influence line for AD

Figure E2.2

Answer: Take a vertical section between AE and BD.

$$V_{right} = x/120, \quad V_{left} = (1 - x/120)$$

(a) Influence line for AB.
   (i) Load to the left of A: $0 \leqslant x \leqslant 45$. Taking moments about D of the free body to the right of AE,

$$F_{AB}5 + V_{right}60 = 0.$$

Therefore

$$F_{AB} = -(x/120)12 = -x/10$$

and when $x = 45$,

$$F_{AB} = -4.5.$$

   (ii) Load to the right of B: $60 \leqslant x \leqslant 120$. Taking moments about D of the free body to the left of BD,

$$F_{AB}5 + V_{left}60 = 0.$$

Therefore

$$F_{AB} = -(1 - x/120)12$$

and when $x = 60$,

$$F_{AB} = -6.0.$$

(b) Influence line for AD: It is convenient to take moments about C where AB and ED when produced meet. The perpendicular distance h from C to AD is AC sin DAB. Tan DAB = 5/15, DAB = 18.44° and AC = 30. Therefore $h = 9.49$.
   (i) Load to the left of A: $0 \leqslant x \leqslant 45$. Taking moments about C of the free body to the right of AE,

$$F_{AD}h = V_{right}45.$$

Therefore

$$F_{AD} = (x/120)4.74 = x/25.31$$

and when $x = 45$,

$$F_{AB} = 1.78.$$

   (ii) Load to the right of B: $60 \leqslant x \leqslant 120$. Taking moments about C of the free body to the left of BD,

$$F_{AD}h + V_{left}75 = 0.$$

Therefore

$$F_{AD} = -(1 - x/120)7.90$$

and when $x = 60$,

$$F_{AD} = -3.95.$$

3. Figure E2.3 shows a simply supported girder of 100 m span. The live loads shown may cross the girder in either direction. Determine the maximum bending moment in the girder and the equivalent uniformly distributed load covering the whole span to produce the same maximum bending moment. Also calculate the maximum value of the shear force at the left-hand quarter section.

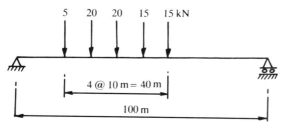

Figure E2.3

Answer: $\Sigma W = 75$, and acts at 22 m from the 5 kN load. Maximum bending moment near the mid span. With three loads to the left and two to the right, the average load on the left is heavier than the average load on the right. With two loads to the left and three loads to the right, the average load on the right is heavier than that on the left. Therefore maximum occurs under the second 20 kN load. The distance between this load and ΣW is $(22 - 20) = 2$ m. Therefore the load position is such that the second 20 kN load is 1 m to the left of mid span when ΣW is 1 m to the right of mid span. The left and right reactions are $\Sigma W(50 - 1)/100 = 36.75$ and $\Sigma W(50 + 1)/100 = 38.25$ respectively. The BM under the load is 1500.75 kN m^{-1}. If $qL^2/8 = 1500.75$, $q = 1.2$ kN m^{-1}.

Maximum negative shear force occurs when the first 20 kN load is at the section, the 5 kN is to the left and the other loads are to the right. Therefore

$$Q_{max} = 5(0.15) + 20(-0.75) + 20(-0.65)$$
$$+ 15(-0.55) + 15(-0.45)$$
$$= -42.25.$$

4. Construct the influence line for the bending moment at the left-hand quarter section of a three-pinned parabolic arch of span 32 m and rise 8 m.

Answer: The influence line for the bending moment at the left-hand quarter section of a simply supported beam of span 32 m is a triangle with a height $h_1 = x(L - x)/L$ where $L = 32$, $x = 8$. Therefore $h_1 = 6$. The influence line for the horizontal reaction at the supports is a symmetrical

triangle with height h_2. When the unit load is at the midspan, the horizontal reaction H is given, for zero moment at the central hinge, by $V_{left}(L/2) = Hr$, where $r = \text{rise} = 8$, $L = 32$, $V = 0.5$. Therefore $H = 1$. Therefore $h_2 = 1$. The influence line for BM at a section in the arch is = influence for BM at the same section in a corresponding simply supported beam − [influence line for H]y. In this case $y = (4r/L^2)x(L - x)$, where $r = 8$, $L = 32$, $x = 8$. Therefore $y = 6$. The influence line for BM is like the influence line for axial forces in a diagonal member of a parallel chord truss with ordinates of 3 at $L/4$, −2 at $(L/2)$, −1 at $(3L/4)$. Positive indicates tension at the bottom.

5. Construct the influence lines for all reactions and the bending moment at C for the propped cantilever with an internal hinge shown in Fig. E2.4.

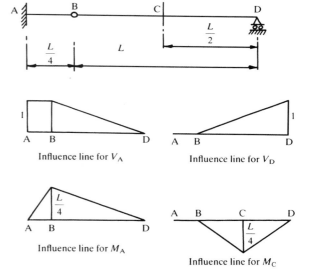

Figure E2.4

Answer: Load in AB: $0 \leqslant x \leqslant 0.25L$, $V_A = 1$, $V_D = 0$, $M_C = 0$, $M_A = -x$. Load in BCD: $0.25L \leqslant x \leqslant 1.25L$, $V_A = 1 - V_D$, $V_D = -0.25 + x/L$, $V_A = 1.25 - x/L$, $M_A = -0.25LV_A = -L\{0.3125 - 0.25x/L\}$

2.12 ● Examples for practice

1. For the continuous beam ABCD with a hinge at C as shown in Fig. E2.5, draw the influence line for the following forces.
  (a) Bending moment at midspan of AB
  (b) Bending moment at support B
  (c) Reaction at support at A
  (d) Reaction at support at B
  (e) Reaction at support at D
  (f) Shear force at midspan of AB
  If the load acting on the beam is a uniformly distributed load of 20 kN m^{-1}, calculate the maximum values of the above forces for both

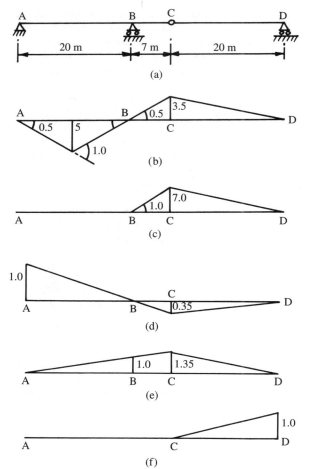

Figure E2.5 ● (a)–(g)

positive and negative values. Note that the load need not cover the entire span.

Answer: The influence lines are shown in Fig. E2.5.
Midspan bending moment in AB = 1000 kN m,
−945 kN m

Moment at B = 1890 kN m
Reaction at A = 200 kN (up), 94.5 kN (down)
Reaction at B = 1645 kN (up)
Reaction at D = 200 kN (up)
Shear force at midspan of AB = 94.5 kN or 50 kN

2. The span of a simply supported beam is 40 m. A train of five concentrated loads in the order of 50, 200, 200, 150 and 150 kN spaced 2.5 m between the loads crosses the beam in any direction. Calculate the bending moment at midspan. How does the moment at midspan compare with the maximum moment anywhere in the span? Also calculate the maximum shear force at the left-hand quarter-span.

Answer: Midspan, M = 6562.5 kN m,
M_{max} = 6563.7 kN m. With 50 kN at 7.5 m from left support, Q = −456.25 kN.

3. Draw the influence line for members DF, DE and CE of the pin-jointed truss shown in Fig. E1.19.

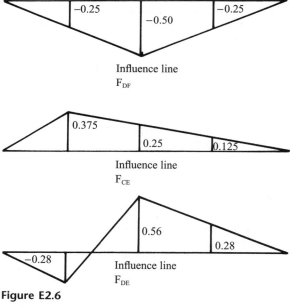

Influence line
F_{DF}

Influence line
F_{CE}

Influence line
F_{DE}

Figure E2.6

Answer: Influence lines are shown in Fig. E2.6.

Chapter three

Simple stress systems

In Chapter 1 it was shown that members of structures assembled from bars, beams and shafts are subjected to forces such as axial and shear forces, bending and twisting moments, and so on. The designer has to ensure that these forces at a section are safely resisted by the material from which members are fabricated. Designers use terms such as 'stress' and 'strain' as measures of how well the member can resist the applied forces. This chapter discusses these terms and explains their use in the design of members.

3.1 ● Concept of stress

Consider the two-dimensional body in the x–y plane as shown in Fig. 3.1(a). It is acted upon by external forces F_1, F_2, F_3, F_4 and F_5 such that the body is in equilibrium, satisfying the equilibrium requirements of Section 1.4.13.

Imagine the body being cut into two parts, I and II, along the section A–A as shown in Fig. 3.1(b). The two parts are free bodies, i.e. parts of the structure imagined to be separated from the rest of the structure with all the external and internal forces acting on them. Part I of the structure is acted upon by external forces F_1 and F_2 only and is therefore unlikely to be in equilibrium under the action of these two forces only. For part I to be in equilibrium it is necessary that forces act along the cut section, reflecting the effect of external forces acting on part II of the structure. In a similar fashion, part II of the structure is acted upon by external forces F_3, F_4 and F_5 only and for equilibrium requires that additional forces act along the cut section reflecting the effect of forces acting on part I of the structure. According to Newton's third law of motion, action and reaction are equal and opposite. In other words, the forces acting at the cut section A–A on part II of the structure are exactly equal and opposite to the forces acting on part I at the cut section A–A. It is important to observe that although the external forces might be concentrated forces, the forces at the cut section are invariably distributed over the cut section in a non-uniform manner, both in magnitude and direction, but have such a resultant as to maintain the equilibrium of the free bodies.

In order to describe the distribution of the forces at the cut face, it is useful to consider a quantity called stress. Stress is defined as follows. Consider an element of infinitesimal area ΔA at the cut section. Let the resultant force acting on this

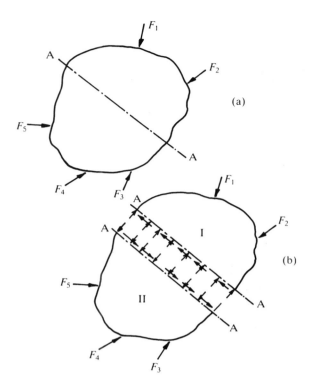

Figure 3.1 ● (a), (b)
2-D body under
external forces

element of area be ΔF. The force ΔF is in general neither normal nor tangential to the area ΔA. Stress is defined as the ratio

Stress = Limit $\{\Delta F/\Delta A\}$ as $\Delta A \rightarrow 0$

In the international system of units, the unit for stress is the Pascal. 1 mega Pascal is equal to 1 N mm^{-2}.

3.1.1 Normal and shear stress

For design purposes, the general definition of stress given in the previous section is not very useful. Therefore a more specialized definition of stress considering the force components normal and tangential to the area ΔA is adopted as follows.

Normal stress σ = Limit $\{$Normal component of $\Delta F/\Delta A\}$ as $\Delta A \rightarrow 0$

Tangential or shear stress τ = Limit $\{$Tangential component of $\Delta F/\Delta A\}$ as $\Delta A \rightarrow 0$.

As a simple illustration of these concepts, consider a rectangular strip of width B and of unit thickness acted upon by a tensile force F as shown in Fig. 3.2. Consider the free body separated by a section inclined at an angle θ to the horizontal. Evidently, for equilibrium, a resultant vertical force F must act at the cut section. If, for the sake of simplicity, it is *assumed* that the distributed force at the cut section is uniformly distributed, then 'f', the vertical force per unit area at the cut section of length $B \sec \theta$, is equal to

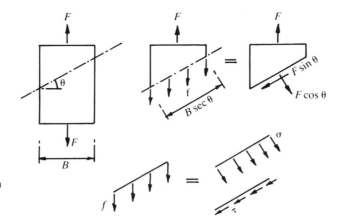

Figure 3.2 ● Normal and shear stresses on an inclined plane in a strip under tensile force

$$f = F/(B \sec \theta) = (F/B) \cos \theta$$

However, this distributed force f is acting in the vertical direction and is therefore neither normal to nor tangential to the cut section. Thus although f is a stress, it is neither a normal nor a shear stress. The normal and shear stresses at the section are determined by considering the normal and tangential components of the force acting on the area. The normal component of the force at the cut section is $F \cos \theta$ and the tangential component of the force is $F \sin \theta$. The area of the cut section is $B \sec \theta$ because the thickness is unity. Therefore *assuming* that the normal and shear stress are uniformly distributed over the width $B \sec \theta$, the stresses can be determined as the appropriate force component divided by the area. Therefore

Normal stress $\sigma = F \cos \theta/(B \sec \theta) = (F/B) \cos^2 \theta = (F/B)\{1 + \cos 2\theta\}/2$

Tangential stress $\tau = F \sin \theta/(B \sec \theta) = (F/B) \cos \theta \sin \theta = (F/B)\{\sin 2\theta\}/2$

It is important to note that because in the definition of stress three quantities are involved – magnitude and direction of the force and the orientation of the area on which the force acts – stress is not a vector but is known as a tensor. The normal law which a vector has to obey – such as the resolution of a force into x and y components – *does not* apply to stress.

The above simple example shows that the stress at a point depends on the orientation of the area at that point.

3.2 ● Stress components in three dimensions

Normal stress components

If, at a point in a body, the normal stress is calculated on an infinitesimal area oriented normal to the x-axis, then that stress is denoted by σ_x. Similarly if the normal stress is calculated considering infinitesimal areas normal to the y- and z-axes respectively, then these normal stresses are denoted by σ_y and σ_z respectively. Figure 3.3(a) shows the three normal stress components. The sign convention used is that tensile stresses are positive and therefore compressive stresses are negative.

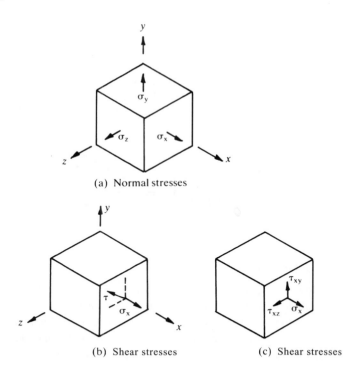

(a) Normal stresses

Figure 3.3 ● (a)–(c)
Normal and shear
stresses in 3-D structure

(b) Shear stresses

(c) Shear stresses

Shear stress components

Figure 3.3(b) shows an element of area ΔA normal to the x-axis. As can be seen, the resultant tangential force on the element is in general parallel neither to the y-axis nor to the z-axis. In order to specify the shear stress without having to also specify its inclination say to the y-axis, a more restricted definition of shear stress is adopted. As shown in Fig. 3.3(c), on the face normal to the x-axis two shear stresses act as follows:

Shear stress τ_{xy} = Limit {y component of the tangential force acting on an element of area ΔA normal to the x-axis/ΔA} as $\Delta A \to 0$.

Shear stress τ_{xz} = Limit {z component of the tangential force acting on an element of area ΔA normal to the x-axis/ΔA} as $\Delta A \to 0$.

Note that as opposed to the specification of normal stress, which requires only one subscript, shear stress requires two subscripts. The first subscript specifies the direction of the normal to the area and the second subscript specifies the direction of the shear stress.

In a similar manner considering the shear stresses on areas normal to the y- and z-axes, one can define shear stresses $\{\tau_{yx}, \tau_{yz}\}$ and $\{\tau_{zx}, \tau_{zy}\}$ respectively. Figure 3.4(a) shows the nine components of stresses at a point, i.e. the three normal stresses (σ_x, σ_y and σ_z) and the six shear stress components (τ_{xy}, τ_{xz}, τ_{yx}, τ_{yz}, τ_{zx}, τ_{zy}). It was stated before that the sign convention used for the normal stress is that tensile stress is positive. The sign convention for shear stress is a bit more complicated. In this book the convention adopted is that if the outward normal to the face

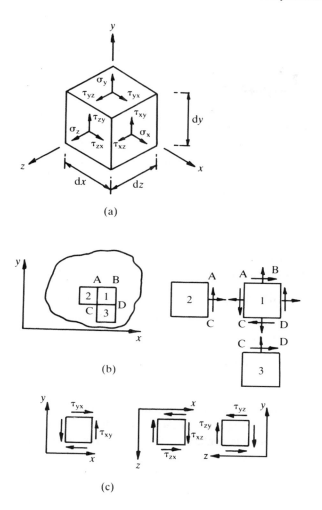

Figure 3.4 ● (a)–(c) Definition of positive shear stress directions

is in the positive direction of the co-ordinate axis (say the x-axis), then the shear stresses on that face are positive in the positive direction of the orthogonal axes, i.e. τ_{xy} and τ_{xz} act along the positive y- and z-axes respectively. The positive directions of the shear stresses on the positive faces are shown in Fig. 3.4(a). The positive direction of the shear stresses on the negative face, i.e. the face on which the outward normal is in the negative direction of the co-ordinate axis, is determined as follows.

Consider the three adjacent free bodies as shown in Fig. 3.4(b). For free body 2, face AC is the positive face. Similarly for free body 3, face CD is the positive face. Clearly on faces AC and CD the shear stresses act in the positive direction of the co-ordinate axes. However, from Newton's third law of motion, action and reaction are equal and opposite. Therefore the stresses on negative faces AC and CD of free body 1 are equal and opposite to the stresses on positive faces AC and CD of free bodies 2 and 3 respectively. Therefore it can be concluded that if the normal to a face is in the negative direction of the co-ordinate axes, then the positive directions of the shear stresses are along the negative directions of the co-ordinate axes.

Figure 3.4(c) shows the positive directions for the shear stresses acting on the faces normal to x-, y- and z-axes.

3.2.1 Complementary shear stress

It was stated in the previous section that in general there are six shear stress components. It will now be shown that in practically all situations it is possible to reduce the number of independent shear stress components to only three. In order to do this, consider the stresses acting on an infinitesimal element as shown in Fig. 3.4(c). Assuming that the stresses are uniformly distributed and that the stresses on the opposite faces are the same because of the small dimensions of the element involved, the equations of equilibrium requiring the sum of the forces in the x-, y- and z-directions to be equal to zero are satisfied. However, considering the equations of equilibrium with respect to the moments of all the forces about the x-, y- and z-axes we have:

Considering the moments of all the forces about the x-axis,

$$(\tau_{yz} . \mathrm{d}x . \mathrm{d}z)\mathrm{d}y - (\tau_{zy} . \mathrm{d}x . \mathrm{d}y)\mathrm{d}z = 0, \text{ therefore } \tau_{yz} = \tau_{zy}.$$

Similarly, considering the moments of all the forces about the y- and z-axes respectively, we have

$$(\tau_{zx} . \mathrm{d}x . \mathrm{d}y)\mathrm{d}z - (\tau_{xz} . \mathrm{d}z . \mathrm{d}y)\mathrm{d}x = 0, \text{ therefore } \tau_{zx} = \tau_{xz}$$

$$(\tau_{xy} . \mathrm{d}z . \mathrm{d}y)\mathrm{d}x - (\tau_{yx} . \mathrm{d}x . \mathrm{d}z)\mathrm{d}y = 0, \text{ therefore } \tau_{xy} = \tau_{yx}$$

This indicates that there are only three independent shear stress components (τ_{xy}, τ_{yz}, τ_{zx}) and the other shear stress components (τ_{yx}, τ_{zy}, τ_{xz}) are known as shear stresses complementary to the shear stresses (τ_{yx}, τ_{zy}, τ_{xz}) or as complementary shear stresses.

The only case when the complementary shear stresses are not equal is the uncommon case of a couple about the co-ordinate axes acting on an infinitesimal volume of the element. Fortunately this situation hardly ever occurs.

3.2.2 Stress components in 2-D problems

If a body is in the vertical x–y plane and the loads also act in the same plane, then the independent stress components to be considered are the two normal stresses σ_x and σ_y and the shear stress τ_{xy}. It should be remembered that the shear stress τ_{xy} is accompanied by the complementary shear stress τ_{yx}.

3.2.3 Summary

From the discussions in the previous two sections it can be concluded that:

(a) In two-dimensional problems in the x–y plane one has to consider two independent normal stresses (σ_x, σ_y) and one independent shear stress (τ_{xy}) and its complement (τ_{yx}). Because there is only one independent component of shear stress, it is common practice to drop the subscripts and denote the shear stress simply as τ.

(b) In three-dimensional problems one has to consider the three independent normal stresses $(\sigma_x, \sigma_y, \sigma_z)$ and three independent shear stresses $(\tau_{xy}, \tau_{yz}, \tau_{zx})$ together with their respective complements $(\tau_{yx}, \tau_{zy}, \tau_{xz})$.

3.2.4   Simple examples

Before proceeding further with the discussion of stresses, it is useful to consider some simple examples to get a 'feel' for normal and shear stress calculation.

Example 1

A strip of uniform cross-section of width B and thickness t is subjected to a pull of F as shown in Fig. 3.5(a). Calculate the resulting stress in the bar.

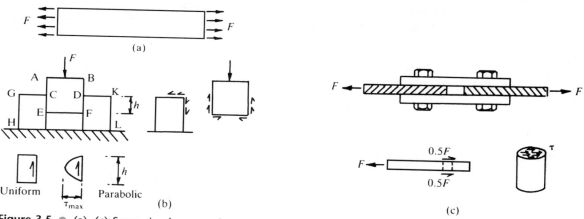

Figure 3.5 ● (a)–(c) Some simple cases of stress distribution

Solution   Assuming that σ_x is the only stress at a vertical section and it is uniformly distributed, then for equilibrium

$$\sigma_x(\text{Area of cross-section} = Bt) = F$$

Therefore

$$\sigma_x = F/(Bt)$$

Example 2

Three blocks of uniform thickness t glued over a length h are loaded as shown in Fig. 3.5(b). Calculate the average shear stress over the glued length h.

Solution   Using a free-body diagram for the top block, as shown in Fig. 3.5(b), the force F acting on the top block is kept in equilibrium by the shear stress acting on the glued edges. If the shear stress is *assumed* to be uniformly distributed, then for equilibrium

$$\tau_{xy}(\text{Total area of glued surface}) = F$$

Total area of glued surface $= 2(ht)$, where $t =$ thickness of the block. The term 2 appears because the block is glued on the edges CE and DF. Therefore

$$\tau_{xy} = F/(2ht) = 0.5F/(ht)$$

Note that the assumption that the shear stress is uniformly distributed violates the requirement of complementary shear stress. This is because there are no shear stresses acting on the edge EF, and so shear stress at E and F must be zero. Similarly, because the edges CG and DK are free from shear stress, shear stress must be zero at C and D as well. These conditions can be met by *assuming* that the shear stress is parabolically distributed. If the maximum shear stress is τ_{max}, then for equilibrium we have

$$(\text{shear stress} \times \text{area over which it acts}) = F$$

Because the shear stress is parabolically distributed, using the well-known result that the area enclosed by a parabola is equal to 2/3 the area of the enclosing rectangle, we have

$$(2/3)\{\tau_{max}\ 2ht\} = F$$

Therefore

$$\tau_{max} = 0.75F/(ht)$$

Example 3

Two plates are connected using bolts and cover plates as shown in Fig. 3.5(c). What is the type of stress acting on the bolt cross-section?

Solution Using the free-body diagrams shown in Fig. 3.5(c), it is clear that the stress acting on the bolt cross-section is a shear stress. As there are two bolt cross-sections resisting the pull F, the shear stress on each cross-section must balance a force of $0.5F$.

If the diameter of the bolt is d, then the average shear stress in the bolt is

$$\tau = 0.5F/\{\pi d^2/4\}$$

Example 4

A steel bar of diameter d is embedded in concrete for a length L as shown in Fig. 3.5(d). The bar is subjected to a pull of F. Calculate the 'bond stress' between the bar and concrete.

Solution As shown in Fig. 3.5(e), the bar is held in equilibrium by the shear stress (also called the bond stress) between the bar and the concrete. The area over which the shear stress acts is given by the product of the perimeter of the bar and the embedded length L. Therefore

$$\pi dL\ \tau = F, \tau = F/\{\pi dL\}$$

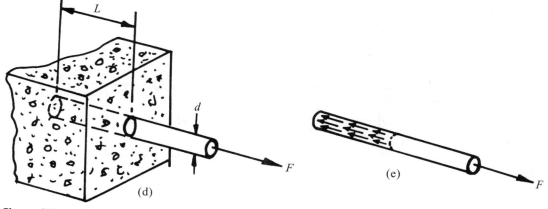

Figure 3.5 ● (d) A reinforcing bar embedded in concrete, (e) free body of the bar

Example 5

Figure 3.5(f) shows a bracket bolted onto the flange of a steel column using two bolts. The bracket has to support a load of F. What is the stress induced in the bolts?

Figure 3.5 ● (f)
A steel bracket bolted
to a column

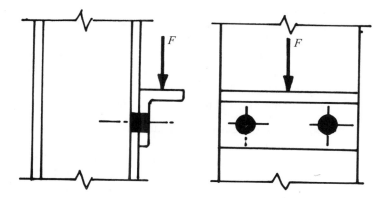

Solution   The load F induces shear stress in the bolts. The area resisting shear is the cross-sectional area of two bolts. Therefore

$$(2 \text{ bolts} \times \pi d^2/4)\tau = F, \ \tau = 2F/\{\pi d^2\}$$

Example 6

Figure 3.5(g) shows the arrangement for punching a hole of 20 mm diameter in a sheet of metal of thickness 6 mm. What is the relationship between the applied load to the punch and the shear stress induced in the plate?

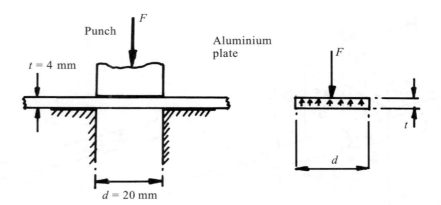

Figure 3.5 ●
(g) An arrangement for
punching a hole in a
metal sheet

Solution As shown in the free-body diagram, Fig. 3.5(g), the applied load F is maintained in
equilibrium by the shear stress τ. The area over which the shear stress acts is equal
to the perimeter of the punched hole × thickness of the plate:

$$(\pi dt)\tau = F, \; \tau = F/\{\pi dt\}$$

3.2.5 Example of biaxial stress state

In the examples in the previous sections, the stress at a point was acting in one
direction only. This is called the uniaxial state of stress. However, as discussed in
Section 3.2, more complicated states of stress occur in practice. As a simple example, consider a long, closed cylinder of diameter D subjected to internal pressure
p as shown in Fig. 3.5(h). Clearly under the influence of the pressure, the cylinder
expands and its diameter as well as its length increases. Considering how equilibrium in the longitudinal direction is maintained, consider the two basic situations:

(a) The pressure acts on the circular ends. The total force F acting on the end is
given by pressure times the end area: $F = p \, \pi D^2/4$
The force F is held in equilibrium by the normal stresses σ_x, acting between the
end plate and the cylinder. The area over which the stress σ_x acts is the product
of the circumference of the cylinder and the thickness of the plate t. Therefore

$$F = p \, \pi D^2/4 = (\pi Dt) \, \sigma_x$$

$$\sigma_x = pD/(4t)$$

(b) The pressure also causes a circumferential or hoop stress. Consider the equilibrium of one half of the cylinder of unit length. Because the pressure always
acts normal to the area, the force acting in the radial direction on an element of
area subtending an angle $d\theta$ at the centre is $pR \, d\theta$, where R is the radius of the
cylinder, $2R = D$. The vertical component of the force is $pR \, d\theta \sin \theta$. Integrating this force from $\theta = 0$ to π, the total vertical force is

$$V = \int_0^\pi pR \, d\theta \sin \theta = -pR \cos \theta \, |_0^\pi = -pR(-1 - 1) = p2R = pD$$

Longitudinal stress

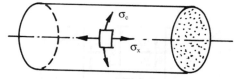

Circumferential stress

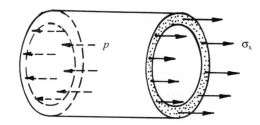

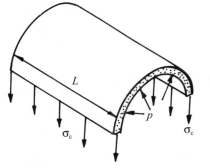

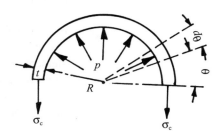

Figure 3.5 ● (h)
A closed cylinder under
internal pressure

This force is resisted by the circumferential stress σ_c. The area over which the circumferential stress acts is the product of unit length and two thicknesses each of t.

$$V = pD = (2t)\,\sigma_c, \ \sigma_c = pD/(2t)$$

Thus the cylinder is subjected at every point to two stresses. In the longitudinal direction to a stress σ_x and to a stress σ_c in the circumferential direction. This is a case of biaxial state of stress.

3.3 ● Stresses on a general plane in 2-D problems

In the previous sections, stress was calculated on planes normal to the co-ordinate axes. The orientation of a given body with respect to the co-ordinate axes is determined purely for mathematical convenience. For design purposes, it may be of interest to determine stresses on planes inclined to the co-ordinate axes because the stresses on these planes might be greater than the stresses on planes normal to the co-ordinate axes.

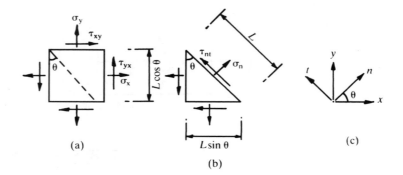

Figure 3.6 ● (a)–(c)
Normal and shear
stresses on an
inclined plane

For example, for the example shown in Fig. 3.2, it was shown that

$$\sigma = (F/B)\{1 + \cos 2\theta\}/2, \quad \tau = (F/B)\{\sin 2\theta\}/2$$

Evidently σ is a maximum when $\cos 2\theta = 1$ or $\theta = 0$. Therefore $\sigma_{max} = (F/B)$. Similarly τ is a maximum when $\sin 2\theta = 1$ or $\theta = 45$. Therefore $\tau_{max} = 0.5(F/B)$. Thus σ_{max} acts on a horizontal plane and τ_{max} acts on a plane inclined at 45° to the horizontal. The designer is interested in ensuring that the maximum stresses, whether normal or shear, do not exceed 'safe' values. Therefore it is necessary to explore the state of stress at a point in all directions so that the maximum values are not missed. This can be done as follows.

Consider an infinitesimal element shown in Fig. 3.6. Let the stresses on this element be given by σ_x, σ_y, τ_{xy} and the complementary shear stress $\tau_{yx} = \tau_{xy}$. Consider a plane whose normal makes an angle θ with respect to the x-axis. Let the normal and shear stress on this plane be (σ_n, τ_{nt}), where the subscripts n and t denote normal and tangent to the given plane. The relationship between the stresses in the (x, y) system and the (n, t) system can be obtained by considering the forces on the triangular wedge shown in Fig. 3.6(b). Let the length of the hypotenuse be L and the thickness of the element be T. Considering the forces acting on the wedge we have, treating the stresses $(\sigma_x, \sigma_y, \tau_{xy})$ as applied stresses,

Applied resultant force in the x-direction $= F_x$
$= -\{$stress $\sigma_x \times$ (area on which it acts $= L \cos \theta T)\}$
$\quad -\{($stress $\tau_{yx} = \tau_{xy}) \times$ (area on which it acts $= L \sin \theta T)\}$

Therefore

$$F_x = \{-\sigma_x \cos \theta - \tau_{xy} \sin \theta\}LT$$

Similarly,

Applied resultant force in the y-direction $= F_y$
$= -\{$stress $\sigma_y \times$ (area on which it acts $= L \sin \theta T)\}$
$\quad -\{($stress $\tau_{yx} = \tau_{xy}) \times$ (area on which it acts $= L \cos \theta T)\}$

Therefore

$$F_y = \{-\tau_{xy} \cos \theta - \sigma_y \sin \theta\}LT$$

Considering the equilibrium of the forces in the n and t directions we have

In the n-direction $\sigma_n(LT) + F_x \cos \theta + F_y \sin \theta = 0$

In the t-direction $\tau_{nt}(LT) - F_x \sin \theta + F_y \cos \theta = 0$

Substituting for F_x and F_y in terms of the applied stresses we have

$$\sigma_n = \sigma_x \cos^2 \theta + \sigma_y \sin^2 \theta + 2\tau_{xy} \sin \theta \cos \theta$$

$$\tau_{nt} = (\sigma_y - \sigma_x) \sin \theta \cos \theta + \tau_{xy}(\cos^2 \theta - \sin^2 \theta)$$

Similarly, considering the stresses acting on a plane whose normal is inclined at an angle $(90 + \theta)$ to the x-axis it can be shown that

$$\sigma_t = \sigma_y \cos^2 \theta + \sigma_x \sin^2 \theta - 2\tau_{xy} \sin \theta \cos \theta$$

Noting that $\cos 2\theta = \cos^2 \theta - \sin^2 \theta$ and $\sin 2\theta = 2\sin \theta \cos \theta$, we can express $(\sigma_n, \sigma_t, \tau_{nt})$ as follows as this leads to some easier mathematical manipulations.

$$\sigma_n = (\sigma_x + \sigma_y)/2 + \{(\sigma_x - \sigma_y)/2\} \cos 2\theta + \tau_{xy} \sin 2\theta$$

$$\sigma_t = (\sigma_x + \sigma_y)/2 - \{(\sigma_x - \sigma_y)/2\} \cos 2\theta - \tau_{xy} \sin 2\theta$$

$$\tau_{nt} = -\{(\sigma_x - \sigma_y)/2\} \sin 2\theta + \tau_{xy} \cos 2\theta$$

3.3.1　Stress invariants

From the expressions for the stresses derived in the previous section, it is interesting to note that

$$\sigma_x + \sigma_y = \sigma_n + \sigma_t$$

This indicates that for a given state of stress, the sum of the normal stresses on any two orthogonal planes is a constant. This is known as the first stress invariant.
  Similarly, it can be shown that

$$\{\sigma_x\sigma_y - \tau_{xy}^2\} = \{\sigma_n\sigma_t - \tau_{nt}^2\}$$

This is known as the second stress invariant.
  The concept of stress invariants is extensively used in the Theory of Plasticity.

3.4　●　Maximum and minimum stresses

It was shown in Section 3.3 that the normal stress σ_n and the tangential stress τ_{nt} acting on a plane whose normal is inclined at an angle θ to the x-axis are dependent on Cartesian stresses $(\sigma_x, \sigma_y, \tau_{xy})$ and the inclination θ. The question naturally arises as to the value of θ which yields maximum and minimum normal and shear stresses. This can be determined by setting $d\sigma_n/d\theta = 0$ for maximum or minimum values of σ_n and $d\tau_{nt}/d\theta = 0$ for maximum or minimum values of τ_{nt}.

Maximum or minimum normal stresses

Differentiating σ_n with respect to θ we have

$$d\sigma_n/d\theta = -(\sigma_x - \sigma_y)\sin 2\theta + \tau_{xy}\, 2\cos 2\theta$$

For maximum (or minimum) σ_n we have $d\sigma_n/d\theta = 0$. Therefore

$$\tan 2\theta = 2\tau_{xy}/(\sigma_x - \sigma_y)$$

therefore

$$\sin 2\theta = 2\tau_{xy}/D \text{ and } \cos 2\theta = (\sigma_x - \sigma_y)/D$$

where $D = \sqrt{\{(\sigma_x - \sigma_y)^2 + 4\tau_{xy}^2\}}$. Substituting for $\cos 2\theta$ and $\sin 2\theta$ in the expressions for σ_n, σ_t and τ_{nt} and simplifying we have

$$\sigma_n = (\sigma_x + \sigma_y)/2 + \sqrt{[\{(\sigma_x - \sigma_y)/2\}^2 + \tau_{xy}^2]}$$

$$\sigma_t = (\sigma_x + \sigma_y)/2 - \sqrt{[\{(\sigma_x - \sigma_y)/2\}^2 + \tau_{xy}^2]}$$

$$\tau_{nt} = 0$$

These maximum and minimum normal stresses are called principal normal stresses σ_1 and σ_2 respectively and on the planes on which the normal stresses are either a maximum or a minimum, the shear stress is zero. Therefore the principal normal stresses are given by

$$\sigma_1 = (\sigma_x + \sigma_y)/2 + \sqrt{[\{(\sigma_x - \sigma_y)/2\}^2 + \tau_{xy}^2]}$$

$$\sigma_2 = (\sigma_x + \sigma_y)/2 - \sqrt{[\{(\sigma_x - \sigma_y)/2\}^2 + \tau_{xy}^2]}$$

Directions of the planes on which σ_1 and σ_2 act

It should be noted that because of the fact that $\tan 2\theta = \tan 2(\theta \pm 90)$, from the expression $\tan 2\theta = 2\tau_{xy}/(\sigma_x - \sigma_y)$ it is not possible to determine whether the value of θ yields the inclination of the plane on which σ_1 acts or σ_2 acts. The best way to resolve the difficulty is to calculate σ_1, σ_2 and the smallest value of θ from the given values of $(\sigma_x, \sigma_y, \tau_{xy})$. Using the value of θ calculated, calculate σ_n on that plane. If σ_n corresponds to σ_1 then the calculated value of θ gives the inclination of σ_1 to the x-axis and σ_2 is inclined at $\theta \pm 90$ to the x-axis. On the other hand, if σ_n is equal to σ_2, then the calculated value of θ gives the inclination of σ_2 to the x-axis and σ_1 acts on a plane inclined at $\theta \pm 90$ to the x-axis. As indicated before, the shear stress is zero on the planes on which the principal stresses act. An example is given later in this section, illustrating the use of the above rules.

Maximum shear stress

Differentiating τ_{nt} with respect to θ we have

$$d\tau_{nt}/d\theta = -(\sigma_x - \sigma_y)\cos 2\theta - \tau_{xy}\, 2\sin 2\theta$$

For maximum (or minimum) τ_{nt}, $d\tau_{nt}/d\theta = 0$. Therefore

$$\tan 2\theta = -(\sigma_x - \sigma_y)/(2\tau_{xy})$$

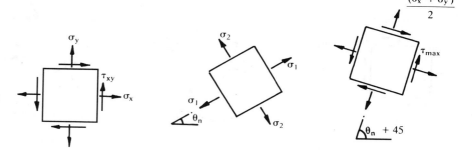

Figure 3.7 ● Cartesian stresses, principal normal stresses and maximum shear stresses acting on an element

Directions of planes on which maximum τ_{nt} acts

It was shown that for maximum τ_{nt}, $\tan 2\theta = -(\sigma_x - \sigma_y)/(2\tau_{xy})$ and that maximum or minimum normal stress act on planes given by $\tan 2\theta_n = 2\tau_{xy}/(\sigma_x - \sigma_y)$. Because

$$\tan 2(\theta_n \pm 45) = -\cot 2\theta_n = \tan 2\theta$$

then

$$\theta_n \pm 45 = \theta$$

In other words, as shown in Fig. 3.7, the maximum shear stress acts on planes which bisect the planes on which the principal normal stresses act. The maximum shear stress is given by

$$\tau_{max} = -\{(\sigma_x - \sigma_y)/2\} \sin 2\theta + \tau_{xy} \cos 2\theta, \text{ where } \theta = \theta_n + 45$$

The normal stress on the planes on which maximum shear stress acts is easily calculated and is given by

$$\sigma_n = (\sigma_x + \sigma_y)/2 + (\sigma_x - \sigma_y)/2\cos 2\theta + \tau_{xy} \sin 2\theta$$

where $\sin 2\theta = -(\sigma_x - \sigma_y)/D$, $\cos 2\theta = 2\tau_{xy}/D$,

$$D = \pm\sqrt{\{(\sigma_x - \sigma_y)^2 + 4\tau_{xy}^2\}}.$$

Simplifying

$$\sigma_n = (\sigma_x + \sigma_y)/2$$

This shows that the normal stress on planes on which the maximum shear stress acts is *not* equal to zero.

A graphical method called Mohr's Circle, used both for visualizing and for calculating the principal stresses, maximum shear stresses and stresses on a given plane is shown in Appendix 3.

Example

Calculate the principal normal stresses and their directions and also the value of the maximum shear stress and the inclination of the plane on which it acts given that at a point

$$\sigma_x = 50 \text{ N mm}^{-2}, \sigma_y = -40 \text{ N mm}^{-2} \text{ and } \tau_{xy} = 30 \text{ N mm}^{-2}$$

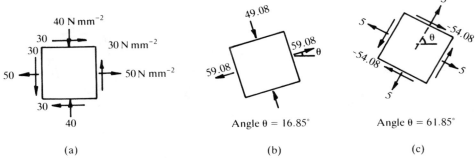

Figure 3.8 ● (a)–(c) Cartesian stresses, principal normal stresses and maximum shear stresses

Solution The solution consists of several steps as follows.

(a) Calculate the principal normal stresses given by

$$\sigma_1 = (\sigma_x + \sigma_y)/2 + \sqrt{[\{(\sigma_x - \sigma_y)/2\}^2 + \tau_{xy}^2]} = 59.08$$
$$\sigma_2 = (\sigma_x + \sigma_y)/2 - \sqrt{[\{(\sigma_x - \sigma_y)/2\}^2 + \tau_{xy}^2]} = -49.08$$

(b) Direction of principal stresses. Calculate θ given by

$$\tan 2\theta = 2\tau_{xy}/(\sigma_x - \sigma_y) = 2(30)/\{50 - (-40)\} = 0.667$$

Therefore

$$\theta = 16.85°$$

(c) Calculate σ_n on the plane whose normal is inclined at $\theta = 16.85$ to the x-axis.

$$\sigma_n = \sigma_x \cos^2 \theta + \sigma_y \sin^2 \theta + 2\tau_{xy} \sin \theta \cos \theta$$
$$= 50 \cos^2 16.85 + (-40) \sin^2 16.85 + 2(30) \sin 16.85 \cos 16.85$$
$$= 59.08$$

Because $\sigma_n = \sigma_1$, σ_1 is inclined at 16.85° to the x-axis and σ_2 is inclined at $90 + 16.85 = 106.85°$ to the x-axis, as shown in Fig. 3.8(b).

(d) Maximum shear stress.
Acts on planes $\theta = \theta_n + 45$ where $\theta_n = 16.85°$.

$$\theta = 16.85 + 45 = 61.85°$$
$$2\theta = 2(61.85) = 123.7, \sin 2\theta = 0.832, \cos 2\theta = -0.5548,$$
$$\tau_{max} = -\{(\sigma_x - \sigma_y)/2\} \sin 2\theta + \tau_{xy} \cos 2\theta, \text{ where } \theta = 61.85$$
$$= -\{50 - (-40)\}/2 \,(0.832) + (30)(-0.5548) = -54.1 \text{ N mm}^{-2}$$

The normal stress on these planes $= \sigma_n = (\sigma_x + \sigma_y)/2 = \{50 + (-40)\}/2 = 5 \text{ N mm}^{-2}$

The state of stress on the planes on which the maximum shear stress acts is shown in Fig. 3.8(c).

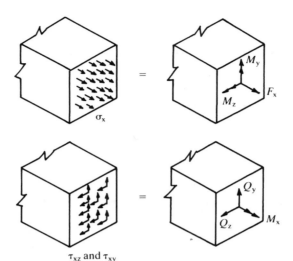

Figure 3.9 ● Stress
resultants

3.5 ● Stress resultants

Figure 3.9 shows a member with its axis orientated along the x-axis. At each point in the cross-section, the stresses which can act are the normal stress σ_x and the shear stresses τ_{xy} and τ_{xz} in the y- and z-directions respectively. In normal design situations, it is difficult to specify the stresses applied at each point because they are generally not known. In addition, many of the testing procedures yield the overall strength of a member rather than the stress distribution at each point. Engineers therefore commonly deal with the resultant forces due to distributed stresses rather than the stresses themselves. The resultant forces and couples due to the distributed stresses can be calculated by integrating the distributed stresses over the whole cross-section as follows.

Assuming the positive direction of the forces coincide with the positive direction of the co-ordinate axes,

$$\text{Resultant axial force in the } x\text{-direction} = \int \sigma_x \, dA = F_x$$

$$\text{Resultant shear force in the } y\text{-direction} = \int \tau_{xy} \, dA = Q_y$$

$$\text{Resultant shear force in the } z\text{-direction} = \int \tau_{xz} \, dA = Q_z$$

where dA = an infinitesimal area of the cross-section.

In a similar manner, adopting the convention that couples are positive if the double-headed arrow corresponding to the couple as a vector is directed along the positive direction of the co-ordinate axes as shown in Fig. 1.75, the resultant couples due to the distributed forces are given by

$$\text{Resultant couple about the } z\text{-axis} = \int \sigma_x y \, dA = -\text{bending moment } M_z$$

$$\text{Resultant couple about the } y\text{-axis} = \int \sigma_x z \, dA = \text{bending moment } M_y$$

$$\text{Resultant couple about the } x\text{-axis} = \int \{\tau_{xz} y - \tau_{xy} z\} \, dA = \text{twisting moment } M_x$$

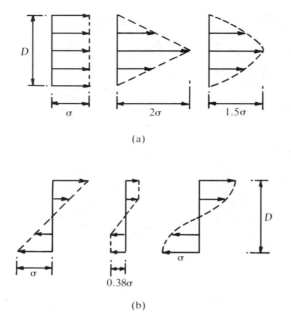

Figure 3.10 ●
(a), (b) Various stress
distributions yielding
identical stress resultants

The resultant forces and couples due to the distributed stresses acting on a cross-section are called stress resultants.

The above discussion repeats the fact discussed in Chapter 1, that in three dimensions, at a cross-section of a member, in general there are six stress resultants consisting of three forces and three couples. In the case of two-dimensional structures in the x–y plane, there are only three stress resultants at a cross-section, i.e. axial force F_x, shear force Q_y and bending moment M_z.

It should be appreciated that different stress distributions can give rise to the same stress resultant. To emphasize this point consider the three cases of axial stress distribution – uniform, triangular and parabolic – shown in Fig. 3.10(a). The three distributions yield the same axial force equal to $bd\sigma$, where b and d are the dimensions of the rectangle. Similarly, the three cases of stress distribution shown in Fig. 3.10(b) result in the same value $(-bd^2\sigma/6)$ for the couple M_z about the z-axis. This emphasizes that a given stress resultant *does not* specify the stress distribution. When the engineer uses the stress resultants in design calculations, he or she already has an idea of the specific stress distribution that they are dealing with.

3.6 ● Strain

When forces act on a structure, the structure deforms. The designer has to ensure that not only is the structure capable of resisting the applied forces but also that the resulting deformations are not large. In the previous sections, the concept of stress was introduced. Stress was used as a measure of the distributed internal force between the particles. In a similar manner we are interested in developing some measure of the way the overall deformation is 'spread' over the infinitesimal areas

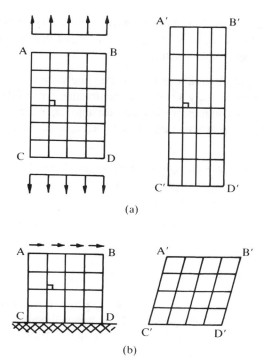

Figure 3.11 ● Normal and shear strains: (a) normal strains only, (b) shear strains

of the structure in response to the stresses acting on those areas. The overall deformation can then be calculated by 'summing up' the deformations of individual infinitesimal areas. Engineers use the term 'strain' as a measure of the distributed deformation.

Before explaining the concept of strain, it is useful to record two simple experimental observations.

1. If a sheet of low modulus rubber is marked with an orthogonal grid of lines as shown in Fig. 3.11(a) and is stretched uniformly across the width, it can be seen that:

   (a) the angle between the grid lines remains unaltered;
   (b) lines in the direction of the pull, stretch;
   (c) lines normal to the direction of pull, contract.

   Evidently a measure of the deformation at any point is given by how much a short length of the structure in a given direction changes its length when the forces are applied to the structure.

2. If a short sheet of rubber is fixed at the base and a shear force is applied at the top as shown in Fig. 3.11(b), then the type of deformation which results is quite different from the case when the sheet of rubber is pulled. Observations show that:

   (a) the angle between the grid lines is altered;
   (b) there is little stretching of the lines themselves.

In this case a measure of the deformation at a point is the change in the angle between two lines at right angles at the point that resulted from the application of the forces to the structure.

These simple observations have led engineers to distinguish between two types of deformations, namely a stretching/contracting type of deformation, which occurs when tensile or compressive stresses are applied, and an angular type of deformation, which occurs when shear stresses are applied. Engineers define two types of strain as follows.

1. Normal strain ε = change in length/original length.
2. Shear strain γ = change in the angle between lines which were initially at a right angle.

It should be noted that strains are non-dimensional quantities.

3.7 ● Displacements

In a two-dimensional structure in the x–y plane, the displacement of any point can be described by its displacement u in the x-direction and v in the y-direction. Because the displacement is a function of the position of the point in the structure, the displacement components u and v are functions of x and y. This is denoted by

$$u = u(x, y) \quad \text{and} \quad v = v(x, y)$$

3.7.1 Strain-displacement relationship

The normal and shear strains can be expressed in terms of the derivatives of displacements as follows.

Consider an element of length $\mathrm{d}x$ orientated parallel to the x-axis as shown in Fig. 3.12(b). Due to the deformation of the structure under the applied forces, if the displacement at one end of the element is u then the displacement at the other end is $\{u + (\partial u/\partial x)\,\mathrm{d}x\}$. Note that the partial derivative of u with respect to x, $\partial u/\partial x$, is used because u is a function of both x and y. The change in length is therefore $(\partial u/\partial x)\,\mathrm{d}x$. The strain in the x-direction is therefore given by

$$\varepsilon_x = \text{change in length/original length} = \{(\partial u/\partial x)\,\mathrm{d}x\}/\mathrm{d}x = \partial u/\partial x$$

Similarly, considering an element orientated parallel to the y-axis, Fig. 3.12(a), the strain in the y-direction is given by

$$\varepsilon_y = \partial v/\partial y$$

To establish the relationship between the displacement derivatives and the corresponding shear strain we proceed as follows.

Consider an element of sides $\mathrm{d}x$ and $\mathrm{d}y$ as shown in Fig. 3.12(b). In order to determine the shear strain, we have to calculate the rotation of the sides from their original position. Considering the side parallel to the x-axis, if the displacement in the y-direction at one end of the line is v, then the displacement at the other end is $\{v + (\partial v/\partial x)\,\mathrm{d}x\}$. Provided the displacements are small, the change in the angle

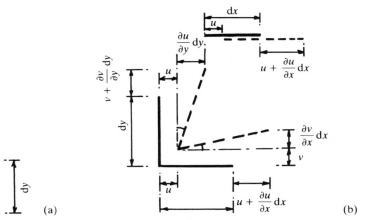

Figure 3.12 ● (a), (b) Calculation of Cartesian strains in terms of Cartesian displacements

$= \partial v / \partial x$. Similarly, considering the side parallel to the y-axis, the displacement in the x-direction at the 'near' and 'far' ends are respectively u and $\{u + \partial u / \partial y) \, dy\}$. Therefore the rotation of the line is $\partial u / \partial y$. The total change in angle is the shear strain γ_{xy}. Therefore

$$\gamma_{xy} = \partial u / \partial y + \partial v / \partial x$$

Summarizing, the relationship between strains and displacements is given by

$$\varepsilon_x = \partial u / \partial x, \ \varepsilon_y = \partial v / \partial y, \ \gamma_{xy} = \partial u / \partial y + \partial v / \partial x$$

3.7.2 Normal strain in a general direction

Very often it is necessary to calculate the normal strain in a given direction inclined at an angle θ to the x-axis as shown in Fig. 3.13. Consider an element of length ds where $\cos \theta = dx/ds$ and $\sin \theta = dy/ds$. The displacements at the two ends of the line are (u, v) at end 1 and $[\{u + (\partial u / \partial x) \, dx + (\partial u / \partial y) \, dy\}, \{v + (\partial v / \partial y) \, dy + (\partial v / \partial x) \, dx\}]$ at the other end. Note that because u and v are functions of both x and y, it is necessary to consider the change in displacements u and v due to changes in both x and y. Thus $\partial u / \partial x$ represents the change in u of two points unit distance apart along the x-axis. Similarly, $\partial u / \partial y$ is the change in u of two points unit distance apart along the y-axis. The projected change in length of the line on the x- and y-axes are respectively

$$\Delta_x = \{(\partial u / \partial x) \, dx + (\partial u / \partial y) \, dy\}$$

$$\Delta_y = \{(\partial v / \partial x) \, dx + (\partial v / \partial y) \, dy\}$$

Therefore

$$\text{Change in length} = \Delta_x \cos \theta + \Delta_y \sin \theta$$

$$\varepsilon_n = \text{the strain in the } \theta \text{ direction} = \text{change of length}/ds$$

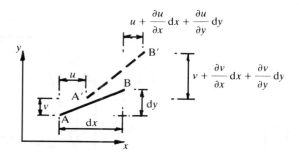

Figure 3.13 ●
Deformations of an
inclined element

Therefore

$$\varepsilon_n = (\Delta_x \cos\theta + \Delta_y \sin\theta)/ds$$

Since $dx/ds = \cos\theta$ and $dy/ds = \sin\theta$, we have, after substituting for Δ_x and Δ_y and simplifying,

$$\varepsilon_n = (\partial u/\partial x) \cos^2\theta + (\partial v/\partial y) \sin^2\theta + (\partial u/\partial y + \partial v/\partial x) \cos\theta \sin\theta$$

Since $\varepsilon_x = \partial u/\partial x$, $\varepsilon_y = \partial v/\partial y$, $\gamma_{xy} = \partial u/\partial y + \partial v/\partial x$, ε_n can be expressed as

$$\varepsilon_n = \varepsilon_x \cos^2\theta + \varepsilon_y \sin^2\theta + \gamma_{xy} \sin\theta \cos\theta$$

$$= (\varepsilon_x + \varepsilon_y)/2 + \{(\varepsilon_x - \varepsilon_y)/2\} \cos 2\theta + (0.5\gamma_{xy}) \sin 2\theta$$

Note the similarity with

$$\sigma_n = (\sigma_x + \sigma_y)/2 + \{(\sigma_x - \sigma_y)/2\} \cos 2\theta + (\tau_{xy}) \sin 2\theta$$

3.7.3 Shear strain of an element rotated by an angle θ with respect to the x-axis

The anticlockwise rotation of the line AB is given by

$$\theta_1 = \Delta_n/ds, \text{ where } \Delta_n = \Delta_y \cos\theta - \Delta_x \sin\theta$$

Substituting for Δ_x and Δ_y and simplifying

$$\theta_1 = \partial v/\partial x \cos^2\theta + \partial v/\partial y \sin\theta \cos\theta - \partial u/\partial x \cos\theta \sin\theta - \partial u/\partial y \sin^2\theta$$

Similarly, the anticlockwise rotation θ_2 of a line inclined at $(\theta + 90)$ to the x-axis is given by replacing $\cos\theta$ by $-\sin\theta$ and $\sin\theta$ by $\cos\theta$ in the above expression for θ_1. Therefore

$$\theta_2 = \partial v/\partial x \sin^2\theta - \partial v/\partial y \sin\theta \cos\theta + \partial u/\partial x \cos\theta \sin\theta - \partial u/\partial y \cos^2\theta$$

The shear strain γ_{nt} is given by the change in the angle $(\theta_1 - \theta_2)$. Therefore

$$\gamma_{nt} = (\theta_1 - \theta_2)$$

$$= (\partial v/\partial x + \partial u/\partial y)(\cos^2\theta - \sin^2\theta) + 2(\partial v/\partial y - \partial u/\partial x) \sin\theta \cos\theta$$

$$= \gamma_{xy}(\cos^2\theta - \sin^2\theta) - (\varepsilon_x - \varepsilon_y) \sin 2\theta$$

For use in a later section it is convenient to write the above expression as

$$(0.5\gamma_{nt}) = -0.5(\varepsilon_x - \varepsilon_y) \sin 2\theta + (0.5\gamma_{xy}) \cos 2\theta$$

Note the similarity with the expression

$$(\tau_{nt}) = -0.5(\sigma_x - \sigma_y) \sin 2\theta + (\tau_{xy}) \cos 2\theta$$

Alternative derivation of expression for strain in general direction

Referring to Fig. 3.6, consider a pair of orthogonal axes (x, y) and (n, t) where the second pair are rotated with respect to the first pair by an angle θ in the anticlockwise direction. The co-ordinates of any point in the two sets of axes are given by

$$n = x \cos \theta + y \sin \theta, \ t = -x \sin \theta + y \cos \theta$$

Solving for x and y, we have

$$x = n \cos \theta - t \sin \theta, \ y = n \sin \theta + t \cos \theta$$

The displacements in the n and t directions are given by

$$u_n = u \cos \theta + v \sin \theta, \ v_t = -u \sin \theta + v \cos \theta$$

The extensional strain in the n direction is given by

$$\varepsilon_n = \partial u_n / \mathrm{d}n = \partial / \partial n \{u \cos \theta + v \sin \theta\}$$

Because n is a function of x and y, we have

$$\varepsilon_n = \partial u_n / \partial n = \partial u_n / \partial x \ \partial x / \partial n + \partial u_n / \partial y \ \partial y / \partial n$$

Since $\partial x / \partial n = \cos \theta$ and $\partial y / \partial n = \sin \theta$, the expression for ε_n can be written as

$$\varepsilon_n = \cos \theta \ \partial u_n / \partial x + \sin \theta \ \partial u_n / \partial y$$

Substituting for $u_n = u \cos \theta + v \sin \theta$ and carrying out the differentiation, we have

$$\varepsilon_n = \cos \theta \{\partial u / \partial x \cos \theta + \partial v / \partial x \sin \theta\} + \sin \theta \{\partial u / \partial y \cos \theta + \partial v / \partial y \sin \theta\}$$
$$= \partial u / \partial x \cos^2 \theta + \partial v / \partial y \sin^2 \theta + (\partial u / \partial y + \partial v / \partial x) \cos \theta \sin \theta$$

Since $\varepsilon_x = \partial u / \partial x$, $\varepsilon_y = \partial v / \partial y$, $\gamma_{xy} = \partial u / \partial y + \partial v / \partial x$, ε_n can be expressed as

$$\varepsilon_n = \varepsilon_x \cos^2 \theta + \varepsilon_y \sin^2 \theta + \gamma_{xy} \sin \theta \cos \theta$$
$$= (\varepsilon_x + \varepsilon_y)/2 + \{(\varepsilon_x - \varepsilon_y)/2\} \cos 2\theta + 0.5\gamma_{xy} \sin 2\theta$$

The shear strain γ_{nt} can be expressed as

$$\gamma_{nt} = \partial u_n / \partial t + \partial v_t / \partial n$$

As before, since n and t are functions of t,

$$\gamma_{nt} = \partial u_n / \partial x \ \partial x / \partial t + \partial u_n / \partial y \ \partial y / \partial t + \partial v_t / \partial x \ \partial x / \partial n + \partial v_t / \partial y \ \partial y / \partial n$$

Since $\partial x / \partial n = \cos \theta$, $\partial y / \partial n = \sin \theta$, $\partial x / \partial t = -\sin \theta$ and $\partial y / \partial t = \cos \theta$, the expression for γ_{nt} can be written as

$$\gamma_{nt} = -\sin \theta \ \partial u_n / \partial x + \cos \theta \ \partial u_n / \partial y + \cos \theta \ \partial v_t / \partial x + \sin \theta \ \partial v_t / \partial y$$

Substituting for u_n and v_t in terms of u and v and differentiating, we have

$$\gamma_{nt} = -\sin\theta\{\partial u/\partial x\,\cos\theta + \partial v/\partial x\,\sin\theta\} + \cos\theta\{\partial u/\partial y\,\cos\theta + \partial v/\partial y\,\sin\theta\}$$
$$+ \cos\theta\{-\partial u/\partial x\,\sin\theta + \partial v/\partial x\,\cos\theta\} + \sin\theta\{-\partial u/\partial y\,\sin\theta + \partial v/\partial y\,\cos\theta\}$$

Therefore

$$\gamma_{nt} = -2\partial u/\partial x\,\sin\theta\,\cos\theta + 2\partial v/\partial y\,\sin\theta\,\cos\theta$$
$$+ \{\partial u/\partial y + \partial v/\partial x\}\{\cos^2\theta - \sin^2\theta\}$$

which leads to

$$\gamma_{nt} = \gamma_{xy}(\cos^2\theta - \sin^2\theta) - 2(\varepsilon_x - \varepsilon_y)\sin\theta\,\cos\theta$$

3.7.4 Principal normal strains

As in the case of stresses, we are interested in calculating the maximum and minimum normal strains at a point. Following steps similar to the ones used in calculating the principal stresses, the principal normal strains can be calculated. However, it can be seen that because the expression for ε_n can be obtained from the equation for σ_n by replacing (σ_x, σ_y and τ_{xy}) respectively by (ε_x, ε_y and $0.5\gamma_{xy}$), the principal normal strains are given by

$$\varepsilon_1 = (\varepsilon_x + \varepsilon_y)/2 + \sqrt{[\{(\varepsilon_x - \varepsilon_y)/2\}^2 + (0.5\gamma_{xy})^2]}$$

$$\varepsilon_2 = (\varepsilon_x + \varepsilon_y)/2 - \sqrt{[\{(\varepsilon_x - \varepsilon_y)/2\}^2 + (0.5\gamma_{xy})^2]}$$

and their directions by $\tan 2\theta = \gamma_{xy}/(\varepsilon_x - \varepsilon_y)$

3.8 ● Material properties

Engineers use various materials like steel, timber, reinforced concrete, plastics, stones, bricks, etc. to construct structures. The types of stresses to which the material is subjected to are a combination of normal tensile/compressive stresses and shear stresses. Obviously engineers are interested in studying the behaviour of these materials under the stresses to which they are likely to be subjected. A common type of test for materials like steel which are strong in tension is a tensile test, or for materials like concrete or brick which are weak in tension but strong in compression a compressive test. In a tension test either a long round bar or a long and narrow strip of steel is subjected to a pull and the tensile stress (= tensile force/cross-sectional area) is plotted vs. tensile strain (= change in length/original length). This is known as a stress–strain diagram. Figure 3.14(a) shows a slightly idealized plot for mild steel. From the stress–strain diagram it is generally noticed that

(a) Up to a certain stress level σ_0, generally known as the yield stress of the material, the relationship between stress and strain is linear.
(b) If the stress is less than the yield stress σ_0 and the material is unloaded, then the unloading path follows the loading path.

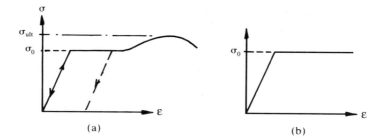

Figure 3.14 ● (a), (b) Stress–strain relationship for ductile materials

(c) If an attempt is made to increase the stress above the yield stress σ_0, the material 'flows', i.e. there is a considerable increase in strain without any noticeable increase in stress. This horizontal portion of the stress–strain diagram is generally known as the 'plastic plateau'.

(d) If after the material has yielded, the material is unloaded at any stage, the unloading path is parallel to the initial linear portion.

(e) Beyond the plastic plateau, there is an increase in stress due to an increase in strain. This is generally attributed to realigning of the particles and is known as the strain hardening range.

(f) The material reaches a maximum stress called σ_{ult} and finally snaps.

This behaviour is fairly typical of 'ductile' materials like mild steel. Other ductile materials such as steel used for prestressing do not show a clearly defined stress at which yielding begins. Brittle materials do not show any plastic plateau and the linear portion continues right up to the ultimate stress. Engineers avoid subjecting materials to stresses which might result in brittle failure, because there is no warning of impending failure. For convenience, the stress–strain diagram for steel is idealized as shown in Fig. 3.14(b) and used as a basis for mathematical calculations. It consists of a linear portion known as the elastic range and a long plastic plateau called the plastic range. The majority of the problems treated in this book assume that the stress range is limited to the linear part. Although the above description assumed that the material was tested in tension/compression, it is generally assumed that similar behaviour is observed if the material is tested in shear.

If the stress and strain in the material are assumed to be within the linear range, the most important property of the material, which is a measure of its deformability, is the slope of the linear portion of the stress–strain curve. In the case of normal stresses this is known as Young's modulus E and in the case of shear stresses it is known as shear or rigidity modulus G. In addition, as was observed (see Section 3.6) in the simple tension test on low modulus rubber, an extension in the pull direction is accompanied by a contraction in the direction normal to the direction of the pull. The numerical value of the ratio of strain in the direction perpendicular to the pull to the strain in the direction of the pull is called Poisson's ratio ν. The material property used in elementary stress analysis is based on these three fundamental concepts, i.e. Young's modulus, shear modulus and Poisson's ratio.

If the state of stress and strain is confined to the linear range only, then the loading and unloading paths coincide. In such a case the behaviour is said to be linearly elastic.

3.8.1 Isotropic and anisotropic materials: linear elastic stress–strain relationship

If a material has the same material properties in all directions, then it is known as an isotropic material. Steel is for all practical purposes an isotropic material. However, there are other materials which do not possess the same properties in all directions. Such materials are called anisotropic materials. For example, a typical anisotropic material is timber, which possesses different properties along the grain and normal to the grain. Another anisotropic material is reinforced plastic. For normally encountered materials, both isotropic and anisotropic, the relationship between stresses and strains can be expressed as follows:

$$\varepsilon_x = C_1\sigma_x + C_2\sigma_y, \; \varepsilon_y = C_2\sigma_x + C_3\sigma_y, \; \tau_{xy} = C_4\gamma_{xy}$$

where C_1 to C_4 are material constants.

Note: If in addition to the stresses the structure experiences an increase in the ambient temperature of T, then additional strains given by $\varepsilon_x = \alpha T$ and $\varepsilon_y = \alpha T$, where α = coefficient of linear thermal expansion, are caused. Thus the total strains are given by

$$\varepsilon_x = C_1\sigma_x + C_2\sigma_y + \alpha T, \; \varepsilon_y = C_2\sigma_x + C_3\sigma_y + \alpha T, \; \tau_{xy} = C_4\gamma_{xy}$$

Isotropic material

In the case of an isotropic material the stress–strain relationships are given by

$$\varepsilon_x = (\sigma_x - v\sigma_y)/E + \alpha T, \; \varepsilon_y = (\sigma_y - v\sigma_x)/E + \alpha T, \; \gamma_{xy} = \tau_{xy}/G$$

or expressing the stresses in terms of strains,

$$\sigma_x = [E/(1 - v^2)] \{\varepsilon_x + v\varepsilon_y\} - [E/(1 - v)] \, \alpha T$$

$$\sigma_y = [E/(1 - v^2)] \{\varepsilon_y + v\varepsilon_x\} - [E/(1 - v)] \, \alpha T$$

$$\tau_{xy} = G\gamma_{xy}, \; G = E/\{2(1 + v)\}$$

where E = Young's modulus, v = Poisson's ratio and $G = E/\{2(1 + v)\}$ as will be shown in the next section. As can be seen, only two material constants, E and v, are required to establish the stress–strain relationships.

Anisotropic material

In the case of a simple anisotropic material with different properties in the *x*- and *y*-directions, the stress–strain relationships are as follows:

$$\varepsilon_x = \sigma_x/E_x - v_x(\sigma_y/E_y) + \alpha T, \; \varepsilon_y = \sigma_y/E_y - v_y(\sigma_x/E_x) + \alpha T,$$

$$\gamma_{xy} = \tau_{xy}/G \text{ or expressing stresses in terms of strains,}$$

$$\sigma_x = [E_x/(1 - v_xv_y)] \{\varepsilon_x + v_x\varepsilon_y\} - [E_x/(1 - v_xv_y)] \{(1 + v_x) \, \alpha T\}$$

$$\sigma_y = [E_y/(1 - v_xv_y)] \{\varepsilon_y + v_y\varepsilon_x\} - [E_y/(1 - v_xv_y)] \{(1 + v_y) \, \alpha T\}$$

$$\tau_{xy} = G\gamma_{xy}$$

Table 3.1

Material	Young's modulus, E (kN mm^{-2})	Poisson's ratio, v
Structural steel	210	0.27
Aluminium alloy	70	0.32
Concrete	21	0.15
Wood (fir)	14	–

where E_x = Young's modulus in the x-direction, E_y = Young's modulus in the y-direction, v_x, v_y = Poisson's ratio in the x- and y-directions respectively. v_x and v_y are not independent but are related by $v_x E_x = v_y E_y$, G = shear modulus. Therefore in this case four material constants are required to establish the stress–strain relationship.

Table 3.1 shows typical material constants for some commonly used engineering materials. The stress–strain relationship given above assumes a linear relationship between stress and strain. Such a relationship is said to obey Hooke's law, after Robert Hooke (1635–1703) who noticed that for many materials 'extension is proportional to force'.

3.8.2 Shear modulus G in isotropic materials

Consider the state of stress given by

$$\sigma_x = 0, \ \sigma_y = 0 \quad \text{and} \quad \tau_{xy} = \tau$$

This is known as a state of pure shear. The principal stresses σ_1 and σ_2 are given by

$$\sigma_1 = (\sigma_x + \sigma_y)/2 + \sqrt{[\{(\sigma_x - \sigma_y)/2\}^2 + (\tau_{xy})^2]} = \tau$$

$$\sigma_2 = (\sigma_x + \sigma_y)/2 - \sqrt{[\{(\sigma_x - \sigma_y)/2\}^2 + (\tau_{xy})^2]} = -\tau$$

Using the stress–strain relationship for isotropic materials, the strains in the x–y system are given by

$$\varepsilon_x = 0, \ \varepsilon_y = 0 \text{ since } \sigma_x = \sigma_y = 0, \ \gamma_{xy} = \tau_{xy}/G = \tau/G$$

Similarly, strain in the direction of σ_1 is given by

$$\varepsilon_1 = \sigma_1/E - v\sigma_2/E = \tau(1 + v)/E$$

However, using the expression derived in Section 3.7.4, the principal strain is given by

$$\varepsilon_1 = (\varepsilon_x + \varepsilon_y)/2 + \sqrt{[\{(\varepsilon_x - \varepsilon_y)/2\}^2 + (0.5\gamma_{xy})^2]} = 0.5\gamma_{xy} = 0.5\tau/G$$

In the case of isotropic materials, the directions of principal stresses and strains coincide because there are no 'strong' and 'weak' directions. This is not true in the case of anisotropic materials because a higher stress in a 'strong' direction may in fact produce a smaller strain than a lower stress in the 'weak' direction. Since we are considering an isotropic material, the two expressions for ε_1 must be equal. Therefore

$$\varepsilon_1 = \tau(1 + v)/E = 0.5\tau/G$$

$$G = E/\{2(1 + v)\}$$

3.8.3 Volumetric strain and bulk modulus

Consider an infinitesimal rectangular prism of sides dx, dy and dz. Let the normal strains in the x-, y- and z-directions be ε_x, ε_y and ε_z respectively. The deformed sides of the prism are now $dx(1 + \varepsilon_x)$, $dy(1 + \varepsilon_y)$ and $dz(1 + \varepsilon_z)$. The original volume dV is given by

$$dV = dx\, dy\, dz$$

The change in volume is given by

$$dx(1 + \varepsilon_x)\, dy(1 + \varepsilon_y)\, dz(1 + \varepsilon_z) - dx\, dy\, dz$$

$$= dx\, dy\, dz\, \{\varepsilon_x + \varepsilon_y + \varepsilon_z + \varepsilon_x\varepsilon_y + \varepsilon_y\varepsilon_z + \varepsilon_z\varepsilon_x + \varepsilon_x\varepsilon_y\varepsilon_z\}$$

Because the strains are small, the products of strains can be neglected. Therefore change of volume is given by

$$\text{change of volume} = dx\, dy\, dz(\varepsilon_x + \varepsilon_y + \varepsilon_z)$$

Volumetric strain is defined as

$$\text{volumetric strain} = \text{change of volume/original volume}$$
$$= \varepsilon_x + \varepsilon_y + \varepsilon_z$$

If the state of stress at a point is given by $\sigma_x = \sigma_y = \sigma_z = \sigma$, then this state of stress is called a hydrostatic state of stress. If the material is isotropic, then

$$\varepsilon_x = (\sigma - v\sigma - v\sigma)/E = \sigma(1 - 2v)/E = \varepsilon_y = \varepsilon_z$$

Therefore volumetric strain is given by

$$\text{volumetric strain} = \sigma\, 3(1 - 2v)/E = \sigma/K$$

where $K = E/\{3(1 - 2v)\}$. K is known as the bulk modulus.

It is interesting to note that if $v = 0.5$, then volumetric strain is zero under a hydrostatic state of stress. The reader should appreciate that in the expression for volumetric strain, only normal strain components appear. This indicates that under a pure shear state of stress, there is no volume change, only a change of shape.

3.9 ● General assumptions in engineers' theory of axial and bending stress analysis

The problem of determination of the 'exact' stresses in a given structure under a given loading is extremely complex. Although the easy availability of computers has made it possible to carry out the 'exact' stress analysis, generally the expense, effort and inherent approximations involved in practical design do not justify such an approach for normal design purposes. Instead, based on assumptions justified on the basis of comparison with more accurate methods of stress analysis and experimental observations, engineers have developed simple procedures on the basis of which, given the stress resultants, it is possible to determine the corresponding stress distribution. This procedure will be elaborated in the succeeding sections of this chapter.

The assumptions made in an engineer's theory of stress analysis are:

(a) the material is linearly elastic, i.e. the material never reaches the plastic plateau;

(b) sections which were plane before stressing remain plane after stressing.

Assumption (a) is easy to follow. Assumption (b) is explained in the following sections.

3.9.1 Engineers' theory – simple axial tension

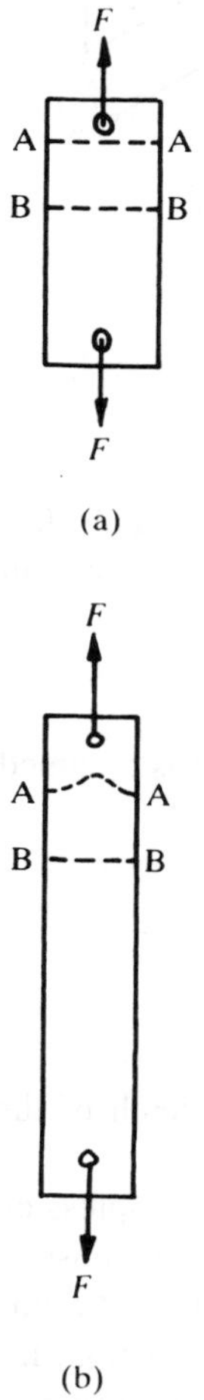

(a)

(b)

Figure 3.15 ● (a), (b) Deformation of a flat strip in tension

Consider, for example, a rectangular strip of constant thickness under a tensile load as shown in Fig. 3.15(a). If a line normal to the load is marked on the strip before the load is applied and if the deformation of this line is observed after the load, it is assumed that it still remains a line parallel to the original line. This indicates that all the points at a section normal to the tensile load suffer the same deformation. Therefore the displacement u in the x-direction and the axial strain ε_x are a function of x only. Therefore $\varepsilon_x = \partial u/\partial x = du/dx$ because u is a function of x only. Furthermore, assuming that only stress σ_x is present, then the stress σ_x is the same at all points at a section normal to the axis. In other words, the stress σ_x is constant across the cross-section. The relationship between the stress resultant F_x and the stress σ_x is given by

$$F_x = \int \sigma_x \, dA = \sigma_x \int dA = \sigma_x A, \text{ i.e. } \sigma_x = F_x/A$$

where A = area of the cross-section.

$$\varepsilon_x = \sigma_x/E + \alpha T = F_x/(AE) + \alpha T$$

$$\varepsilon_x = \partial u/\partial x = du/dx = F_x/(AE) + \alpha T$$

Therefore

$$u = F_x x/(AE) + \alpha T x + C$$

where C = constant of integration to be determined from the boundary conditions.

One important limitation of this approach must be obvious. If the tensile force is applied as a concentrated load, simple observation with a low modulus rubber sheet will indicate, as shown in Fig. 3.15(b), that near the point of load application, plane sections are far from remaining plane after the load application. In other words, the engineers' theory is applicable to sections 'far removed' from points of load application. Other situations where the simple equation is not applicable are discussed in Section 3.17.

3.9.2 Engineers' theory – simple symmetrical bending

Consider a member cross-section with a vertical axis of symmetry as shown in Fig. 3.16. Let a bending moment be applied so that the beam bends in the vertical plane. As can be confirmed by bending a beam made from low modulus rubber, the extreme top face will be in compression and the bottom face will be in tension. In other words, over the depth of the beam the stress σ_x changes from tension to

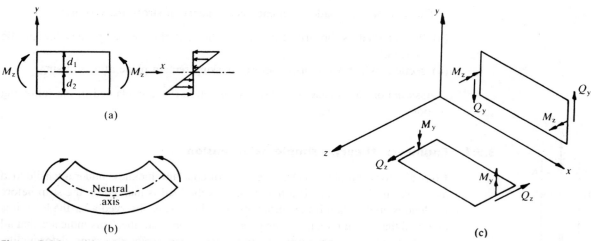

Figure 3.16 ● (a)–(c) Bending deformation and bending stresses

compression. If it is assumed that plane sections remain plane before and after bending, then the axial displacement at any point is a linear function of y from an unspecified origin. Therefore

$u = Cy$, where C is a function of x only

The strain in the x-direction $\varepsilon_x = \partial u / \partial x = C'y$, where $C' = dC/dx$. If it is assumed that no stress other than σ_x is acting, then

$\sigma_x = E\varepsilon_x$, where $E = $ Young's modulus.

Therefore

$\sigma_x = EC'y$

indicating that the bending stress σ_x is linearly distributed over the depth of the beam.

Since the stress σ_x has to change from tension at the bottom face to compression at the top face, there is a point in the cross-section at which the stress σ_x must be zero. It is useful to use this point as the origin for the measurement y. The line joining the points of zero stress is called the neutral axis. At a cross-section, the point of zero bending stress can be determined as follows. At any cross-section normal to the axis, because no resultant axial force F_x has been applied, it is necessary that it should be zero. Therefore

$$F_x = \int \sigma_x \, dA = C'E \int y \, dA = 0$$

$C' \neq 0$ as otherwise σ_x will be zero everywhere. Therefore the condition that $F_x = 0$ can be satisfied only if $\int y \, dA = 0$.

The axis in the cross-section about which $\int y \, dA = 0$ is called the centroidal axis. Incidentally $\int y \, dA$ is called the first moment of area A about the horizontal centroidal axis. In other words, the zero bending stress axis or the neutral axis coincides with the centroidal axis of the section.

The moment stress resultant on the positive face is given by

$$\int \sigma_x y \, dA = C'E \int y^2 \, dA = C'EI_{zz} = -M_z$$

where $I_{zz} = \int y^2 \, dA$. I_{zz} is known as the second moment of area of the cross-section about the z-axis. Therefore

$$C'E = -M_z/I_{zz}$$

and because $C'Ey = \sigma_x$

$$\sigma_x = -(M_z/I_{zz})y$$

where y is measured from the centroid of the cross-section.

In a similar manner if a bending moment is applied in the horizontal x–z plane, then the stress σ_x is given by

$$\sigma_x = +(M_y/I_{yy})z$$

where the origin for z is again at the centroid of the cross-section and $I_{yy} = \int z^2 \, dA$. I_{yy} is known as the second moment of area about the yy axis.

It will be shown in Section 3.12 that the above formulae are applicable as long as the y- or z-axis is an axis of symmetry or a principal axis. The term principal axis is explained later, in Section 3.12.5.

The determination of second moment of area I_{yy} and I_{zz} is treated in the next section.

3.9.3 Composite beams

In the previous section it was assumed that the beam is made up of homogeneous material. In practice timber beams are frequently strengthened by steel plates resulting in non-homogeneous sections.

Similarly it is common to use composite beams which consist of steel beams with a concrete slab connected to the top flange of the steel beam. Reinforced concrete can be considered as a composite material where steel is used to improve the tension-carrying capacity of concrete. In such cases, although $\varepsilon_x = C'y$ (based on the assumption that plane sections remain plane before and after bending is valid) $\sigma_x = E(y)\varepsilon_x$, where the Young's modulus E is a function of y. Therefore stress σ_x is not linearly distributed in the cross-section.

$$F_x = \int \sigma_x \, dA = C' \int E(y)y \, dA = C'E \int y\{E(y)/E\} \, dA$$

where E is a representative Young's modulus. Similarly,

$$-M_{zz} = C'E \int y^2\{E(y)/E\} \, dA$$

In the case of homogeneous beams, $E(y)/E = 1$. Therefore, in dealing with composite beams, it is convenient to consider a modified beam of homogeneous material by adjusting the width at any level to correspond to the ratio $E(y)/E$. All calculations are carried out on the modified beam. This is illustrated by simple examples at the end of this chapter.

3.10 ● Calculation of second moment of area

The calculation of the stress due to bending involves the calculation of the second moment of area. The majority of cross-sections can be treated as an assemblage of rectangular sections but triangular and circular cross-sections also occur. In this section some simple examples will be given of the calculation of the second moment of area.

Example 1

Calculate the second moment of area of the rectangular section about the axis a–a passing through the centroid and the axis b–b passing through the bottom edge as shown in Fig. 3.17(a).

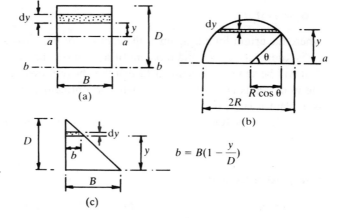

Figure 3.17 ● (a)–(c) Typical cross-sections for calculating the second moment of area

$I_{aa} = \int y^2 \, dA$, where $dA = B \, dy$ is the shaded area shown in Fig. 3.17(a). Therefore

$$I_{aa} = \int By^2 \, dy \text{ where the limits for } y = \pm 0.5D$$

$$= By^3/3 \text{ between the limits } y = \pm 0.5D \text{ gives}$$

$$= BD^3/12$$

$I_{bb} = \int y^2 \, dA$ where y is the distance from the b–b axis to the element of area.

$$= \int By^2 \, dy \text{ where the limits for } y = 0 \text{ to } D$$

$$= By^3/3 \text{ where the limits are for } y = 0 \text{ and } D$$

$$= BD^3/3$$

Example 2

Calculate the second moment of area of the semi-circular section of radius R about an axis coinciding with the diameter as shown in Fig. 3.17(b).

It is convenient to use polar co-ordinates. As shown in Fig. 3.17(b), $y = R \sin \theta$, $dy = R \cos \theta \, d\theta$,

$$dA = (2R \cos \theta) \, dy = (2R \cos \theta)R \cos \theta \, d\theta = 2R^2 \cos^2 \theta \, d\theta,$$

$$I_{aa} = \int y^2 \, dA = \int (R^2 \sin^2 \theta)(2R^2 \cos^2 \theta \, d\theta)$$

$$= 2R^4 \int \sin^2 \theta \cos^2 \theta \, d\theta$$

Since $2\sin \theta \cos \theta = \sin 2\theta$, $\cos 2\theta = 1 - 2\sin^2 \theta$, we have

$$I_{aa} = (R^4/2) \int \sin^2 2\theta \, d\theta = (R^4/4) \int (1 - \cos 4\theta) \, d\theta$$

$$= (R^4/4)(\theta - \sin 4\theta/4)$$

where the limits for $\theta = 0$ and $\pi/2$

$$= \pi R^4/8.$$

Example 3

Calculate I_{aa} for the triangle shown in Fig. 3.17(c). The width b of the triangle at a height y from the base is, from similar triangles, equal to $B(D - y)/D$. Therefore $dA = \{B(D - y)/D\} \, dy$

$$I_{aa} = \int y^2 \, dA = \int B(1 - y/D)y^2 \, dy \text{ where the limits for } y = 0 \text{ and } D$$

$$= B\{y^3/3 - y^4/(4D)\} \; |_0^D$$

$$= BD^3/12$$

3.10.1   Parallel axis theorem

A very useful theorem which enables the calculation of the second moment of area of a section about an axis parallel to the centroidal axis in terms of the corresponding second moment of area about the centroidal axis is the parallel axis theorem. Consider an axis bb which is parallel to the axis zz passing through the centroid and at a distance h_y from it, as shown in Fig. 3.18. Clearly

$$I_{bb} = \int (y - h_y)^2 \, dA = \int y^2 \, dA + h_y^2 \int dA - 2h_y \int y \, dA$$

where $\int dA = A$, and since the axis zz passes through the centroid, $\int y \, dA = 0$ by definition and $\int y^2 \, dA = I_{zz}$. Therefore

$$I_{bb} = I_{zz} + Ah_y^2$$

Similarly, if the second moment of area I_{cc} about an axis cc parallel to the yy axis and at a distance h_z from it is required, then

$$I_{cc} = \int (z - h_z)^2 \, dA = \int z^2 \, dA + h_z^2 \int dA - 2h_z \int z \, dA$$

As before, $\int z \, dA = 0$ because the y-axis passes through the centroid and $\int z^2 \, dA = I_{yy}$. Therefore

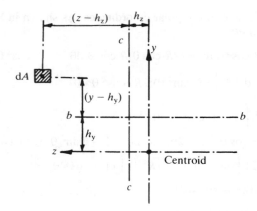

Figure 3.18 ● A set of parallel axes

$$I_{cc} = I_{yy} + Ah_z^2$$

Note that the signs of h_y and h_z are irrelevant because they always appear as squared quantities. The above equations can be stated as follows.

The parallel axis theorem states that the second moment of area about an axis is equal to the second moment of area about the *parallel* axis passing through the centroid of the section + area of the section multiplied by the square of the distance between the two axes.

Note: The parallel axis theorem leads to a simple rule for determining I of some special sections with 'holes'. When the section is considered as (solid section – hole), and if the relevant centroidal axis of the solid and the hole coincide, then, about the common centroidal axis,

I of section with hole = I of the solid section – I of the 'hole'

See examples in the next section.

3.10.2 Examples of the use of the parallel axis theorem

Example 1

For the cross-section shown in Fig. 3.19, calculate the second moment of area about the *zz* and *yy* axes passing through the centroid of the section.

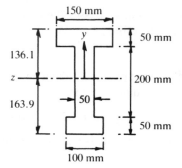

Figure 3.19 ● An *I* section

(a) Calculate the position of the centroid. The centroid always lies on the axis of symmetry. To calculate the position of the centroid on the y-axis, take moments about an arbitrary axis parallel to the axis with respect to which the second moment of area is required. In this case, to calculate I_{zz}, take moments about, say, the top flange. The distance $\bar{y}$ of the centroid from the chosen arbitrary axis is given by definition by

$$\bar{y} = \sum A_i y_i / \left(\sum A_i \right)$$

where A_i = area of cross-section of the ith of the N areas into which the section is divided, and y_i = distance to the centroid of the ith area from the chosen arbitrary axis.

For manual calculation purposes, it is useful to set out the calculations in tabular form as shown in Table 3.2.

Table 3.2

Description	Dimensions	A_i	y_i	$A_i y_i$
Top flange	150 × 50	7.5 × 10³	25	1.875 × 10⁵
Web	200 × 50	10.0 × 10³	150	15.0 × 10⁵
Bottom flange	100 × 50	5.0 × 10³	275	13.75 × 10⁵
		Σ22.5 × 10³		Σ30.625 × 10⁵

$\bar{y} = 30.625 \times 10^5 / (22.5 \times 10^3) = 136.1$ mm

(b) Calculation of I_{zz}. Use the parallel axis theorem

$$I_{zz} = \sum I_{zzi} + \sum A_i h_{yi}^2$$

where I_{zzi} = second moment of area of the ith area about its centroidal axis parallel to the zz axis, and h_{yi} = perpendicular distance between the zz axis and the parallel axis through the centroid of the ith area. Note that $h_{yi} = (y_i - \bar{y})$.

The calculations can be set out as below. It should be noted that I_{zzi} for the three sections are as follows:

$$\text{Top flange} = (1/12) \times 150 \times 50^3 = 1.56 \times 10^6$$

$$\text{Web} = (1/12) \times 50 \times 200^3 = 33.33 \times 10^6$$

$$\text{Bottom flange} = (1/12) \times 100 \times 50^3 = 1.04 \times 10^6$$

Table 3.3

Description	A_i	$h_{yi} = (y_i - \bar{y})$	$A_i h_{yi}^2$	I_{zzi}
Top flange	7.5 × 10³	−111.1	92.57 × 10⁶	1.56 × 10⁶
Web	10.0 × 10³	+13.9	1.93 × 10⁶	33.33 × 10⁶
Bottom flange	5.0 × 10³	+138.9	96.47 × 10⁶	1.04 × 10⁶
			Σ190.97 × 10⁶	Σ35.93 × 10⁶

$I_{zz} = (35.93 + 190.97) \times 10^6 = 226.9 \times 10^6$ mm⁴

(c) Calculation of I_{yy}. Because the section is symmetric with respect to the y-axis, the centroid lies on the y-axis. Since the centroids of the individual areas A_i also lie on the y-axis, $h_{zi} = 0$. Therefore

$$I_{yy} = \sum I_{yyi}$$

$$= 50(150)^3/12 \text{ for top flange } + 200(50)^3/12 \text{ for web } + 50(100)^3/12 \text{ for bottom flange}$$

$$= 20.31 \times 10^6 \text{ mm}^4$$

Example 2

Figure 3.20(a) shows a compound section built from two channels connected by 300×12 mm plates. Calculate I_{yy} and I_{zz}.

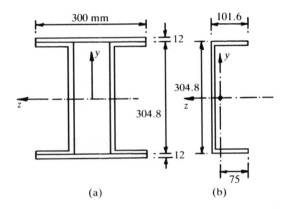

Figure 3.20 ● (a), (b)
A composite section

(a)   (b)

It is given that the channel has the following section properties. The centroid lies on the axis of symmetry at a distance of 75.0 mm from the tip of the flanges as shown in Fig. 3.20(b). The second moment of area about the zz and yy axes passing through its own centroid are $I_{zzi} = 82.14 \times 10^6 \text{ mm}^4$, $I_{yyi} = 4.995 \times 10^6 \text{ mm}^4$, $A_i = 5.883 \times 10^3 \text{ mm}^2$. The overall depth of the channel is 304.8 mm.

Solution   The compound section is doubly symmetric. Therefore, the centroid lies on the intersection of the axes of symmetry. The section properties of the two plates about their own centroidal axes are $A_i = 300 \times 12 = 3.6 \times 10^3$, $I_{zzi} = 300(12)^3/12 = 0.0432 \times 10^6$, $I_{yyi} = 12(300)^3/12 = 27 \times 10^6$.

Note that h_y for the two plates is given by (half the depth of channel + half the plate thickness) $= 0.5 \times 304.8 + 0.5 \times 12 = 152.4 + 6 = 158.4$ mm.

(a) Calculation of I_{zz} (see Table 3.4). Using the formula $I_{zz} = \sum I_{zzi} + \sum A_i h_{zi}^2$
(b) Calculation of I_{yy}. Using the formula $I_{yy} = \sum I_{yyi} + \sum A_i h_{zi}^2$.

Note that $h_{zi} = 0$ for the two plates because their centroidal zz axis and the centroidal yy axis of the compound section coincide. The distance h_{zi} for the two channels is given by $300/2 - 75 = 75$ mm as shown in Fig. 3.20(a).

Table 3.4

Description	A_i	h_{yi}	$A_i h_{yi}^2$	I_{zzi}
Top plate	3.60×10^3	158.4	90.32×10^6	0.04×10^6
channel left	5.88×10^3	0	0	82.14×10^6
channel right	5.88×10^3	0	0	82.14×10^6
Bottom plate	3.60×10^3	−158.4	90.32×10^6	0.04×10^6
			$\Sigma 180.64 \times 10^6$	$\Sigma 164.36 \times 10^6$

$I_{zz} = (164.36 + 180.64) \times 10^6 = 345.0 \times 10^6 \text{ mm}^4$

Table 3.5

Description	A_i	h_{zi}	$A_i h_{yi}^2$	I_{yyi}
Top plate	3.60×10^3	0	0	27.00×10^6
channel left	5.88×10^3	75	33.09×10^6	4.99×10^6
channel right	5.88×10^3	−75	33.09×10^6	4.99×10^6
Bottom plate	3.60×10^3	0	0	27.00×10^6
			$\Sigma 66.18 \times 10^6$	$\Sigma 63.98 \times 10^6$

$I_{yy} = (63.98 + 66.18) \times 10^6 = 130.16 \times 10^6 \text{ mm}^4$

It is worth noting that in the case of 'hollow' sections, *provided that the centroidal axis of the 'hole' and the 'solid' coincide*, then the

I for hollow section = *I* for solid section − *I* for hole

This idea is illustrated by a few examples.

(a) Rectangular hollow section of dimension $b \times d$ and wall thickness t:
Solid $= b \times d$, hole $= (b − 2t)(d − 2t)$
$I_{zz} = \{bd^3 − (b − 2t)(d − 2t)^3\}/12$

(b) Circular hollow section of diameter D and wall thickness t;
Solid $=$ diameter $= D$, hole $=$ diameter $= (D − 2t)$
$I_{zz} = \pi\{D^4 − (D − 2t)^4\}/64$

(c) Symmetrical *I*-section: Flange width $= b$, overall depth $= d$, flange thickness $= t_f$, web thickness $= t_w$
Solid $= b \times d$, hole $= (b − t_w)(d − 2t_f)$
$I_{zz} = \{bd^3 − (b − t_w)(d − 2t_f)^3\}/12$

(d) Symmetrical channel-section web vertical: Flange width $= b$, overall depth $= d$, flange thickness $= t_f$, web thickness $= t_w$
Solid $= b \times d$, hole $= (b − t_w)(d − 2t_f)$
$I_{zz} = \{bd^3 − (b − t_w)(d − 2t_f)^3\}/12$

3.10.3 Example of the calculation of bending stress

The section shown in Fig. 3.19 is used as a simply supported beam of span $L = 5$ m to carry a uniformly distributed load of $q = 100 \text{ kN m}^{-1}$. Calculate the maximum compressive and tensile bending stress in the section.

Solution In Section 3.10.1, Example 1, the second moment of area of the section was shown to be $I_{zz} = 225.9 \times 10^6$ N mm^{-2}. The maximum bending moment occurs at midspan and is given by

$$M_{max} = qL^2/8 = 100(5)^2/8 = 312.5 \text{ kN m} = 312.5 \times 10^6 \text{ N mm}$$

Therefore

$$M_z = M_{max}$$

The bending stress is linearly distributed and is given by

$$\sigma_x = -(M_z y)/I_{zz}$$

The stress σ_x is compressive at the top face and tensile at the bottom face. The distance from the neutral axis to the top and bottom face is 136.1 mm and −163.9 mm respectively. Therefore

$$\sigma_{top} = -312.5 \times 10^6 (136.1)/(225.9 \times 10^6) = -188 \text{ N mm}^{-2} \text{ compressive}$$

$$\sigma_{bottom} = -312.5 \times 10^6 (-163.9)/(225.9 \times 10^6) = 227 \text{ N mm}^{-2} \text{ tensile}$$

3.10.4 Section modulus

As was seen in the last example, the stress due to bending is a maximum at the top or the bottom fibre.

The expression for stress at the top is

$$\sigma_{top} = (M/I)\, y_{top}$$

The above expression can be written as

$$\sigma_{top} = M/Z_{top}$$

where $Z_{top} = I/y_{top}$
Similarly

$$\sigma_{bottom} = (M/I)\, y_{bottom}$$

The above expression can be written as

$$\sigma_{bottom} = M/Z_{bottom}$$

where $Z_{bottom} = I/y_{bottom}$

The values for Z_{bottom} and Z_{top} are known as section moduli for top and bottom respectively. Only in the case of sections symmetrical about the z-axis is the value of Z_{bottom} equal to that of Z_{top}.

3.10.5 Comparative strength of sections

In order to appreciate that, depending upon the distribution of the material in a cross-section, widely different strengths in bending can be achieved, consider five different cross-sections all having a total cross-sectional area of $2a^2$ but different distributions of material in the cross-section.

(a) Rectangular section: depth, $d = 2a$, width, $b = a$
 Area of cross-section $= b \times d = 2a^2$
 Second moment of area, $I = bd^3/12 = 0.667a^4$
 Section modulus, $Z_{bottom} = Z_{top} = I/(0.5d) = 0.667a^3$

(b) Square section: depth = width = S, placed with *sides* vertical and horizontal.
 $S = \sqrt{2}a$
 Area of cross-section $= S^2 = 2a^2$
 Second moment of area, $I = S^4/12 = 0.333a^4$
 Section modulus, $Z_{bottom} = Z_{top} = I/(0.5S) = 0.471a^3$

(c) Square section: depth = width = S, placed with the *diagonals* vertical and horizontal.
 $S = \sqrt{2}\ a$
 Area of cross-section $= S^2 = 2a^2$
 Second moment of area, I: Consider as two triangles of base $b = 2a$ and height $h = S/\sqrt{2} = a$
 $I = 2(bh^3/12) = 0.333a^4$
 Section modulus, $Z_{bottom} = Z_{top} = I/h = 0.333a^3$

(d) Solid circle: diameter = $D = 1.596a$
 Area of cross-section $= \pi D^2/4 = 2a^2$
 Second moment of area, $I = \pi D^4/64 = 0.319a^4$
 Section modulus, $Z_{bottom} = Z_{top} = I/(0.5D) = 0.40a^3$

(e) Equilateral triangular section: sides, $L = 2.149a$, height $h, = 0.5\sqrt{3}a$
 Area of cross-section $= 0.5Lh = 2a^2$
 Second moment of area, I: I about the centroidal axis $= bh^3/36 = 0.385a^4$
 Section modulus, $Z_{bottom} = Z_{top} = I/0.6667h = 0.310a^3$

If the allowable bending stress σ is the same for all the cross-sections analysed, then the bending capacity is proportional to the minimum section modulus, because bending capacity, $M = (Z_{bottom}$ or $Z_{top})\ \sigma$ where σ = maximum allowable stress.

The relative bending capacities, in increasing order, as a proportion of the capacity of a rectangular section are: Rectangle: 1.0, square: 0.71, circle: 0.60, square (diagonal vertical) = 0.50 and equilateral triangle = 0.47.

Clearly the rectangular section is the most efficient. The reason for this is that for a large value of second moment of area, it is necessary to have maximum material positioned away from the centroidal axis. Thus a square section placed with the diagonal vertical is inefficient, because the amount of material reduces as one moves away from the centroidal axis. *I*-sections, which have the maximum material away from the centroidal axis, are the most efficient and this accounts for their popularity.

3.11 ● Shear stresses in beams

Consider two identical beams one placed on top of another and loaded as shown in Fig. 3.21(a). Experimental observation shows that the two beams share the applied

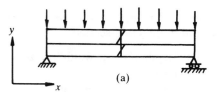

(a)

Figure 3.21 ●
(a), (b) Illustration of
development of
horizontal and vertical
shear stress in beam
bending

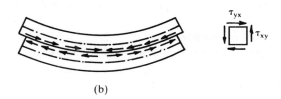

(b)

load equally. At the junction between the two beams, due to bending stresses the top beam extends and the bottom beam contracts. This leads to a difference in the length at the interface as shown exaggerated in Fig. 3.21(b). If the two beams are joined together at the interface by welding, gluing or by some other means, then shear stresses will develop at the interface to prevent slip. Evidently the presence of horizontal shear stress τ_{yx} will mean the presence of shear stress τ_{xy} in the vertical direction due to the fact that complementary shear stresses are equal.

3.11.1 Calculation of shear stresses in beams

The calculation of stresses due to axial force in the previous sections was carried out assuming that the average stress is given by load divided by area of cross-section. Unfortunately this assumption cannot be used for calculating shear stresses in beams because the assumption leads to the wrong values of shear stresses. Consider the simple rectangular beam shown in Fig. 3.22. Let a shear force Q be applied to the section in the y-direction. If the shear strain is uniform throughout the cross-section of the beam, then the shear stress is also uniformly distributed in the cross-section. This leads to the existence of a shear stress at the top and bottom boundaries which are free from shear stresses. In other words, the assumption leads to the wrong shear stress distribution in the cross-section. In addition, as shown in Section 1.12, bending moment M and shear force Q are related by the equilibrium equation

$$dM/dx + Q = 0$$

In a similar way bending and shear stresses cannot be calculated independently but have to satisfy equilibrium considerations. Consider an element dx of a simple beam with the bending moment applied in the vertical plane as shown in Fig. 3.23(a). Let the bending moment at a section be M. The bending moment at a neighbouring section at a distance of dx is given by $\{M + (dM/dx)\,dx\}$. At any level y from the neutral axis, the bending stress σ_x at the two sections is given by σ_x and $\{\sigma_x + (\partial\sigma_x/\partial x)\,dx\}$ respectively. Evidently the difference in normal stress between the two faces is given by $(\partial\sigma_x/\partial x)\,dx$. Let it be required to determine the shear stress τ_{yx} at a level $y = h$ from the neutral axis. Considering the free body

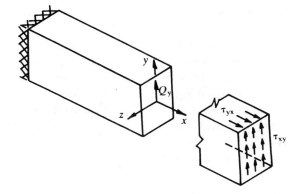

Figure 3.22 ●
Shear force Q_y and corresponding uniform distribution of shear stress τ_{xy}

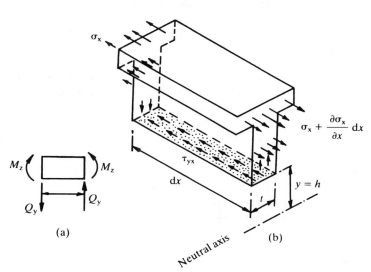

Figure 3.23 ●
(a), (b) Free body for calculating horizontal shear stress τ_{yx}

shown in Fig. 3.23(b), the net horizontal force due to difference in bending stresses is equal to $\int \{(\partial\sigma_x/\partial x)\,dx\}\,dA$, where the integration is carried out over the area *above* the section where it is required to determine the shear stress. In order to balance this force, a horizontal force is required which can be provided by the shear stress $\tau = \tau_{yx}$ at the level $y = h$. Assuming that the shear stress is uniformly distributed over the thickness t, we have

$$\int (\partial\sigma_x/\partial x\,dx)\,dA = \tau t\,dx$$

But $\sigma_x = -(M_z/I_{zz})y$, provided the symmetric bending theory is applicable. Therefore

$$\partial\sigma_x/\partial x = -\{(dM_z/dx)y\}/I_{zz}.$$

Since shear force $Q = -dM_z/dx$, $\partial\sigma_x/\partial x = -(Qy)/I_{zz}$,

$$\tau = \frac{Q}{I_{zz}t}\int y\,dA$$

where $\int y\,dA =$ first moment about the neutral axis of the area above the level where the shear stress τ is required. Because shear stress $\tau_{xy} = \tau_{yx}$, a vertical shear stress equal to τ acts at the cross-section such that the equilibrium condition $\int \tau \, dA = Q$ is satisfied.

Note: The above formula has been obtained from equilibrium considerations and on the assumption that the shear stress τ is uniformly distributed over the thickness t. This is acceptable in the case of sections with reasonably parallel sides. However, as shown in Fig. 3.25(e), in the case of say circular and triangular sections, the shear stresses at the edges of the cross-section are parallel to the edges and not to the vertical. The calculated shear stress τ at any level y must be treated as an average value only over the width and does not reflect the true distribution of the shear stress in the cross-section.

3.11.2 Examples of the calculation of shear stress in beams

Example 1

As a simple example, consider a rectangular section shown in Fig. 3.24. It is required to determine the shear stress at a distance y from the neutral axis. The shaded area A above the level y is equal to $B(0.5D - y)$ as shown in Fig. 3.24(a). The centroid $\bar{y}$ of this area is at a distance $\bar{y} = y + 0.5(D/2 - y) = 0.5(D/2 + y)$. The first moment of this area about the neutral axis is given by

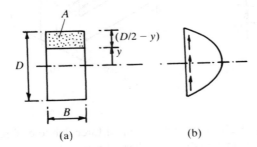

Figure 3.24 ● (a), (b) Shear stress distribution in a rectangular beam cross-section

$$\int y \, dA = \text{area } A \times \bar{y} = \{B(0.5D - y)\}0.5(D/2 + y)$$

$$= \{B/2\}\{D^2/4 - y^2\} = \frac{BD^2}{8}\left\{1 - \left(\frac{2y}{D}\right)^2\right\}$$

Therefore

$$\tau_{xy} = [Q/(I_{zz}t)](BD^2/8)\{1 - (y/0.5D)^2\}$$

However, $t = B$ and $I_{zz} = BD^3/12$, therefore

$$\tau_{xy} = 1.5(Q/BD)\{1 - (y/0.5D)^2\}$$

Expressing the average shear stress $Q/BD = \tau_{average}$, we have

$$\tau_{xy} = 1.5\tau_{average}\{1 - (y/0.5D)^2\}$$

This shows that the shear stress varies parabolically over the depth of the beam with a maximum at the neutral equal to 1.5 times the average shear stress. It is interesting to note that the bending stress is a maximum tension or compression at the top or bottom faces of the beam, while shear stress is a maximum at the neutral axis where the bending stress is zero.

Example 2

Calculate the shear stresses $\tau = \tau_{xy}$ in the I beam shown in Fig. 3.19. Since the thickness is not constant, it is necessary to consider three cases as shown in Fig. 3.25.

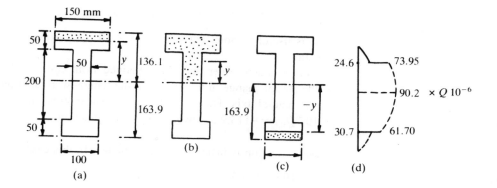

Figure 3.25 ● (a)–(d) Shear stress calculation in a beam of I cross-section

Case 1. Section is inside the top flange. In this case as shown in Fig 3.25(a), $t = 150$, $136.1 \geqslant y > 86.1$.

Calculation of $\int y\, dA$: the distance from the neutral axis to the top fibre is 136.1 mm. For a section distance y from the neutral axis, the shaded area A above the section is given by $A = 150(136.1 - y)$. The distance to the centroid $\bar{y}$ of this area from the neutral axis is

$$\bar{y} = y + 0.5\{136.1 - y\} = 0.5(136.1 + y)$$

Therefore

$$\int y\, dA = A\,\bar{y} = 150(136.1 - y)0.5(136.1 - y) = 1.30 \times 10^6\{1 - (y/136.1)^2\}$$

Since $I_{zz} = 225.9 \times 10^6$, $\tau = \{Q/(It)\} \int y\, dA = 41.0 \times 10^{-6}\{1 - (y/136.1)^2\}Q$

It is useful to note that $\tau = 0$, $y = 136.1$, at top surface

$\tau = 24.59 \times 10^{-6}Q$, $y = 86.1$ at the web junction of top flange.

Case 2. Section is inside the web. In this case $t = 50$, $86.1 \geqslant y \geqslant -113.9$. The shaded area A in Fig. 3.25(b) $= A_1$ which is the area of top flange $+ A_2$ which is the area of web above y

$$A_1 = 150 \times 50, \quad A_2 = 50\{(136.1 - 50) - y\} = 50(86.1 - y)$$

From the neutral axis, the centroid of top flange is at a distance $\bar{y}_1$ equal to $\bar{y}_1 = (136.1 - 50/2) = 111.1$ mm from the neutral axis.

From the neutral axis, the centroid of the area of web A_2 is at a distance $\bar{y}_2 = y + 0.5(136.1 - 50 - y) = 0.5(86.1 + y)$

$$\int y \, dA = A_1 \bar{y}_1 + A_2 \bar{y}_2$$
$$= 150 \times 50 \times 111.1 + 50(86.1 - y)\{0.5(86.1 + y)\}$$
$$= 0.833 \times 10^6 + 0.185 \times 10^6\{1 - (y/86.1)^2\}$$
$$= 1.019 \times 10^6 \{1 - 0.18(y/86.1)^2\}$$

Substituting in the expression for $\tau = \{Q/(It)\} \int y \, dA$, $I_{zz} = 225.9 \times 10^6$, $t = 50$ we have

$$\tau = 90.18 \times 10^{-6}\{1 - 0.18(y/86.1)^2\}Q$$

It is useful to note that $\tau = 73.95 \times 10^{-6}Q$, $y = 36.1$ at the upper flange–web junction.

This is much larger than that obtained from case 1. The main reason for this is the abrupt change in thickness t at the flange–web junction. $\tau = 90.18 \times 10^{-6}Q$, $y = 0$ at neutral axis. This is the maximum shear stress in the section; and $\tau = 61.77 \times 10^{-6}Q$, $y = -113.9$ at lower flange–web junction.

Case 3. Section is inside the bottom flange. In this case

$$t = 100, \quad -113.9 > y \geqslant -163.9$$

In this case y is negative. The distance from the neutral axis to the bottom face is 163.9 mm. Therefore area A of the shaded portion in Fig. 3.25(c) is

$$A = 100\{163.9 - (-y)\} = 100\{163.9 + y\}$$

The distance to the centroid of this area from the neutral axis is

$$\bar{y} = -y + 0.5\{163.9 - (-y)\} = 0.5\{163.9 - y\}$$

Therefore

$$\int y \, dA = A \bar{y} = 100(163.9 + y)\{0.5(163.9 - y)\}$$
$$= 1.343 \times 10^6\{1 - (y/163.9)^2\}$$

Substituting in the expression for $\tau = \{Q/(It)\} \int y \, dA$, $I_{zz} = 225.9 \times 10^6$, $t = 100$ we have

$$\tau = 59.6 \times 10^{-6}\{1 - (y/163.9)^2\}Q$$

$$\tau = 30.74 \times 10^{-6}Q \text{ at } y = -113.9 \text{ at lower flange–web junction}$$

$$\tau = 0, \quad y = -163.9 \text{ at bottom surface.}$$

The above shear stress $\tau = \tau_{xy}$ acts in the vertical direction at a cross-section. As shown in Fig. 3.25(d) the shear stress distribution in each portion is parabolic but due to the sudden changes in the thickness at flange–web junctions, the distribution of shear stress is discontinuous. As in the case of the rectangular section, the major portion of the shear force is carried by the web rather than by the flange. The small vertical shear stress in the flanges is of little design significance. In practice the discontinuity of shear stresses shown will not occur and an exact analysis of the problem will not show this discontinuity (see Section 3.17), but the difference between the exact and approximate solution obtained is generally of little consequence for most design purposes.

Position of maximum shear stress in a web

In the previous examples it was shown that the maximum shear stress in the cross-section was at the neutral axis. The reason for this is that in every cross-section, the maximum difference in the resultant axial force due to the bending stress between two sections (see Fig. 3.22(d)), occurs at the neutral axis. If the web is of constant thickness, then the maximum shear stress also occurs at the neutral axis. However, if the web is of variable thickness, then the maximum shear stress might not always be at the neutral axis. Figure 3.25(e) shows a triangular cross-section with 'tapering webs'.

The shear stress distribution is calculated using the formula

$$\tau = \{Q/(I_{zz}t)\} \int y \, dA$$

where

$$I_{zz} = bh^3/36$$

$$t = (b/h)\{2h/3 - y\}$$

Area of the section above $y = 0.5t(2h/3 - y) = (t/6)(2h - 3y)$

The centroid of this area from the neutral axis of the section

$$= y + (1/3)\{2h/3 - y\}$$

$$= (2/9)(h + 3y)$$

$$\therefore (1/t) \int y \, dA = (1/27)(2h - 3y)(h + 3y)$$

Substituting the above expressions,

$$\tau = \{Q/I_{zz}\}(2h - 3y)(h + 3y)/27$$

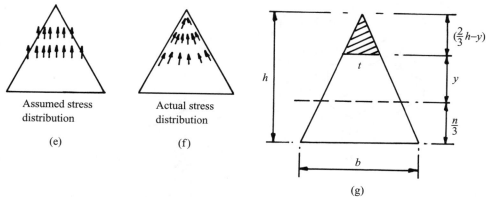

Figure 3.25 ● (e)–(g) Distribution of shear stresses in a beam of triangular cross-section

For maximum τ, $d\tau/dx = 0$. Therefore maximum τ occurs when $(2h - 3y)(3) + (h + 3y)(-3) = 0$, solving $y = h/6$. $\tau_{max} = 3Q/(bh)$ when $y = h/6$

For $\tau_{neutral\ axis}$, $y = 0$

$\therefore \tau_{neutral\ axis} = 8Q/(3bh) = 2.67Q/(bh)$

As can be seen, the ratio between the shear stress at the neutral axis and the maximum shear stress is $2.67/3 = 0.89$. Thus in most cases it can be accepted that the shear stress at the neutral axis should be a very good approximation to the maximum shear stress in the cross-section.

Horizontal shear stress distribution in flanges

Example 1

For the I beam shown in Fig. 3.26(a), calculate the horizontal shear stress $\tau = \tau_{xz}$ in the flanges.

A free body of the flange shown in Fig. 3.26(b) shows clearly that horizontal shear stress τ_{zx} is necessary to maintain equilibrium. Because of the fact that the thickness of the flange is small, it is reasonable to assume that the shear stress τ_{xz} is uniformly distributed over the thickness. The average bending stress in the flange is the same as the bending stress at the mid-depth of the flange and is uniform across the flange width. Therefore the equilibrium equation can be established for the free body as

$$\tau_{xz}t_f\,dx + [\sigma_x - \{\sigma_x + (\partial\sigma_x/\partial x)\,dx\}](0.5B - z)t_f = 0$$

where t_f = thickness of flange, B = width of flange, σ_x = bending stress at mid-depth of flange = $-(M/I_{zz})y_f$, y_f = distance from neutral axis to mid-depth of flange.

Simplifying,

$$\tau_{xz} = (\partial\sigma_x/\partial x)(0.5B - z)$$

$$\partial\sigma_x/\partial x = -\{(dM/dx)y_f\}/I_{zz}.$$

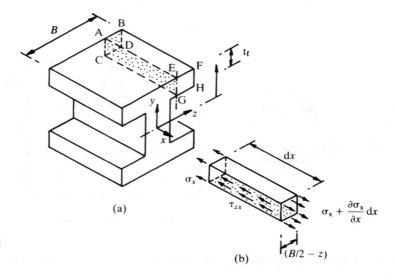

Figure 3.26 ●
(a), (b) Free body for the calculation of the horizontal shear stress in the flange of an I beam

(a)

(b)

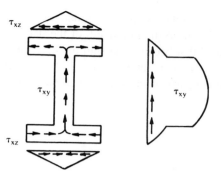

Figure 3.27 ● Shear stress distribution in the web and flanges of an *I* beam

Since

$$dM/dx = -Q, \quad \partial\sigma_x/\partial x = (Q/I_{zz})y_f$$

then

$$\tau_{xz} = (Q/I_{zz})(0.5B - z)y_f$$

This shows that τ_{xz} is linearly distributed. Figure 3.27 shows the shear stress distribution in the flanges and web of the *I* beam.

3.11.3   Some simple applications of shear stress calculation in design

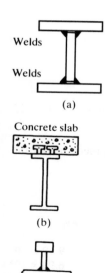

Figure 3.28 ● (a)–(c) Methods of connecting flanges and web in an *I* beam

In practice when steel beams larger than those that can be rolled are needed, beams are fabricated by welding steel plates as shown in Fig. 3.28(a). Similarly when composite steel–concrete beams are required then the reinforced concrete slab is connected to the steel beams using either friction grip bolts or shear connectors which are headed studs welded to the top flange of the steel beam as shown in Fig. 3.28(b). The shear stress at the flange–web interface is

$$\tau = \{Q/(I_{zz}t)\} \int y \, dA$$

where $\int y \, dA$ is the first moment of area of the top flange about the neutral axis.

The shear flow = $\tau \times t$ = horizontal shear force per unit length at the flange–web interface which has to be carried by the welds or shear connectors. Therefore

$$\text{shear flow} = (Q/I_{zz}) \int y \, dA$$

In the case of beams, the shear force is generally at a maximum near supports. This means that either larger welds or more shear connectors are required near the supports than near the midspan. However, very often it is convenient to use the same size of weld or the same number of shear connectors per unit length throughout the span and aim to provide sufficient 'average' strength. In this case the total shear force to be carried by the welds over a length O to L_1 is given by

$$\int_0^{L_1} \tau t \, dx = \left\{\left(\int y \, dA\right)\Big/I_{zz}\right\}\int_0^{L_1} Q \, dx$$

where $\int y \, dA$ is calculated only for the top flange = area of top flange × depth from neutral axis of the beam to the mid-depth of the flange.

It should be noted that because the sign of shear force changes over the span of a beam, it is necessary that integration $\int Q \, dx$ is carried out separately for the positive and negative parts of the shear force diagram.

3.11.4 Example

Example 1

Figure 3.29 shows an I beam reinforced by a plate welded to the top flange. The beam is used as a simply supported beam of span 6 m to carry a uniformly distributed load of 25 kN m^{-1}. Calculate the force per unit length carried by the welds. It is given that for the I beam

$$I_{zz} = 44.73 \times 10^6 \text{ mm}^4, \ A = 6.0 \times 10^3 \text{ mm}^2$$

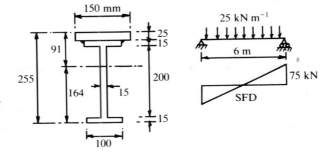

Figure 3.29 ●
Calculation of shear flow in the welds connecting the reinforcing plate to the beam

Solution Calculate the I_{zz} of the modified beam. For the reinforcing plate $A_i = 150 \times 25 = 3.75 \times 10^3$ mm^2, $I_{zzi} = 150 \times 25^3/12 = 0.195 \times 10^6$ mm^4.

(a) Calculation of the centroid of the composite beam. Take moments about the bottom flange.

Table 3.6

Description	Dimensions	A_i	y_i	$A_i y_i$
Top plate	150 × 25	3.75×10^3	242.5	9.09×10^5
I beam	–	6.00×10^3	115.0	6.90×10^5
		$\Sigma 9.75 \times 10^3$		$\Sigma 15.99 \times 10^5$

$\bar{y} = 15.99 \times 10^5/9.75 \times 10^3 = 164.04$ mm from bottom face. The distance to the top face from the neutral axis is $(230 + 25 - \bar{y}) = 90.96$ mm.

Table 3.7

Description	A_i	$h_{yi} = (y_i - \bar{y})$	$A_i h_y^2$	I_{zzi}
Top plate	3.75×10^3	$(242.5 - 164) = 78.5$	23.11×10^6	0.195×10^6
I beam	6.00×10^3	$(115 - 164) = -49$	14.41×10^6	44.73×10^6
			$\Sigma 37.52 \times 10^6$	$\Sigma 44.93 \times 10^6$

(b) Calculate I_{zz} of composite section (see Table 3.7):

$$I_{zz} = (44.93 + 37.52) \times 10^6 = 82.45 \times 10^6$$

The distance from the neutral axis to the centroid of the reinforcing plate is $(230 + 25/2 - 164) = 78.5$ mm. Therefore $\int y \, dA$ for the top flange is given by

$$\int y \, dA = \text{(area of top flange)} \times 78.5$$

$$= (150 \times 25)78.5 = 0.294 \times 10^6 \text{ mm}^3$$

Shear flow at the weld level is given by

$$\text{shear flow} = \tau t = (Q/I_{zz}) \int y \, dA$$

Substituting the known values

$$\text{shear flow} = Q \times 0.294 \times 10^6/(82.45 \times 10^6) = 3.57 \times 10^{-3} Q$$

In the case of a simply supported beam, the reaction at the support $= 25(6/2) = 75$ kN. The shear force Q at a section x from the left-hand support is

$$Q = -\{75 - 25x\}$$

$$\int Q \, dx = -\{75x - 25x^2/2\} \text{ with limits for } x = 0 \text{ and } 3 \text{ m}$$

$$= -112.5 \text{ kN m} = -112.5 \times 10^6 \text{ N mm}$$

Total shear flow over half span $= 3.57 \times 10^{-3} \int Q \, dx$

$$= 3.57 \times 10^{-3}(112.5 \times 10^6) = 0.40 \times 10^6 \text{ N}$$

This shear flow is resisted by the two welds each of length 3000 mm. Therefore shear flow to be resisted per unit length of weld is

$$0.40 \times 10^6/(2 \times 3000) = 66.94 \text{ N mm}^{-1}$$

As an exercise, the bending stress in the beam can be calculated. The maximum bending moment is given by $M_{zz} = qL^2/8 = 25 \times 6^2/8 = 112.5$ kN m $= 112.5 \times 10^6$ N mm. Using the formula $\sigma_x = -M_{zz}y/I_{zz}$, we have

$$\sigma_x \text{ top} = -112.5 \times 10^6(90.96)/(82.45 \times 10^6)$$

$$= -124.1 \text{ N mm}^{-2} \text{ compression}$$

$$\sigma_x \text{ bottom} = -112.5 \times 10^6(-164.04)/(82.45 \times 10^6)$$

$$= 223.8 \text{ N mm}^{-2} \text{ tension}$$

3.11.5 Distribution of bending and shear stresses

It was noticed in the several examples on bending and shear stress calculation that the bending stress is a maximum at the extreme fibres of the cross-section, but the shear stress is a maximum at the neutral axis. In the case of an I section, the bending moment is resisted mainly by the flanges and the shear stress is resisted mainly by the webs. To emphasize this point, consider an I section corresponding to 914 × 305 mm × 289 kg m^{-1} universal beam. The beam has the following dimensions:

Area of top and bottom flanges = 307.8 × 32 mm = 9849.6 mm^2

Distance between the centre line of flanges = 894.6 mm

Web = 862.6 × 19.6 mm

$I_{zz} = 5045.94 \times 10^6$ mm^4

Bending stress calculation

I_{zz} of section = contribution of flanges + contribution of the web

Contribution of the flanges = 2 {area of each flange × (894.6/2)^2}

$$= 3941.6 \times 10^5 \text{ mm}^4 = 0.78 I_{zz}$$

Since maximum $\sigma_x = -M_z y_{max}/I_{zz}$, then

$$M_z = -I_{zz}(\max \sigma_x/y_{max}) = -\{I_{zz} \text{ from web} + I_{zz} \text{ from flanges}\} \{\max \sigma_x/y_{max}\}$$

Since I_{zz} from flanges = 0.78 I_{zz} of the whole section, the flanges alone carry 78 per cent of the total bending moment and the webs carry only 22 per cent of the total bending moment applied to the cross-section.

Shear stress calculation

Ignoring the flanges and assuming that all the shear force is resisted by the web of dimensions 862.6 × 19.6 mm, the maximum shear stress in the web is given by (see Section 3.11.1)

$$\tau_{max} = 1.5\tau_{average} = 1.5Q/(862.6 \times 19.6) \text{ or } Q = 1.13(\tau_{max} \times 10^4)$$

Using the more exact calculation given by $\tau = Q A \bar{y}/(I_{zz}t)$ (see Section 3.11.1), we have

$A\bar{y}$ = (Area of flange = 9849.6) × (distance to centroid of flange = 894.6/2)
      + (area of half the web = 862.6 × 19.6) × (distance to its centroid = 0.5 × 862.6/2)

$$= 8.05 \times 10^6 \text{ mm}^3$$

$I_{zz} = 5045.94 \times 10^6$ mm^4, t = web thickness = 19.6 mm

Therefore $Q = 1.23(\tau_{max} \times 10^4)$

Therefore for the same τ_{max},

$$\frac{Q_{web}}{Q_{total}} = \frac{1.13}{1.23} = 0.92$$

This shows that the web alone carries 92 per cent of the total shear force. Therefore at the preliminary design stage an I beam can be proportioned using the shear force to calculate the depth and thickness of the web and the bending moment to calculate the area of flanges. This procedure is widely used in design calculations involving I beams.

3.12 ● Unsymmetrical bending

In Section 3.9.2 the equation for calculating bending stress σ_x given by

$$\sigma_x = -(M_z y)/I_{zz} + (M_y z)/I_{yy}$$

was developed. It was pointed out that this was developed on the assumption that plane sections remain plane and further that the y- and/or the z-axis are the axes of symmetry of the cross-section. Figure 3.30(a) shows some typical sections where this is satisfied and Fig. 3.30(b) some commonly used sections where this is not satisfied.

In order to demonstrate the anomaly that would develop if the formula is applied to a section where neither the y- nor the z-axis is an axis of symmetry, consider the Z-section shown in Fig. 3.31. The second moments of area for this section can be shown to be

$$I_{zz} = 95.72 \times 10^6 \text{ mm}^4, \quad I_{yy} = 23.94 \times 10^6 \text{ mm}^4$$

Let the section be subjected to a moment of $M_z = 10$ kN m $= 10 \times 10^6$ N mm. The bending stress $\sigma_x = -(M_z y)/I_{zz} = -0.105y$ N mm^{-2}. The average stress in the top flange at $y = +144$ is σ_x top $= -15.0$ N mm^{-2} (compressive stress). Similarly, the

Figure 3.30 ●
Symmetrical and unsymmetrical cross-sectional shapes:
(a) cross-sections with a vertical or horizontal axis of symmetry, (b) cross-sections with no axis of symmetry

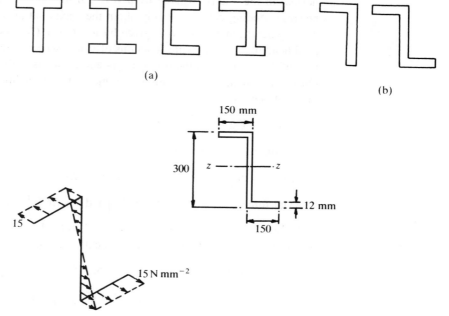

Figure 3.31 ●
Incorrect bending stress distribution obtained by using $\sigma = -My/I_{zz}$ for a z cross-section

average bottom flange stress when y = −144 is given by σ_x bottom = +15.0 N mm^{-2} (tensile). If the stress resultants due to this stress distribution are calculated, then F_x = axial force = $\int \sigma_x \, dA$ = 0 because the stress distribution is antisymmetrical with respect to the z-axis.

It can be seen that the total force F in the two flanges are equal and opposite and given by

F = flange area (= 150 × 12 mm^2) × average stress (= 15.0 N mm^{-2})

   = 0.027 × 10^6 N (compressive in top flange and tensile in bottom flange)

The forces in the two flanges form a couple about the y-axis and the normal distance between the forces is 150 mm. Therefore

$M_y = F \times 150 = 0.027 \times 10^6 \times 150 = 4.06$ kN m

But for equilibrium M_y must be zero as there is no applied couple about the y-axis.

The above example shows that in the case of the Z-section, although the applied force is a couple about the z-axis, the section must bend not only about the z-axis but also about the y-axis to ensure that the stress distribution satisfies the equations of equilibrium. A general approach for doing this is developed in the next section. When a section is subjected to couples about the y- and z-axis which are not axes of symmetry of the cross-section, it is said to be a case of unsymmetrical bending.

3.12.1 Unsymmetrical bending – general expression for bending stress

Bending stress in the case of symmetrical bending was calculated on the assumption that plane sections remain plane. When the applied couple was M_z only, the assumption of plane sections means that the axial displacement at any section normal to the x-axis is given by $u = Cy$. In the case of unsymmetrical bending, because bending occurs not only about the z-axis but also about the y-axis, then a more general function for u is given by $u = C_1 y + C_2 z$. As in the case of symmetrical bending, C_1 and C_2 are functions of x only where the x-axis coincides with the axis of the section and the origin for y- and z-axes is at the centroid of the section. Once this fundamental assumption is made, the calculations proceed as for symmetrical bending.

$u = C_1 y + C_2 z$

$\varepsilon_x = \partial u/\partial x = C'_1 y + C'_2 z$

where $C'_1 = dC_1/dx$, $C'_2 = dC_2/dx$ and $\sigma_x = E\varepsilon_x$

Axial force = $F_x = \int \sigma_x \, dA = EC'_1 \int y \, dA + EC'_2 \int z \, dA = 0$

If the origin for y and z is at the centroid, then $\int y \, dA = \int z \, dA = 0$ by definition and the equilibrium requirement is automatically satisfied.

Moment about the z-axis = $-M_z = \int \sigma_x y \, dA$

$= EC'_1 \int y^2 \, dA + EC'_2 \int yz \, dA$

Moment about the y-axis $= M_y = \displaystyle\int \sigma_x z \, dA$

$$= EC_1' \int yz \, dA + EC_2' \int z^2 \, dA$$

Define

$\displaystyle\int y^2 \, dA = I_{zz}$, second moment of area about the z-axis.

$\displaystyle\int z^2 \, dA = I_{yy}$, second moment of area about the y-axis.

$\displaystyle\int yz \, dA = I_{yz}$, cross second moment of area about the y–z axis.

The cross second moment of area is also called the product moment of area or product inertia. Therefore

$$-M_z = EC_1' I_{zz} + EC_2' I_{yz}$$

$$M_y = EC_1' I_{yz} + EC_2' I_{yy}$$

Solving for C_1' and C_2' we have

$$EC_1' = -(M_z I_{yy} + M_y I_{yz})/(I_{yy} I_{zz} - I_{yz}^2)$$

$$EC_2' = (M_z I_{yz} + M_y I_{zz})/(I_{yy} I_{zz} - I_{yz}^2)$$

$$\sigma_x = -\frac{(M_z I_{yy} + M_y I_{yz})}{(I_{yy} I_{zz} - I_{yz}^2)} y + \frac{(M_z I_{yz} + M_y I_{zz})}{(I_{yy} I_{zz} - I_{yz}^2)} z$$

As can be seen, if $I_{yz} = 0$, then $\sigma_x = -(M_z/I_{zz})y + (M_y z)/I_{yy}$. This is exactly the same as the formula derived for the calculation of bending stress in symmetric bending. The formula derived for the case of unsymmetrical bending is a more general one of which the symmetrical bending is a particular case.

3.12.2 Calculation of cross second moment of area

The calculation of cross moment of area or product inertia is similar to the calculation of second moment of area. However, one important property of symmetrical sections will be found useful. As an example consider the I section with one axis of symmetry as shown in Fig. 3.32. Let the cross-section be divided into a large number of small areas. Let it be required to calculate the cross second moment of area or product inertia about the centroidal axes. If the section is divided into a sufficiently large number of small rectangles, then it is reasonable to assume that $I_{yz} = \int yz \, dA \approx \Sigma y_i z_i \, dA_i$ where dA_i is an infinitesimal element of area and y_i and z_i are the distance to the centroid of dA_i from the z- and y-axis respectively. It is immediately apparent that, because of symmetry, for every element of area dA_i with positive y_i and z_i there is a corresponding dA_i with positive y_i and negative z_i. In other words, the total contribution of these two small areas dA_i to the total I_{yz} is equal to zero. Extending similar arguments to other elements of area dA_i symmetrically situated with respect to the co-ordinate axes, it becomes evident that for a

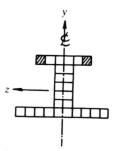

Figure 3.32 ●
Example to show that I_{yz} is zero about an axis of symmetry

section with at least one axis of symmetry, I_{yz}, when calculated about its centroidal axes, is equal to zero. This is an important result and explains why it was possible to develop such simple expressions for the calculation of bending stress in the case of symmetrical bending.

3.12.3 Parallel axis theorem for cross second moment of area or product inertia

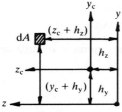

y_c
y
$(z_c + h_z)$
dA
h_z
z_c
$(y_c + h_y)$ h_y
z

Figure 3.33 ● A set of parallel axes

Consider the pair of orthogonal co-ordinate axes (y_c, z_c) and (y, z) shown in Fig. 3.33. Axes (y_c, z_c) pass through the centroid of the section. Then

$$I_{zz} = \int y^2 \, dA = \int (y_c + h_y)^2 \, dA$$

$$I_{yy} = \int z^2 \, dA = \int (z_c + h_z)^2 \, dA$$

$$I_{yz} = \int yz \, dA = \int (y_c + h_y)(z_c + h_z) \, dA$$

where (h_y, h_z) are the co-ordinates of the centroid of the area A with respect to the (y, z) system. Expanding the products inside the integrals, we have

$$I_{zz} = \int y_c^2 \, dA + h_y^2 \int dA + 2h_y \int y_c \, dA$$

$$I_{yy} = \int z_c^2 \, dA + h_z^2 \int dA + 2h_z \int z_c \, dA$$

$$I_{yz} = \int y_c z_c \, dA + h_y \int z_c \, dA + h_z \int y_c \, dA + h_y h_z \int dA$$

Evidently $\int dA$ = area of cross-section A and $\int z_c \, dA = \int y_c \, dA = 0$ because they are the first moments of area about the centroidal axes and are by definition equal to zero. $\int y_c^2 \, dA = I_{zz}$ about the z_c axis passing through the centroid of the area. $\int z_c^2 \, dA = I_{yy}$ about the y_c axis passing through the centroid of the area. $\int y_c z_c \, dA = I_{yz}$ about the y_c and z_c axes passing through the centroid. Therefore

$$I_{zz} = I_{zz} \text{ about the centroidal axis} + Ah_y^2$$

$$I_{yy} = I_{yy} \text{ about the centroidal axis} + Ah_y^2$$

$$I_{yz} = I_{yz} \text{ about the centroidal axis} + Ah_y h_z$$

For a symmetric section, I_{yz} about the centroidal axis is equal to zero. Therefore $I_{yz} = h_y h_z A$.

One important point to remember is that in the calculation of I_{yz}, h_y and h_z do not appear as squared quantities. It is therefore important to ensure that the correct signs are used in the calculation of h_y and h_z. If a section is composed of a number of individual rectangles, then the above equations are applied to each rectangle in turn so that for the whole section

$$I_{yy} = \sum I_{zzi} \text{ about the centroidal axis} + \sum A_i h_{yi}^2$$

$$I_{yy} = \sum I_{yyi} \text{ about the centroidal axis} + \sum A_i h_{zi}^2$$

$$I_{yz} = \sum A_i h_{yi} h_{zi}$$

where the summation is over $i = 1$ to N, $N =$ total number of individual rectangles in the section considered.

3.12.4   Examples of unsymmetrical bending stress calculation

Example 1

The z-section shown in Fig. 3.34 is used as a simply supported beam of span 5 m to carry a uniformly distributed load of 20 kN m^{-1}. Calculate the bending stress distribution at midspan.

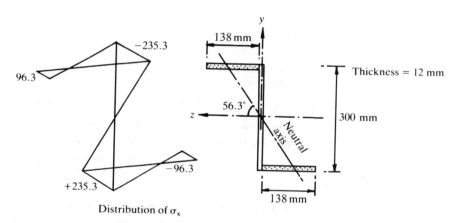

Figure 3.34
Stress distribution in a
z-section

Solution   (a) Calculate the section properties. The position of the centroid can be determined by inspection. The calculation of the section properties is set out in Table 3.8.

Table 3.8

Section	$b \times d$	A_i	h_{yi}	h_{zi}	I_{zzi}	I_{yyi}	$A_i h_{yi}^2$	$A_i h_{zi}^2$	$A_i h_{yi} h_{zi}$
Top flange	12×138	1656	144	75	0.02×10^6	2.63×10^6	34.34×10^6	9.32×10^6	17.89×10^6
Web	12×300	3600	0	0	27.0×10^6	0.04×10^6	0	0	0
Bottom flange	12×138	1656	-144	-75	0.02×10^6	2.63×10^6	34.34×10^6	9.32×10^6	17.89×10^6
Σ		6912			27.04×10^6	5.30×10^6	68.68×10^6	18.64×10^6	35.78×10^6

$I_{zz} = (27.04 + 68.68) \times 10^6 = 95.72 \times 10^6 \text{ mm}^4$
$I_{yy} = (5.30 + 18.64) \times 10^6 = 23.94 \times 10^6 \text{ mm}^4$
$I_{yz} = 35.78 \times 10^6 \text{ mm}^4$

(b) Calculation of applied 'forces' on the section. Since the loads are applied in the x–y plane, $M_{yy} = 0$ and $M_{zz} = qL^2/8$ at midspan, where $L = 5$ m and $q = 20$ kN m^{-2}. Therefore $M_{zz} = 20 \times 5^2/8 = 62.5$ kN m $= 62.5 \times 10^6$ N mm.

Substituting for the section properties and the applied forces in the equation, we have

$$\sigma_x = -\{(M_z I_{yy} + M_y I_{yz})/(I_{yy}I_{zz} - I_{yz}^2)\}y + \{(M_z I_{yz} + M_y I_{zz})/(I_{yy}I_{zz} - I_{yz}^2)\}z$$

$$\sigma_x = -1.48y + 2.21z$$

The equation to neutral axis is given by

$$\sigma_x = 0, \text{ i.e. } -1.48y + 2.21z = 0$$

$$y = 1.49z, \tan\theta = 1.49, \theta = 56.3° \text{ to the } z\text{-axis.}$$

Note that the neutral axis is inclined to the z-axis. In other words, although the applied force is only M_z, the section bends about the z- *as well as about the y-axis*.

The stress σ_x at the upper flange tip for which $y = 150$, $z = 144$, is given by

$$\sigma_x = -1.48 \times 150 + 2.21 \times 144 = 96.3 \text{ N mm}^{-2}$$

The upper flange tip is in tension, which is quite contrary to what one would expect from using inappropriately the equation $\sigma_x = -M_z y/I_{zz}$ applicable for symmetric bending.

Similarly, the stress at the top of the web for which $y = 150$, $z = -6$ is given by

$$\sigma_x = -1.48 \times 150 + 2.22 \times (-6) = -235.3 \text{ N mm}^{-2}$$

This is the maximum compressive stress at the section. It is instructive to compare this with the value given by

$$\sigma_x = -M_z y/I_{zz} = -(62.5 \times 10^6)(150)/(95.7 \times 10^6)$$

$$= -97.9 \text{ N mm}^{-2}$$

In other words, if the equation derived for the symmetric bending is applied to sections undergoing unsymmetrical bending, then very large errors result. In the specific case of the z-section considered here, the ratio of the exact to inexact stress is $-235.3/(-97.9) = 2.40$, an unacceptably large error.

Example 2

The angle section shown in Fig. 3.35 is used as a simply supported beam of 2 m span to carry a uniformly distributed load of q kN m^{-1}. Calculate the permissible load q in order to limit the maximum stress (tensile or compressive) to 200 N mm^{-2}.

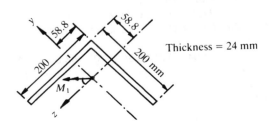

Thickness = 24 mm

Figure 3.35 ● An angle section bent about a horizontal axis

Solution (a) Calculate the section properties. For convenience, take the co-ordinate axes y and z as parallel to the sides of the angle section.

(b) Calculation of centroid. Taking moments about the top flange

Table 3.9

Section	$b \times d$	A_i	y_i	$A_i y_i$
Top flange	176 × 24	4.22 × 10³	12	0.51 × 10⁵
Web	200 × 24	4.80 × 10³	100	4.80 × 10⁵
		Σ9.02 × 10³		Σ5.31 × 10⁵

$$\bar{y} = 5.31 \times 10^5/(9.02 \times 10^3) = 58.81 \text{ mm}$$

Because of symmetry, $\bar{z} = 58.81$ mm

(c) Calculation of second moments of area. From Table 3.10

$$I_{zz} = (16.20 + 17.40) \times 10^6 = 33.60 \times 10^6$$

$$I_{yy} = (11.13 + 22.17) \times 10^6 = 33.30 \times 10^6$$

$$I_{yz} = 19.78 \times 10^6$$

Table 3.10

Section	$b \times d$	A_i ($\times 10^3$)	h_{yi}	h_{zi}	I_{zzi} ($\times 10^6$)	I_{yyi} ($\times 10^6$)	$A_i h_{yi}^2$ ($\times 10^6$)	$A_i h_{zi}^2$ ($\times 10^6$)	$A_i h_{yi} h_{zi}$ ($\times 10^6$)
Flange	24 × 176	4.22	46.81	53.19	0.20	10.90	9.26	11.95	10.52
Web	24 × 200	4.80	−41.19	−46.81	16.0	0.23	8.14	10.52	9.26
Σ					16.20	11.13	17.40	22.17	19.78

$I_{zz} = (16.20 + 17.40) \times 10^6 = 33.60 \times 10^6$
$I_{yy} = (11.13 + 22.17) \times 10^6 = 33.30 \times 10^6$
$I_{yz} = 19.78 \times 10^6$

(d) Calculation of forces on the section. The applied bending moment is about the horizontal axis. The moment $M_1 = q2^2/8 = 0.5q$ kN m $= 0.5q \times 10^6$ N mm. Its components along the zz and yy axes are $M_z = M_1 \cos 45 = 0.354q \times 10^6$ N mm and $M_y = M_1 \sin 45 = 0.354q \times 10^6$ N mm.

(e) Calculate σ_x. Using the formula

$$\sigma_x = -\{(M_z I_{yy} + M_y I_{yz})/(I_{yy}I_{zz} - I_{yz}^2)\}y + \{(M_z I_{yz} + M_y I_{zz})/(I_{yy}I_{zz} - I_{yz}^2)\}z$$

$$\sigma_x = (-y + z)2.562 \times 10^{-2}q \text{ N mm}^{-2}$$

The maximum stress is at the tips of the flange where $y = 58.81$, $z = 141.2$. This gives $\sigma_x = 2.11q$. If $\sigma_x = 200$ N mm^{-2}, then $q = 200/2.11 = 94.8$ kN m^{-1}.

3.12.5 Principal axes of a section

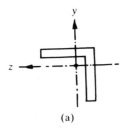

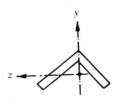

Figure 3.36 ● (a), (b) An equal angle section in two different orientations

The formula for the bending stress in the case of unsymmetric bending shows that if $I_{yz} = 0$, then the formula reduces to the case of symmetric bending. Figure 3.36 shows an equal angle section. The angle is orientated in two different ways. In Fig. 3.36(a), neither the y- nor the z-axis is an axis of symmetry and therefore $I_{yz} \neq 0$. Therefore, for this orientation, the bending stress has to be calculated using the formula for unsymmetric bending. On the other hand, for the orientation in Fig. 3.36(b), the y-axis is an axis of symmetry and therefore $I_{yz} = 0$ and σ_x can be calculated using the formula for symmetry bending. This example prompts one to ask the question whether every section possesses a set of orthogonal axes for which the cross second moment of area is zero. This can be answered by studying how the cross second moment of area changes for different orientation of axes in the cross-section.

As shown in Fig. 3.37, let (y, z) and (y_1, z_1) represent two sets of orthogonal axes such that (y_1, z_1) is obtained by rotating the (y, z) set about the x-axis by an angle α in the clockwise direction. The co-ordinates of any point in the two systems are related by the equations

$$y_1 = y \cos \alpha - z \sin \alpha, \quad z_1 = y \sin \alpha + z \cos \alpha$$

so that

$$I_{z_1 z_1} = \int y_1^2 \, dA = \int (y^2 \cos^2 \alpha + z^2 \sin^2 \alpha - 2yz \cos \alpha \sin \alpha) \, dA$$

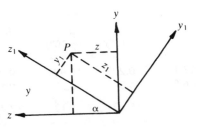

Figure 3.37 ● Rotation of axes

Therefore

$$I_{z_1 z_1} = \cos^2 \alpha \int y^2 \, dA + \sin^2 \alpha \int z^2 \, dA - 2\cos \alpha \sin \alpha \int yz \, dA$$

$$= \cos^2 \alpha \, I_{zz} + \sin^2 \alpha \, I_{yy} - 2\cos \alpha \sin \alpha \, I_{yz}$$

$$= (I_{zz} + I_{yy})/2 + \{(I_{zz} - I_{yy})/2\} \cos 2\alpha - I_{yz} \sin 2\alpha$$

$$I_{y_1 y_1} = \int z_1^2 \, dA = \int (y^2 \sin^2 \alpha + z^2 \cos^2 \alpha + 2yz \cos \alpha \sin \alpha) \, dA$$

Therefore

$$I_{y_1 y_1} = \sin^2 \alpha \int y^2 \, dA + \cos^2 \alpha \int z^2 \, dA + 2\cos \alpha \sin \alpha \int yz \, dA$$

$$= \sin^2 \alpha \, I_{zz} + \cos^2 \alpha \, I_{yy} + 2\cos \alpha \sin \alpha \, I_{yz}$$

$$= (I_{zz} + I_{yy})/2 - \{(I_{zz} - I_{yy})/2\} \cos 2\alpha + I_{yz} \sin 2\alpha$$

$$I_{y_1 z_1} = \int y_1 z_1 \, dA = \int \{y^2 \cos \alpha \sin \alpha + z^2 \cos \alpha \sin \alpha + yz(\cos^2 \alpha - \sin^2 \alpha)\} \, dA$$

Therefore

$$I_{y_1 z_1} = (I_{zz} - I_{yy})\cos \alpha \sin \alpha + I_{yz}(\cos^2 \alpha - \sin^2 \alpha)$$
$$= \{(I_{zz} - I_{yy})/2\}\sin 2\alpha + I_{yz} \cos 2\alpha$$

Summarizing, we have

$$I_{z_1 z_1} = (I_{zz} + I_{yy})/2 + \{(I_{zz} - I_{yy})/2\} \cos 2\alpha - I_{yz} \sin 2\alpha$$
$$I_{y_1 y_1} = (I_{zz} + I_{yy})/2 - \{(I_{zz} - I_{yy})/2\} \cos 2\alpha + I_{yz} \sin 2\alpha$$
$$I_{y_1 z_1} = \{(I_{zz} - I_{yy})/2\} \sin 2\alpha + I_{yz} \cos 2\alpha$$

Naturally $I_{y_1 z_1} = 0$, when $\tan 2\alpha = -2I_{yz}/(I_{zz} - I_{yy})$. In other words, there is always a certain orientation for which $I_{y_1 z_1} = 0$ irrespective of whether the section has an axis of symmetry or not. It can be further noticed that for maximum or minimum $I_{z_1 z_1}$

$$d(I_{z_1 z_1})/d\alpha = +(I_{zz} - I_{yy}) \sin 2\alpha + 2I_{yz} \cos 2\alpha = 0$$

i.e. $\tan 2\alpha = -2I_{yz}/(I_{zz} - I_{yy})$

In other words, the orientation for which $I_{y_1 z_1} = 0$ is also the orientation for which $I_{z_1 z_1}$ is stationary, i.e. a maximum or a minimum. The maximum and minimum second moments of areas are called principal second moments of area and the corresponding axes principal axes of the section. Substituting for $\tan 2\alpha = -2I_{yz}/(I_{zz} - I_{yy})$, i.e. $\sin 2\alpha = -I_{yz}/D$, $\cos 2\alpha = 0.5(I_{zz} - I_{yy})/D$ where $D = \sqrt{[\{(I_{zz} - I_{yy})/2\}^2 + I_{yz}^2]}$, in the expressions for $I_{z_1 z_1}$ and $I_{y_1 y_1}$, the maximum and minimum second moments of areas are given by

$$I_{11} = (I_{zz} + I_{yy})/2 + \sqrt{[\{(I_{zz} - I_{yy})/2\}^2 + I_{yz}^2]}$$
$$I_{22} = (I_{zz} + I_{yy})/2 - \sqrt{[\{(I_{zz} - I_{yy})/2\}^2 + I_{yz}^2]}$$

The above expressions will not be used for calculating the bending stress because they are fairly inconvenient to use. However, the concept of the minimum second moment of area is of importance in the study of instability of columns which will be discussed in Chapter 8.

3.13 ● Combined force actions – principle of superposition

In the previous sections, equations were derived for the calculation of the stresses in a section due to axial force, bending moment and shear force. In a general design situation, it is necessary to calculate the stresses induced not by a single force action but by a combination of several forces. In the case of elastic structures which respond linearly to the application of forces, a useful principle known as the principle of superposition can be used to calculate the response to combined forces as simply as the algebraic/vectorial addition of individual forces. As a simple example consider a structure subjected to forces W_1 and W_2. It is assumed that the response of the structure (for example, the displacement or stress at a point) to the individual loads is linear, as shown in Fig. 3.38(a) and (b). If the loads W_1 and W_2 are applied sequentially, the corresponding response will be as shown in Fig. 3.39(c). In fact in the particular class of structures considered here – linearly elastic structures – the

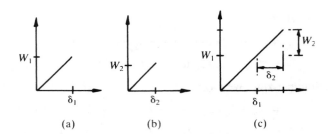

Figure 3.38 ● (a)–(c)
Illustration of the
principle of
superposition

(a)   (b)   (c)

response is governed purely by the final value of the loads on the structure. The order of load application – whether the loads are applied simultaneously or sequentially or in some other manner – is irrelevant.

3.14 ● Equivalent force resultants due to an eccentric axial force

Consider the column shown in Fig. 3.39 subjected to an eccentric axial force W with an eccentricity of e_y with respect to the z-axis passing through the centroid. Let the state of stress induced at any point in the beam be σ_x. Calculating the stress resultants about the centroid, we have

$$\int \sigma_x \, dA = \text{axial force } F_x$$

$$\int \sigma_x y \, dA = \text{couple } (-M_z)$$

For equilibrium, the stress resultants must balance the applied forces. Therefore $F_x = W$ and $M_z = -We_y$. In other words, *a force W eccentric with respect to the centroid can be considered as equivalent to an axial force equal to the value of the eccentric force plus a moment about the centroidal axes equal to the value of the eccentric force multiplied by the eccentricity with respect to the particular axes of bending.*

Assuming that plane sections remain plane, at least at sections 'far' removed from the point of application of concentrated forces, then

Due to axial force W, $\sigma_x = W/A$, where A = area of cross-section

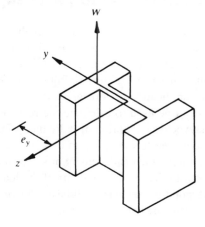

Figure 3.39 ● An *I*
section under the action
of an eccentric axial
load

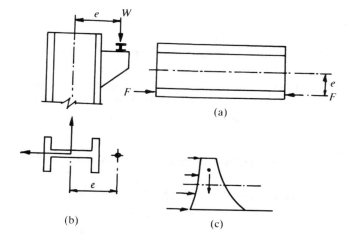

Figure 3.40 ●
Examples of sections subjected to eccentric axial load: (a) a pre-stressed concrete beam, (b) a column carrying an external load on a bracket, (c) a gravity dam with varying width

(a)

(b)

(c)

Due to bending moment $M_z = -We_y$, $\sigma_x = -(M_z/I_{zz})y = (We_y/I_{zz})y$
The total stress is therefore equal to

$$\sigma_x = \frac{W}{A} + \frac{We_y}{I_{zz}}y = \frac{W}{A}\left[1 + \frac{e_y A y}{I_{zz}}\right]$$

Similarly, if the load is eccentric with respect to both the y- and the z-axes, then the beam will be under an additional bending moment of $M_y = We_z$. Therefore

$$\sigma_x = \frac{W}{A} + \frac{We_y}{I_{zz}}y + \frac{We_z}{I_{yy}}z$$

The load situation considered, namely the application of an eccentric axial force, occurs very commonly in the design of structures. For example, the application of an eccentric force near the bottom face of a beam as in Fig. 3.40(a) causes the beam to curve upwards, inducing tension at the top face and compression at the bottom face. This has the beneficial effect of compensating for the compression at the top face and tension at the bottom face caused by the gravity forces. This effect is made use of in prestressed concrete structures by applying an eccentric axial force using prestressing tendons. Another common situation is the eccentric force on a column due to the action of a crane resting on a bracket welded to the column as shown in Fig. 3.40(b). Other examples include the gravity dam such as the one shown in Fig. 3.40(c), where at any horizontal section the resultant gravity forces due to the weight of the structure above the section acts eccentrically with respect to the centroid of the section.

3.14.1 Stress distribution due to combined axial load and bending moment

Because both the axial force and the bending moment induce a normal stress σ along the axis of the member, the stresses can be algebraically added. The following examples illustrate the procedure.

Example 1

The built-up column section shown in Fig. 3.20(a) is used as a short column. It is subjected to an axial load of 1200 kN and is also subjected to bending moment about the zz-axis of 170 kN m inducing tension in the bottom face and to bending moment about the yy-axis of 50 kN m inducing tension to the left. Calculate the maximum tensile and compressive stresses in the cross-section. The cross-sectional properties are:

$$\text{Area } A = 18.96 \times 10^3 \text{ mm}^2, \ I_{zz} = 345.0 \times 10^6 \text{ mm}^4, \ I_{yy} = 130.1 \times 10^6 \text{ mm}^4$$

Solution (a) Axial force $P = 1200$ kN: This force induces a uniform compression in the cross-section. The stress is given by

$$\sigma = -P/A = -1200 \times 10^3/(18.96 \times 10^3) = -63.3 \text{ N mm}^{-2}$$

(b) Bending moment $M_{zz} = 170$ kN m: The bending moment induces a maximum tensile stress in the bottom fibres and a maximum compressive stress in the top fibres. The value of the stress is given by

$$\sigma = \pm(M_{zz}/I_{zz}) \ (y = 328.8/2) = \pm\{170 \times 10^6/(345.0 \times 10^6)\} \times 164.4$$

$$= \pm81.0 \text{ N mm}^{-2}$$

Note: The total depth of the section
        = 304.8 (depth of channel) + 2 × 12 (plate thickness) = 328.8 mm
(c) Bending moment $M_{zz} = 50$ kN m: The bending moment induces a maximum tensile stress in the flanges to the left and a maximum compressive stress in the flanges to the right. The value of the stress is given by

$$\sigma = \pm(M_{yy}/I_{yy}) \ (z = 300.0/2) = \pm\{50 \times 10^6/(130.1 \times 10^6)\} \times 150.0$$

$$= \pm57.6 \text{ N mm}^{-2}$$

Note: The total width of plates = 300 mm.

(i)   Top flange, top right-hand tip:
      $\sigma = -63.3$ due to $P - 81.0$ due to $M_{zz} - 57.6$ due to M_{yy}
      $= -201.9 \text{ N mm}^{-2}$
(ii) Top flange, top left-hand tip:
      $\sigma = -63.3$ due to $P - 81.0$ due to $M_{zz} + 57.6$ due to $M_{yy} = -86.7 \text{ N mm}^{-2}$
(iii) Bottom flange, bottom right-hand tip:
      $\sigma = -63.3$ due to $P + 81.0$ due to $M_{zz} - 57.6$ due to $M_{yy} = -39.9 \text{ N mm}^{-2}$
(iv) Bottom flange, bottom left-hand tip:
      $\sigma = -63.3$ due to $P + 81.0$ due to $M_{zz} + 57.6$ due to $M_{yy} = 75.3 \text{ N mm}^{-2}$
      (tensile)

The maximum tensile stress of 75.3 N mm^{-2} occurs in the bottom flange at the left-hand tip and the maximum compressive stress of 201.9 N mm^{-2} occurs in the top flange at the right-hand tip.

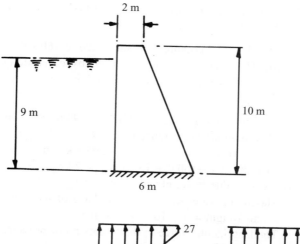

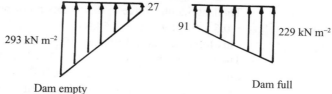

293 kN m⁻² 27

Dam empty

91 229 kN m⁻²

Dam full

Figure 3.41 ● A small concrete dam

Example 2

Figure 3.41 shows the cross-section of a small concrete dam. Calculate the maximum stresses at the base of the dam due to:

(a) the self-weight of the dam
(b) pressure of water on the vertical face.

Assume the height of water, when the dam is full, is 9 m.
Take the unit weight of concrete as 24 kN m^{-3} and the unit weight of water as 10 kN m^{-3}.

Solution (a) Stresses due to self-weight. Consider 1 m length of the dam. The cross-sectional area of the dam A is

A = average width × height
Average width = $0.5(2 + 6) = 4$ m
Height = 10 m
Area $A = 4 \times 10 = 40$ m^2
Self-weight = volume × unit weight = $\{40 \times 1\} \times 24 = 960$ kN
The centroid of the weight is at a distance of $\bar{z}$ from the vertical face.
$\bar{z} = [10 \times 1 \times 2 \times (2/2)$ for the rectangular portion $+ 0.5 \times 10 \times 1 \times (6 - 2)$
$\times \{2 + (6 - 2)/3\}$ for the triangular portion]$/A = 2.167$ m
Considering the base of the dam, the distance between the centroid of the base and the centroid of the load is given by
$e = (6/2) - 2.167 = 0.833$ m to the left of the centroid.

Thus, at the base, there is an axial load of 960 kN acting at an eccentricity of 0.833 m.

(i) The axial compressive stress due to 960 kN = $960/(6 \times 1)$ = 160 kN m^{-2}

(ii) The bending moment induced by eccentricity is M_{zz} = 960 kN × 0.833 = 799.68 kN m. I_{zz} of the section = $1 \times 6^3/12$ = 18 m^4, section modulus $Z_{zz} = 1 \times 6^2/6 = 6$ m^3

The stress induced = M_{zz}/Z_{zz} = 799.68 × 3/6 = 133.3 N m^{-2}

Maximum stresses induced are $(-P/A \pm M_{zz}/Z_{zz})$

At left, σ = −160.0 − 133.3 = −293.3 kN m^{-2}

At right, σ = −160.0 + 133.3 = −26.7 kN m^{-2}

(b) Pressure due to weight of water:

Maximum water pressure at the base of the dam = Depth below water level × unit weight = 9 × 10 = 90 kN m^{-2}

Total horizontal force = 0.5 × maximum pressure × depth of water × length = 0.5 × 90 × 9 × 1 = 405 kN

This force acts at a distance of 9/3 = 3 m from the base.

The resultant moment M = 405 × 3 = 1215 kN m

Stresses due to water pressure = $\pm M/Z_{zz}$ = ±1215/6 = ±202.5 kN m^{-2}

This induces tension on the left side (upstream face) and compression on the right side (downstream face).

(c) Net stresses:

When the dam is empty, only stresses due to self weight exist.

At left, σ = −293.3 kN m^{-2}

At right, σ = −26.7 kN m^{-2}

When the dam is full, total stresses are the sum of stresses due to self weight and stresses due to water pressure.

At left, σ = (−293.3 + 202.5) = −90.8 kN m^{-2}

At right, σ = (−26.7 − 202.5) = −229.2 kN m^{-2}

3.14.2 Middle third rule

In the previous section it was shown that the stress at a section due to the combined action of an axial force and a bending moment is given by

$$\sigma_x = W/A + (We_y/I_{zz})y = (W/A)\{1 + (Ay/I_{zz})e_y\}$$

Masonry structures for example are very weak in tension. One common form of loading is an eccentric axial compressive load. It is therefore usual to design such structures such that the stresses due to axial load and bending moment are never tensile. If σ_x is not to be tensile, then $-W/A(1 - e_y Ay/I_{zz})$ must be always negative. For a rectangular section for which A = BD and I_{zz} = BD^3/12, σ_x is a maximum tension when y = 0.5D. Therefore Ay/I_{zz} = 6/D. Therefore for σ_x not to be tensile

$$-W/A[1 - e_y 6/D] \leqslant 0$$

$$1 - e_y 6/D) \geqslant 0$$

$$e_y \leqslant D/6$$

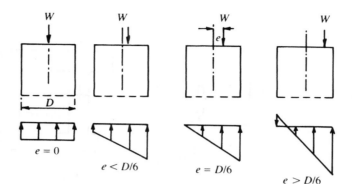

Figure 3.42 ● Stress distribution under varying eccentricity

In other words, if the eccentricity of the axial load is kept within the middle third of the depth, then tension will not develop in the cross-section. This is commonly known as the middle third rule. Figure 3.42 shows typical stress distributions when the eccentricity of the force is varied. In the case of non-rectangular sections, the limit within which the eccentric force must lie in order that no tension develops is easily established from the equation $1 - e_y(Ay/I) \geqslant 0$.

3.14.3 Core of a section

In the last section it was shown that in the case of a rectangular section, as long as the eccentric axial load moving along an axis of symmetry lies within the middle third region, then no tension develops in the section. Similar expressions can be derived for sections other than rectangular and in fact establish the region with which an eccentric axial load, which is eccentric with respect to both the principal axes about which the beam bends, should lie in order that no tension is introduced in the cross-section. This region is generally known as the CORE of the section. The following examples demonstrate the procedure for calculating the core of the section.

Example 1

For the unsymmetrical *I*-section shown in Fig. 3.43(a), determine the region in which an eccentric axial load should lie in order to induce no tension anywhere in the cross-section. The dimensions of the cross-section are:

Top flange: 150 mm wide × 50 mm thick
Bottom flange: 100 mm wide × 50 mm thick
Web: 200 mm deep × 50 mm thick
Overall depth = 300 mm
The section properties as calculated previously are:
Area of cross-section A = 150 × 50 top flange + 100 × 50 bottom flange
+ 200 × 50 web
= 22.5×10^3 mm^2
Second moment of area $I_{zz} = 225.9 \times 10^6$ mm^4
Second moment of area $I_{yy} = 20.3 \times 10^6$ mm^4
Position of centroid $\bar{y}$ = 136.1 mm from top or 163.9 mm from below.

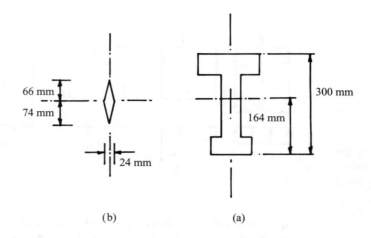

(b)   (a)

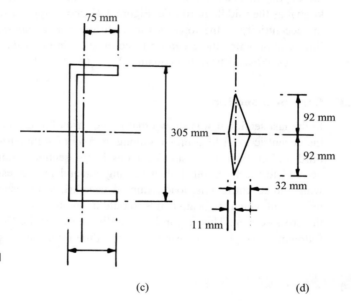

Figure 3.43 ●
(a)–(d) Core of an I and
channel section used in
a column

(c)   (d)

Solution **(a)** Consider the eccentricity about the zz-axis. Let the load P be above the centroidal zz-axis at an eccentricity of $e_{y\ top}$. Tension will then be induced in the bottom flange. The maximum tensile stress is given by:

$$\sigma = -P/A + (P\ e_{y\ top}/I_{zz})\ (y = 163.9)$$

For σ to be zero, $e_{y\ top} = I_{zz}/(A\ .\ 163.9) = 61$ mm

(b) Consider the eccentricity about the zz-axis. Let the load P be below the centroidal zz-axis at an eccentricity of $e_{y\ bot}$. Tension will then be induced in the top flange. The maximum tensile stress is given by:

$$\sigma = -P/A + (P\ e_{y\ bot}//I_{zz})\ (y = 136.1)$$

For σ to be zero, $e_{y\ bot} = I_{zz}/(A\ .\ 136.1) = 74$ mm

(c) Consider the eccentricity about the yy-axis. Let the load P be eccentric about the centroidal yy- axis at an eccentricity of e_z. Tension will then be induced in the top flange to the left if the load is to the right and vice versa. The maximum tensile stress is given by:

$$\sigma = -P/A + (P\ e_{z\ \text{left}}/I_{yy})\ (z = 150/2)$$

For σ to be zero, $e_{z\ \text{left}} = e_{z\ \text{right}} = I_{yy}/(A\ .\ 75.0) = 12$ mm
Thus the above calculations give the maximum eccentricities possible without inducing any tension in section when the load is eccentric about any one axis. If the load is acting eccentrically by e_y about the zz-axis and by e_z about the yy-axis, then say for tension not to be induced in the top flange, the condition is that, if the load is above the zz-axis and to the left of the yy-axis, then

$$\sigma = -P/A + (P\ e_y/I_{zz})\ (y = 136.1) + (P\ e_z/I_{yy})\ (y = 150/2) = 0$$

Therefore, dividing throughout by (P/A) and simplifying,

$$1 = (A\ e_y/(I_{zz})\ 136.1 + (A\ e_z/(I_{yy})75.0$$

but $I_{zz}/(A \times 136.1) = e_{y\ \text{bot}}$ and $I_{yy}/(A \times 75.0) = e_{z\ \text{left}}$
Therefore

$$e_y/e_{y\ \text{bot}} + e_z/e_{z\ \text{left}} = 1$$

This is the equation for a straight line connecting the variables e_z and e_y. Similar equations can be established for other combinations of load position. Thus the region in which an eccentric load should lie without inducing tension in the cross-section is diamond-shaped as shown in Fig. 3.43(b).

Example 2

For the symmetrical channel section shown in Fig. 3.43(c), determine the region in which an eccentric axial load should lie in order to induce no tension anywhere in the cross-section.
The section properties are:
Flanges: 101.6 mm wide × 14.8 mm thick
Web = 10.2 mm thick
Overall depth = 304.8 mm
Area of cross-section $A = 5.88 \times 10^3$ mm^2
Second moment of area $I_{zz} = 82.14 \times 10^6$ mm^4
Second moment of area $I_{yy} = 4.995 \times 10^6$ mm^4
Position of centroid $\bar{y} = 152.4$ mm from top or bottom and $\bar{z}$ at 75 mm from the tips of the flange.

(a) Consider the eccentricity about the zz-axis. Let the load P be above the centroidal zz-axis at an eccentricity of $e_{y\ \text{top}}$. Tension will then be induced in the bottom flange. The maximum tensile stress is given by:

$$\sigma = -P/A + (P\,e_{y\,top}/I_{zz})\;(y = 152.4)$$

For σ to be zero, $e_{ytop} = I_{zz}/(A\,.\,152.4) = 92$ mm
Because of symmetry about the zz-axis, $e_{y\,top} = e_{y\,bottom}$
(b) Consider the eccentricity about the yy-axis. Let the load P be eccentric about
the centroidal yy-axis at an eccentricity of $e_{z\,left}$. Tension will then be induced
in the top flange to the right. The maximum tensile stress is given by:

$$\sigma = -P/A + (P\,e_{z\,left}/I_{yy})\;(z = 75)$$

$e_{z\,left} = I_{yy}/(A \times 75) = 11$ mm
(c) Consider the eccentricity about the yy-axis. Let the load P be eccentric about
the centroidal yy-axis at an eccentricity of $e_{z\,right}$. Tension will then be induced
in the web at left. The maximum tensile stress is given by:

$$\sigma = -P/A + (P\,e_{z\,right}/I_{yy})\;(z = (101.6 - 75.0) = 26.6)$$

$e_{z\,right} = I_{yy}/(A \times 26.6) = 32$ mm

Figure 3.43(d) shows the region in which an eccentric load should lie in order not
to induce any tension in the cross-section.

Example 3

A symmetrical I-section is used as a short column. It is subjected to an eccentric
axial load of 3000 kN. The eccentricities are 100 mm along the y-axis above the
horizontal centroidal axis and 20 mm along the z-axis to the right. Calculate the maximum tensile and compressive stresses in the cross-section. Also calculate the 'core
of the section'.
Given: Flanges: 300×50 mm, Thickness of web $= 50$ mm, Overall depth $= 600$ mm

Solution Cross-sectional properties:
Area $A = 300 \times 600 - (300 - 50) \times (600 - 50 - 50)$
$\qquad = 300 \times 600 - 250 \times 500 = 55 \times 10^3$ mm^2
$I_{zz} = 300 \times 600^3/12 - 250 \times 500^3/12 = 2795.83 \times 10^6$ mm^4
$I_{yy} = 2\{300^3 \times 50/12\} + 50^3 \times 500/12 = 230.21 \times 10^6$ mm^4
Forces: $P = 3000$ kN, $M_{zz} = P\,e_y = 3000 \times 100 \times 10^{-3} = 300$ kN m,
$M_{yy} = P\,e_z = 3000 \times 20 \times 10^{-3} = 60$ kN m,
Stresses:
$-P/A = -3000 \times 10^3/(55 \times 10^3) = -54.6$ N mm^{-2}
$\pm(M_{zz}/I_{zz})y = \pm\{300 \times 10^6/(2795.83 \times 10^6)\}(600/2) = \pm32.2$ N mm^{-2}
$\pm(M_{yy}/I_{yy})z = \pm\{60 \times 10^6/(230.21 \times 10^6)\}(300/2) = \pm39.1$ N mm^{-2}
At top right corner $\sigma = -54.6 - 32.2 - 39.1 = -125.9$ N mm^{-2}
At top left corner $\sigma = -54.6 - 32.2 + 39.1 = -47.7$ N mm^{-2}
At bottom right corner $\sigma = -54.6 + 32.2 - 39.1 = -61.5$ N mm^{-2}
At bottom left corner $\sigma = -54.6 + 32.2 + 39.1 = +16.7$ N mm^{-2}
Maximum stresses are 125.9 N mm^{-2} compressive and 16.7 N mm^{-2} tensile.

Core of the section:

(a) Assuming that the eccentricity exists about the zz-axis only, let the load P be eccentric about the centroidal zz-axis at an eccentricity of $e_{y\ top}$. Tension will then be induced in the bottom flange. The maximum tensile stress is given by

$$\sigma = -P/A + (P\ e_{y\ top}/I_{zz})\ (y = 600/2)$$

Because of symmetry about the zz-axis, $e_{y\ top} = e_{y\ bottom}$

For σ to be zero, $e_{y\ top} = e_{y\ bot} = I_{zz}/(A\ .\ 300) = 169.4$ mm

(b) Consider the eccentricity about the yy-axis. Let the load P be eccentric about the centroidal yy-axis at an eccentricity of $e_{z\ left}$. Tension will then be induced in the top flange to the right. The maximum tensile stress is given by:

$$\sigma = -P/A + (P\ e_{z\ left}/I_{yy})\ (z = 300/2)$$

Because of symmetry about the yy-axis, $e_{z\ left} = e_{z\ right}$

$e_{z\ left} = e_{z\ right} = I_{yy}/(A \times 150) = 27.9$ mm

The shape of the core is therefore a doubly symmetric diamond with a height equal to $2 \times 169.4 = 338.8$ mm and width equal to $2 \times 27.9 = 55.8$ mm.

3.15 ● Equivalent force resultants due to an eccentric shear force

Consider a simple cantilever subjected to an end load as shown in Fig. 3.44(a). If the section has an axis of symmetry, then evidently if the load is parallel to the axis of symmetry and passes through the centroid as in Fig. 3.44(b), then the shear stress in the cross-section is symmetrically distributed with respect to the axis of symmetry and the section does not twist.

Now consider the case of a channel section shown in Fig. 3.44(c) with the load applied normal to the axis of symmetry. The shear stress distribution in the web and flanges will be as shown in Fig. 3.44(c). Obviously the shear flows in the flanges are such that their resultant force in the horizontal direction is equal to zero.

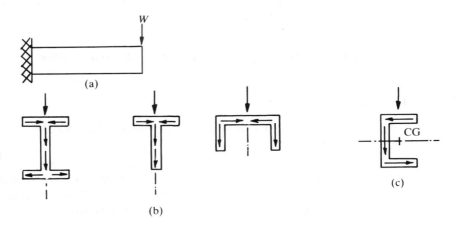

Figure 3.44 ●
A cantilever beam of various cross-sections subjected to an end load: (a) cantilever beam, (b) cross-sections with a vertical axis of symmetry, (c) cross-section unsymmetrical about the vertical axis

However, they do produce a resultant torque equal to the shear force in the flange multiplied by the distance between the centre line of the flanges. If the section is not to twist, then the shear force in the web together with the applied force must produce a torque of an opposite nature to that produced by the shear force in the flanges. In other words, in this particular case, for the section not to twist, then the applied load must be displaced to the left with respect to the centroid as shown in Fig. 3.45. The point through which the applied load must pass through for the section not to twist is called the shear centre.

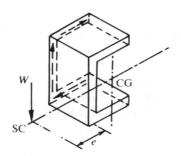

Figure 3.45 ●
A channel section
with shear centre
and centroid shown

From the above qualitative description of the existence of the shear centre, we can deduce a few simple principles for the location of the shear centre.

1. In the case of sections with two axes of symmetry, the shear centre and centroid of the section coincide.
2. In the case of sections with one axis of symmetry, the shear centre lies on the axis of symmetry.
3. For sections such as angle and T sections shown in Fig. 3.46, which can be divided into a series of individual rectangles with their centre lines meeting at a point, the shear centre is the point of intersection of the centre lines.

This discussion leads to the conclusion that if the applied shear force at a section is 'eccentric' with respect to the shear centre, then the applied shear force is equivalent to a shear force through the shear centre plus a torque equal to the applied force times the distance between the point of application of the shear force and the shear centre of the section.

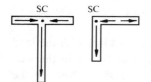

Figure 3.46 ● Shear
centre position for T and
angle sections

3.15.1 Examples of the calculation of shear centre

Example 1

Calculate the position of the shear centre for the channel section shown in Fig. 3.47(a).

Solution   Assuming that the channel section is used as a cantilever beam with a concentrated load W at the tip, then the shear flow in the flange, as shown in Fig. 3.46(b) and as explained in Section 3.10.1, Example 3, is given by

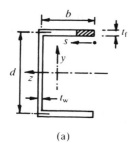

(a)

(b)

Figure 3.47 ●
(a), (b) Calculation of
the position of shear
centre for a channel
section

$$\tau_{xz}t_f = QA\bar{y}/I_{zz}$$

where $A\bar{y} = st_f(0.5d)$ and s is measured from the tip of the flange and shear force $Q = W$. The shear flow in the flange is a linear function of s. The total shear force Q_f in the flange is therefore

$$Q_f = \int \tau_{xz}t_f\,ds = (W/I_{zz})0.5dt_f \int s\,ds$$

where the limits for s are 0 and b. Therefore

$$Q_f = (W/I_{zz})0.25b^2dt_f$$

The torque produced by the shear forces in the flanges is given by

$$Q_f d = (W/I_{zz})0.25b^2d^2t_f$$

The torque produced by having the shear force W acting eccentrically with respect to the web is We, where e = distance of shear centre from the centre line of the web. Therefore for equilibrium $We = Q_f d$. This gives

$$e = 0.25b^2d^2t_f/I_{zz}$$

$I_{zz} = 2\{bt_f^3/12 + bt_f\,(0.5d)^2\}$ for the two flanges $+ t_w d^3/12$ for the web. Since the contribution to I_{zz} is mainly from the flanges, ignoring the second moment of area of the flanges about their own centroidal axis, i.e. $bt_f^3/12$ and the second moment of area of the web, I_{zz} is approximately given by $I_{zz} \approx 2(bt_f)(0.5d)^2 = 0.5bt_f d^2$. Using this approximate value of I_{zz}, $e \approx 0.5b$.

Example 2

Calculate the position of the shear centre for the z-section shown in Fig. 3.48.

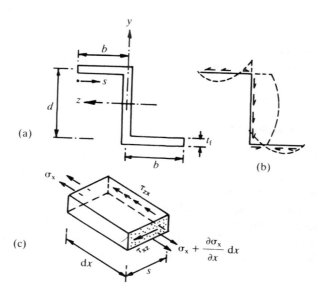

(a)

(b)

(c)

Figure 3.48 ● (a)–(c)
Calculation of the
position of the shear
centre in a z-section

Solution Considering the z-section used as a cantilever beam fixed at $x = 0$ and loaded at $x = L$ by a concentrated load W, then $M_z = -W(L - x)$ and $M_y = 0$. Note that $dM_z/dx = W$. Using the unsymmetrical bending formula, as shown in Section 3.12.1, the bending stress σ_x is given by

$$\sigma_x = M_z(-yI_{yy} + zI_{yz})/(I_{yy}I_{zz} - I_{yz}^2)$$

The difference in σ_x between two neighbouring sections dx apart is

$$(\partial\sigma_x/\partial x)\,dx = \{(dM_z/dx)(-yI_{yy} + zI_{yz})/(I_{yy}I_{zz} - I_{yz}^2)\}$$

Substituting for $dM_z/dx = W$,

$$(\partial\sigma_x/\partial x)\,dx = \{W(-yI_{yy} + zI_{yz})/(I_{yy}I_{zz} - I_{yz}^2)\}$$

Consider the upper flange, for which $y = 0.5d$ and $z = (b - s)$, where the origin for s is at the tip of the flange. Considering the equilibrium of the free body of the flange shown in Fig. 3.48(c),

$$\int \{(\partial\sigma_x/\partial x)\,dx\}(t_f\,ds) = dx\; t_f\,\tau_{xz}$$

Substituting for $(\partial\sigma_x/\partial x)\,dx$ in the above expression and simplifying we have

$$\tau_{xz} = \{W/(I_{zz}I_{yy} - I_{yz}^2)\}\int \{-0.5dI_{yy} + (b - s)I_{yz}\}\,ds$$

$$= C\{-0.5dI_{yy}s + (bs - s^2/2)I_{yz}\}$$

where $C = W/(I_{yy}I_{zz} - I_{yz}^2)$. Note that the shear stress in the flange varies parabolically with respect to s as shown in Fig. 3.48(b). This contrasts with the linear distribution with respect to s in the case of a channel section considered in Example 1. The total shear force in the flange is

$$Q_f = \int_0^b \tau_{xz}t_f\,ds = C\{-0.25db^2I_{yy} + 0.333b^3I_{yz}\}$$

For the z-section, $I_{yy} = 2(t_fb^3/3)$ and $I_{yz} = 2(bt_f0.5d0.5b) = 0.5b^2dt_f$. When the above values of I_{yy} and I_{yz} are substituted in the expression for Q_f, Q_f is found to be zero. In other words, the net shear force in the flanges is zero and hence does not result in a net torque. Therefore the shear centre coincides with the centroid.

3.16 ● Measurement of strain

The designer is generally interested in the state of stress at a point. Unfortunately it is not easy to measure stress. However, in the elastic state, the state of stress at a point can be inferred from the state of strain at the same point. The state of strain at a point is completely defined if the three quantities ε_x, ε_y and γ_{xy} at a point are specified. Of the three quantities specifying the state of strain, it is very difficult to measure γ_{xy}. However, it is known that the extensional strain in any direction inclined at an angle θ to the x-axis is given, as derived in Section 3.7.2, by

$$\varepsilon_n = \varepsilon_x \cos^2\theta + \varepsilon_y \sin^2\theta + \gamma_{xy} \sin\theta \cos\theta$$

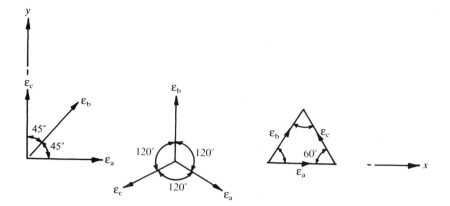

Figure 3.49 ● Strain gauge rosettes

Therefore by measuring the extensional strains in three specified directions, one can determine the three components of strain at a point. In practice, the arrangements shown in Fig. 3.49, known as rectangular, equiangular and delta or triangular arrangements, are used. Normal strains are measured using strain gauges. These can be mechanical, electrical, electronic, etc. Electrical strain gauges based on the principle of measuring the extension of a piece of wire by measuring the corresponding change in resistance are the most popular. Strain gauges arranged in groups as shown in Fig. 3.49 are called strain gauge rosettes. The necessary equations for the calculation of ε_x, ε_y and γ_{xy} can be derived as follows.

Rectangular rosette. Let the measured strains be as follows: ε_a at $\theta = 0$, ε_b at $\theta = 45$ and ε_c at $\theta = 90$ to the x-axis. Then $\varepsilon_a = \varepsilon_x$, $\varepsilon_c = \varepsilon_y$ and $\varepsilon_b = \varepsilon_x \cos^2 45 + \varepsilon_y \sin^2 45 + \gamma_{xy} \sin 45 \cos 45$.
    Solving, we have

$$\varepsilon_x = \varepsilon_a, \; \varepsilon_y = \varepsilon_c, \; \gamma_{xy} = 2\varepsilon_b - \varepsilon_a - \varepsilon_c$$

Equiangular rosette. Let the measured strains be as follows. ε_a at $\theta = -30$, ε_b at $\theta = 90$ and ε_c at $\theta = 210$ to the x-axis. Therefore

    For $\theta = -30$,

$$\varepsilon_a = \varepsilon_x \cos^2 (-30) + \varepsilon_y \sin^2 (-30) + \gamma_{xy} \sin (-30) \cos (-30)$$

    For $\theta = 90$, $\varepsilon_b = \varepsilon_y$

    For $\theta = 210$, $\varepsilon_c = \varepsilon_x \cos^2 210 + \sin^2 210 + \gamma_{xy} \sin 210 \cos 210$

Solving,

$$\varepsilon_x = (2\varepsilon_a - \varepsilon_b + 2\varepsilon_c)/3, \; \varepsilon_y = \varepsilon_b, \; \gamma_{xy} = 2(\varepsilon_c - \varepsilon_a)/\sqrt{3}$$

Triangular or delta rosette. Let the measured strains be as follows. ε_a at $\theta = 0$, ε_b at $\theta = 60$ and ε_c at $\theta = 120$ to the x-axis. Therefore

$$\varepsilon_a = \varepsilon_x$$

$$\varepsilon_b = \varepsilon_x \cos^2 60 + \varepsilon_y \sin^2 60 + \gamma_{xy} \sin 60 \cos 60$$

$$\varepsilon_c = \varepsilon_x \cos^2 120 + \varepsilon_y \sin^2 120 + \gamma_{xy} \sin 120 \cos 120$$

Solving, we have

$$\varepsilon_x = \varepsilon_a, \ \varepsilon_y = (2\varepsilon_c + 2\varepsilon_b - \varepsilon_a)/3, \ \gamma_{xy} = 2(\varepsilon_b - \varepsilon_c)\sqrt{3}$$

Once the state of strain at a point is known, then the principal strains and their direction can be easily determined.

3.16.1 Example of strain rosettes

Example

In a rectangular rosette, the measured strains were as follows:

$$\varepsilon_a = 900 \times 10^{-6}, \ \varepsilon_b = -500 \times 10^{-6}, \ \varepsilon_c = -570 \times 10^{-6}$$

Assuming that the material is elastic with isotropic properties of $E = 210 \times 10^3$ N mm^{-2} and $v = 0.3$, calculate the principal stresses and their directions.

Solution (a) Use the equations for a rectangular rosette, $\varepsilon_x = \varepsilon_a = 900 \times 10^{-6}$, $\varepsilon_y = \varepsilon_c = -570 \times 10^{-6}$, $\gamma_{xy} = 2\varepsilon_b - \varepsilon_a - \varepsilon_c = -1330 \times 10^{-6}$

(b) Calculate the stresses using the stress–strain relationship:

$$\sigma_x = \{E/(1 - v^2)\}(\varepsilon_x + v\varepsilon_y) = 168.2 \text{ N mm}^{-2}$$

$$\sigma_y = \{E/(1 - v^2)\}(v\varepsilon_x + \varepsilon_y) = -53.1 \text{ N mm}^{-2}$$

$$\tau_{xy} = G\gamma_{xy} = \{E/2(1 + v)\}\gamma_{xy} = -107.42 \text{ N mm}^{-2}$$

(c) Calculate the principal stresses:

$$\sigma_1 = (\sigma_x + \sigma_y)/2 + \sqrt{[\{(\sigma_x - \sigma_y)/2\}^2 + \tau_{xy}^2]} = 211.8 \text{ N mm}^{-2}$$

$$\sigma_2 = (\sigma_x + \sigma_y)/2 - \sqrt{[\{(\sigma_x - \sigma_y)/2\}^2 + \tau_{xy}^2]} = -96.7 \text{ N mm}^{-2}$$

$$\tan 2\theta = 2\tau_{xy}/(\sigma_x - \sigma_y) = -0.971$$

$$\theta = -22.07$$

(d) Calculate σ_n, when $\theta = -22.07$

$$\sigma_n = \sigma_x \cos^2 \theta + \sigma_y \sin^2 \theta + 2\tau_{xy} \sin \theta \cos \theta = 211.8$$

Since $\sigma_n = \sigma_1$, σ_1 is inclined at $-22.07°$ to the x-axis and σ_2 is inclined at $(-22.08 + 90) = 67.93°$ to the x-axis.

3.17 ● Concept of stress flow

Engineers' theory of stress calculation for an element loaded by axial force and bending moment is based on the fundamental assumption that plane sections remain plane. It was pointed out that this assumption is invalid near the point of application of concentrated forces as in Fig. 3.15. There are many other situations where the fundamental assumption becomes invalid. Some of these cases are discussed in this section.

Example 1

Consider the stepped column shown in Fig. 3.50(a). Let the area of cross-section of the wider section be $2A$ and that of the narrower section be A. If an axial force F acts on the column, then engineers' theory predicts a stress of $\sigma_x = F/A$ in the

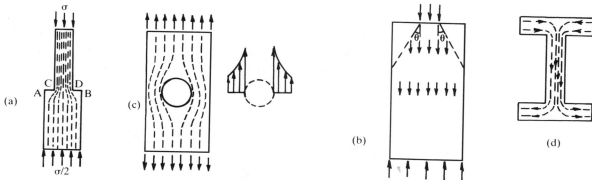

Figure 3.50 ● (a)–(d) Some examples of stress flow

narrow section and a stress σ_x of $F/(2A)$ in the wider section. These calculations indicate abrupt changes in stress. But as shown in Fig. 3.50(a), in practice stress changes take place gradually. The stress from one section 'flows' or 'diffuses' into the next section in a smooth manner. Engineers visualize the 'flow' of stress in terms of the flow of water in a channel. As is well known, water tends to 'cut corners'. A river at a bend tends to flow faster at the convex edge and slow down at the concave edge. In fact these concepts are applicable to 'stress flow' as well. Thus for example, stress tends to concentrate at the re-entrant corner such as corners C and D but becomes zero at corners A and B. Engineers sometimes allow for the gradual 'flow' of stresses by assuming that stress 'diffuses' at 30° as shown in Fig. 3.50(b). The fact that stress tends to concentrate at re-entrant corner is an important idea. As far as possible, one should avoid abrupt changes of section and try to 'round the corners' to aid smooth 'flow' of stress.

Example 2

Figure 3.50(c) shows a plate with a hole which is subjected to uniform tension at a section remote from the hole. If one imagines the hole as an obstruction to the free flow of stress, it is clear that stress tends to concentrate near the edge of the hole. In fact 'exact' calculations show that in a very wide plate the axial stress adjacent to the hole is nearly three times as large as for points away from the hole as shown in Fig. 3.50(c). Holes and re-entrant corners are known as stress raisers for obvious reasons.

Example 3

Figure 3.50(d) shows the shear stress 'flow' in the case of an *I*-section. Again one should note the stress concentration in the re-entrant corners and virtual absence of vertical shear stress in the flange areas far removed from the web. The reader should contrast this smooth 'flow' of stresses with the abrupt stress changes calculated in Examples 2 and 3, Section 3.11.2, using engineers' stress theory.

Example 4

Figure 3.51(a) shows a corner of a rigid-jointed frame. If the corner is subjected to a bending moment, then it is clear that the stresses have to 'flow' round the corner. In fact the stress distribution in the corner 'block' is quite complex and cannot be determined by engineers' theory. The presence of stress concentration at the inner corner is obvious.

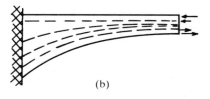

Figure 3.51 ●
(a), (b) Additional
examples of stress flow

(a)   (b)

Example 5

Figure 3.51(b) shows a beam with a haunch. Again the stress 'flow' pattern shown indicates that the stress state does not consist simply of bending stress because the stress path is inclined to the horizontal. It should be noted that although the bending moment is constant, due to the variable depth of the beam, the bending stress σ_x at two similar points at two adjacent sections is not equal. As was shown in Section 3.11, if $d\sigma_x/dx$ is not zero, then for equilibrium shear stress τ_{xy} is required.

The examples could be multiplied. The important lessons to bear in mind are

(a) The stress 'flow' is always smooth and not abrupt as indicated by simple calculations involved in engineers' theory of stress calculation.
(b) Stress raisers such as re-entrant corners and holes should be avoided and where they are unavoidable, it is important to 'round the corners' to aid smooth 'flow' of stress.

3.17.1 Saint Venant's principle

The intuitive ideas presented in the previous section were formulated by Saint Venant in the form of a famous principle which can be stated as follows.

The stress distribution at sections far removed from the point of application of concentrated forces depends purely on stress resultants and not on the actual distribution of forces.

The principle is best explained using simple examples.

Example 1

Consider the column shown in Fig. 3.52(a) subjected to an eccentric force F at an eccentricity of e. As is well known (see Section 3.14), the eccentric force is statically equal to an axial force F and a couple Fe. Saint Venant's principle asserts that, although near the load point the state of stress is complex, away from the concentrated load, the state of stress is the sum of a uniform stress due to the axial force F and a linear stress distribution due to the bending moment equal to Fe. It is generally found that beyond a distance approximately equal to the width of the column the stress distribution is practically that given by engineers' theory.

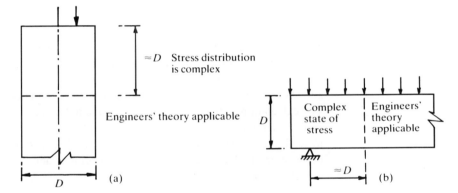

Figure 3.52 ● (a), (b) Examples of regions where engineers' theory is inapplicable

Example 2

Consider the end of a beam supported on a bearing which is assumed to supply a concentrated vertical reaction as shown in Fig. 3.52(b). It is found from 'exact' analysis that at a distance approximately equal to the depth of the beam, the state of stress, i.e. bending and shear stress, is that given by engineers' theory. However, near the bearing itself, the state of stress is extremely complex. Therefore if a simply supported beam has a span of less than twice the depth of the beam, engineers' theory cannot be used to calculate the state of stress in such a beam. Such beams are known as deep beams and show complex stress distributions which are very different from those given by engineers' theory.

3.18 ● Torsion

In the previous sections of this chapter commonly occurring cases of axial and bending stress distributions were determined on the assumption that plane sections

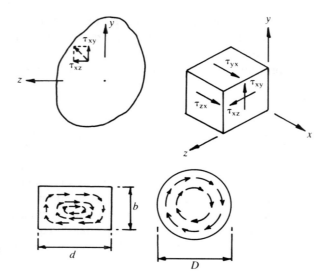

Figure 3.53 ● Shear stress distribution due to torsion

remain plane. The distribution of shear stress in beam problems due to shear force was determined on the basis of equilibrium requirements. Apart from the axially loaded members such as struts and ties and beams subjected to bending moment and shear force, engineers have to design members subjected to twisting moments. The calculation of the stress distribution due to torsion is complex and except in the case of circular cross-sections cannot be handled by simple theory. Therefore, instead of developing equations which are valid only for circular sections, a general description of the state of stress due to torsion and a summary of results from 'exact' analysis will be given.

3.18.1 Open and closed sections

Consider the cross-section shown in Fig. 3.53. If a member orientated along the x-axis is subjected to a twisting moment, then as discussed in Section 3.5, shear stresses τ_{xy} and τ_{xz} are required to resist the applied torsional moment about the x-axis. Figure 3.53 shows the qualitative distribution of shear stress in rectangular and circular cross-section shafts.

From the point of view of the behaviour of members of different cross-sections subjected to torsion, it is convenient to divide them into two basic groups. They are

1. 'Solid' and closed form or 'hollow' cross-sections. Figure 3.54(a) shows typical cross-sections. Closed form cross-sections include circular and rectangular hollow sections.
2. 'Open' cross-sections. This category includes typical rolled steel sections as shown in Fig. 3.54(b) and a split circular tube.

If the torque is constant along the length of the shaft, then the stress distribution at any point consists of stresses τ_{xy} and τ_{xz} only. The resultant stress distribution in the case of rectangular and circular cross-sections is as shown in Fig. 3.53. In

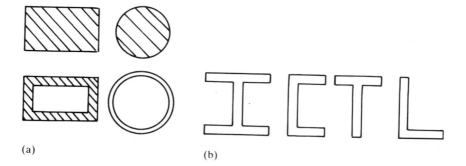

Figure 3.54 ● (a), (b) 'Closed' and 'open' cross-sections

(a)

(b)

circular sections, the resultant shear stress at any point varies linearly with the distance from the centre and the maximum shear stress acts normal to the radius R of the tube. On the other hand, in rectangular sections, the distribution is more complex, mainly due to the presence of corners which the 'flow' of shear stress tends to bypass. The shear stress is small towards the centroid of the rectangle and increases non-linearly towards the edges. The maximum shear stress is at the middle of the *longer* side.

The relationship between the applied twisting moment T and the corresponding maximum shear stress is given by

(a) Circular sections: $\tau_{max} = 2T/(\pi R^3)$
(b) Rectangular sections ($b \times d$, $d > b$): $\tau_{max} = kT/(b^2 d)$ where k = function of d/b ratio. Numerical values of k are given in Table 3.11.

Table 3.11

d/b	k	d/b	k
1.0	4.801	3.0	3.745
1.2	4.566	4.0	3.546
1.5	4.329	5.0	3.436
2.0	4.065	10.0	3.205
2.5	3.876	∞	3.000

The classification of sections into 'closed' and 'open' sections is important because the way in which torsional shear stresses are distributed in the cross-section in the two types is quite different. As shown in Fig. 3.55, the shear stress in 'closed' sections forms a closed loop. Because the larger shear stress acts at a 'larger' lever arm, for a given value of the maximum stress, the section is able to resist a 'large' torque. In the case of 'open' sections such as rolled steel structural sections which can be divided into a number of thin individual rectangles, the resistance to torsion of the 'open' section is for all practical purposes the same as the sum of the resistance of the individual rectangles. Since the individual rectangles are thin, the total resistance to torsion of 'open' sections tends to be very small. Thus from the point of view of torsional resistance, for example, a 'closed' type of section like a rectangular hollow section is much preferable to an 'open' type of section like an *I* section.

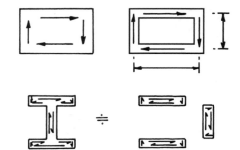

Figure 3.55 ● Shear stress distribution in 'closed' and 'open' cross-sections

3.18.2 Torsional resistance of closed thin-walled tubes

Consider the torsional shear stress distribution in a rectangular section. As pointed out before, the shear stress increases away from the centroid and thus 'larger' shear stress acts at a 'larger' lever arm. If the section is hollowed out, the torsional resistance will decrease but because in the hollow part only 'small' shear stresses act at 'small' lever arm, the decrease in the torsional resistance is not proportional to the decrease in the cross-sectional area. Thus for efficient utilization of material, a thin closed tube is preferable to a solid section. If the thickness of the tube is small, then one can disregard the variation of shear stress across the thickness and assume that shear stress is uniform. Exact analysis has shown that in all parts of the tube (except near the corners) it is reasonable to assume that the product of uniform shear stress and thickness is constant. With this assumption, it is possible to derive some simple expressions connecting applied torque and the shear stress in the walls of the tube.

Example 1

Calculate the relationship between the applied torque T and the uniform shear stress τ in the wall of the thin-walled circular tube shown in Fig. 3.56(a).

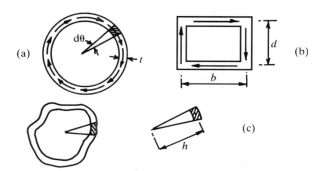

Figure 3.56 ● (a)–(c) Closed form sections

Solution   Let the shear stress in the tube be τ. Consider an element of area dA of the tube. $dA = R\,d\theta t$, where R = radius of the centre line of the tube, t = thickness. The force acting on this area is τdA and it acts at a lever arm R. Therefore

$$T = \int \tau \, dA \ R = \int \tau R \ d\theta t R = \tau R^2 t \int d\theta \text{ where limits for } \theta = 0 \text{ and } 2\pi. T = 2\pi R^2 t \tau$$

Since area A_0 enclosed by the centre line is πR^2, we have

$$\tau t = T/(2A_0)$$

Example 2

Calculate the relationship between the applied torque T and the uniform shear flow q, where $q = \tau t$ in the wall of the thin-walled rectangular tube shown in Fig. 3.56(b).

Solution *Assuming* that the shear flow q = wall thickness × shear stress is constant, we have

(a) The shear force in the flanges = qb. The force qb in the two flanges acts at a distance d and therefore forms a couple $(qb)d$. Therefore the torque resisted by the flanges is equal to qbd.

(b) The shear force in the webs = qd. The force qd in the two webs acts at a distance b and therefore forms a couple $(qd)b$. Therefore the torque resisted by the webs is equal to qdb. Therefore

$$T = \text{torque resisted by flanges and webs} = q(bd + db)$$

$$q = T/(2bd)$$

Since area A_0 enclosed by the centre line is bd, we have

$$q = T/(2A_0)$$

Shear stress τ in the wall = q/thickness of wall

Example 3

Calculate the relationship between the applied torque T and the uniform shear stress τ in the wall of the thin-walled single-cell tube shown in Fig. 3.56(c).

Solution Let q = shear stress × wall thickness. It is *assumed* that q is constant in the wall of the tube. As shown in Fig. 3.56, the force F acting on a small element of length ds and wall thickness t is

$$F = \tau(ds \ t) = q \ ds$$

The torque dT resisted by this force is

$$dT = Fh$$

where h = lever arm to the force F. Therefore

$$dT = q \ ds \ h$$

$$ds \ h = \text{twice the area of triangle ABC}$$

Therefore

$$T = q \int ds \ h = q(2A_0)$$

where A_0 = area enclosed by the centre line of the tube.

$$q = T/(2A_0)$$

The above examples show that in a thin-walled single-cell tube, irrespective of the shape of the tube, the torsional shear flow q is given by

$$q = T/(2A_0)$$

3.19 ● References

Gere J M, Timoshenko S P 1987 *Mechanics of Materials* (2nd edn) Van Nostrand Reinhold (Has a large number of examples for practice. Somewhat old fashioned.)

Pilkey W D, Pilkey O H 1974 *Mechanics of Solids* Quantum (A good book.)

Popov E P 1978 *Mechanics of Materials* (2nd edn) Prentice-Hall (A good book with plenty of examples for practice.)

3.20 ● Problems

1. A stepped column carries loads as shown in Fig. E3.1. Calculate the normal stress in the column.

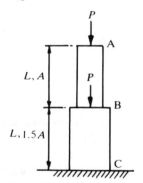

Figure E3.1

Answer: $-P/A$ and $-1.33P/A$.

2. Two bars are connected by a pin-joint as shown in Fig. E3.2. If the tensile force in the bars is P, calculate the average shear stress in the pin of diameter d.

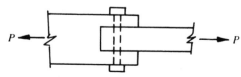

Figure E3.2

Answer: Case of double shear, $\tau = 2P/(\pi d^2)$.

3. A flanged shaft coupling shown in Fig. E3.3 has to transmit a torque T. The eight bolts are arranged on a circle of diameter D. Calculate the average shear stress in the bolts of diameter d.

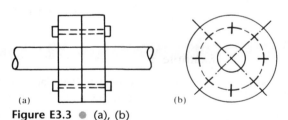

Figure E3.3 ● (a), (b)

Answer: If Q is the shear force per bolt, then the torque resisted by one bolt is $Q(0.5D)$. Therefore $T = 8Q(0.5D)$, $Q = 16T/D$, $\tau = 4Q/(\pi d^2)$.

4. At a point in a structure, the stresses are $\sigma_x = -50 \text{ N mm}^{-2}$, $\sigma_y = 80 \text{ N mm}^{-2}$, $\tau_{xy} = 70 \text{ N mm}^{-2}$. Determine the principal stresses and their directions. Also determine the maximum shear stress and the directions of the planes on which it acts. If $E = 210 \text{ kN mm}^{-2}$ and $\nu = 0.3$, calculate the principal strains.

Answer: $\sigma_1 = 110.5$ and acts at $66.4°$ and $\sigma_2 = -80.53$ and acts at $-23.6°$ to the x-axis. $\tau_{max} = 95.53 \text{ N mm}^{-2}$. It is accompanied by a normal stress of 15 N mm^{-2}. They act on planes inclined at 21.4° and 11.4° to the horizontal. $\varepsilon_1 = 641 \times 10^{-6}$, $\varepsilon_2 = -541 \times 10^{-6}$.

5. At a point in a structure made of an anisotropic material, the measured strains at a point are $\varepsilon_x = 800 \times 10^{-6}$, $\varepsilon_y = -680 \times 10^{-6}$ and $\gamma_{xy} = -1200 \times 10^{-6}$. Given that $E_x = 150$ kN mm^{-2}, $E_y = 250$ kN mm^{-2}, $G = 105.7$ kN mm^{-2}, $v_x = 0.3$ and $v_y = 0.18$, calculate the principal strains, stresses and their directions. Also determine the maximum shear stress and the directions of the planes on which it acts.

Answer: $\varepsilon_1 = 1013 \times 10^{-6}$ inclined at $-19.5°$ to the x-axis. $\varepsilon_2 = -893 \times 10^{-6}$ inclined at 70.5° to the x-axis. $\sigma_y = -141.7$, $\sigma_x = 94.5$, $\tau_{xy} = -126.8$ N mm^{-2}, $\sigma_1 = 149.7$ inclined at $-23.5°$ to the x-axis, $\sigma_2 = -196.9$ inclined at 66.5° to the x-axis. Note that because the material is anisotropic, the directions of principal stresses and strains do not coincide. $\tau_{max} = -173.3$ N mm^{-2}. It is accompanied by a normal stress of -23.6 N mm^{-2}. They act on planes inclined at 21.5° and 11.5° to the horizontal.

6. Calculate the total change in the length of the stepped column shown in Fig. E3.1 due to given loads and a temperature change of T. Use the strain-displacement relationship of Section 3.7.

Answer:

(a) BC: $\sigma_y = -1.33P/A$, $\varepsilon_y = \partial v/\partial y = \sigma_y/E + \alpha T = -1.33P/(EA) + \alpha T$, $v = -1.33Py/(EA) + \alpha Ty + C$, $v = 0$ at the fixed base, i.e. $v = 0$ at $y = 0$. Therefore $v = -1.33Py/(EA) + \alpha T$. At B, $y = L$. Therefore at B, $v = -1.33PL/(EA) + \alpha TL$

(b) AB: $\sigma_y = -P/A$, $\varepsilon_y = \partial v/\partial y = \sigma_y/E + \alpha T = -P/(EA) + \alpha T$, $v = -Py/(EA) + \alpha Ty + C$. But at B, $v = -1.33PL/(AE) + \alpha TL$ at $y = L$. Therefore $C = -0.33Py/(EA)$. At A, $y = 2L$. Therefore at A, $v = -P2L/(EA) + \alpha T2L - 0.33PL/(AE) = -2.33PL/(AE) + \alpha T2L$.

7. What is the state of stress in the stepped column in Fig. E3.1 if it is fully fixed at A and C and subjected to a temperature rise of $T°$?

Answer: Let an unknown axial compressive load P be applied to restrain the column from expansion. BC: $\sigma_y = -P/(1.5A) = -0.67P/A$, $\partial v/\partial y = \varepsilon_y = \sigma_y/E + \alpha T = -0.67P/(AE) + \alpha T$, $v = -0.67Py/(AE) + \alpha Ty + C$, $v = 0$ at $y = 0$, $C = 0$. v at B where $y = L$ is given by $-0.67PL/(AE) + \alpha TL$. AB: $\sigma_y = -P/A$, $\partial v/\partial y = \varepsilon_y = \sigma_y/E + \alpha T = -P/(AE) + \alpha T$, $v = -Py/(AE) + \alpha Ty + C$. From the previous calculation, $v = -0.67PL/(AE) + \alpha TL$ at $y = L$, thus

$C = 0.33PL/(AE)$. Therefore $v = -Py/(AE) + \alpha TL + 0.33PL/(AE)$. Since A is fixed, $v = 0$ at A where $y = 2L$. Therefore $v = 0 = -P(2L)/(AE) + \alpha T2L + 0.33PL/(AE)$. Therefore $P/A = 1.2E\alpha T$. In AB, $\sigma_y = -1.2E\alpha T$, and in BC, $\sigma_y = -0.8E\alpha T$.

8. Calculate the I of a solid circle about a diameter.

Answer: Using I for a semicircle $= \pi R^4/8$, I for circle $= 2(\pi R^4/8) = \pi R^4/4 = \pi D^4/64$, where $D = 2R$.

9. Calculate the I of a rectangle about a diagonal.

Answer: As shown in Fig. E3.4, $D^2 = b^2 + d^2$, area of rectangle $= bd = 2(0.5Dh) = dh$, therefore $h = bd/D$. I of rectangle about a diagonal $= I$ of two triangles about their common base $= 2(Dh^3/12) = b^3d^3(6D^2)$.

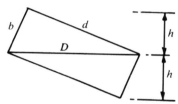

Figure E3.4

10. Calculate I_{yy} and I_{zz} for the rectangular section with two circular holes of 100 mm diameter as shown in Fig. E3.5.

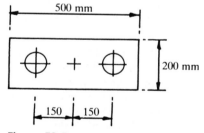

Figure E3.5

Answer: $I_{zz} = I$ of solid rectangle $500 \times 200 - I$ of two circular holes of 100 mm diameter. This is correct as the horizontal centroidal axis of the solid rectangle and holes coincide. $I_{zz} = 500 \times 200^3/12 - 2\{\pi100^4/64\} = 323.5 \times 10^6$ mm^4. For I_{yy} use the parallel axis theorem. The holes are at a distance of 150 mm from the centroidal axis of the hollow section. Therefore $I_{yy} = 200 \times 500^3/12$ for the solid $-2\{\pi100^4/64 + (\pi/4)100^2150^2\}$ for the two holes of 100 mm diameter, $I_{yy} = 1720.1 \times 10^6$ mm^4.

11. Calculate the I_{zz} and I_{yy} of the I section, channel section and the composite section shown in Fig. E3.6. Determine the maximum moment about

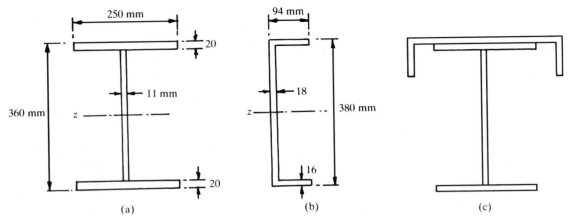

Figure E3.6 ● (a)–(c)

the *zz* axis that the composite section can resist if the maximum stress is not to exceed 250 N mm^{-2}.

Answer:

(a) *I* section:
$$A = 250 \times 360 - (250 - 11)(360 - 20 - 20)$$
$$= 13520 \text{ mm}^2$$

$$I_{zz} = I \text{ solid} - I \text{ hole}$$
$$= 250 \times 360^3/12$$
$$- (250 - 11)(360 - 20 - 20)^3/12$$
$$= 319.4 \times 10^6 \text{ mm}^4$$

$$I_{yy} = (360 - 20 - 20)11^3/12 \text{ for web}$$
$$+ 2\{20 \times 250^3/12\} \text{ for the two flanges}$$
$$= 52.1 \times 10^6 \text{ mm}^4$$

(b) Channel: $I_{zz} = 94 \times 380^3/12$ for the 'solid' $(94 - 18)(380 - 16 - 16)^3/12$ for the 'hole' $= 162.9 \times 10^6$ mm^4. Before I_{yy} can be determined, the position of the centroid has to be established. $A = 18 \times (300 - 16 - 16)$ $+ 2 \times 94 \times 16 = 9272$ mm^2. Taking moments about the left-hand edge,

$$A\bar{Z} = 348 \times 18 \times (18/2) + 2 \times 94 \times 16$$
$$\times (94/2)$$
$$= 197.75 \times 10^3 \text{ mm}^3$$

Therefore the centroid is at $197.75 \times 10^3/9272$ $= 21.3$ mm from the left edge.

$$I_{yy} = \{348 \times 18^3/12 + 348 \times 18 \times$$
$$(21.3 - 18/2)^2\} \text{ for the web}$$
$$+ 2\{16 \times 94^3/12 + 94 \times 16$$
$$\times (21.33 - 94/2)^2\} \text{ for the two flanges}$$
$$= 5.3 \times 10^6 \text{ mm}^4$$

(c) Composite section: Determine the common centroid. Net area = 13 520 for *I* section

+ 9272 for channel = 22 772 mm^2. Taking moments about the top face of the composite section

$$A\bar{y} = 9252 \times 21.3 \text{ for channel}$$
$$+ 13\,520 \times (18 + 360/2) \text{ for the } I \text{ section}$$
$$= 2.87 \times 10^6 \text{ mm}^3$$

Therefore

$$\bar{y} = 2.87 \times 10^6/22\,772 = 126.2 \text{ mm}$$

Distance from the neutral axis to the bottom fibre = $18 + 360 - 126.2 = 251.8$ mm

$$I_{zz} = \{319.4 \times 10^6 + 13\,520$$
$$\times (18 + 360/2 - 126.2)^2 \text{ for } I \text{ section}$$
$$+ \{5.3 \times 10^6 + 9252 \times (126.2 - 21.3)^2\}$$
$$\text{for channel}$$
$$= 496.15 \times 10^6 \text{ mm}^4$$

Note that in the above calculation I_{yy} of the channel about its own centroid is used as this is the axis parallel to the composite *z*-axis.

$$I_{yy} = 52.12 \times 10^6 \text{ for } I + 162.9 \times 10^6 \text{ for}$$
$$\text{channel}$$
$$= 215.0 \times 10^6 \text{ mm}^4$$

Note that the centroidal axes of the two sections coincide with the common centroidal axis. Therefore there is no need to use the parallel axis theorem.

(d) $\sigma_{x\text{max}} = (M_z/I_{zz})y_{\text{max}}$
In this case $y_{\text{max}} = 251.8$ to the bottom fibre. Therefore $\sigma_{x\text{max}} = 250 = (M_z/496.15 \times 10^6)251.8$, and

$$M_z = 492.6 \times 10^6 \text{ N mm} = 492.6 \text{ kN m.}$$

12. Determine the maximum bending moment and shear force that the section shown in Fig. E3.7 can

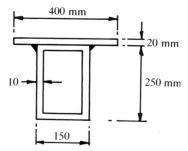

Figure E3.7

resist if the maximum bending and shear stress are not to exceed 300 N mm^{-2} and 180 N mm^{-2} respectively. Also determine the average shear flow that the welds connecting the top plate to the beam have to resist if the section is used as a simply supported beam to carry a uniformly distributed load of 15 kN m^{-1} over a span of 10 m.

Answer:

(a) Determine the section properties of component parts.

Area of rectangular hollow section (RHS) = A_1
= $250 \times 150 - (250 - 10 - 10)(150 - 10 - 10)$
= 7600 mm^2

I_{zz} of RHS = I_{zz1} = $150 \times 250^3/12$ for solid
$\quad - (150 - 10 - 10)(250 - 10$
$\quad - 10)^3/12$ for the hole
$\quad = 63.5 \times 10^6$ mm^4

Area of top plate = A_2 = 400×20 = 8000 mm^2

I_{zz} of top plate = I_{zz2} = $400 \times (20)^3/12$
$\quad = 0.27 \times 10^6$ mm^4

(b) Determine the centroid of the composite section:

Total area = $A_1 + A_2$ = 7600 + 8000
$\quad = 15\ 600$ mm^2

Take moments about the top fibre:

$A\bar{Y}$ = $A_1(20 + 250/2) + A_2 \times (20/2)$
$\quad = 1.18 \times 10^6$ mm^3

$\bar{Y}$ = $1.18 \times 10^6/15\ 600$ = 75.8 mm

Distance to bottom fibre = 20 + 250 − 75.8
$\quad = 194.2$ mm

(c) Composite I_{zz}:

$I_{zz} = I_{zz1} + A_1(20 + 250/2 - 75.8)^2 + I_{zz2}$
$\quad + A_2(20/2 - 75.8)^2$
$\quad = 134.8 \times 10^6$ mm^4

(d) σ_{xmax} = 300 = $(M_z/134.8 \times 10^6)y_{max}$, where y_{max} = 194.2 mm. Therefore

M_z = 208.2 kN m

(e) Maximum shear stress calculation:
Shear stress is a maximum at the neutral axis:

$\tau_{max} = [Q/(I_{zz}t)] \int y\ dA$

$\int y\ dA$ = first moment about the neutral axis
of the area above the neutral axis
= $400 \times 20 \times (75.8 - 20/2)$ for the
flange plate
+ $150 \times (75.8 - 20) \times (75.8$
$- 20)/2$ for the 'solid' RHS
+ $(150 - 10 - 10) \times (55.8 - 10)$
$\times (55.8 - 10)/2$ for the 'hole' RHS
= 0.62×10^6 mm^3

t = 2×10 for two webs = 20 mm

Therefore

τ_{max} = 180 = $Q \times 0.62 \times 10^6/(134.8 \times 10^6 \times 20)$

Q = 782.7×10^3 N = 782.7 kN

(f) Shear flow at the flange plate–RHS junction:

$\tau t = \{Q/I_{zz}\} \int y\ dA$

$\int y\ dA$ = $400 \times 20 \times (75.8 - 20/2)$ for flange
plate only
= 0.526×10^6 mm^3

τt = $Q \times 0.526 \times 10^6/134.8 \times 10^6$
$\quad = 3.90 \times 10^{-3} \times Q$ N mm^{-1}

In the simply supported beam supporting u.d.l., the SFD is linear with the maximum at the supports of 75 kN. Total shear flow over half the span is given by $\int \tau t\ dx$ = $3.90 \times 10^{-3} \int Q\ dx$, where
$\int Q\ dx$ = $0.5 \times 75 \times 5$ m = 187.5 kN m
$\quad = 187.5 \times 10^6$ N mm

Therefore

$\int \tau t\ dx$ = $3.90 \times 10^{-3} \times 187.5 \times 10^6$
$\quad = 731.3 \times 10^3$ N

This shear force is resisted by two welds 5000 mm long. Therefore *average* shear flow in the welds is equal to

$731.3 \times 10^3/(2 \times 5000)$ = 73.1 N mm^{-1}

13. Figure E3.8 shows an unequal angle section. Determine the maximum bending moment that can be applied about the horizontal zz axis without the maximum bending stress exceeding 250 N mm^{-2}.

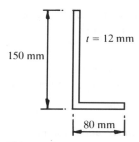

Figure E3.8

Answer: Determine section properties:

$$A_{web} = A_1 = 150 \times 12 = 1800$$

$$A_{flange} = A_2 = 12(80 - 12) = 816$$

$$A = A_1 + A_2 = 2616 \text{ mm}^2$$

Taking moments about the bottom face,

$$A\bar{Y} = A_1 \times 150/2 + A_2 \times 12/2 = 139.896 \times 10^3$$

$$\bar{Y} = 53.5 \text{ mm from bottom face.}$$

Taking moments about left face,

$$A\bar{Z} = A_1 \times 12/2 + A_2 \times \{12 + (80 - 12)/2\}$$
$$= 48.34 \times 10^3 \text{ mm}^3$$

$$\bar{Z} = 18.48 \text{ mm}$$

With respect to the centroidal axes, the co-ordinates (h_y, h_z) of the centroids of A_1 and A_2 are

$$A_1: h_{y1} = 150/2 - \bar{Y} = 21.5,$$

$$h_{z1} = \bar{Z} - 12/2 = 12.48 \text{ mm}$$

$$A_2: h_{y2} = 12/2 - \bar{Y} = -47.5 \text{ mm}$$

$$h_{z2} = \bar{Z} - \{12 + (80 - 12)/2\} = -27.52$$

$$I_{zz1} = 12 \times 150^3/12 = 3.375 \times 10^6$$

$$I_{yy1} = 150 \times 12^3/12 = 0.022 \times 10^6$$

$$I_{zz2} = (80 - 12) \times 12^3/12 = 0.0098 \times 10^6$$

$$I_{yy2} = 12 \times (80 - 12)^3/12 = 0.3144 \times 10^6$$

Composite second moments of area:

$$I_{zz} = I_{zz1} + A_1 h_{y1}^2 + I_{zz2} + A_2 h_{y2}^2 = 6.06 \times 10^6 \text{ mm}^4$$

$$I_{yy} = I_{yy1} + A_1 h_{z1}^2 + I_{yy2} + A_2 h_{z2}^2 = 1.23 \times 10^6 \text{ mm}^4$$

$$I_{yz} = A h_{y1} h_{z1} + A_2 h_{y2} h_{z2} = 1.55 \times 10^6 \text{ mm}^4$$

Using the formula for unsymmetrical bending,

$$\sigma_x = M_{zz}(-0.243y + 0.306z) \times 10^{-6}$$

The neutral axis is given by $-0.243y + 0.306z = 0$, i.e. $y = 1.26z$. Therefore the neutral axis is inclined

at $51.6°$ to the horizontal. Maximum stress occurs at the top of the web on the inside for which $y = 150 - \bar{Y} = 96.52$, $z = \bar{Z} - 12 = 6.48$, $\sigma_x = 250$ $= M_{zz} (-0.243 \times 96.52 + 0.306 \times 6.48) \times 10^{-6}$, $M_{zz} = 11.64 \text{ kN m.}$

14. Determine the position of the shear centre for the channel section shown in Fig. E3.9. Assume thickness = 12 mm.

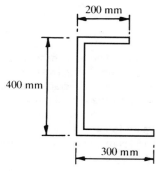

Figure E3.9

Answer: This is an unsymmetrical section. Therefore unsymmetrical bending should be considered.

(a) Section properties:

Area of top flange = $A_1 = (200 - 12)12 = 2252$

Area of web = $A_2 = 400 \times 12 = 4800$

Area of bottom flange = $A_3 = (300 - 12)12$
$$= 3456$$

Total area $A = 10\,512 \text{ mm}^2$

$$I_{zz1} = (200 - 12)12^3/12 = 0.03 \times 10^6$$

$$I_{yy1} = 12 \times (200 - 12)^3/12 = 6.65 \times 10^6$$

$$I_{zz2} = 12 \times 400^3/12 = 64.0 \times 10^6$$

$$I_{yy2} = 400 \times 12^3/12 = 0.06 \times 10^6$$

$$I_{zz3} = (300 - 12)12^3/12 = 0.04 \times 10^6$$

$$I_{yy3} = 12 \times (300 - 12)^3/12 = 23.689 \times 10^6$$

(b) Position of centroid:
Taking moments about bottom face:

$$\bar{Y} = \{A_1(400 - 6) + A_2 \times 400/2 + A_3(12/2)\}/A$$
$$= 177.85 \text{ mm}$$

Taking moments about left face:

$$\bar{Z} = \{A_1(188/2 + 12) + A_2 \times 12/2$$
$$+ A_3(288/2 + 12)\}/A$$
$$= 76.77 \text{ mm}$$

(c) Co-ordinates (h_y, h_z) of the position of the centroids of individual rectangles with respect to common centroidal axes:

$$h_{y1} = 400 - 6 - \bar{Y} = 216.15$$

$$h_{z1} = \bar{Z} - (12 + 188/2) = -29.23$$

$$h_{y2} = 400/2 - \bar{Y} = 22.15,$$

$$h_{z2} = \bar{Z} - (12/2) = 70.77$$

$$h_{y3} = 12/2 - \bar{Y} = -171.85,$$

$$h_{z3} = \bar{Z} - (12 + 288/2) = -79.23$$

(d) Section properties:

$$I_{zz} = \sum(I_{zzi} + A_i h_{yi}^2) = 273.9 \times 10^6 \text{ mm}^4$$

$$I_{yy} = \sum(I_{yyi} + A_i h_{zi}^2) = 78.26 \times 10^6 \text{ mm}^4$$

$$I_{yz} = \sum A_i h_{yi} h_{zi} = 40.35 \times 10^6 \text{ mm}^4$$

(e) Substituting in the unsymmetric bending formula:

$$\sigma_x = M_z \times 10^{-9}\{-3.954y + 2.039z\} + M_y \times 10^{-9}\{-2.039y + 13.829z\}$$

(f) Determine the stress at the centre line of flanges and web:
  (i) Top flange: $y = 216.23$,
$$-123.23 \leqslant z \leqslant 70.77$$

$$\sigma_x = M_z \times 10^{-9}\{-854.97 + 2.037z\} + M_y \times 10^{-9}\{-440.89 + 13.829z\}$$

  (ii) Bottom flange: $y = -171.77$,
$$-223.23 \leqslant z \leqslant 70.77$$

$$\sigma_x = M_z \times 10^{-9}\{679.18 + 2.0397z\} + M_y \times 10^{-9}\{350.24 + 13.829z\}$$

  (iii) Web: $z = 70.77$, $-171.23 \leqslant y \leqslant 216.23$

$$\sigma_x = M_z \times 10^{-9}\{-3.954y + 144.30\} + M_y \times 10^{-9}\{-2.039y + 978.68\}$$

If forces W_y and W_z are applied in the y- and z- directions at the tip of a cantilever fixed at $x = 0$, then $M_z = W_y(L - x)$ and $M_y = -W_z(L - x)$. Therefore $dM_z/dx = -W_y$ and $dMy/dx = W_z$.

(g) Determine the shear stress distribution and the shear force in the flanges and web: as in examples in Section 3.15.1,
  (i) Top flange:

$$\partial\sigma_x/\partial x = -W_y \times 10^{-9}\{-854.97 + 2.039z\} + W_z \times 10^{-9}\{-440.89 + 13.829z\}$$

It is useful to use the variable s starting from the tip of the flange where the shear stress is zero. Therefore $s = z + 123.23$. Substituting for z,

$$\partial\sigma_x/\partial x = -W_y \times 10^{-9}\{-1106.24 + 2.039s\} + W_z \times 10^{-9}\{-2145.04 + 13.829s\}$$

$$\tau_{xz} = \int \partial\sigma_x/\partial x \, ds$$
$$= -W_y \times 10^{-9}\{-1106.24s + 2.039s^2/2\} + W_z \times 10^{-9}\{-2145.04s + 13.829s^2/2\}$$

Q_t flange $= \int \tau_{xz} t \, ds$, where t = thickness of flange.
$$= -W_y \times 10^{-9}\{-1106.24s^2/2 + 2.039s^3/6\}t + W_z \times 10^{-9}\{-2145.804s^2/2 + 13.829s^3/6\}t$$

The limits are $0 \leqslant s \leqslant 194$;
Q_t flange $= [0.22W_y - W_z0.28]$

(ii) Bottom flange: Similar to above

$$\partial\sigma_x/\partial x = [-W_y \times 10^{-9}\{679.18 + 2.039z\} + W_z \times 10^{-9}\{350.24 + 13.829z\}]$$

Let $s = 223.23 + z$ where s starts from the tip of the flange where the shear stress is zero.

$$\partial\sigma_x/\partial x = [-W_y \times 10^{-9}\{224.01 + 2.039s\} + W_z \times 10^{-9}\{-2736.81 + 13.829s\}]$$

$$\tau_{xz} = \int \partial\sigma_x/\partial x \, ds$$
$$= [-W_y \times 10^{-9}\{224.01s + 2.039s^2/2\} + W_z \times 10^{-9}\{-2736.81s + 13.829s^2/2\}]$$

Q_b flange $= \int \tau_{xz} t \, ds$, where t = thickness of flange
$$= -W_y \times 10^{-9}\{224.01s^2/2 + 2.039s^3/6\}t + W_z \times 10^{-9}\{-2736.81s^2/2 + 13.829s^3/6\}t$$

The limits are $0 \leqslant s \leqslant 294$;
$Q_b = [-0.22W_y - 0.72W_z]$

Note that the shear forces in the flanges corresponding to W_y sum to zero as should be expected from equilibrium considerations. Similarly, the shear forces in the flange corresponding to W_z sum to W_z again as expected from equilibrium considerations.

(a) W_y only: taking moments about the lower flange–web junction, anticlockwise torque by the shear forces in the top flange = $Q_t(400 - 12) = 0.22W_y \times 388 = 85.36W_y$. The shear centre is therefore located to the left of the web so that the shear force in the web and the applied force produce a clockwise torque of $e_x W_y$. Therefore $e_x = 85.36$ mm.

(b) W_z only: taking moments about the lower flange–web junction, shear force in the top flange $\times (400 - 12) = W_z \times$ distance to W_z.
   For equilibrium, W_z should act at $0.28 \times (400 - 12) = 108.64$ mm from the centre line of the bottom flange. Therefore the shear centre is at $108.64 + 6 = 114.64$ mm from the bottom face.

15. Figure E3.10 shows a timber–steel composite beam. The steel plates are 12 mm thick. Determine the maximum moment that the beam can withstand when bent about
(a) the z-axis;
(b) the y-axis.
Assume that the steel plates are 12 mm thick.

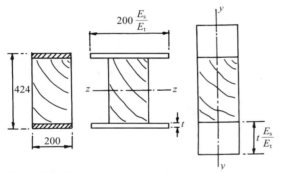

Figure E3.10

If the beam is used as a simply supported beam of 10 m span, calculate the uniformly distributed load that the beam can support. What is the maximum shear flow at the steel plate–timber junction when the beam is resisting the maximum bending about the z-axis? It is given that $E_{steel} = 200$ kN mm^{-2}, $E_{timber} = 14$ kN mm^{-2}. Maximum permissible stress in timber $= 10$ N mm^{-2}.

Answer:
(a) Bending about the z-axis: Replace steel plates by equivalent timber plate of width $200 \times E_{steel}/E_{timber} = 2857$ mm. The total depth remains unaltered.
I_{zz}composite $= 200 \times 400^3/12$ for timber
$+ 2\{2857 \times 12^3/12 + 2857$
$\times 12 \times (400/2 + 12/2)^2\}$ for the two equivalent timber plates

$I_{zz} = 3977.2 \times 10^6$ mm^4

$\sigma_x = 10 = (M_{zz}/I_{zz})200$

Note $y = 200$ mm at the top face of timber beam, and $M_z = 198.9$ kN m. Stress at the top face of the composite beam $= (M_z/I_{zz})(200 + 12) = 10.6$ N mm^{-2}. This is the stress in the timber. Since the strain in steel and timber must be the same, the *real* stress in the steel is $10.6(E_{steel}/E_{timber}) = 151.4$ N mm^{-2}. If q is the u.d.l. that the beam is supporting, then $ql^2/8 = M_z$, $L = 10$ m, $q = 15.9$ kN mm^{-1}.

Shear flow

$$= Q \int (y \, dA)/I_{zz}$$
$$= Q[2857 \times 12 \times (200 + 12/2)]/I_{zz}$$
$$= 1.78 \times 10^{-3}Q$$

Q is a maximum at the supports
$= 15.9 \times 10/2 = 79.5$ kN;
maximum shear flow $= 141.5$ N mm^{-1}.

(b) Bending about the y-axis:

$$I_{yy} = 400 \times 200^3/12 \text{ for timber}$$
$$+ 2[(12E_{steel}/E_{timber})200^3/12]$$
$$= 495.2 \text{ mm}^4$$

Note how only the 'equivalent width' of steel plates is altered in calculating both I_{zz} and I_{yy}.

16. Figure E3.11 shows a rectangular hollow section. It is used as a beam to resist at a section a bending moment M_z of 40 kN m so as to produce tension on the bottom face and a downward shear force of 120 kN. In addition, it is subjected to a clockwise torque of 30 kN m. Calculate the principal stresses in the middle of the flange and web.

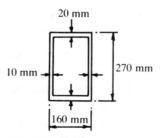

Figure E3.11

Answer:

$I_{zz} = 160 \times 270^3/12 - 140 \times 230^3/12$
$= 120.5 \times 10^6$ mm^4

$\int y \, dA$ of the area above the neutral axis
$= 160 \times 135 \times 135/2$ solid area $- 140 \times 115$
$\times 115/2$ for the hole $= 0.53 \times 10^6$ mm^3
(a) σ_x in flange $= (40 \times 10^6/I_{zz})270/2$
$= 44.8$ N mm^{-2}

(b) τ_1 due to shear force at the middle of the web
$= Q \int y \, dA/(I_{zz}t)$
$= 120 \times 10^3 \times 0.53 \times 10^6/(120.5 \times 10^6 \times 2 \times 10)$
$= 26.5 \text{ N mm}^{-2}$.
Note that $t = 2 \times 10$ due to two webs each 10 mm thick.

There is no shear stress at the middle of the flange because of symmetry.

(c) Shear flow q due to torsion $= T/(2A_0)$, $A_0 = 150 \times 250 = 37\,500 \text{ mm}^2$, $T = 30 \times 10^6 \text{ N mm}$, shear flow $q = 400 \text{ N mm}^{-1}$, τ_3 shear stress in flange due to torsion $= q/t = 400/20 = 20 \text{ N mm}^{-2}$, τ_2 in the web $= q/10 = 40 \text{ N mm}^{-2}$.

(d) Net stresses in the flange: $\sigma_x = 44.8$, $\tau = \tau_3 = 20$, $\sigma_1 = 52.8$, $\sigma_2 = -7.2$.

(e) Net stresses in the web: $\sigma_x = 0$, $\tau = \tau_1 + \tau_2 = 26.5 + 40 = 66.5$, $\sigma_1 = 66.5$, $\sigma_2 = -66.5$, a case of pure shear.

17. A symmetrical *I*-section with the following dimensions is used as a short column to support an axial load of 2500 kN and bending moments of 100 kN m and 80 kN m about the zz (major) and yy (minor) axis respectively. Calculate the maximum tensile and compressive stresses in the cross-section. Also calculate the dimensions of the core of the section.
Flanges: 375 mm × 27 mm
Web: 17 mm thick
Overall depth = 375 mm

Answer:
(a) Stress calculation

$A = 25.71 \times 10^3 \text{ mm}^2$, $I_{xx} = 661.2 \times 10^6 \text{ mm}^4$

$Z_{zz} = I_{zz}/(375/2) = 3.53 \times 10^6 \text{ mm}^3$

$I_{yy} = 237.4 \times 10^6 \text{ mm}^4$, $Z_{yy} = 1.27 \times 10^6 \text{ mm}^3$

$P/A = 97 \text{ N mm}^{-2}$, $M_{zz}/Z_{zz} = \pm28.3 \text{ N mm}^{-2}$, $M_{yy}/Z_{yy} = \pm63.0 \text{ N mm}^{-2}$

$\sigma = (-188 \text{ and } -6) \text{ N mm}^{-2}$

The core is a symmetrical diamond shape with overall height = 276 mm and overall width = 98 mm.

18. A T-section with the following dimensions is used as a short column to support an axial load of 1000 kN applied at eccentricities of 50 mm along the y-axis (vertical) and 20 mm along the z-axis (horizontal) respectively. Calculate the maximum tensile and compressive stresses in the cross-section. Also calculate the dimensions of the core of the section.
Flange: 500 mm × 100 mm
Web: 100 mm thick
Overall depth = 600 mm

Answer:
(a) Stress calculation
$A = 100 \times 10^3 \text{ mm}^2$, Centroidal axis = 400 mm from below,
$I_{zz} = 3330.0 \times 10^6 \text{ mm}^4$

$Z_{zz \text{ top}} = 16.65 \times 10^6 \text{ mm}^3$,
$Z_{zz \text{ bottom}} = 8.33 \times 10^6 \text{ mm}^3$

$I_{yy} = 1083.0 \times 10^6 \text{ mm}^4$

Z_{yy} for flange $= 4.33 \times 10^6 \text{ mm}^3$, Z_{yy} for web $= 21.65 \times 10^6 \text{ mm}^3$

$P/A = -1000 \times 10^3/(100 \times 10^3) = -10 \text{ N mm}^{-2}$

$M_{zz} = P \, e_y = 1000 \text{ kN} \times 50 \text{ mm} = 50 \text{ kN m}$

$M_{zz}/Z_{zz} \text{ Top} = -50 \times 10^6/(16.65 \times 10^6) = -3.0 \text{ N mm}^{-2}$

$M_{zz}/Z_{zz} \text{ Bottom} = +50 \times 10^6/(8.33 \times 10^6) = +6.0 \text{ N mm}^{-2}$

$M_{yy} = P \, e_z = 1000 \text{ kN} \times 20 \text{ mm} = 20 \text{ kN m}$

$M_{yy}/Z_{yy} \text{ flange} = \pm20 \times 10^6/(4.33 \times 10^6) = \pm4.6 \text{ N mm}^{-2}$

$M_{yy}/Z_{yy} \text{ web} = \pm20 \times 10^6/(21.65 \times 10^6) = \pm0.9 \text{ N mm}^{-2}$

σ compressive $= -17.6 \text{ N mm}^{-2}$ in the flange at the top right-hand tip

σ tensile $= -3.1 \text{ N mm}^{-2}$ in the web at the bottom left-hand tip.

The core is a symmetrical diamond shape with the height above = 83 mm and height below = 167 mm and overall width = 86 mm.

3.21 ● Examples for practice

1. At a point in a 2-D structure, the following stresses were recorded. Calculate (i) principal stresses and their directions and (ii) Maximum shear stress and the directions of the planes on which it acts.

(a) $\sigma_x = -240$ N mm^{-2}, $\sigma_y = 120$ N mm^{-2},
$\tau_{xy} = 100$ N mm^{-2}

Answer: $\sigma_1 = 146$ N mm^{-2} at 75.5° to the horizontal and $\sigma_2 = -266$ N mm^{-2} at -14.5° to the horizontal. $\tau_{max} = 206$ N mm^{-2} accompanied by a normal stress of -60 N mm^{-2} acting on planes inclined at 30.5° and 120.5° to the horizontal.

(b) $\sigma_x = 120$ N mm^{-2}, $\sigma_y = -80$ N mm^{-2},
$\tau_{xy} = -140$ N mm^{-2}

Answer: $\sigma_1 = 192$ N mm^{-2} at -27.2° to the horizontal and $\sigma_2 = -152$ N mm^{-2} at 62.8° to the horizontal. $\tau_{max} = 172$ N mm^{-2} accompanied by a normal stress of 20 N mm^{-2} acting on planes inclined at 17.8° and 107.8° to the horizontal.

2. A symmetrical I beam with the following dimensions:
Flange: width = 400 mm, thickness = 15 mm
Web: thickness = 12 mm
Overall depth = 450 mm
has a 20 mm thick × 400 mm wide plate welded to the top flange.
Calculate the second moments of area about the horizontal and vertical axes and the corresponding section moduli. If the section is subjected to a bending moment of 500 kN m about the zz-axis so as to cause tension on the bottom fibres, calculate the tensile and compressive stresses in the cross-section.

Answer: Centroid = 300 mm from below,
$I_{zz} = 943 \times 10^6$ mm^4, $I_{yy} = 267 \times 10^6$ mm^4,
$Z_{zz\ bottom} = 3.14 \times 10^6$ mm^3, $Z_{zz\ top} = 5.55 \times 10^6$ mm^3, $Z_{yy} = 1.33 \times 10^6$ mm^3, stresses: top = -90 and bottom $+159$ N mm^{-2}.

3. An unsymmetrical I beam has the following dimensions:
Top flange: width = 250 mm, thickness = 16 mm
Bottom flange: width = 350 mm, thickness = 16 mm
Web: thickness = 12 mm
Overall depth = 680 mm.
Calculate the second moments of area about the horizontal and vertical axes and the corresponding section moduli. If the section is subjected to a bending moment of 900 kN m about the zz-axis so as to cause tension on the top fibres, calculate the tensile and compressive stresses in the cross-section.

Answer: Centroid = 309 mm from below,
$I_{zz} = 1314 \times 10^6$ mm^4, $I_{yy} = 78 \times 10^6$ mm^4,
$Z_{zz\ bottom} = 4.25 \times 10^6$ mm^3,
$Z_{zz\ top} = 3.54 \times 10^6$ mm^3, $Z_{yy} = 0.446 \times 10^6$ mm^3,
stresses: top = 254 and bottom -212 N mm^{-2}.

4. A T-beam has the following dimensions:
Flange: width = 1200 mm, thickness = 300 mm
Web: thickness = 300 mm
Overall depth = 1000 mm
Calculate the second moments of area about the horizontal and vertical axes and the corresponding section moduli.

Answer: Centroid = 666 mm from below,
$I_{zz} = 4.44 \times 10^{10}$ mm^4, $I_{yy} = 4.48 \times 10^{10}$ mm^4,
$Z_{zz\ bottom} = 66.68 \times 10^6$ mm^3, $Z_{zz\ top} = 133 \times 10^6$ mm^3, $Z_{yy} = 74.67 \times 10^6$ mm^3.

5. A Z-section has the following properties. Calculate the stress in the cross-section when subjected to bending moments of 70 and 40 kN m about the horizontal and vertical axes respectively. Also calculate the principal second moments of area of the cross-section.
Flanges: 150 mm, overall depth = 300 mm, thickness = 12 mm uniform.

Answer: $I_{zz} = 95.72 \times 10^6$ mm^4, $I_{yy} = 23.93 \times 10^6$ mm^4, $I_{yz} = 35.76 \times 10^6$ mm^4, $\sigma_x = -3.07$ y $+ 6.255$ z, At flange tips $\sigma_x = 440$ N/mm mm^2, $I_{11} = 110.5 \times 10^6$ mm^4, $I_{22} = 9.16 \times 10^6$ mm^4.

6. A symmetrical I section in question 2 above is used as a column. If the section is subjected to an axial load of 850 kN and bending moments of 350 kN m about the zz-axis so as to cause tension on the bottom fibres and 120 kN m about the yy-axis so as to cause tension on the right-hand side fibres, calculate the maximum tensile and compressive stresses in the cross-section. Also calculate the core of the section.

Answer: Area = 17.04×10^3 mm^2, $I_{zz} = 642 \times 10^6$ mm^4, $I_{yy} = 160 \times 10^6$ mm^4, $Z_{zz} = 2.85 \times 10^6$ mm^3, $Z_{yy} = 0.80 \times 10^6$ mm^3, stresses: top left-hand tip of flange = -323 N mm^{-2} and bottom right-hand tip of flange $+173$ N mm^{-2}. Core is a symmetrical diamond shape, e_y maximum = 167 mm, e_z maximum = 47 mm.

7. A T-section has the following dimensions:
Flange: width = 100 mm, thickness = 10 mm
Web: thickness = 10 mm
Overall depth = 100 mm
If the section is subjected to an axial load of 50 kN and bending moments of 1.50 kN m about the zz-axis so as to cause tension on the bottom fibres and 1.25 kN m about the yy-axis so as to cause

tension on the right-hand side fibres, calculate the maximum tensile and compressive stresses in the cross-section. Also calculate the core of the section.

Answer: Area = 1.9×10^3 mm^2, centroidal axis 71.3 mm from below, $I_{zz} = 1.80 \times 10^6$ mm^4, $I_{yy} = 0.84 \times 10^6$ mm^4, Z_{zz} top = 62.72×10^3 mm^3, Z_{zz} bottom = 25.25×10^3 mm^3, Z_{yy} = 16.82 $\times 10^3$ mm^3, stresses: top right-hand tip of flange = -148 N mm^{-2} and bottom left-hand tip of web = $+17$ N mm^{-2}. Core is an unsymmetrical diamond shape, e_y maximum top = 13 mm, e_y maximum bottom = 33 mm, e_z maximum = 9 mm.

8. A T-section in question 7 above is used as a beam. If the cross-section is subjected to a vertical shear force of 50 kN, calculate the maximum shear stresses in the cross-section. Compare this with the average shear stress in the web.

Answer: Centroidal axis 71.3 mm from below, $I_{zz} = 1.80 \times 10^6$ mm^4, $\int y\, dA = 25.4 \times 10^3$ mm^3, $\tau = 70.6$ N mm^{-2} at the neutral axis in the web. Average shear stress = 50 N mm^{-2}, maximum shear stress = $1.41 \times$ average shear stress.

9. Determine expressions for the shear stress distribution and the maximum shear stress in the following cross-sections when used as a beam.

(i) a solid circular cross-section of radius R.

Answer: Measuring the angle θ from the horizontal, width = $2R \cos \theta$, $y = R \sin \theta$,

$$I_{zz} = \frac{\pi R^4}{4}, \int y\, dA = \int_{\theta}^{\pi/2} 2R^3 \cos^2 \theta \sin \theta\, d\theta = \frac{2R^3 \cos^3 \theta}{3},$$

$$\tau = \frac{4Q \cos^2 \theta}{3\pi R^2}$$

Maximum shear stress = $1.33 \times$ average shear stress.

(ii) A square section of side a, placed with a diagonal horizontal.

Answer: Diagonal length = $\sqrt{2}\, a$. Consider the cross-section as made up of two triangles of height h and base $2h$, where $h = a/\sqrt{2}$, $I_{zz} = a^4/12$,

$\tau = \{Q/a^2\}(1 - y/h)[1 + 2(y/h)]$, y = distance from the neutral axis to the level at which the shear stress is required, $\tau_{max} = \{Q/a^2\} = \tau_{average}$

10. An unsymmetrical I beam in question 3 above is used as a beam.
If the cross-section is subjected to a shear force of 800 kN, calculate the maximum and average shear stress in the web. Also calculate the maximum strength of the welds required to connect the top and bottom flanges to the web.

Answer: Centroid = 309 mm from below, and 371 mm from above, $I_{zz} = 1314 \times 10^6$ mm^4, $\int y\, dA = 2.208 \times 10^6$ mm^3 at neutral axis, $\tau = 112$ N mm^{-2}, average shear stress $\tau = 98$ N mm^{-2}, $\int y\, dA = 1.45 \times 10^6$ mm^3 for top flange and $\int y\, dA = 1.58 \times 10^6$ mm^3 for bottom flange, weld strength for top flange = 441 N mm^{-1} per weld and for bottom flange 481 N mm^{-1} per weld.

11. A plate 60 mm wide and 10 mm thick is attached to a support through a bolt of diameter d. The bolt is in double shear. The plate is subjected to a pull of P kN. The permissible stress of plate material in tension is equal to 300 N mm^{-2} and the permissible stress of bolt material in shear is equal to 170 N mm^{-2}. What is the optimum diameter of the bolt in order to obtain maximum value of the load P?

Answer: $d = 19.73$ mm, in practice a 20 mm diameter bolt will be used.

12. A rectangular beam is cut from a log of diameter D. Determine the dimensions of the beam in terms of D in order to get a beam of maximum strength. Also determine the ratio of the section modulus for the rectangular beam to that of the log.

Answer: $b = D/\sqrt{3} = 0.5774\, D$, therefore $h = \sqrt{2}\, D/\sqrt{3} = 0.8165 D$
$Z = bh^2/6 = 0.0642\, D^3$
The section modulus of a circular section = $\pi D^3/32$ = $0.0982\, D^3$
The ration of $Z_{beam}/Z_{log} = 0.6539$.

Chapter four

Deformation of elements

In the design of structures the designer has to ensure that under working loads two important criteria are satisfied:

1. the structure is capable of resisting the applied forces without the stresses anywhere exceeding the permissible stresses;
2. the structure must not undergo unacceptably large deformations.

The calculation of stresses at a section due to a given set of stress resultants was discussed in Chapter 3. This chapter is concerned with the calculation of the displacements due to applied forces.

4.1 ● Differential equation for axial deformation

Consider the bar subjected to a distributed axial load $q(x)$ as shown in Fig. 4.1(a). Because of the distributed axial load, the axial force in the bar is not constant. Let the axial force at two neighbouring sections dx apart be F and $\{F + (dF/dx)dx\}$ as shown in Fig. 4.1(b). If $q(x)$ as a function of x is the distributed axial force per unit length, then the equilibrium equation is given by

$$-F + \{F + (dF/dx)dx\} + q(x)dx = 0$$

Simplifying

$$dF/dx + q(x) = 0 \tag{4.1}$$

This differential equation is in terms of forces and is independent of material properties.

In order to determine the displacements, we have to express the axial force F in terms of the axial displacement $u(x)$. The axial strain $\varepsilon_x = \partial u/\partial x$. However, because it is assumed that plane sections remain plane, i.e. the strain ε_x is independent of y, then $\varepsilon_x = du/dx$. From Hooke's law, the axial stress $\sigma_x = E\varepsilon_x$, where E = Young's modulus. The axial force $F = A\sigma_x$, where A = cross-sectional area. Therefore

$$F = AE \, du/dx$$

Substituting for F in terms of the displacement u, the differential equation of equilibrium [4.1] becomes

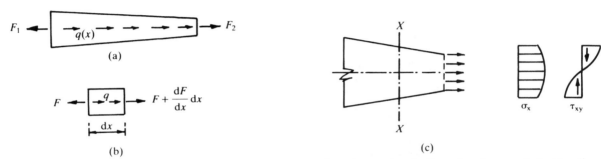

Figure 4.1 ● A bar subjected to tensile forces: (a) a non-uniform bar under distributed and concentrated tensile load, (b) an infinitesimal element in equilibrium, (c) stress distribution in a non-uniform bar in tension

$$d/dx\,(AE\,du/dx) + q(x) = 0 \tag{4.2}$$

If it is assumed that the axial rigidity AE is constant, then the differential equation is simplified to

$$AE\,d^2u/dx^2 + q(x) = 0 \tag{4.3}$$

This is a fundamental differential equation for an axially loaded bar with constant properties along the axis.

Note: In the derivation of the equation it was assumed that irrespective of the variation of AE along the x-direction, σ_x is the only stress present and that it is uniformly distributed in the cross-section. From the discussion of 'stress flow' in Section 3.17, we would expect the 'stress flow' to produce normal and shear stress distribution in a cross-section to be non-uniform as shown in Fig. 4.1(c). Exact analysis shows that if the taper is less than 20° to the horizontal, the error in assuming a uniform distribution of σ_x and zero shear stress is not significant.

4.1.1 Examples of the calculation of axial displacement

Example 1

Calculate the axial displacement of the bar shown in Fig. 4.2(a).

Figure 4.2 ●
A uniform bar under tensile forces:
(a) a concentrated load at the end, (b) uniform distributed load,
(c) linearly varying load,
(d) a fixed ended bar under uniformly distributed load

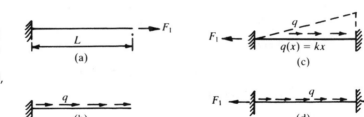

Solution In this case $q = 0$. Therefore, from equation [4.3],

$$d^2u/dx^2 = 0$$

Integrating successively with respect to x

$$du/dx = C_1$$

$$u = C_1x + C_2$$

where C_1 and C_2 are constants of integration.

The boundary conditions are $u = 0$ at $x = 0$ because of fixity and $F = F_1$ at $x = L$. However, because $F = AE \, du/dx$, $du/dx = F_1/(AE)$ at $x = L$. Using the boundary conditions and solving for C_1 and C_2, we get

$$C_2 = 0, \; C_1 = F_1/(AE)$$

Therefore $u = F_1x/(AE)$, $du/dx = F_1/(AE)$ and $F = AE \, du/dx = F_1$. The axial displacement Δ at the tip is given by u when $x = L$. Therefore $\Delta = F_1L/(AE)$.

The results can be summarized as follows.

For an axially loaded bar of constant cross-section fixed at one end and subjected to an axial force at the other end:

(a) The displacement varies linearly, i.e. $u = F_1x/(AE)$.
(b) The strain and therefore the axial force is constant, i.e. $F = F_1$.
(c) The axial extension is given by $\Delta = F_1L(AE)$. The axial stiffness of the bar, defined as the force F_1 required for unit extension Δ, is therefore given by AE/L. AE is often called the axial rigidity of the cross-section.

Example 2

Calculate the axial displacement in a bar subjected to a uniformly distributed axial load of constant intensity q per unit length as shown in Fig. 4.2(b).

Solution From differential equation [4.3],

$$d^2u/dx^2 = -q/(AE)$$

where $q = $ constant. Integrating successively with respect to x

$$du/dx = -\{q/(AE)\}x + C_1$$

$$u = -qx^2/(2AE) + C_1x + C_2$$

The boundary conditions are $u = 0$ at $x = 0$ because of fixity and $F = AE \, du/dx = 0$ at $x = L$ because the end $x = L$ is free from force. Therefore at $x = L$, $AE \, du/dx = 0$ or $du/dx = 0$. Solving for C_1 and C_2,

$$C_2 = 0 \text{ and } C_1 = qL/(AE)$$

Substituting for C_1 and C_2 in the expression for u,

$$u = qx(2L - x)/(2AE)$$

Therefore

$$F = AE \, du/dx = q(L - x)$$

The extension Δ of the bar at the free end is given by u at $x = L$. Therefore

$$\Delta = qL^2/(2AE)$$

If the *total* applied force is F_1, then $F_1 = qL$. Therefore $\Delta = 0.5F_1L/(AE)$. The results indicate that:

(a) The displacement of the bar varies parabolically. It is zero at the fixed end and has a value $qL^2/(2AE)$ at the free end.

(b) The strain and therefore the axial force varies linearly. It is a maximum at the fixed end and zero at the free end.

(c) The total axial extension $\Delta = 0.5F_1L/(AE)$ is only half the extension that would have occurred if a concentrated force equal to F_1 had been applied at the tip.

Example 3

Determine the displacement and stresses if q varies linearly with respect to x and the ends are fixed as shown in Fig. 4.2(c).

Solution In this case let $q(x) = kx$, where k = constant and from equation [4.3],

$$AE \, d^2u/dx^2 = -kx$$

$$AE \, du/dx = -kx^2/2 + C_1$$

$$AEu = -kx^3/6 + C_1x + C_2$$

The boundary conditions are $u = 0$ at $x = 0$ and $x = L$ because both ends are fixed. Using the boundary conditions, the constants of integration are given by $C_2 = 0$ and $C_1 = kL^2/6$. Substituting for C_1 and C_2,

$$AEu = k[-x^3 + L^2x]/6$$

$$F = AE \, du/dx = k(L^2 - 3x^2)/6$$

The axial forces at the ends of the bar are given at $x = 0$ by $F_1 = F = kL^2/6$ and at $x = L$ by $F_2 = -F = kL^2/3$; where because F_2 is a compressive force, it is treated as a negative force.

Example 4

Calculate the forces at the ends of a prismatic bar fixed at the ends and carrying a uniformly distributed axial load q as shown in Fig. 4.2(d).

Solution As in Example 2,

$$u = -qx^2/(2AE) + C_1x + C_2$$

Because the ends are fixed, $u = 0$ at $x = 0$ and $x = L$. Therefore $C_2 = 0$ and $C_1 = qL/(2AE)$. Therefore

$$u = \{q/(2AE)\}(Lx - x^2)$$

$$\mathrm{d}u/\mathrm{d}x = \{q/(2AE)\}(L - 2x)$$

Axial force $F = AE\,\mathrm{d}u/\mathrm{d}x = q(L - 2x)/2$

Axial force at $x = 0$ is $qL/2$ and at $x = L$ is $-qL/2$. The negative sign in the force at $x = L$ indicates it is a compressive force. Note that the axial force is tensile in the left half of the bar, zero at the middle and compressive in the right half of the bar.

Example 5

Calculate the forces at the ends of a prismatic bar fixed at the ends and subjected to a concentrated force F_c at $x = a$ as shown in Fig. 4.3(a).

Figure 4.3 ● (a), (b)
A fixed ended bar
subjected to an off-
centre axial force

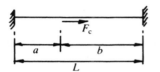
(a)

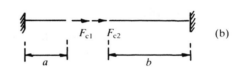
(b)

Solution Let the load F_c be shared between the left and right parts of the bar as shown in Fig. 4.3(b). The left part of the bar is in uniform tension of F_{c1} and the right part is in uniform compression of F_{c2}. Using the result of Example 1, Section 4.1.1, the extension Δ_1 of the left part of the bar is

$$\Delta_1 = F_{c1}a/(AE)$$

Similarly the contraction Δ_2 of the right part of the bar is

$$\Delta_2 = F_{c2}b/(AE)$$

where $b = L - a$. But $\Delta_1 = \Delta_2$ in order that no gap develops at the junction of the left and right sections of the bar. Therefore

$$F_{c1}a = F_{c2}b$$

But

$$F_{c1} + F_{c2} = F_c$$

$$\therefore F_{c1} = F_c b/L \text{ and } F_{c2} = F_c a/L$$

Example 6

A rigid block weighing 200 kN is supported by three short columns as shown in Fig. 4.4. The cross-sectional areas of the middle column and the outer columns are respectively $2A$ and A. Calculate the force in each column.

Solution This is a statically indeterminate structure, because the block is supported by three vertical reactions which cannot be determined by statics. Because of symmetry the

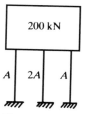

Figure 4.4 ●
A statically
indeterminate structure

outer columns will have the same force. Further, because the block is rigid and therefore cannot deform, it must remain level at all times. This requires that all the three columns must contract by the same amount.

Let the force in the outer columns be F_1 and in the inner column be F_2. Clearly for equilibrium,

$$2F_1 + F_2 = 200 \text{ kN}$$

Let the axial contraction of all three columns be Δ. Then

$$\Delta = F_1L/(AE) = F_2L/(2AE)$$

Therefore

$$F_1 = 0.5F_2$$

Solving $F_1 = 50$ kN and $F_2 = 100$ kN.

This example shows a very important principle, namely that a total force will be distributed among members in proportion to their 'stiffness'. In the example considered, the axial stiffness of the end columns was AE/L and that of the middle column was $(2AE)/L$. Because the central column was twice as stiff as the outer columns, it absorbed twice the force of the outer columns. It has to be remembered that stiffness is *directly* proportional to cross-sectional area A and *inversely* proportional to the length L of the member.

Example 7

A hollow steel tube is filled with concrete and used as a short column. The steel section has a diameter of 200 mm and a wall thickness of 10 mm. The Young's moduli for steel and concrete are $E_s = 210$ kN mm^{-2}, $E_c = 30$ kN mm^{-2}. If a total axial load of P is applied to the column, what proportion of the load is taken by each material?

Solution   Let P_s and P_c be the load carried by steel and concrete respectively. For equilibrium,

$$P = P_s + P_c$$

Because both materials contract by the same amount, for compatibility,

$$P_sL/(A_sE_s) = P_cL/(A_cE_c)$$

where L = length of column and A_s and A_c are the areas of cross-section of steel and concrete respectively. Therefore,

$$P_s = P_c\{(A_sE_s)/(A_cE_c)\}$$

Substituting for

$$A_s = \pi(200^2 - 180^2)/4 = 5.97 \times 10^3 \text{ mm}^2, \quad A_c = \pi(180^2)/4 = 25.45 \times 10^3 \text{ mm}^2$$

$$A_s/A_c = 0.235, \quad E_s/E_c = 7.0$$

$$P_s = 1.642P_c$$

Therefore

$$P_s = 0.622P, \ P_c = 0.378P$$

This is another example of the basic idea that the stiffer the section, the more load it attracts. Thus although the steel section has only a quarter of the area of the concrete section, it attracts 62 per cent of the applied load. The reason for this is the higher Young's modulus of steel compared to that of concrete.

In practice it often happens that a particular member is found to be over-stressed. Then there are two options open to reduce the level of stress in the member. The first option is to increase the area of cross-section of the over-stressed member. This inevitably increases the stiffness of the member and therefore *increases* the share of the total force and therefore might *not* really solve the original problem of an over-stressed member. The second option is to increase the area of cross-section of other members so that the share of the force by the over-stressed member is decreased. Designers routinely use these options to 'optimize', i.e. minimize, the total consumption of material in a structure.

4.1.2 Stiffness matrix for a bar element

Example 1

For the axially loaded bar in Fig. 4.5, establish the relationship between the axial forces and the corresponding axial displacements at the ends of the bar. Assume that q is constant. It is given that the axial force and displacement at $x = 0$ and $x = L$ are respectively (F_{a1}, Δ_{a1}) and (F_{a2}, Δ_{a2}).

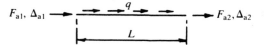

Figure 4.5 ● A bar element

Solution As in Example 2 of Section 4.1.1,

$$u = -qx^2/(2AE) + C_1x + C_2$$

The boundary conditions are $u = \Delta_{a1}$ at $x = 0$ and $u = \Delta_{a2}$ at $x = L$. Therefore $C_2 = \Delta_{a1}$, $C_1 = (\Delta_{a2} - \Delta_{a1})/L + qL/(2AE)$.
This leads to

$$u = (1 - x/L)\Delta_{a1} + (x/L)\Delta_{a2} + qx(L - x)/(2AE)$$

$$F = AE \ du/dx = (AE/L)\{\Delta_{a2} - \Delta_{a1}\} + q(L - 2x)/2$$

At $x = 0$, $F_{a1} = -F = -AE/L(\Delta_{a2} - \Delta_{a1}) - 0.5qL$

Note that F_{a1} is a compressive force.

At $x = L$, $F_{a2} = F = AE/L(\Delta_{a2} - \Delta_{a1}) - 0.5qL$

The above two equations can be displayed in matrix format as

$$\begin{bmatrix} F_{a1} \\ F_{a2} \end{bmatrix} = AE/L \begin{bmatrix} 1 & -1 \\ -1 & 1 \end{bmatrix} \begin{bmatrix} \Delta_{a1} \\ \Delta_{a2} \end{bmatrix} + \begin{bmatrix} -0.5qL \\ -0.5qL \end{bmatrix} \tag{4.4}$$

In the above equation if $\Delta_{a1} = \Delta_{a2} = 0$, then

$$\begin{bmatrix} F_{a1} \\ F_{a2} \end{bmatrix} = \begin{bmatrix} -0.5qL \\ -0.5qL \end{bmatrix}$$

The forces which result at the ends due to the load on the bar when the ends are prevented from displacing are called fixed end forces and the corresponding force vector as the fixed end force vector.

Similarly, if $q = 0$, then

$$\begin{bmatrix} F_{a1} \\ F_{a2} \end{bmatrix} = AE/L \begin{bmatrix} 1 & -1 \\ -1 & 1 \end{bmatrix} \begin{bmatrix} \Delta_{a1} \\ \Delta_{a2} \end{bmatrix}$$

The above is a relationship between the forces at the ends of the bar and the corresponding displacements at the ends. Because the relationship in a spring between the force and the corresponding extension is known as the stiffness of the spring, by analogy we have,

{Force vector} = {Stiffness matrix} {Displacement vector}

where

$$\{\text{Stiffness matrix}\} = AE/L \begin{bmatrix} 1 & -1 \\ -1 & 1 \end{bmatrix}$$

$$\{\text{Displacement vector}\} = \begin{bmatrix} \Delta_{a1} \\ \Delta_{a2} \end{bmatrix}$$

$$\{\text{Force vector}\} = \begin{bmatrix} F_{a1} \\ F_{a2} \end{bmatrix}$$

Therefore the relationship between the forces and displacements can be displayed as

{Force vector} = {Stiffness matrix} {Displacement vector}
+ {Fixed end force vector}

In general therefore for a bar element

$$\begin{bmatrix} F_{a1} \\ F_{a2} \end{bmatrix} = AE/L \begin{bmatrix} 1 & -1 \\ -1 & 1 \end{bmatrix} \begin{bmatrix} \Delta_{a1} \\ \Delta_{a2} \end{bmatrix} + \begin{bmatrix} F_{fa1} \\ F_{fa2} \end{bmatrix}$$

where F_{fa1} and F_{fa2} are the fixed end forces at the ends 1 and 2 respectively due to axial loads applied to the bar along its length.

The stiffness matrix for a bar element derived above forms the basis of a popular computer-oriented method of truss analysis. The use of this equation for the analysis of trusses is discussed in detail in Chapter 6.

Loading	F_{fa1}	F_{fa2}
u.d.l. q	$-0.5qL$	$-0.5qL$
Off-centre load F_c	$-F_c b/L$	$-F_c a/L$

4.1.3 Fixed end forces for a bar element

The results of Examples 4 and 5 of Section 4.1.1 can be summarized in Table 4.1 for future use.

4.2 ● Differential equation for bending deformations

It was shown in Section 3.9.2 that for sections with the principal axes parallel to the y- and z-axes, the stress due to bending, assuming that plane sections remain plane, is given by

$$\sigma_x = -M_z y/I_{zz} + M_y z/I_{yy}$$

If σ_x is the only stress present, then the strain in the x-direction is

$$\varepsilon_x = \sigma_x/E = -M_z y/(EI_{zz}) + M_y z/(EI_{yy})$$

From the strain–displacement relationship $\varepsilon_x = \partial u/\partial x$, therefore

$$\partial u/\partial x = -M_z y/(EI_{zz}) + M_y z/(EI_{yy})$$

Assuming for simplicity that $M_y = 0$, and integrating partially with respect to x, $u = -y \int (M_z/EI_{zz})\,dx + C(y)$ where $C(y) = $ a 'constant' of integration which because of integrating partially with respect to x can be a function of y.

The equation gives only the axial displacement. However, as can be verified by applying a bending moment to a simple beam, the major displacement of design interest is the displacement in the plane of bending, v. Therefore an attempt must be made to obtain an expression relating the applied moment M_z and the resulting displacement v. If σ_x is the only stress present, then shear stress τ_{xy} and shear strain γ_{xy} must be zero. Therefore $\gamma_{xy} = \partial u/\partial y + \partial v/\partial x = 0$. Then $\partial v/\partial x = -\partial u/\partial y$. $\partial u/\partial y$ can be calculated by partially differentiating u with respect to y. Therefore

$$\partial u/\partial y = -\int (M_z/EI_{zz})\,dx + dC(y)/dy$$

and

$$\partial v/\partial x = -\partial u/\partial y = \int (M_z/EI_{zz})\,dx - dC(y)/dy$$

Differentiating the above expression with respect to x in order to eliminate $C(y)$, we have

$$\partial^2 v/\partial x^2 = M_z/(EI_{zz})$$

Note that because $C(y)$ is a function of y only, when differentiated with respect to x, $\partial C(y)/\partial x = 0$. Because both M_z and EI_{zz} are functions of x only we have, replacing $\partial^2 v/\partial x^2$ by $d^2 v/dx^2$

$$EI_{zz}\, d^2 v/dx^2 = M_z \tag{4.5}$$

Proceeding in an identical manner but assuming that the shear strain $\gamma_{xz} = \partial u/\partial z + \partial w/\partial x = 0$ where $w =$ displacement in the z-direction, it can be shown that

$$EI_{yy}\, d^2 w/dx^2 = -M_y \tag{4.5a}$$

These are fundamental differential equations connecting the bending moments M_z and M_y and the corresponding second derivative of displacements v and w. Assuming that only M_z is present, instead of relating the derivative of displacement v to the bending moment, which can be calculated in the case of statically determinate structures only, it is possible to relate it directly to the external load acting on the beam. This can be done as follows.

Referring to Section 1.12 and Fig. 1.62, the equilibrium relationship between the applied loading q, the shear force Q and bending moment M_z is given by

$$dQ/dx = -q \text{ and } dM_z/dx = -Q$$

Substituting $EI_{zz}\, d^2 v/dx^2 = M_z$ we have

$$Q = -d/dx\{EI_{zz}\, d^2 v/dx^2\}$$

$$q = -dQ/dx = d^2/dx^2\{EI_{zz}\, d^2 v/dx^2\}$$

If the flexural rigidity EI_{zz} is constant, then

$$Q = -EI\, d^3 v/dx^3$$

$$q = EI_{zz}\, d^4 v/dx^4$$

where q is positive in the positive y-direction.

4.2.1 Interpretation of $d^2 v/dx^2$ in terms of curvature of the deformed beam

Consider an element of length dx of a beam subjected to a bending moment of M_z as shown in Fig. 4.6(a). Let the neutral axis of the element of the beam be assumed to deform into an arc of a circle of radius R as shown in Fig. 4.6(b). The neutral axis remains unstretched. Therefore $dx = R\, d\theta$, where $d\theta =$ subtended angle. The deformed length of any fibre at a distance y from the neutral axis is given by

$$ds = (R - y)\, d\theta$$

The strain in the fibre is given by

$$\varepsilon_x = (ds - dx)/dx = -y/R$$

However, according to the strain calculated from the stress σ_x, ε_x is given by $\varepsilon_x = \sigma_x/E$. Substituting $\sigma_x = -M_z y/I_{zz}$ we have

$$\varepsilon_x = -M_z y/(EI_{zz})$$

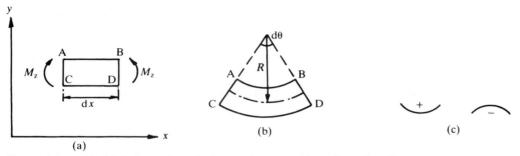

Figure 4.6 ● (a)–(c) Deformation of a beam element subjected to a bending moment

Equating the two expressions for e_x,

$$1/R = M_z/(EI_{zz})$$

$1/R$ is the curvature. Curvature is a measure of how 'curved' a given curve is. A straight line is not curved and therefore has zero curvature. It can be assumed to be a part of a circle of infinite radius R for which $1/R$ is therefore zero. It is shown in elementary texts on calculus that

$$1/R = d^2v/dx^2/\{1 + (dv/dx)^2\}^{3/2}$$

When deformation is assumed to be small, dv/dx remains small and $(dv/dx)^2$ can be disregarded, leading to

$$1/R \approx d^2v/dx^2 = M_z/(EI_{zz})$$

The sign convention adopted for curvature is shown in Fig. 4.6(c). The main advantage of this interpretation is that it enables us to relate bending moment distribution to the corresponding deflected form. For example, if there is no change in the sign of the bending moment, as in a simply supported beam, then the corresponding curvature also does not change in sign.

4.2.2 Application of $M_z = EI_{zz}\, d^2v/dx^2$ in the presence of bending moment and shear force

The equation $M_z = EI_{zz}\, d^2v/dx^2$ is strictly applicable only if shear strain is zero. In general for beam problems, shear force and bending moment coexist. Obviously the equation $EI_{zz}\, d^2v/dx^2 = M_z$ is not strictly applicable in such cases. However, experimental evidence and more rigorous analysis indicate that in the vast majority of cases the error involved in assuming that the shear strain is zero, even in the presence of shear forces, is very small and can be safely ignored. Further comments on this aspect are given in Section 4.13.

4.3 ● Examples of the calculation of bending deformations in cantilever beams

In the case of cantilever beams, only one end is supported and the beam is generally built into something 'solid' as shown in Fig. 1.44. The forces developed over the embedded end are complex and for simplicity the beam is assumed to be fixed at the edge of the 'foundation' as shown in Fig. 1.44. The boundary conditions, i.e. the constraints on the displacement at the fixed end, are taken to be $v = 0$ and $dv/dx = 0$. From now on it is assumed that bending takes place in the vertical x–y plane. Therefore M_z and I_{zz} are used without subscripts as M and I respectively.

Example 1

Calculate the bending deformations in the cantilever beam shown in Fig. 4.7(a).

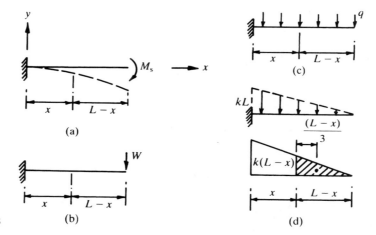

Figure 4.7 ● (a)–(d)
A cantilever under
various external loadings

Solution The bending moment M at a section x is equal to $-M_s$, the moment at the tip of the cantilever. The reason for the negative sign is that bending moment is considered positive here if it causes tension in the bottom face. Therefore

$$M = EI\ d^2v/dx^2 = -M_s$$

Integrating successively with respect to x,

$$EI\ dv/dx = -M_s x + C_1$$

$$EIv = -M_s x^2/2 + C_1 x + C_2$$

The constants of integration C_1 and C_2 can be determined from the boundary conditions at the support, i.e. $v = dv/dx = 0$ at $x = 0$. This gives $C_1 = C_2 = 0$. Therefore

$$EI\ dv/dx = -M_s x$$

$$EIv = -M_s x^2/2$$

The deflection and slope at the tip $x = L$ are given by

$$v_{(x=L)} = -M_s L^2/(2EI)$$

$$(dv/dx)_{(x=L)} = -M_s L/(EI)$$

The minus sign indicates that the deflection is downwards in the negative direction of y and the rotation is clockwise.

Example 2

Calculate the bending deformations in the cantilever beam shown in Fig. 4.7(b).

Solution In this case the bending moment M at a section x from the origin is given by $M = -W(L - x)$. Therefore

$$EI\,d^2v/dx^2 = -W(L - x)$$

Successively integrating with respect to x

$$EI\,dv/dx = -W(Lx - x^2/2) + C_1$$

$$EIv = -W(Lx^2/2 - x^3/6) + C_1x + C_2$$

The boundary conditions are $v = dv/dx = 0$ at $x = 0$. Therefore $C_1 = C_2 = 0$ and

$$EI(dv/dx) = -W(Lx - x^2/2)$$

$$v = -W(3Lx^2 - x^3)/(6EI)$$

The deflection and slope at the tip at $x = L$ are

$$v_{(x=L)} = -WL^3/(3EI)$$

$$dv/dx_{(x=L)} = -WL^2/(2EI)$$

Example 3

Calculate the bending deformations in the cantilever beam shown in Fig. 4.7(c).

Solution In this case bending moment M at a section x is given by $M = -q(L - x)^2/2$. Therefore proceeding as in the previous examples,

$$EI\,d^2v/dx^2 = -q(L^2/2 - Lx + x^2/2)$$

$$EI\,dv/dx = -q(L^2x/2 - Lx^2/2 + x^3/6) + C_1$$

$$EIv = -q(L^2x^2/4 - Lx^3/6 + x^4/24) + C_1x + C_2$$

Using the boundary conditions $v = dv/dx = 0$ at $x = 0$, we have $C_1 = C_2 = 0$. Therefore

$$EI\,dv/dx = -q(L^2x/2 - Lx^2/2 + x^3/6)$$

$$EIv = -q(L^2x^2/4 - Lx^3/6 + x^4/24)$$

The deflection and slope at the tip at $x = L$ are given by

$$v_{(x=L)} = -qL^4/(8EI)$$

$$dv/dx_{(x=L)} = -qL^3/(6EI)$$

Example 4

Calculate the bending deflections in the cantilever beam subjected to linearly distributed load $q = k(L - x)$, where $k = $ constant, as shown in Fig. 4.7(d).

This situation commonly occurs when designing cantilever walls retaining soil or water.

Solution   The bending moment at a section x from the fixed support is given by the area of the hatched triangle times the distance of the centroid of the triangle from the section. Therefore

$$EI\, d^2v/dx^2 = -[0.5k(L-x)(L-x)](L-x)/3 = -k\{L^3 - 3L^2x + 3Lx^2 - x^3\}/6$$

$$EI\, dv/dx = -k\{L^3x - 3L^2x^2/2 + Lx^3 - x^4/4\}/6 + C_1$$

$$EIv = -k\{L^3x^2/2 - L^2x^3/2 + Lx^4/4 - x^5/20\}/6 + C_1x + C_2$$

Because at $x = 0$, $v = 0$ and $dv/dx = 0$, $C_1 = C_2 = 0$. The deflection and slope at the tip $x = L$ is given by

$$v_{(x=L)} = -kL^5/(30\,EI)$$

$$dv/dx_{(x=L)} = -kL^4/(24\,EI)$$

4.3.1   Standard cases of cantilever loading

The results of the examples discussed above are very useful and can be used to solve a variety of other cases of loading on cantilevers. The results are summarized in Table 4.2. In the table the negative signs for deflection and slope are omitted because from the given type of loading and support conditions it is obvious that the deflection is downwards and the rotation at the tip is clockwise.

It should be noted that if a concentrated load W is applied at the tip, then the tip deflection is $WL^3/(3\,EI)$. Similarly if the same load is uniformly distributed such that $q = W/L$, then the corresponding tip deflection is $qL^4/(8\,EI) = WL^3/(8\,EI)$. Therefore for the same value of the total load W, a concentrated load at the tip causes a deflection at the tip 2.67 times as large as that produced if the load W had been uniformly distributed. In the case of triangularly distributed loading, the total load on the cantilever is $W = 0.5kL^2$. Thus a cantilever carrying a concentrated load W at the tip deflects 5 times as much as a cantilever carrying the same load as a triangularly distributed load with the apex at the fixed end. Distributing the total load reduces the intensity of the bending moment and this reduces the bending deflection.

Table 4.2 ● Tip deflection and rotation in cantilever beams

Load case	v	dv/dx
End couple	$M_sL^2/(2EI)$	$M_s/(EI)$
End concentrated load	$WL^3/(3EI)$	$WL^2/(2EI)$
Uniformly distributed load	$qL^4/(8EI)$	$qL^3/(6EI)$
Triangular load with the apex at the fixed end	$kL^5/(30EI)$	$kL^4/(24EI)$

4.3.2 Solution of non-standard cases of cantilever loading using standard cases

The values of slopes and deflections calculated for standard cases of loading on a cantilever and shown in Table 4.2 can be used to solve more complicated cases of loading. The procedure used is illustrated by some simple examples.

Example 1

Calculate the deflection and slope at the tip of the cantilever shown in Fig. 4.8(a).

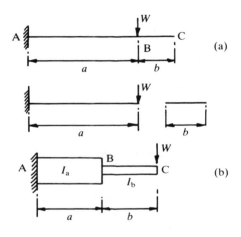

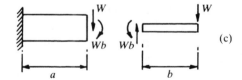

Figure 4.8 ●
(a)–(c) A cantilever
under external
concentrated load

Solution From the given loading, the bending moment is zero in the portion BC. Therefore

$$EI \, d^2v/dx^2 = 0$$

$$EI \, dv/dx = C_1$$

$$EIv = C_1x + C_2$$

Therefore the slope of portion BC is constant and it deflects as a straight line. However, for continuity of slope and deflection at its junction with AB, the portion BC must have at B the same slope and deflection as AB at B. From the summary of results for the standard loading cases given in Table 4.2, the deflection and slope of AB at B are

$$v = Wa^3/(3EI), \ dv/dx = Wa^2/(2EI)$$

Because the deflection and slope are small quantities,

Deflection at C = Deflection at B + {Slope at B}b

$$= Wa^3/(3EI) + Wa^2/(2EI)b$$

$$= Wa^3/(3EI)\{1 + 3b/(2a)\}$$

The deflection is downwards. The slope at C is the same as the slope at B = $Wa^2/(2EI)$.

Example 2

Calculate the deflection at the tip of the stepped cantilever shown in Fig. 4.8(b).

Solution Because the cantilever has different second moments of area for portions AB and BC, it is necessary to consider the free bodies as shown in Fig. 4.8(c).

(a) For portion AB, the slope and deflection at B are given by
  (i) Due to load W, $v = Wa^3/(3EI_a)$, $dv/dx = Wa^2/(2EI_a)$
  (ii) Due to moment $M_s = Wb$, $v = (Wb)a^2/(2EI_a)$, $dv/dx = (Wb)a/(EI_a)$
  The total deflection and slope are

$$v = [Wa^3/(3EI_a)]\{1 + 1.5b/a\}$$

$$dv/dx = [Wa^2/(2EI_a)]\{1 + 2b/a\}$$

(b) For the cantilever BC, the deflection and slope due to the load W, *if* the end B is held fixed, are given by

$$v = Wb^3/(3EI_b), \quad dv/dx = Wb^2/(2EI_b)$$

(c) The net deflection and slope at C are given by

Deflection at C

= deflection at C due to cantilever BC + deflection at B of cantilever AB + {slope at B of cantilever AB}b

$$v_c = \frac{Wb^3}{3EI_b} + \frac{Wa^3(1 + 1.5b/a)}{3EI_a} + \frac{Wa^2b(1 + 2b/a)}{2EI_a}$$

Slope at C = slope at B + slope at C

$$\frac{dv}{dx} = \frac{Wa^2(1 + 2b/a)}{2EI_a} + \frac{Wb^2}{2EI_b}$$

Example 3

Determine the slope and deflection of the partially loaded cantilever beam shown in Fig. 4.9(a).

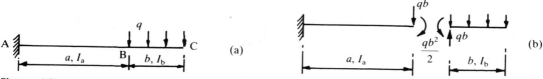

Figure 4.9 ● (a), (b) A non-uniform cantilever under patch loading

Solution Using the two free bodies shown in Fig. 4.9(b), the problem can be solved as in the previous example.

(a) For portion AB, the slope and deflection at B are given by
  (i) Due to point load $W = qb$, $v = (qb)a^3/(3EI_a)$, $dv/dx = (qb)a^2/(2EI_a)$
  (ii) Due to moment $M_s = qb^2/2$, $v = (qb^2/2)a^2/(2EI_a)$, $dv/dx = (qb^2/2)a/(EI_a)$
  The total deflection and slope at B are

$$v = [(qb)a^3/(3EI_a)]\{1 + 3b/(4a)\}$$

$$dv/dx = [(qb)a^2/(2EI_a)]\{1 + b/a\}$$

(b) For the cantilever BC, the deflection and slope due to the load q, *if* the end B is held fixed, are given by

$$v = qb^4/(8EI_b), \quad dv/dx = qb^3/(6EI_b)$$

(c) The net deflection and slope at C are given by

Deflection at C

  = deflection at C due to cantilever BC + deflection at B of cantilever AB + {slope at B of cantilever AB}b

  = $qb^4/(8EI_b) + [qba^3/(3EI_a)]\{1 + 0.75b/a\} + [qba^2/(2EI_a)]\{1 + b/a\}]b$

$$v = \frac{qb^4}{8EI_b} + \frac{qba^3}{3EI_a}(1 + 0.75b/a) + \frac{qb^2a^2}{2EI_a}(1 + b/a)$$

  Slope at C = slope at B + slope at C = $[qba^2/(2EI_a)]\{1 + b/a\} + qb^3/(6EI_b)$

4.4 ● Examples of the calculation of bending deformations in simply supported beams

In the case of simply supported beams, one end of the beam rests on a 'pinned' support and the other end on a 'roller' support as shown in Fig. 4.10. The 'roller'

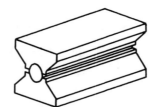
Knuckle pin bearing

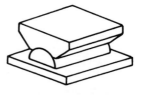
Cylindrical knuckle bearing

Knuckle leaf bearing (Hinge)

Elastomeric laminated bearing

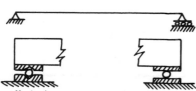
(V-plate + roller) pin support   Simple roller support

Figure 4.10 ● Various types of pinned or hinged supports and roller supports

support allows horizontal translation and rotation about the z-axis perpendicular to the plane of the paper but translation in the vertical directions is restrained. The 'pinned' support is similar to the 'roller' support except that the translations in both the horizontal and the vertical direction are restrained. For the purposes of the calculation of bending deformation, the boundary conditions at both the 'pinned' and 'roller' supports are $v = 0$ but $dv/dx \neq 0$.

Example 1

Calculate the bending deflections in the simply supported beam subjected to end couples M_s as shown in Fig. 4.11(a).

Figure 4.11 ● (a), (b) A simply supported beam subjected to end couples

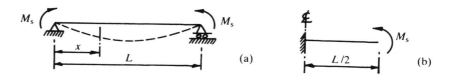

Solution The vertical reactions at the supports are zero. The bending moment at x from the left-hand support is M_s. Therefore

$$EI \, d^2v/dx^2 = M_s$$

$$EI \, dv/dx = M_s x + C_1$$

$$EIv = M_s x^2/2 + C_1 x + C_2$$

Since $v = 0$ at $x = 0$ and $x = L$, $C_2 = 0$ and $C_1 = -M_s L/2$. Therefore

$$EIv = M_s x(x - L)/2$$

$$EI \, dv/dx = M_s(2x - L)/2$$

The maximum slope occurs at the ends and is given at $x = 0$ by $dv/dx = -M_s L/(2EI)$ and at $x = L$ by $dv/dx = M_s L/(2EI)$. From symmetry, the maximum deflection occurs at the centre where $x = 0.5L$ and is given by $v(x = 0.5L) = -M_s L^2/(8EI)$.

It is interesting to note that the above values of maximum slope and deflection could have been derived from the standard cases of cantilever loading. By symmetry, the slope is zero at the centre. Therefore, considering only one half of the beam as shown in Fig. 4.11(b), and using the standard results in Table 4.2 for a cantilever subjected to an end couple, we have

$$\text{Maximum slope} = -(-M_s)(0.5L)/EI = M_s L/(2EI)$$

$$\text{Maximum deflection} = -(-M_s)(0.5L)^2/(2EI) = M_s L^2/(8EI)$$

The change in sign of the deflection is due to the fact that the deflection given is the deflection of the support with respect to the 'fixed' support.

Example 2

Calculate the bending deflections in the simply supported beam subjected to a concentrated load W as shown in Fig. 4.12(a).

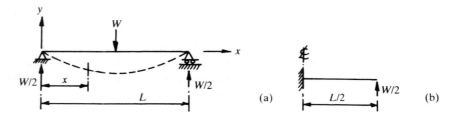

Figure 4.12 ● (a), (b)
A simply supported
beam under midspan
concentrated load

Solution By symmetry, the vertical reactions at the supports are $0.5W$. The bending moment for the left half of the beam is $0.5Wx$, where $0 \leqslant x \leqslant 0.5L$. Therefore

$$EI \, \mathrm{d}^2v/\mathrm{d}x^2 = Wx/2$$

$$EI \, \mathrm{d}v/\mathrm{d}x = Wx^2/4 + C_1$$

$$EIv = Wx^3/12 + C_1x + C_2$$

where $0 \leqslant x \leqslant 0.5L$. The boundary conditions are at $x = 0$, $v = 0$ and, because of symmetry, at $x = 0.5L$, $\mathrm{d}v/\mathrm{d}x = 0$. Therefore $C_2 = 0$ and $C_1 = -WL^2/16$ and

$$v = Wx(4x^2 - 3L^2)/(48EI), \ 0 \leqslant x \leqslant 0.5L$$

The maximum deflection occurs under the load where $\mathrm{d}v/\mathrm{d}x = 0$ and is given by

$$v_{(x=0.5L)} = -WL^3/(48EI)$$

The maximum slope occurs at the ends and is given by, at $x = 0$,

$$\mathrm{d}v/\mathrm{d}x_{(x=0)} = -WL^2/(16EI)$$

By symmetry, the slope at the right-hand support is $WL^2/(16EI)$.

The above values could also have been derived using the results for cantilevers. As shown in Fig. 4.12(b), considering only one half of the beam

$$v_{(x=0.5L)} = (0.5W)(0.5L)^3/(3EI) = WL^3/(48EI)$$

$$\mathrm{d}v/\mathrm{d}x_{(x=0.5L)} = (0.5W)(0.5L)^2/(2EI) = WL^3/(16EI)$$

Example 3

Calculate the bending deflections in the simply supported beam subjected to uniformly distributed load as shown in Fig. 4.13(a).

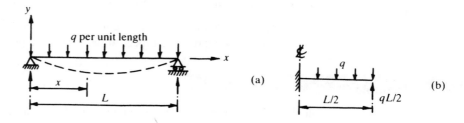

Figure 4.13 ● (a), (b)
A simply supported
beam under uniformly
distributed load

Solution The reactions at the supports are, by symmetry, $0.5qL$. The bending moment at a section x from the LHS is

$$EI \, d^2v/dx^2 = qLx/2 - qx^2/2$$

$$EI \, dv/dx = qLx^2/4 - qx^3/6 + C_1$$

$$EIv = qLx^3/12 - qx^4/24 + C_1x + C_2$$

The boundary conditions are $v = 0$ at $x = 0$ and $x = L$. It is also possible, if desired, to use the condition that $dv/dx = 0$ at $x = 0.5L$ because of symmetry. In other words, any two of the three conditions can be used. Solving for C_1 and C_2, $C_2 = 0$ and $C_1 = -qL^3/24$. Therefore

$$v = q(2Lx^3 - x^4 - L^3x)/(24EI)$$

$$dv/dx = q(6Lx^2 - 4x^3 - L^3)/(24EI)$$

The maximum deflection is at $x = 0.5L$ and the maximum slope is at the supports $x = 0$ and $x = L$. The values are given by

$$v_{(x=0.5L)} = -5qL^4/(384EI)$$

$$dv/dx_{(x=0 \text{ or } x=L)} = \mp qL^3/(24EI)$$

Again it is interesting to see that the above values could have been derived on the basis of results for cantilevers given in Table 4.2. Considering one half of the beam, as shown in Fig. 4.13(b),

$$v_{(x=0.5L)} = \text{due to uniform load } q + \text{tip load } W = 0.5qL$$

$$= -q(0.5L)^4/(8EI) + 0.5qL(0.5L)^3/(3EI)$$

$$= 5qL^4/(384EI)$$

$$dv/dx_{(x=0)} = -q(0.5L)^3/(6EI) + 0.5qL(0.5L)^2/(2EI)$$

$$= qL^3/(24EI)$$

Example 4

Calculate the slope and deflection of the simply supported beam subjected to a couple at one end only as shown in Fig. 4.14.

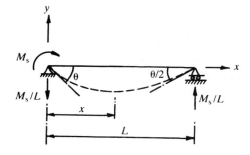

Figure 4.14 ● A simply supported beam under an end couple

Solution The reactions at the two ends are M_s/L and act so as to balance the applied couple M_s. The bending moment at a section x from the LHS is given by

$$EI\, d^2v/dx^2 = M_s - M_s x/L$$

$$EI\, dv/dx = M_s x - M_s x^2/(2L) + C_1$$

$$EIv = M_s x^2/2 - M_s x^3/(6L) + C_1 x + C_2$$

The boundary conditions are $v = 0$ at $x = 0$ and $x = L$. This gives $C_2 = 0$ and $C_1 = -M_s L/3$. Therefore

$$v = M_s\{3x^2 L - x^3 - 2L^2 x\}/(6EIL)$$

$$dv/dx = M_s\{6xL - 3x^2 - 2L^2\}/(6EIL)$$

The slopes at the supports are given at $x = 0$ by $dv/dx = -M_s L/(3EI)$, and at $x = L$ by $dv/dx = M_s L/(6EI)$. It can be seen that the slope at the unloaded support is exactly half the slope at the loaded support. The deflection at the centre is $v(x = 0.5L) = ML^2/(16EI)$. This is not necessarily the maximum deflection because dv/dx is not zero at $x = 0.5L$. For dv/dx to be zero, $\{6xL - 3x^2 - 2L^2\} = 0$. Solving the quadratic equation, $x = (1 - 1/\sqrt{3})L = 0.423L$. Substituting this value of x in the equation for v, $v_{max} = ML^2/(15.59EI)$. This shows that the maximum deflection is only about 2.5 per cent larger than the deflection at the centre.

Example 5

Calculate the deflection in the simply supported beam carrying trapezoidal load as shown in Fig. 4.15(a).

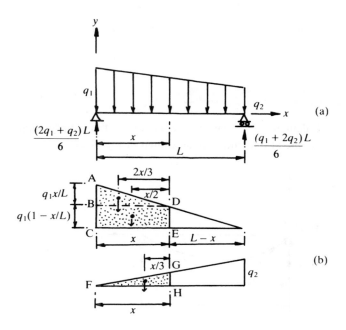

Figure 4.15 ● (a), (b) Simply supported beam under trapezoidal loading

Solution It is mathematically convenient to treat the trapezoidal load as two triangular loads as shown in Fig. 4.15(b). Using the result of Example 5, Section 1.11.2, the left- and right-hand side reactions are respectively $(2q_1 + q_2)L/6$ and $(q_1 + 2q_2)L/6$. The bending moment at a section x from the left-hand side is given by

(a) Due to LHS reaction: $M_1 = (2q_1 + q_2)Lx/6$

(b) Due to q_1: (Because the trapezoidal load ACDE can be split into a uniformly distributed load BCDE of $q_1(1 - x/L)$ and a triangular load ABD with an intensity at the apex of q_1x/L)

$$M_2 = -q_1(1 - x/L)\, x^2/2 - 0.5\,(q_1x/L)x(2x/3) = q_1\{-x^2/2 + x^3/(6L)\}$$

(c) Due to q_2:

$$M_3 = -0.5(q_2x/L)x(x/3) = -q_2x^3/(6L)$$

$$EI\, \mathrm{d}^2v/\mathrm{d}x^2 = M_1 + M_2 + M_3$$

Therefore

$$EI\frac{\mathrm{d}^2v}{\mathrm{d}x^2} = q_1\left\{2Lx - 3x^2 + \frac{x^3}{L}\right\}\bigg/6 + q_2\left\{Lx - \frac{x^3}{L}\right\}\bigg/6$$

$$EI\frac{\mathrm{d}v}{\mathrm{d}x} = q_1\left\{Lx^2 - x^3 + \frac{x^4}{4L}\right\}\bigg/6 + q_2\left\{\frac{Lx^2}{2} - \frac{x^4}{4L}\right\}\bigg/6 + C_1$$

$$EIv = q_1\left\{\frac{Lx^3}{3} - \frac{x^4}{4} + \frac{x^5}{20L}\right\}\bigg/6 + q_2\left\{\frac{Lx^3}{6} - \frac{x^5}{20L}\right\}\bigg/6 + C_1x + C_2$$

Using the boundary conditions that $v = 0$ at $x = 0$ and $x = L$, $C_2 = 0$ and $C_1 = -(8q_1 + 7q_2)L^3/360$

$$v = \left[q_1\left\{20Lx^3 - 15x^4 + \frac{3x^5}{L} - 8L^3x\right\} + q_2\left\{10Lx^3 - \frac{3x^5}{L} - 7L^3x\right\}\right]\bigg/(360EI)$$

The deflection at the centre is given by

$$v(x = 0.5L) = -5(q_1 + q_2)L^4/(768EI)$$

The rotations at the supports are $-(8q_1 + 7q_2)L^3/(360EI)$ at the left support and $+(7q_1 + 8q_2)L^3/(360EI)$ at the right support.

Note that if $q_1 = q_2 = q$, then the load is uniformly distributed, and the results coincide with those derived in Example 3.

4.4.1 Standard cases of loading on simply supported beams

It is useful to record for future reference some results of standard cases of loading on simply supported beams derived in the previous section. They can be used to solve a variety of practical problems.

Table 4.3 ● Standard
cases of loading on
simply supported beams

Load case	v at centre	dv/dx at supports
Equal end couples	$M_sL^2/(8EI)$	$-M_sL/(2EI)$, $M_sL/(2EI)$
Couple at one end only	$M_sL^2/(16EI)$	$-M_sL/(3EI)$, $M_sL/(6EI)$
Central concentrated load	$WL^3/(48EI)$	$-WL^2/(16EI)$, $WL^2/(16EI)$
Uniformly distributed load	$qL^45/(384EI)$	$-qL^3/(24EI)$, $qL^3/(24EI)$

Note: In all these cases, the beam deflects vertically down. If $W = qL$, then the
maximum deflection in the case of a central concentrated load is $8/5 = 1.6$ times
the maximum deflection if the same load had been uniformly distributed.

4.4.2 Application to simple problems

Example 1

Calculate the deflection at the tip of the cantilever and at the centre of the span BC
of the beam shown in Fig. 4.16(a).

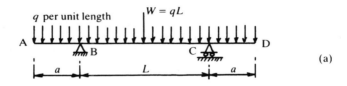

(a)

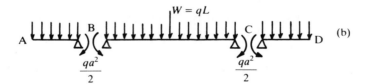

(b)

Figure 4.16 ● (a), (b)
A simply supported
beam with overhangs

Solution Separate the structure into simply supported and cantilever segments as shown in
Fig. 4.16(b). Using the data in Tables 4.2 and 4.3, deflection and slope in the
various segments can be calculated as follows.

(a) Consider the simply supported segment BC first. The deflection at the centre
and the slope at the ends due to the three loads are
  (i) Concentrated load $W = qL$ at the centre:

$$v = (qL)L^3/(48EI), \quad dv/dx = (qL)L^2/(16EI)$$

  (ii) Uniformly distributed load q:

$$v = qL^45/(384EI), \quad dv/dx = qL^3/(24EI)$$

  (iii) End moments $M_s = qa^2/2$:

$$v = -(qa^2/2)L^2/(8EI), \quad dv/dx = -(qa^2/2)L/(2EI)$$

The total central deflection and support rotation are therefore $v = qL^4\{13 - 24a^2/L^2\}/(384EI)$, $dv/dx = qL^3\{5 - 12a^2/L^2\}/(48EI)$. It should be noted that the rotation at the left support is a clockwise rotation.

(b) Consider the cantilever segment AB:

Deflection at the tip = deflection due to q − {clockwise rotation at B}a

$$v = qa^4/(8EI) - [qL^3\{5 - 12a^2/L^2\}/(48EI)]a$$

$$v = qL^3a\{-5 + 12a^2/L^2 + 6a^3/L^3\}/(48EI)$$

Clearly, $v = 0$ if

$$\{-5 + 12(a/L)^2 + 6(a/L)^3\} = 0$$

By solving the above cubic equation by trial and error, $v = 0$ when $(a/L) = 0.57$. Thus if $(a/L) < 0.57$, then the cantilever deflects up and if $(a/L) > 0.57$, then the cantilever deflects down.

Example 2

Calculate the horizontal deflection at the point of load application in the U-frame shown in Fig. 4.17(a).

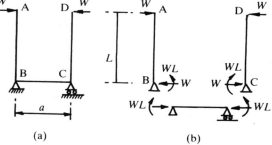

Figure 4.17 ● (a), (b) A simple U-frame bent under equal and opposite concentrated loading

(a)   (b)

Solution Separating the structure into cantilever and simply supported segments as shown in Fig. 4.17(b), we have

(a) Simply supported segment BC: The rotation at the ends due to moments WL is given by

$$dv/dx = \pm(WL)a/(2EI)$$

(b) Cantilever segment AB: Deflection at the load point is given by

Deflection = deflection due to W + {clockwise rotation at support B}L

$$v = WL^3/(3EI) + \{WLa/(2EI)\}L$$

$$= WL^3\{1 + 1.5a/L\}/(3EI)$$

4.5 ● Application to simple statically indeterminate problems

Statically indeterminate problems are those problems where the stress resultants at a section in the structure cannot be determined by consideration of statics only. It becomes necessary to use constraints on displacements to give additional equations to solve the problem. This aspect is illustrated by a few simple examples.

Example 1

Calculate the reactions of the beam on three supports shown in Fig. 4.18(a).

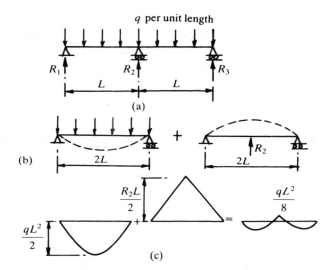

Figure 4.18 ●
(a)–(c) A statically
indeterminate
continuous beam

Solution This is a simple example of a statically indeterminate structure. In this case there are three vertical reactions. However, only two equations of statics involving the vertical reactions are available, i.e. $\Sigma F_y = 0$ and $\Sigma M_z = 0$. Before the problem can be solved, it is necessary to derive an additional equation. This can be done as follows.

Let the central reaction be R_2. If we consider the given structure as a simply supported beam of span $2L$ and loaded by uniformly distributed load q and a central load R_2, as shown in Fig. 4.18(b), then the value of R_2 should be such as to make the deflection at the centre zero. The deflection at the centre of the simply supported beam of span $2L$ due to q and R_2 is given, using values from Table 4.3, by

(a) Due to q, downward deflection = $q(2L)^4 5/(384EI)$

(b) Due to R_2, upward deflection = $R_2(2L)^3/(48EI)$

For zero deflection at the centre, upward and downward deflections must be equal. This is the additional equation. This gives $R_2 = 1.25qL$. The end reactions must be, by symmetry, $R_1 = R_3 = (2qL - R_2)/2 = 0.375qL$. Superposing the BM due to q and R_z, the final bending moment distribution is as shown in Fig. 4.18(c). As is to be expected, the provision of a central support considerably reduces the BM in the structure.

In theory, the above procedure can be applied to a beam resting on any number of supports. However, the calculations become very cumbersome and a more efficient procedure will be discussed in Chapter 6.

Example 2

Calculate the reactions in the 'propped cantilever' shown in Fig. 4.19(a).

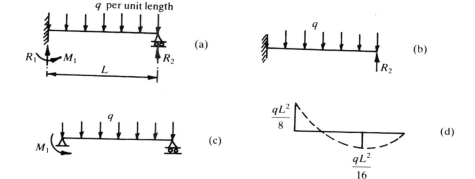

Figure 4.19 ● (a)–(d)
A propped cantilever
under uniformly
distributed load

Solution (a) *Approach 1.* There are three reactions to be determined: two vertical reactions and a couple at the fixed end. Let the reaction at the prop be R_2. Considering the cantilever loaded by q and a concentrated load R_2 at the tip, as shown in Fig. 4.19(b), then for the deflection at the tip we have

(i) Due to q, downward deflection = $qL^4/(8EI)$
(ii) Due to R_2, upward deflection = $R_2L^3/(3EI)$
Therefore for zero deflection at the prop, $R_2 = 3qL/8$. From statics $R_1 = qL - R_2$ = $5qL/8$ and the couple M_1 at the fixed end is

$$M_1 = qL^2/2 - R_2L = qL^2/8$$

The final bending moment is a superposition of that due to q and R_2 at the tip of the cantilever.

(b) *Approach 2.* In the solution given above, the problem was solved by treating the cantilever as the 'basic' structure modified by the provision of a prop so as to reduce the deflection at the tip to zero. It is also possible to start with the simply supported structure as the basic structure, modified by the application of a couple at the fixed end so as to ensure that the rotation at the fixed end is zero. From this point of view, considering the structure shown in Fig. 4.19(c), for the rotation at the 'fixed' end we have

(i) Due to q, clockwise rotation = $qL^3/(24EI)$
(ii) Due to M_1, anticlockwise rotation = $M_1L/(3EI)$
Therefore for zero rotation at the fixed end, $M_1 = qL^2/8$. The other vertical reactions can be determined from statics.

The final bending moment is due to q and end couple M_1 at the left support of the simply supported beam. The final BM shown in Fig. 4.19(d) is a superposition of the two values of bending moment for the two cases of loading on the statically determinate structure. Again there is a considerable reduction in the final BM of the structure.

(c) *Approach 3*. Because this is a statically indeterminate structure, the bending moment distribution cannot be found from statics. Therefore $EI\, d^2v/dx^2 = M$ cannot be used as the basic equation. However, because the beam has constant flexural rigidity EI,

$$EI\, d^4v/dx^4 = -q$$

The negative sign before q indicates that it is a downward load. Integrating:

$$EI\, d^3v/dx^3 = -qx + C_1$$

$$EI\, d^2v/dx^2 = -qx^2/2 + C_1x + C_2$$

$$EI\, dv/dx = -qx^3/6 + C_1x^2/2 + C_2x + C_3$$

$$EIv = -qx^4/24 + C_1x^3/6 + C_2x^2/2 + C_3x + C_4$$

The boundary conditions are:
At $x = 0$, $v = dv/dx = 0$. This requires that $C_4 = C_3 = 0$. At $x = L$, $v = 0$ and $M = EI\, d^2v/dx^2 = 0$. This requires that $C_1 = 5qL/8$, $C_2 = -qL^2/8$. Therefore

$$EIv = -qx^4/24 + 5qLx^3/48 - qL^2x^2/16 = -q[2x^4 - 5Lx^3 + 3L^2x^2]/48$$

$$EI\, dv/dx = -q[8x^3 - 15Lx^2 + 6L^2x]/48$$

$$EI\, d^2v/dx^2 = -q[4x^2 - 5Lx + L^2]/8$$

$$EI\, d^3v/dx^3 = -q[8x - 5L]/8$$

At $x = 0$, $M = EI\, d^2v/dx^2 = -qL^2/8$

and

$$Q = -EI\, d^3v/dx^3 = -5qL/8$$

At $x = L$, $M = EI\, d^2v/dx^2 = 0$

and

$$Q = -EI\, d^3v/dx^3 = 3qL/8$$

The above values agree with the values calculated using the principle of superposition.

In most cases it will be found simpler to use approaches 1 or 2 based on the principle of superposition.

Example 3

For the propped cantilever shown in Fig. 4.20(a), determine the couple required to cause unit rotation at the propped end.

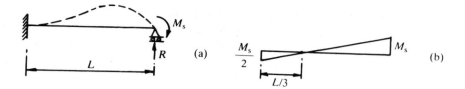

Figure 4.20 ●
(a), (b) A propped
cantilever under an end
couple

Solution   Let the value of the couple be M_s and the prop reaction be R. Because this is a statically indeterminate structure, the prop force R has to be determined so as to maintain zero defection at the prop. Considering the structure as a cantilever, the deflection and rotation at the prop end due to R and M_s are, using the values in Table 4.2,

(a) Due to R: $v = RL^3/(3EI)$ upwards, $dv/dx = RL^2/(2EI)$ anticlockwise
(b) Due to M_s: $v = M_sL^2/(2EI)$ downwards, $dv/dx = M_sL/(EI)$ clockwise

For zero deflection at the prop end, $R = 1.5M_s/L$ and the rotation θ at the prop is given by

$$\theta = M_sL/(EI) - RL^2/(2EI) = M_sL/(4EI) \text{ or } M_s = 4(EI)\theta/L$$

Therefore the couple required for rotation θ equal to unity is $4EI/L$. This is known as the rotational stiffness at the propped end of a propped cantilever. This is an important result which will be used in Chapters 6 and 7.

The moment M at the fixed support is easily calculated and is given by $M = RL - M_s = 0.5M_s$. As can be seen, this couple acts in the same direction as the applied couple M_s. Figure 4.20(b) shows the resulting bending moment diagram. It can be seen that the bending moment is equal to zero at $L/3$ from the left support.

Example 4

For the cantilever shown in Fig. 4.21(a), calculate the vertical force and the couple required to displace the tip of the cantilever by Δ but the rotation at the tip is prevented.

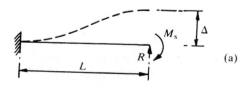

Figure 4.21 ● (a), (b)
A propped cantilever
subjected to support
translation

Solution In the previous example, rotation was allowed at the propped end but not translation. In this example, translation is allowed but rotation is prevented. Using the equations developed in the previous example, for zero rotation

$$M_sL/(EI) - RL^2/(2EI) = 0$$

Therefore $M_s = 0.5RL$. The vertical deflection is given by

$$\Delta = RL^3/(3EI) - M_sL^2/(2EI) = RL^3/(12EI) \text{ or}$$

$$R = 12EI\,\Delta/L^3 \text{ and } M_s = 0.5RL = 6EI\,\Delta/L^2$$

For unit value of Δ, the force required is $R = 12EI/L^3$. This is known as the translational stiffness. The bending moment diagram shown in Fig. 4.21(b) indicates that the bending moment is zero at the centre of the beam.

Example 5

Calculate all the reactions at the supports of the structure shown in Fig. 4.22(a). The axial link BC is undeformable.

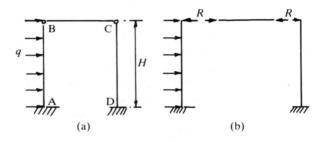

(a)   (b)

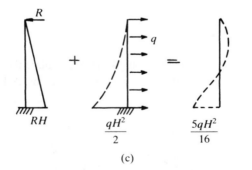

(c)

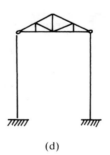

(d)

Figure 4.22 ●
(a)–(d) Two cantilevers connected by a rigid link and subjected to a uniformly distributed load q

Solution In this case two cantilevers are connected by a rigid pin-ended member. Obviously there can only be an axial force in the connecting member. If the force in the connected member is known, then the forces acting on the two cantilevers can be determined. Because the member is rigid, the two cantilevers must deflect at the top by the same amount. Considering the free bodies shown in Fig. 4.22(b), let the force in the connecting member be R. The deflection at the tip of the cantilevers can be calculated as follows.

(a) Cantilever AB:

The total deflection at the top is given by

(i) Due to q, $v = qH^4/(8EI)$

(ii) Due to R, $v = -RH^3/(3EI)$

Net deflection at the tip $= qH^4/(8EI) - RH^3/(3EI)$

(b) Cantilever DC:

Total deflection at the tip is due to R only. Therefore $v = RH^3/(3EI)$.
For the tip deflections of the two cantilevers to be equal we have

$$qH^4/(8EI) - RH^3/(3EI) = RH^3/(3EI)$$

Therefore $R = 3qH/16$ and the deflection at the top is $qH^4/(16EI)$. Once the value of R is known, the forces at the base of the two cantilevers are easily determined. Figure 4.22(c) shows the reactions and the bending moment in the loaded cantilever. It should be noted that the cantilever AB behaves as a partially propped cantilever. Because deflection at the top is permitted, the bending moment at the base is $5qH^2/(16)$, 2.5 times larger than $qH^2/8$ which occurs in a fully propped cantilever as shown in Example 2, but is only $(5/8)$ the value of the bending moment of $0.5qH^2$ in an unpropped cantilever.

This problem commonly occurs in steel structures where two fixed base columns are connected by a truss at the top, as shown in Fig. 4.22(d).

Example 6

Analyse the problem shown in Fig. 4.23(a) of two cantilevers connected by

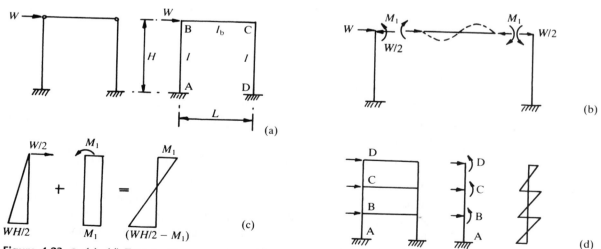

Figure 4.23 ● (a)–(d) Two cantilevers connected by a rigid link and a flexurally deformable beam

(a) a rigid pin-ended axial link,

(b) a link which is axially rigid but deformable in bending and rigidly joined to the two cantilevers.

Solution The problem of the cantilevers connected by a pin-ended rigid link is easily solved because for equal deflection at the top the total load is equally shared by the two cantilevers. Therefore the force in the link is a compressive force of $0.5W$. Under a load of $0.5W$ at the top, the two cantilevers rotate at the top in the clockwise direction by $0.5WH^2/(2EI) = 0.25WH^2/(EI)$.

If the two cantilevers are connected by a bending element, then because the connection between the cantilevers and the beam is rigid, the beam is also forced to undergo clockwise rotations at both ends, as shown in Fig. 4.23(b). Assuming that the couples at the ends of the beam are M_1 clockwise, then considering the free bodies we have

(a) Cantilever AB and DC. The rotation at the top is given by
   (i) Due to load $0.5W$: $dv/dx = 0.5WH^2/(2EI)$ clockwise.
   (ii) Due to couple M_1:

$$dv/dx = -M_1H/(EI)$$

Net rotation at top $= dv/dx = 0.5WH^2/(2EI) - M_1H/(EI)$

(b) Beam BC. Assuming the second moment of area of the beam is I_b, then the rotations at the ends are given by

Couple M_1 at end B:

$$(dv/dx)_B = M_1L/(3EI_b), \quad (dv/dx)_C = -M_1L/(6EI_b)$$

Couple M_1 at end C:

$$(dv/dx)_B = -M_1L/(6EI_b), \quad (dv/dx)_C = M_1L/(3EI_b)$$

Net rotation at B and $C = M_1L/(6EI_b)$

For compatibility, the rotations at the junction of the beam and the cantilever must be equal. Therefore

$$0.5WH^2/(2EI) - M_1H/(EI) = M_1L/(6EI_b)$$

Solving for M_1, $M_1 = 0.25WH/C$, where
$C = \{1 + LI/(6HI_b)\}$.
As can be seen, the greater the value of I_b, the smaller the value of C and the larger the value of M_1. Using the free body of the cantilever, the deflection at the top is given by

$$v = 0.5WH^3/(3EI) - M_1H^2/(2EI) = WH^3\{1 - 0.75/C\}/(6EI)$$

This shows that the smaller the value of C, the smaller the deflection at the top of the cantilever. The bending moment in the cantilever is a superposition of the bending moments due to W and M_1. As can be seen from Fig. 4.23(c), the interaction couple M_1 considerably reduces the bending moment in the cantilever.

This example demonstrates an important reason for adopting rigid-jointed portal frames such as that shown in Fig. 4.23(d). The rigid joint between the beam and the column forces equal rotation of the beam and the column at the joints. This induces anticlockwise couples on the column at the joints which not only reduce the deflections but also the bending moments in the columns.

Example 7

Two simply supported beams as shown in Fig. 4.24(a), are placed one on top of another and are connected so that they are forced to deflect by the same amount at the junction. The top beam is loaded by a uniformly distributed load of intensity q. Calculate the bending moment distribution in the two beams.

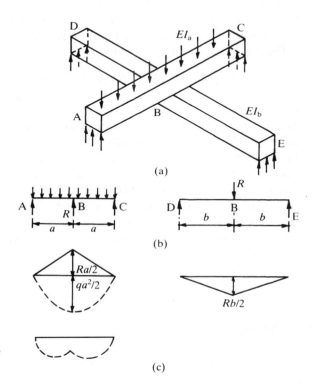

Figure 4.24 ● (a)–(c) Two intersecting beams in the horizontal plane subjected to a uniformly distributed load

Solution The load is applied to the top beam ABC. It will deflect under the load, but because of the connection at the junction, the beam DBE is also forced to deflect. The unloaded beam DBE gives support to the loaded beam ABC. Because the two beams are only connected at B, the interaction force at B can only be a vertical force R. The force R acts downwards on beam DBE and upwards on beam ABC. Considering the two beams in turn and using the data in Table 4.3, we have

(a) Beam ABC.
The deflection at the centre is due to

(i) Load q: $v = \dfrac{5q(2a)^4}{(384 EI_a)}$, a downward deflection

(ii) Force R: $v = -R(2a)^3/(48EI_a)$, an upward deflection.

Total downward deflection

$$v = 5q(2a)^4/(384EI_a) - R(2a)^3/(48EI_a)$$

(b) Beam DBE.

The deflection is due to R only and is given by $v = R(2b)^3/(48EI_b)$, a downward deflection. Equating the two expressions for deflections, we have

$$q(2a)^4 5/(384EI_a) - R(2a)^3/(48EI_a) = R(2b)^3/(48EI_b)$$

Solving for R, $R = 1.25qa/C$, where $C = \{1 + (b/a)^3(I_a/I_b)\}$.

Note that if the beam DBE is infinitely stiff, i.e. $I_b \to \infty$, then $C \to 1$ and $R \to 1.25qa$ which is equal to the central reaction calculated in Example 1, Section 4.4, for a beam resting on three unyielding supports.

The maximum bending moment in the beam ABC is

$$M_{max} = q(2a)^2/8 - R(2a)/4 = 0.5qa^2(1 - 1.25/C)$$

The maximum deflection at the centre is $v = 5q(2a)^4\{1 - 1/C\}/384EI_a)$. It should be noted that, as we would expect, as the 'unloaded' beam DBE is made stiffer by either reducing its span $2b$ or increasing its second moment of area I_b, the value of C decreases and this results in an increase in the value of R. An increase in the value of R will decrease the deflection and will also reduce the bending moment in the loaded beam ABC. Of course this is achieved at the expense of an increase in the bending moment in the 'unloaded' beam DBE. Examples 6 and 7 are simple statically indeterminate structures which show that in such structures the 'stresses' can be controlled by varying the relative second moment of area of the members.

4.6 ● Fixed beams

In practice beams are generally connected to other elements of the structure, as shown in Fig. 4.25. The restraint provided by the other members of the structure prevents free rotation at the ends of the beam. As a simple example, consider the three span beam shown in Fig. 4.26(a). Obviously if the span of the end sections is increased, then the restraint to rotation that the end spans provide at the ends of the central span decreases because the rotational stiffness, i.e. $3EI/L$ (see Table 4.3) is inversely proportional to span. If the process is continued, then in the limit the

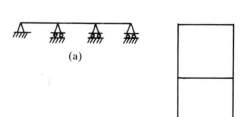

(a)

(b)

(c)

Figure 4.25 ● Beams as part of structures: (a) continuous beam, (b) rigid-jointed frame, (c) plane grid

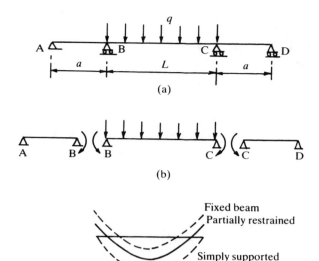

Figure 4.26 ● (a)–(c) Behaviour of a beam in one span affected by its interaction with other members in the neighbouring spans

restraint reduces to zero and the central span behaves as a simply supported beam. On the other hand, as the span of the end sections is decreased, then the restraint to rotation at the ends of the central span increases and in the limit the rotations at the ends of the central span are fully restrained.

The above discussion shows that in most cases of beams joined to other members, the bending moment lies between the corresponding bending moment of beams with no restraint to rotation at supports and beams where the support rotation is zero (Fig. 4.26(c)). The analysis of simply supported beams was discussed in Section 4.4. In the case of simply supported beams, the rotation at the supports is unrestrained. This can be treated as one extreme situation that occurs in beams connected to other members. The other extreme case is when the rotations at the ends are fully restrained. Such beams are called fixed beams. In the next section a few cases of standard loadings on fixed beams will be discussed. The problems will be solved using the slope and deflection values for standard cases of loading on simply supported beams given in Table 4.3.

It should be noted that as opposed to simply supported beams, in the case of fixed beams, the boundary conditions adopted are $v = 0$ and $dv/dx = 0$ at the supports. It is generally assumed that the beam can move freely in the axial direction at the ends. It should be remembered that fixed and simply supported support conditions are idealized conditions rarely achieved in practice.

4.6.1 Fixed beam – examples

Example 1

Calculate the moment at the supports of the fixed beam shown in Fig. 4.27.

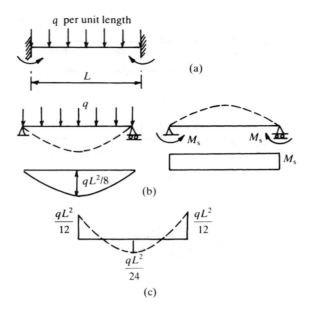

q per unit length

(a)

L

q

M_s M_s

M_s

$qL^2/8$

(b)

$\dfrac{qL^2}{12}$ $\dfrac{qL^2}{12}$

$\dfrac{qL^2}{24}$

(c)

Figure 4.27 ●
(a)–(c) A fixed beam
under uniformly
distributed loading

Solution The loading on the structure can be considered to be made up of applied load q and end moments M_s. Using the information in Table 4.3, the rotation at the support is given by

(a) Due to applied load q:
$\mathrm{d}v/\mathrm{d}x = qL^3/(24EI)$, clockwise at the left and anticlockwise at the right support. Deflection at the centre is $v = 5qL^4/(384EI)$, downwards.

(b) Due to equal values of M_s at the ends:
$\mathrm{d}v/\mathrm{d}x = M_s L/(2EI)$, anticlockwise at the left and clockwise at the right support. Deflection at the centre is $v = M_s L^2/(8EI)$, upwards.
Therefore, net rotation at the support $= \pm\{qL^3/(24EI) - M_s L/(2EI)\}$, where the positive sign refers to clockwise rotation. Therefore for zero rotation at the support

$$\{qL^3/(24EI) - M_s L/(2EI)\} = 0$$

Therefore

$$M_s = qL^2/12$$

Net deflection at centre $= 5qL^4/(384EI) - M_s L^2/(8EI) = qL^4/(384EI)$

The final bending moment is the sum of moments due to q and M_s. As can be seen, while the maximum bending moment in a simply supported beam occurs at midspan causing tension at the bottom surface, in the corresponding fixed beam, the maximum bending moment occurs at the supports and causes tension at the top surface. The maximum bending moment at midspan is $qL^2/24$ which is only a third of the corresponding midspan bending moment of $qL^2/8$ in the case of a simply supported beam. The maximum deflection in a fixed beam is only 20 per cent of the corresponding deflection in a simply supported beam.

The sign of the bending moment varies from tension at the top near the supports to tension at the bottom at midspan.

Example 2

Calculate the moment at the supports of the fixed beam shown in Fig. 4.28.

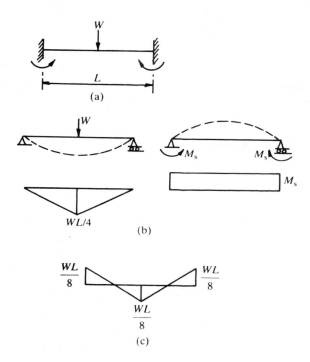

Figure 4.28 ●
(a)–(c) A fixed beam
under midspan
concentrated load

Solution Proceeding as in Example 1, we have

(a) Due to W:
At support $dv/dx = \pm WL^2/(16EI)$, and at centre $v = WL^3/(48EI)$.

(b) Due to M_s:
At support $dv/dx = \mp M_s L/(2EI)$, and at centre $v = M_s L^2/(8EI)$ upwards.
For zero rotation at support $\{WL^2/(16EI) - M_s L/(2EI)\} = 0$, leading to $M_s = WL/8$. The net deflection at the centre is $WL^3/(48EI) - M_s L^2/(8EI) = WL^3/(192EI)$. The maximum bending moment at midspan (Fig. 4.28c) is only 50 per cent of the midspan bending moment of $WL/4$ in a simply supported beam. The maximum deflection in the fixed beam is only 25 per cent of that in the corresponding simply supported beam.

Example 3

Calculate the moment at the supports of the fixed beam shown in Fig. 4.29.

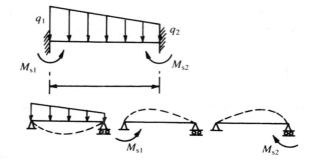

Figure 4.29 ●
A fixed beam under
trapezoidal load

Solution In this case because there is no symmetry in the loading, the fixed end moment at the two ends will not be of the same magnitude. Let the fixed end moments at the left and right supports be M_{s1} and M_{s2} respectively. Then proceeding as in the previous examples, we have

(a) Due to applied loading. Using the results of Example 5, Section 4.4, we have:

At the left support, rotation $dv/dx = -(8q_1 + 7q_2)L^3/(360EI)$

At the right support, rotation $dv/dx = +(7q_1 + 8q_2)L^3/(360EI)$

The deflection at the centre is $v = -5(q_1 + q_2)L^4/(768EI)$

(b) Due to fixed end moment M_{s1} at the left support:

At the left support, rotation $dv/dx = +M_{s1}L/(3EI)$

At the right support, rotation $dv/dx = -M_{s1}L/(6EI)$

The deflection at the centre $v = M_{s1}L^2/(16EI)$

(c) Due to fixed end moment M_{s2} at the right support:

At the left support, rotation $dv/dx = +M_{s2}L/(6EI)$

At the right support, rotation $dv/dx = -M_{s2}L/(3EI)$

The deflection at the centre is $v = M_{s2}L^2/(16EI)$

For zero rotation at the supports

$$-(8q_1 + 7q_2)L^3/(360EI) + M_{s1}L/(3EI) + M_{s2}L/(6EI) = 0$$

$$(7q_1 + 8q_2)L^3/(360EI) - M_{s1}L/(6EI) - M_{s2}L/(6EI) = 0$$

Solving, $M_{s1} = (3q_1 + 2q_2)L^2/60$ and $M_{s2} = (2q_1 + 3q_2)L^2/60$. The deflection at the centre is

$$5(q_1 + q_2)L^4/(768EI) - M_{s1}L^2/(16EI) - M_{s2}L^2/(16EI) = (q_1 + q_2)L^4/(768EI)$$

This shows that the deflection at the middle of the span in a fixed beam is only 20 per cent of that in a corresponding simply supported beam.

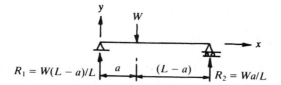

Figure 4.30 ●
A simply supported
beam under off-centre
concentrated load

$R_1 = W(L - a)/L$ a $(L - a)$ $R_2 = Wa/L$

4.7 ● Integration of $EI\,d^2v/dx^2 = M$, when M is discontinuous

Consider the simply supported beam with a concentrated load, as shown in Fig. 4.30. The reactions at the left and the right supports are respectively $R_1 = W(L - a)/L$ and $R_2 = Wa/L$.

The bending moment in the beam is discontinuous and is given by

$$EI\,d^2v/dx^2 = M = R_1 x, \text{ if } 0 \leqslant x \leqslant a$$

$$EI\,d^2v/dx^2 = M = R_1 x - W(x - a), \text{ if } a \leqslant x \leqslant L$$

Integrating successively, we have

$$EI\,d^2v/dx^2 = R_1 x$$

$$EI\,dv/dx = R_1 x^2/2 + C_1$$

$$EIv = R_1 x^3/6 + C_1 x + C_2 \text{ for } 0 \leqslant x \leqslant a$$

Similarly,

$$EI\,d^2v/dx^2 = R_1 x - W(x - a)$$

$$EI\,dv/dx = R_1 x^2/2 - W(x^2/2 - ax) + C_3$$

$$EIv = R_1 x^3/6 - W(x^3/6 - ax^2/2) + C_3 x + C_4, \text{ for } a \leqslant x \leqslant L$$

Because $v = 0$ at $x = 0$, then $C_2 = 0$. Similarly $v = 0$ at $x = L$. Therefore

$$R_1 L^3/6 - W(L^3/6 - aL^2/2) + C_3 L + C_4 = 0 \tag{i}$$

At $x = a$, the deflection and slope from the two expressions must be the same. Therefore for v at $x = a$,

$$R_1 a^3/6 + C_1 a = R_1 a^3/6 - W(a^3/6 - a^2/2) + C_3 a + C_4$$

Therefore

$$C_1 a = C_3 a + C_4 + Wa^3/3 \tag{ii}$$

For dv/dx at $x = a$,

$$R_1 a^2/2 + C_1 = R_1 a^2/2 - W(a^2/2 - a^2) + C_3$$

Therefore

$$C_1 = C_3 + Wa^2/2 \tag{iii}$$

Because $R_1 = W(L - a)/L$, solving equations (i–iii) for the constants we have $C_4 = Wa^3/6$, $C_3 = -Wa(2L^2 + a^2)/(6L)$ and

$$C_1 = -Wa(2L^2 - 3aL + a^2)/(6L)$$

The deflection under the load at $x = a$ is

$$v = Wa^2(L - a)^2/(3L)$$

This example shows that when the bending moment is discontinuous calculation using the standard procedure is cumbersome. Fortunately a simple procedure known as the singularity function method or Macaulay's method is available to handle the cases of integration with discontinuous functions.

4.7.1 Singularity functions

Consider the function defined by $F(x) = \langle x - a \rangle^n$. It is assumed that $F(x) = 0$ for $x \leq a$ and $F(x) = (x - a)^n$ for $x > a$. When integrated

$$\int F(x)\,dx = \int \langle x - a \rangle^n\,dx = \langle x - a \rangle^{n+1}/(n + 1)$$

The brackets $\langle\ \rangle$ are known as Macaulay brackets. Using the singularity function, it becomes possible to represent discontinuous functions and their integrals by a single function. The use of singularity functions is illustrated by some simple examples.

4.7.2 Examples using singularity functions

Example 1

Consider the simply supported beam carrying a concentrated load W at $x = a$ as shown in Fig. 4.30. The equation to be integrated is given by

$$EI\,d^2v/dx^2 = R_1x - W\langle x - a \rangle$$

After integrating successively with respect to x

$$EI\,dv/dx = R_1x^2/2 - W\langle x - a \rangle^2/2 + C_1$$

$$EIv = R_1x^3/6 - W\langle x - a \rangle^3/6 + C_1x + C_2$$

Note that $\langle x - a \rangle^2$ is treated as a single entity and not as $(x - a)^2$ which is equal to $(x^2 - 2ax + a^2)$.

Note that there are only two integration constants, which apply to the entire span irrespective of whether the function to be integrated is continuous or not. The boundary condition $v = 0$ at $x = 0$ requires that $C_2 = 0$ and $v = 0$ at $x = L$ requires that

$$0 = R_1L^3/6 - W(L - a)^3/6 + C_1L$$

Because $R_1 = W(L - a)/L$, $C_1 = -W(L - a)(2L - a)a/(6L)$

$$EI\,dv/dx = W(L - a)x^2/(2L) - W\langle x - a \rangle^2/2 - W(L - a)(2L - a)a/(6L)$$

$$EIv = W(L - a)x^3/(6L) - W\langle x - a \rangle^3/6 - W(L - a)(2L - a)ax/(6L)$$

The slopes at $x = 0$ and $x = L$ are given by

dv/dx at $x = 0$ is $-W(L - a)(2L - a)a/(6EIL)$

$$= -Wab(L + b)/(6EIL)$$

dv/dx at $x = L$ is $W(L - a)(L + a)a/(6EIL)$

$$= +Wab(L + a)/(6EIL)$$

where $(L - a) = b$. The deflection under the load at $x = a$ is given by

$EIv = W(L - a)a^3/(6L) - 0 - W(L - a)(2L - a)a^2/(6L)$

$$= -W(L - a)^2 a^2/(3L)$$

v at $x = a$ is $-Wa^2b^2/(3EIL)$

Note that if a series of N concentrated loads W_i act at a_i from the left support, then the equations are given by

$$EI\,d^2v/dx^2 = R_1 x - \sum W_i \langle x - a_i \rangle$$

$$EI\,dv/dx = R_1 x^2/2 - \sum W_i \langle x - a_i \rangle^2/2 + C_1$$

$$EIv = R_1 x^3/6 - \sum W_i \langle x - a_i \rangle^3/6 + C_1 x + C_2$$

The summation sign Σ extends over $i = 1$ to N, where N = number of concentrated loads acting on the beam. R_1, the reaction at the left support, is given by $R_1 = \Sigma W_i - \Sigma W_i a_i/L$.

Example 2

As a simple application of the above equation, consider the case of a symmetrically loaded simply supported beam subjected to two equal concentrated loads of W each at $L/3$ from the supports.

Solution   In this case $W_1 = W$, $a_1 = L/3$ and $W_2 = W$, $a_2 = 2L/3$ and $R = W$ from symmetry.

$$EIv = Wx^3/6 - \{W\langle x - L/3 \rangle^3 + W\langle x - 2L/3 \rangle^3\}/6 + C_1 x + C_2$$

The constants of integration are determined from the boundary conditions. At $x = 0$, $v = 0$ leading to $C_2 = 0$ and at $x = L$, $v = 0$, $WL^3/6 - \{W(L - L/3)^3 + W(L - 2L/3)^3\}/6 + C_1 L = 0$, leading to $C_1 = -WL^2/9$.

Under the load, deflection is given by v at $x = L/3$. Note that $W\langle x - L/3 \rangle^3 = 0$ because the term inside the bracket is zero but $W\langle x - 2L/3 \rangle^3$ is also zero *because the term inside the bracket is negative.*

$$EIv = W(L/3)^3/6 + C_1 L/3 = -5WL^3/162$$

Deflection at the middle of the span is given by v at $x = L/2$. Note that $W\langle x - L/3 \rangle^3 = W(L/6)^3$ because the term inside the bracket is positive but $W\langle x - 2L/3 \rangle^3$ is zero *because the term inside the bracket is negative.*

$$EIv = W(L/2)^3/6 - W(L/6)^3/6 + C_1 L/2 = -5.75WL^3/162$$

Thus the deflection at midspan is only $5.75/5 = 1.15$, i.e. 15 per cent greater than the deflection under the load.

Example 3

Calculate the fixed end moment for the beam shown in Fig. 4.31.

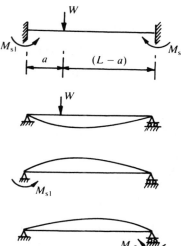

Figure 4.31 ● A fixed beam under off-centre concentrated load

Solution Let the fixed end moments at the left and right supports be M_{s1} and M_{s2} respectively. Then, using the information from Example 1 and that in Table 4.3, the total rotation at the supports can be calculated. Table 4.4 shows deflection and slopes due to individual loads.

Table 4.4

Load	$El(dv/dx)_1$	$El(dv/dx)_2$
W	$-Wab(L + b)/(6L)$	$Wab(L + a)/(6L)$
M_{s1}	$M_{s1}L/3$	$-M_{s1}L/6$
M_{s2}	$M_{s2}L/6$	$-M_{s2}L/3$

For zero rotation at the supports,

$$-Wab(L + b)/(6L) + M_{s1}L/3 + M_{s2}L/6 = 0$$

$$Wab(L + a)/(6L) - M_{s1}L/6 + M_{s2}L/3 = 0$$

Solving, $M_{s1} = Wab^2/L^2$, $M_{s2} = Wa^2b/L^2$. If $a = b = 0.5L$, then $M_{s1} = M_{s2} = WL/8$, agreeing with the result of Example 2, Section 4.6.1.

Example 4

Consider the simply supported beam carrying a concentrated couple M_c at $x = a$ as shown in Fig. 4.32

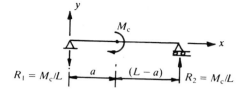

Figure 4.32 ● A simply supported beam under an off-centre couple

The reactions at the left and right supports are $-M_c/L$ and M_c/L respectively. The equation to be integrated is

$$EI\, d^2v/dx^2 = -M_c x/L + M_c\langle x - a\rangle^0$$

Note that the effect of the concentrated couple is written as $M_c\langle x - a\rangle^0$ to ensure that the effect of M_c comes into effect only after $x \geqslant a$. Integrating

$$EI\, dv/dx = -M_c x^2/(2L) + M_c\langle x - a\rangle^1 + C_1$$

$$EIv = -M_c x^3/(6L) + M_c\langle x - a\rangle^2/2 + C_1 x + C_2$$

Using the boundary conditions $v = 0$ at $x = 0$ leads to $C_2 = 0$. C_1 is determined from the condition that $v = 0$ at $x = L$. This leads to

$$0 = -M_c L^2/6 + M_c(L - a)^2/2 + C_1 L$$

or

$$C_1 = M_c(6aL - 3a^2 - 2L^2)/(6L)$$

Therefore

$$EI(dv/dx) \text{ at } x = 0 \text{ is } M_c(6aL - 3a^2 - 2L^2)/(6L)$$

$$= M_c(L^2 - 3b^2)/(6L)$$

$$EI(dv/dx) \text{ at } x = L \text{ is } M_c(L^2 - 3a^2)/(6L)$$

where $b = L - a$.

The deflection at the point of application of the couple $x = a$ is given by

$$EIv = -M_c a^3/(6L) + 0 + M_c a(6aL - 3a^2 - 2L^2)/(6L)$$

$$= M_c a(3aL - 2a^2 - L^2)/(3L)$$

As in the case of concentrated loads, if a series of couples M_{ci} act at a_i from the left support, then the resulting equations are

$$EI\, d^2v/dx^2 = R_1 x + \sum M_{ci}\langle x - a_i\rangle^0$$

$$EI\, dv/dx = R_1 x^2/2 + \sum M_{ci}\langle x - a_i\rangle^1 + C_1$$

$$EIv = R_1 x^3/6 + \sum M_{ci}\langle x - a_i\rangle^2/2 + C_1 x + C_2$$

where R_1, the reaction at the left support $= -\sum M_{ci}/L$.

Example 5

As a simple application of the above equation, consider the case of a **symmetrically** loaded simply supported beam subjected to two equal concentrated couples of M each at $L/3$ from the supports.

Solution   In this case $M_{c1} = M$ (clockwise), $a_1 = L/3$ and $M_{c2} = -M$ (anticlockwise), $a_2 = 2L/3$ and $R = 0$ because $\Sigma M_{ci} = 0$.

$$EIv = \{M\langle x - L/3\rangle^2 - M\langle x - 2L/3\rangle^2\}/2 + C_1x + C_2$$

The constants of integration are determined from the boundary conditions. At $x = 0$, $v = 0$ leading to $C_2 = 0$ and at $x = L$, $v = 0$, $\{M(L - L/3)^2 - M(L - 2L/3)^2\}/2 + C_1L = 0$, leading to $C_1 = -ML/6$.

At the point of application of the couples, deflection is given by v at $x = L/3$. Note that $M\langle x - L/3\rangle^2 = 0$ because the term inside the bracket is zero but $M\langle x - 2L/3\rangle^2$ is zero *because the term inside the bracket is negative.*

$$EIv = C_1L/3 = -ML^2/18$$

Deflection at the middle of the span is given by v at $x = L/2$. Note that $M\langle x - L/3\rangle^2 = W(L/6)^2$ because the term inside the bracket is positive but $M\langle x - 2L/3\rangle^2$ is also zero *because the term inside the bracket is negative.*

$$EIv = M(L/6)^2/2 + C_1L/2 = -5ML^2/72 = -1.25ML^2/18$$

Thus the deflection at midspan is only 25 per cent greater than the deflection under the load.

Example 6

Calculate the fixed end moment for the beam shown in Fig. 4.33.

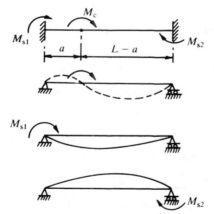

Figure 4.33 ●
A fixed beam under
off-centre couple

Solution   Let the fixed end moments at the left and right supports be M_{s1} and M_{s2} respectively. Then, using the information in Table 4.3, the total rotation at the supports can be calculated. Table 4.5 shows the slopes at the ends due to individual forces. For zero rotation at the supports,

$$M_c(L^2 - 3b^2)/(6L) - M_{s1}L/3 + M_{s2}L/6 = 0$$

$$M_c(L^2 - 3a^2)/(6L) + M_{s1}L/6 - M_{s2}L/3 = 0$$

Table 4.5

Load	$EI(dv/dx)_1$	$EI(dv/dx)_2$
M_c	$M_c(L^2 - 3b^2)/(6L)$	$M_c(L^2 - 3a^2)/(6L)$
M_{s1}	$-M_{s1}L/3$	$M_{s1}L/6$
M_{s2}	$M_{s2}L/6$	$-M_{s2}L/3$

Solving,

$$M_{s1} = M_c\{L^2 - a^2 - 2b^2\}/L^2$$

$$M_{s2} = M_c\{L^2 - 2a^2 - b^2\}/L^2$$

If $a = b = 0.5L$, then $M_{s1} = M_{s2} = 0.25M_c$.

Example 7

Consider the simply supported beam subjected to a uniformly distributed load of q extending from $x = a$ to $x = b$ as shown in Fig. 4.34(a). When using the singularity functions, we have to remember that once the action of a force is brought into play, it cannot be removed except by the introduction of the action of a force of the opposite nature. Therefore the uniformly distributed load q extending from $x = a$ to $x = b$ can be thought of as a combination of q starting at $x = a$ and continuing plus a load of $-q$ starting at $x = b$ and continuing as shown in Fig. 4.34(b). Therefore the equations are given by

$$EI\, d^2v/dx^2 = R_1x - q\langle x - a\rangle^2/2 + q\langle x - b\rangle^2/2$$

$$EI\, dv/dx = R_1x^2/2 - q\langle x - a\rangle^3/6 + q\langle x - b\rangle^3/6 + C_1$$

$$EIv = R_1x^3/6 - q\langle x - a\rangle^4/24 + q\langle x - b\rangle^4/24 + C_1x + C_2$$

Using the boundary conditions $v = 0$ at $x = 0$, $C_2 = 0$. Because $v = 0$ at $x = L$,

$$0 = R_1L^3/6 - q(L - a)^4/24 + q(L - b)^4/24 + C_1L$$

By statics $R_1 = q(b - a)(2L - a - b)/(2L)$, so C_1 can be determined and therefore the deflection and slope throughout the beam is defined.

As in the previous examples, if a series of uniformly distributed loads q_i extending from $x = a_i$ to $x = b_i$ act, then the necessary equations are given by

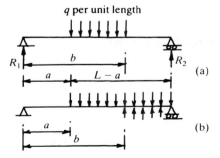

Figure 4.34 $\bullet$ (a), (b) A simply supported beam under uniformly distributed patch loading

$$EI \, d^2v/dx = R_1x - \sum q_i \langle x - a_i \rangle^2/2 + \sum q_i \langle x - b_i \rangle^2/2$$

$$EI \, dv/dx = R_1x^2/2 - \sum q_i \langle x - a_i \rangle^3/6 + \sum q_i \langle x - b_i \rangle^3/6 + C_1$$

$$EIv = R_1x^3/6 - \sum q_i \langle x - a_i \rangle^4/24 + \sum q_i \langle x - b_i \rangle^4/24 + C_1x + C_2$$

where $R_1 = \Sigma q_i(b_i - a_i)(2L - a_i - b_i)/(2L)$.

Example 8

As a simple application of the above equation, consider the case of a symmetrically loaded simply supported beam subjected to two equal uniformly distributed loads of q each extending from the supports to a distance of $L/3$.

Solution In this case $q_1 = q$, $a_1 = 0$, $b_1 = L/3$ and $q_2 = q$, $a_2 = 2L/3$, $b_2 = L$ and $R = qL/3$ from symmetry.

$$EIv = (qL/3)x^3/6 - \{qx^4 + q\langle x - 2L/3 \rangle^4\}/24 + \{q\langle x - L/3 \rangle^4 + q\langle x - L \rangle^4\}/24 + C_1x + C_2$$

The constants of integration are determined from the boundary conditions. At $x = 0$, $v = 0$ leading to $C_2 = 0$ and at $x = L$, $v = 0$,

$$(qL/3)L^3/6 - \{qL^4 + q\langle L - 2L/3 \rangle^4\}/24 + \{q\langle L - L/3 \rangle^4 + q\langle L - L \rangle^4\}/24 + C_1L = 0$$

leading to $C_1 = -7qL^3/324$

At the edge of the load, deflection is given by v at $x = L/3$. Note that $q\langle x - L/3 \rangle^4 = 0$ because the term inside the bracket is zero but $q\langle x - 2L/3 \rangle^4$ and $q\langle x - L \rangle^4$ are also zero *because the terms inside the brackets are negative.*

$$EIv = (qL/3)(L/3)^3/6 - \{q(L/3)^4\}/24 + C_1(L/3) = -11qL^4/1944$$

Deflection at the middle of the span is given by v at $x = L/2$. Note that $q\langle x - L/3 \rangle^3 = q(L/6)^4$ because the term inside the bracket is positive but $q\langle x - 2L/3 \rangle^4$ and $q\langle x - L \rangle^4$ are also zero *because the terms inside the brackets are negative.*

$$EIv = (qL/3)(L/2)^3/6 - \{q(L/2)^4\}/24 + \{q\langle L/2 - L/3 \rangle^4\}/24 + C_1(L/2)$$

$$= -12.5qL^4/1944$$

Thus the deflection at midspan is only $12.5/11.0 = 1.14$, i.e. 14 per cent greater than the deflection under the load.

4.8 ● Table of fixed end moments

The results of fixed end moment evaluation in the previous examples for the load cases in Fig. 4.35 can be usefully collected in Table 4.6.

The fixed end couple in the case of a propped cantilever propped at end 2 is given by

$$M_s = M_{s1} - 0.5M_{s2}$$

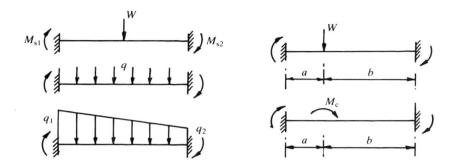

Figure 4.35 ● A fixed beam under various types of loading

Table 4.6 ● Support couples in fixed beams

Load	M_{s1}	M_{s2}
Central load W	$-WL/8$	$WL/8$
u.d.l. q	$-qL^2/12$	$qL^2/12$
Trapezoidal load	$-(3q_1 + 2q_2)L^2/60$	$(2q_1 + 3q_2)L^2/60$
Off-centre load W^*	$-Wab^2/L^2$	Wa^2b/L^2
Couple M_1^*	$M_1\{L^2 - a^2 - 2b^2\}/L^2$	$M_1\{L^2 - 2a^2 - b^2\}/L^2$

*a = distance of the load or couple from the left support; $b = (L - a)$. The negative sign indicates that it is an anticlockwise couple

Table 4.7 ● Support couples in propped cantilevers fixed at the left end

Load	Support couple
Central load W	$-3WL/16$
u.d.l. q	$-qL^2/8$
Trapezoidal load	$-(8q_1 + 7q_2)L^2/120$, with q_2 at the propped end
Off-centre load W^*	$-Wb(L^2 - b^2)/(2L^2)$
Couple M_1^*	$M_1\{L^2 - 3b^2\}/(2L^2)$

*b = distance of load or couple from the *propped* end

where M_{s1} and M_{s2} are the fixed end couples of the corresponding fixed beam.

Table 4.7 shows fixed end moments for some load cases for propped cantilevers.

4.9 ● Beams with variable second moment of area along the span

In previous sections, all the beams studied have had constant second moment of area over the entire span. Quite commonly, designers have to deal with beams having stepped variation of second moment of area along the span. Some simple examples are shown in Fig. 4.36.

Figure 4.36(a) is the case of a steel beam which has been strengthened by the use of additional flange plates over a certain section near the midspan. Figure 4.36(b) is the case of a column where the bottom section of the column has to carry crane

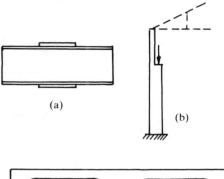

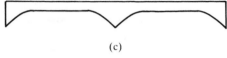

Figure 4.36 ● (a)–(c)
Beams with variable
second moment of area

loads as well as roof loads and is therefore made much stronger than the section which carries roof loads only. These two cases of stepped variation of second moment of area can be handled reasonably simply, as will be shown presently. However, in cases where the variation of second moment of area is more complex, such as that shown in Fig. 4.36(c) for the case of bridge girders which have their depth uniformly increasing towards the supports, the calculations are more complex and they are possibly best handled by numerical integration using a computer program.

Because $EI \, \mathrm{d}^2v/\mathrm{d}x^2 = M$ and if both M and EI are functions of x, then it is preferable to write the equation as

$$\mathrm{d}^2v/\mathrm{d}x^2 = M/(EI)$$

Integrating with respect to x

$$\mathrm{d}v/\mathrm{d}x = \int \{M/(EI)\} \, \mathrm{d}x + C_1$$

$$v = \int (\mathrm{d}v/\mathrm{d}x)\mathrm{d}x = \int \left[\int \{M/(EI)\}\mathrm{d}x\right]\mathrm{d}x + C_1x + C_2$$

The above equations can be given a physical interpretation which assists in the evaluation of the above integrals. Assume that a beam, called a conjugate beam, is loaded by a distributed load $q_c(x)$ equal to $M/(EI)$. The shear force Q_c and bending moment M_c at a section x in the conjugate beam loaded by $M/(EI)$ are related by the equations

$$\mathrm{d}Q_c/\mathrm{d}x = q_c \text{ and } \mathrm{d}M_c/\mathrm{d}x = -Q_c$$

We have

$$Q_c = \int q_c\mathrm{d}x + C_3 = \int \{M/(EI)\} \, \mathrm{d}x + C_3$$

$$M_c = -\int Q_c\mathrm{d}x - C_4 = -\int \left[\int \{M/(EI)\}\mathrm{d}x\right]\mathrm{d}x - C_3x - C_4$$

In the above equation for M_c, the arbitrary constant of integration C_4 has been chosen to be negative so as to make the expression for v in the real beam and $-M_c$ in the conjugate beam remain similar.

Clearly if $C_1 = C_3$ and $C_2 = C_4$, then the shear force Q_c in the conjugate beam is equal to the slope dv/dx in the real beam, and the bending moment M_c in the conjugate beam is equal to the negative deflection in the real beam. All that remains is to establish the boundary conditions for the conjugate beam before the constants of integration C_3 and C_4 can be evaluated. Two cases are of particular interest.

1. If $dv/dx = 0$ at $x = 0$ in the real beam, then $C_1 = 0$. Similarly, if $Q_c = 0$ at $x = 0$ in the conjugate beam, then $C_3 = 0$.
2. If $v = 0$ at $x = 0$ in the real beam, then $C_2 = 0$. Similarly if $M_c = 0$ at $x = 0$ in the conjugate beam, then $C_4 = 0$.

The above observations lead to the following real-conjugate beam pairs.

(a) If the real beam is a cantilever fixed at $x = 0$, then because slope and deflection are zero at $x = 0$, the conjugate beam has a 'free' end at $x = 0$ in order that shear force Q_c (corresponding to slope in the real beam) and bending moment M_c (corresponding to deflection in the real beam) are zero at the end corresponding to the 'fixed' end of the cantilever.
(b) If the real beam is a simply supported beam, then the corresponding conjugate beam is also a simply supported beam in order that at the supports $dv/dx \neq 0$ and $v = 0$ in the real beam and $Q_c \neq 0$ and $M_c = 0$ in the conjugate beam.

4.9.1 Examples using conjugate beams

Example 1

Using the conjugate beam method, calculate the deflection and slope at the tip of the stepped cantilever shown in Fig. 4.37(a).

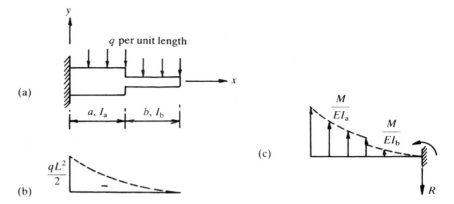

Figure 4.37 ● (a)–(c) Real and conjugate simply supported beams

Solution In the real beam, the bending moment at a distance x from the support as shown in Fig. 4.37(b) is

$$M = -q(L - x)^2/2$$

The distributed load on the conjugate beam as shown in Fig. 4.37(c) is

$$q_c = M/(EI) = -q(L - x)^2/(2EI_a) \quad 0 \leqslant x \leqslant a$$
$$= -q(L - x)^2/(2EI_b) \quad a < x \leqslant L$$

Because the real beam is a cantilever *fixed* at the *left*-hand end, the conjugate beam is a cantilever *fixed* at the *right*-hand end. The slope at the tip of the real cantilever is equal to the shear force at the support of the conjugate beam. Similarly, the deflection at the tip of the real cantilever is equal to the bending moment at the support of the conjugate beam. Note that because only bending moment and shear force are calculated in the conjugate beam, the variation of second moment of area along its length is irrelevant.

(a) The slope at the tip of the real cantilever is equal to the shear force Q_c at the support of the conjugate beam and is given by

$$Q_c = \int q_c \, dx = \int q(L - x)^2 \, dx/(2EI_a) \quad 0 \leqslant x \leqslant a$$
$$+ \int q(L - x)^2 \, dx/(2EI_b) \quad a \leqslant x \leqslant L$$

(b) The deflection at the tip of the real cantilever is equal to the bending moment M_c at the support of the conjugate beam and is given by

$$M_c = \int q_c(L - x) \, dx = \int q(L - x)^3 \, dx/(2EI_a) \quad 0 \leqslant x \leqslant a$$
$$+ \int q(L - x)^3 \, dx/(2EI_b) \quad a \leqslant x \leqslant L$$

The above integrals are evaluated by introducing the variable $r = L - x$; therefore $dx = -dr$.

(i) $Q_c = -\int qr^2 \, dr/(2EI_a) \quad L \geqslant r \geqslant (L - a)$
$$- \int qr^2 \, dr/(2EI_b) \quad L - a \geqslant r \geqslant 0$$
$$= qL^3/(6EI_a) - q(L - a)^3/(6EI_a) + q(L - a)^3/(6EI_b)$$
$$= qL^3/(6EI_a) + q(L - a)^3/6[1/(EI_b) - 1/(EI_a)]$$

(ii) $M_c = -\int qr^3 \, dr/(2EI_a) \quad L \geqslant r \geqslant (L - a)$
$$- \int qr^3 \, dr/(2EI_b) \quad L - a \geqslant r \geqslant 0$$
$$= qL^4/(8EI_a) - q(L - a)^4/(8EI_a) + q(L - a)^4/(8EI_b)$$
$$= qL^4/(8EI_a) + \{q(L - a)^4/8\}[1/(EI_b) - 1/(EI_a)]$$

Note that if $I_a = I_b$, then the results revert to those in Table 4.2.

Example 2

Calculate the slope at the supports and the deflection at the centre of the simply supported stepped beam shown in Fig. 4.38(a).

Solution   The problem is similar to Example 1. It should be noted that because, in the simply supported beam, the slope is not zero at the supports, in the conjugate beam the

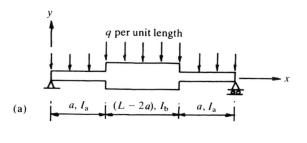

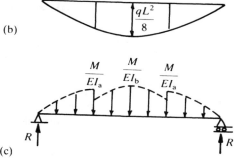

(a)

(b)

(c)

Figure 4.38 ● (a)–(c) Real and conjugate cantilevers

shear force is non-zero at the supports. Similarly, because the deflection is zero at the support of the real beam, the bending moment must also be zero at the support of the conjugate beam. Therefore the real and conjugate beams are both simply supported beams. The bending moment at a distance x from the left support of the real beam as shown in Fig. 4.38(b) is

$$M = qLx/2 - qx^2/2 = qx(L - x)/2$$

As shown in Fig. 4.38(c), the distributed load q_c on the conjugate beam is

$$q_c = M/(EI) = qx(L - x)/(2EI_a) \text{ for } 0 \leqslant x \leqslant a \text{ and } (L - a) \leqslant x \leqslant L$$
$$= qx(L - x)/(2EI_b) \text{ for } a < x < (L - a)$$

The slope at the support of the real beam is equal to the shear force at the support of the conjugate beam. The shear force at the support is equal to the reaction R at the support and is given by $\mp R = \mp \int q_c \, dx$ between the limits 0 and 0.5L:

$$\mp R = \mp \int \{qx(L - x)/(2EI_a)\} \, dx, \, 0 \leqslant x \leqslant a$$
$$\mp \int \{qx(L - x)/(2EI_b)\} \, dx, \, a \leqslant x \leqslant 0.5L$$
$$= \mp \{qa^2(3L - 2a)/12\}[1/(EI_a) - 1/(EI_b)] \mp qL^3/(24EI_b)$$

where the upper sign refers to the left support and the lower sign to the right support.

Similarly, the deflection at the centre of the real beam is equal to the bending moment M_c at the centre of the conjugate beam and is given by

$$M_c = R_1 0.5L - \int q_c(0.5L - x) \, dx$$
$$= R_1 0.5L - \int \{qx(L - x)(0.5L - x)/(2EI_a)\} \, dx, \, 0 \leqslant x \leqslant a$$
$$- \int \{qx(L - x)(0.5L - x)/(2EI_b)\} \, dx, \, a \leqslant x \leqslant 0.5L$$
$$= qa^3(4L - 3a) \{(1/(24EI_a) - 1/(24EI_b)\} - 5qL^4/(384EI_b)$$

If $I_a = I_b$, then the results revert to those given in Table 4.3.

4.10 ● Point of contraflexure

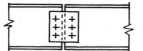

Figure 4.39 ●
Web splice

It was noticed in the case of propped cantilevers and fixed beams that the bending moment does not cause tension on the same face of the beam throughout the span. The bending moment generally produces tension at the top face near the supports and at the bottom face near the middle of the span. Thus the bending moment and hence the curvature is zero somewhere along the span. The point where the bending moment or the curvature changes from +ve to −ve is called a *point of contraflexure*. Points of contraflexure are of some importance in design because at these sections the beam is not stressed in bending and therefore they are the most suitable positions for joining two sections of the beam, e.g. by splicing in the case of steel structures, as shown in Fig. 4.39, or by lapping of tension reinforcement in the case of reinforced concrete beams. In addition, the approximate location of the points of contraflexure enables a complex structure to be isolated into a series of statically determinate segments which are easily analysed. This is particularly valuable at the initial design stage of statically indeterminate structures because a proper analysis of such structures requires information about the relative stiffness of members, which is not available at the preliminary design stage. Further information is given in Appendix 2.

The positions of points of contraflexure are easily determined. This is illustrated by a few simple examples.

Example 1

Determine the point of contraflexure in a propped cantilever carrying a uniformly distributed load q as shown in Fig. 4.40(a).

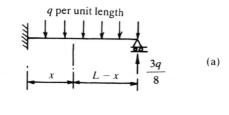

(a)

(b)

Figure 4.40 ● (a)–(c)
Position of contraflexure point in a propped cantilever under uniformly distributed loading

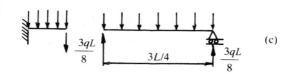

(c)

Solution As was shown in Example 2, Section 4.6, the value of the reaction at the prop is $3qL/8$. Therefore the bending moment at any section x, as shown in Fig. 4.40, is given by

$$M = (3qL/8)(L - x) - q(L - x)^2/2 = q(L - x)(4x - L)/8$$

$$M = 0 \text{ when } (L - x)(4L - x) = 0 \text{ i.e. at } x = L \text{ and at } x = 0.25L$$

Therefore the point of contraflexure is at $0.25L$ from the fixed support.

Therefore by fixing the left support, the point of contraflexure moves towards the centre of the span by $0.25L$. As can be seen from Fig. 4.40(c), the point of contraflexure divides the structure into a cantilever segment and a simply supported segment.

Example 2

Determine the points of contraflexure in the propped cantilever carrying a load W at the centre as shown in Fig. 4.41.

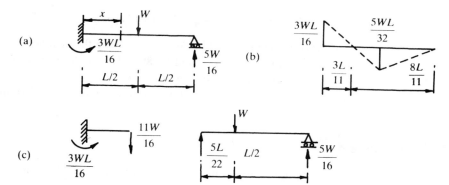

Figure 4.41 ● (a)–(c) Point of contraflexure in a propped cantilever under a central concentrated load

Solution The moment at the fixed end can be calculated as follows. Using the information for a simply supported beam given in Table 4.3, the rotation caused at the left support by W is $WL/(16EI)$ and the rotation caused by the couple M_s at the support is $M_sL/(3EI)$. For zero rotation at the support, $M_s = 3WL/16$. The reaction at the prop is $0.5W - M_s/L = 5W/16$. Therefore the vertical reaction at the left support is $11W/16$.

A point of contraflexure will be between the fixed support and the load. The bending moment at a section x is given by

$$M = 11Wx/16 - M_s = W(11x - 3L)/16, \text{ if } 0 \leqslant x \leqslant 0.5L$$

$$M = 0 \text{ when } x = 3L/11 = 0.273L$$

As compared to the case of uniform loading, the shift in the point of contraflexure is 9 per cent greater.

Example 3

Determine the points of contraflexure in the fixed beam carrying a uniformly distributed load q as shown in Fig. 4.42. The support moments are $M_s = qL^2/12$. The vertical reactions are by symmetry $0.5qL$. The bending moment at a section x is given by

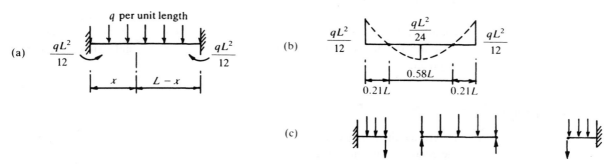

(a) q per unit length   (b)   (c)

Figure 4.42 ● (a)–(c) Contraflexure points in a fixed beam under uniformly distributed loading

$$M = 0.5qLx - M_s - qx^2/2 = (q/12)\{6Lx - L^2 - 6x^2\}$$

$M = 0$ when $(6Lx - L^2 - 6x^2) = 0$. Solving the quadratic equation, $x = L\{3 \pm \sqrt{3}\}/6$. This gives $x = 0.211L$ and $0.789L$. The contraflexure points are at a distance of $0.211L$ from each end (Fig. 4.42(b)).

Fig. 4.42(c) shows the division into cantilevers and simply supported parts.

Example 4

Determine the points of contraflexure in a fixed beam carrying a central concentrated load W as shown in Fig. 4.43.

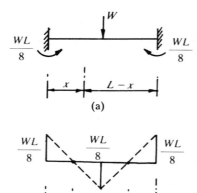

Figure 4.43 ● (a), (b) Contraflexure points in a fixed beam under midspan concentrated load

Solution In this case, the vertical reactions at the support are $0.5W$ and the fixed end moment is $M_s = WL/8$. There are two points of contraflexure symmetrically positioned from the ends and lying between the fixed end and the applied load. The bending moment at a section x is

$$M = 0.5Wx - M_s = (W/8)\{4x - L\}, \ 0 \leqslant x \leqslant 0.5L$$

$$M = 0 \text{ at } x = 0.25L$$

The second point of contraflexure is at $x = 0.75L$ (Fig. 4.43(b)).

4.10.1 General comment on points of contraflexure

From the four examples considered in the previous section it can be seen that:

1. In the case of a uniformly distributed load, fixing both the supports shifts the points of contraflexure by $0.21L$ towards the midspan. In the case of a propped cantilever, fixing one end only shifted the point of contraflexure by $0.25L$ from the fixed end towards the midspan. In other words, for all practical purposes it can be assumed that the fixity at any one end affects the position of the point of contraflexure at that end only.
2. In the case of a central concentrated load, the observations made for uniform loading hold good because the point of contraflexure in the case of a propped cantilever is $0.273L$ from the fixed end and in the case of a fixed beam, it is $0.25L$ from the fixed end.
3. As a general rule, irrespective of the loading on the beam, it can be assumed that the point of contraflexure is approximately $0.2L$ from the fixed ends.

Some examples on the approximate analysis of structures using the above information are given in Appendix 2.

4.11 ● Control of deflection through limiting span/depth ratio

It has been indicated that in the design of structures it is necessary not only to limit the maximum allowable stress in order to prevent permanent deformations at working loads but also to limit the maximum deflections in order to prevent damage to plaster and other finishes. For example, in building structures it is generally specified that the maximum deflection of the beam should not exceed span/300 or thereabouts.

Consider a simply supported beam under uniformly distributed loading q. The maximum bending moment occurs at the centre and is equal to $M_{max} = qL^2/8$. The maximum deflection Δ also occurs at the centre and is equal to $5qL^4/(384EI)$. Therefore (maximum deflection)/span is given by

$$\Delta/L = 5qL^3/(384EI) = (M_{max}/I)L\{5/(48E)\}$$

But $M_{max}y_{max}/I = \sigma_{max}$, where y_{max} = maximum distance from the neutral axis to the top or bottom face where the maximum bending stress σ_{max} occurs.

Therefore

$$\Delta/L = (\sigma_{max}/E)(L/y_{max})(5/48)$$

Similarly, if the load on the beam is a centrally concentrated load, then $\Delta = WL^3/(48EI)$ and $M_{max} = WL/4$. Therefore

$$\Delta/L = (\sigma_{max}/E)(L/y_{max})(4/48)$$

As can be seen, the expression for Δ/L in terms of σ_{max} and L/y_{max} is not particularly sensitive to the type of loading on the beam. Because y_{max} is some function of the depth d of the beam, and further, for a given material, E and σ_{max} are constants, any prescribed limit on Δ/L can be achieved by limiting L/y_{max} or L/d. This limit is very often prescribed in codes of practice for the design of structures.

4.12 ● Differential equations for bending deformations – unsymmetrical bending

In Section 4.2, the differential equations

$$EI_{zz}\frac{d^2v}{dx^2} = M_z \text{ and } EI_{yy}\frac{d^2w}{dx^2} = -M_y$$

were established for the case of symmetrical bending.

In the case of symmetrical bending, the bending stress σ_x is given by

$$\sigma_x = -\frac{M_z}{I_{zz}}y + \frac{M_y}{I_{yy}}z$$

In the case of unsymmetrical bending, σ_x is given by

$$\sigma_x = -\frac{(M_z I_{yy} + M_y I_{yz})}{(I_{zz}I_{yy} - I_{zy}^2)}y + \frac{(M_z I_{yz} + M_y I_{zz})}{(I_{yy}I_{zz} - I_{zy}^2)}z$$

Assuming that σ_x is the only normal stress present, then $\varepsilon_x = \partial u/\partial x = \sigma_x/E$. Integrating partially with respect to x, we have

$$u = -(y/C)\int \{M_z I_{yy} + M_y I_{yz}\}\,dx + (z/C)\int \{M_z I_{yz} + M_y I_{zz}\} + C_1(y, z)$$

where $C_1(y, z)$ is a 'constant' of integration.

$$C = E(I_{yy}I_{zz} - I_{zy}^2)$$

Differentiating u partially with respect to y we have

$$\partial u/\partial y = -(1/C)\int \{M_z I_{yy} + M_y I_{yz}\}\,dx + \partial C_1/\partial y$$

Assuming that shear strain $\gamma_{xy} = \partial u/\partial y + \partial v/\partial x = 0$, we have

$$\partial v/\partial x = -\partial u/\partial y = (1/C)\int \{M_z I_{yy} + M_y I_{yz}\}\,dx - \partial C_1/\partial y$$

Differentiating with respect to x we have

$$\partial^2 v/\partial x^2 = (1/C)\{M_z I_{yy} + M_y I_{yz}\}$$

Quite clearly if $I_{yz} = 0$, then the equation reverts to the same form as the equation in the case of symmetrical bending.

In the case of unsymmetrical bending the neutral axis is inclined to the co-ordinate axes y and z. Therefore, in addition to the displacement v in the vertical x–y plane, there is also a displacement w in the horizontal x–z plane. This can be determined by assuming that the shear strain $\gamma_{zx} = \partial u/\partial z + \partial w/\partial x = 0$. This is equivalent to, for example, assuming that shear strain in the flange in the z-section shown in Fig. 3.34 is zero. Differentiating u partially with respect to z we have

$$\partial u/\partial z = (1/C) \int \{M_z I_{yz} + M_y I_{zz}\} + \partial C_1/\partial z$$

For zero γ_{xz}, $\partial w/\partial x = -\partial u/\partial z$. Therefore

$$\partial w/\partial x = -\partial u/\partial z = -(1/C) \int \{M_z I_{yz} + M_y I_{zz}\} - \partial C_1/\partial z$$

Differentiating partially with respect to x we have

$$\partial^2 w/\partial x^2 = -(1/C)\{M_z I_{yz} + M_y I_{zz}\}$$

If $I_{yz} = 0$, then clearly there is no 'curvature' in the x–z plane, when $M_y = 0$. Given M_z and M_y, integration of the moment–curvature relationships can be carried out as for the symmetrical bending case.

4.13 ● Deformation due to shear force

It was pointed out in Section 4.2 that the equation $EI\,d^2v/dx^2 = M$ applies to the displacement v caused by M under the assumption that shear strain $\gamma_{xy} = \partial u/\partial y + \partial v/\partial x = 0$. This was assumed to be valid even when shear force and bending moment coexist. The object of this section is to calculate the additional deflection due to shear strain and to develop a criterion to decide when it is safe to ignore the deformation caused by shear altogether.

Consider a simple cantilever beam subjected to an end load as shown in Fig. 4.44(a). The deformation due to bending is of the form shown in Fig. 4.44(b) and was shown in Section 4.3, Example 2, to be

$$v_{bending} = -W(3Lx^2 - x^3)/(6EI)$$

At a cross-section, as was shown in Section 3.11.1, shear stress is a maximum at the neutral axis and is zero at the extreme fibres. However, for simplicity it is assumed that an 'average' shear strain characterizes the deformation due to shear as shown in Fig. 4.44(c). The 'average' shear strain is given by

$$dv_s/dx = Q/(GA_s)$$

where Q = shear force at the section, v_s = displacement due to shear, A_s = area of cross-section to resist shear which is less than the area of cross-section A. The ratio A/A_s is generally known as the area shear factor, G = shear modulus.

As a simple application of the above equation, consider the cantilever shown in Fig. 4.44(a). For this case, $Q = -W$. Therefore

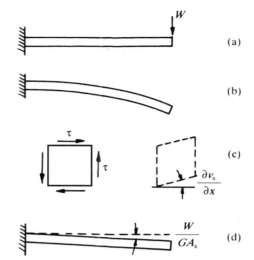

Figure 4.44 ● (a)–(d) Deformation due to bending and shear in a cantilever under a concentrated load at the tip

$$dv_s/dx = -W/(GA_s)$$

Integrating with respect to x,

$$v_s = -Wx/(GA_s) + C_1$$

where C_1 = constant of integration. Because $v_s = 0$ at $x = 0$, then $C_1 = 0$, and $v_s = -Wx/(GA_s)$.

It should be noted that, as shown in Fig. 4.44(d), $dv_s/dx \neq 0$ at the 'fixed' end because of the shear deformation. The tip deflection due to shear is

$$v_s \text{ at } (x = L) = -WL/(GA_s)$$

The total downward deflection due to bending and shear at the tip is

$$v_{total} = v_{bending} + v_s$$

$$= WL^3/(3EI) + WL/(GA_s)$$

$$= WL^3/(3EI)\{1 + 3EI/(GA_sL^2)\}$$

For a rectangular beam of width b and depth d, $I = bd^3/(12)$, $A_s = bd/(1.2)$ (see Section 4.13.1). If the material is assumed to be isotropic, then $G = E/\{2(1 + v)\}$.

Substituting the above values,

$$v_{total} = v_{bending}\{1 + 0.6(1 + v)(d/L)^2\}$$

Obviously the second term in the brackets depends on the span to depth ratio L/d, and the smaller the ratio the greater the contribution due to shear deformation. Assuming that Poisson's ratio $v = 0.3$, the ratio of $v_{total}/v_{bending}$ varies with L/d ratio as shown in Table 4.8.

The shear deformation in the case of a uniformly distributed load can be calculated in a similar manner. Referring to Fig. 4.45, $Q = -q(L - x)$. Therefore

Table 4.8 ● Shear deformation in cantilever beams loaded at the tip

L/d	$v_{total}/v_{bending}$
10	1.0078
5	1.0312
3	1.0867
2	1.1950

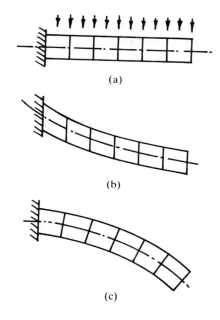

(a)

(b)

Figure 4.45 ● (a)–(c) A cantilever under uniformly distributed load: (a) loading, (b) shear deformation, (c) flexural deformation

(c)

$$dv_s/dx = -q(L - x)/(GA_s)$$

$$v_s = -q(Lx - x^2/2)(GA_s) + C_1$$

As $v_s = 0$ at $x = 0$, $C_1 = 0$.

Therefore

$$v_s = -q(Lx - x^2/2)/(GA_s)$$

The deflection at the tip $x = L$ is

$$v_s = qL^2/(GA_s)$$

From Table 4.1, $v_{bending} = qL^4/(8EI)$. The total deflection at the tip is

$$v_{total} = v_{bending} + v_s$$

$$= qL^4/(8EI) + qL^2/(2GA_s)$$

$$= qL^4/(8EI)\{1 + 4EI/(GA_sL^2)\}$$

Assuming a rectangular beam as before and $v = 0.3$,

Table 4.9 ●
Shear deformation
in cantilevers with
uniformly distributed
loading

L/d	$v_{total}/v_{bending}$
10	1.0104
5	1.0416
3	1.1160
2	1.2600

$$v_{total} = v_{bending}\{1 + 1.04(d/L)^2\}$$

For different values of L/d, the ratios of $v_{total}/v_{bending}$ are shown in Table 4.9. As can be seen, the effect of shear deformation on the total tip deflection is of the same order irrespective of whether a concentrated load or a uniformly distributed load is acting. Quite clearly the shear deformation contribution to total displacement becomes important only in the case of very 'stocky' beams with 'small' span/depth ratio.

4.13.1 Area shear factor

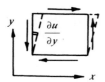

Figure 4.46 ●
(a) Shear deformation of
a beam element

The ratio of the area of cross-section A to 'average' area in shear A_s is called the area shear factor. In the case of a rectangular beam, the value of area shear factor can be derived as follows.

It was shown in Section 3.11.1 that the distribution of shear stress τ_{xy} in a rectangular beam is parabolic and is given by

$$\tau_{xy} = 1.5(Q/A)\{1 - (2y/d)^2\}$$

where Q = shear force at the section, A = cross-sectional area, d = depth of the section and the origin for y is at the centroid.

As shown in Fig. 4.46(a), the rotation of the cross-section is given by

$$\gamma_{xy} = \partial u/\partial y = \tau_{xy}/G$$

where G = shear modulus. Therefore

$$\partial u/\partial y = C\{1 - (2y/d)^2\}$$

where $C = 1.5Q/(GA)$. Integrating partially with respect to y we have

$$u = Cd\{y/d - (4/3)(y/d)^3\} + \text{a function of } x$$

Assuming that $u = 0$ at $y = 0$, then

$$u = Cd\{y/d - (4/3)(y/d)^3\}$$

For simplicity, if it is *assumed* that an average value of shear strain is required, then

$$\gamma_{xy\ average} = (\partial u/\partial y)_{average}$$

or

$$u_{average} = \{\gamma_{xy\ average}\}y$$

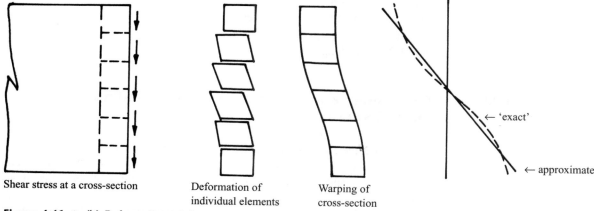

Shear stress at a cross-section   Deformation of   Warping of
                                   individual elements   cross-section

← 'exact'

← approximate

Figure 4.46 ● (b) Deformation of the cross-section of a rectangular beam due to shear stresses

The value of $\gamma_{xy\ average}$ can be determined by minimizing the square of the error (in order to eliminate the sign of the error) between the 'exact' value of u which is cubic in y and the average value of u which is a linear function of y. This shows that $\gamma_{xy\ average} = 0.8\,C = 1.2Q/(GA) = Q/(GA_s)$ where $A_s = A/1.2$.

The error in the value of u between the exact and approximate values is given by

$$\text{Error} = F(y) = Cd\{(y/d) - (4/3)(y/d)^3\} - \gamma_{xy\ average}(y/d)\ dy$$

Total of square of error, S, over the whole cross-section $= \int F^2(y)\ dy$ where the integration is carried over the limits $y = \pm 0.5d$.

Minimizing the total of square of error S over the whole cross-section with respect to $\gamma_{xy\ average}$, we have

$$\frac{dS}{d\gamma_{average}} = 0 = \int F(y)\ dy = 0$$

$$\int [C\{(y/d) - (4/3)(y/d)^3\} - \gamma_{xy\ average}(y/d)]y\ dy = 0$$

Carrying out the integration and substituting the limits,

$$C(1/15) - \gamma_{xy\ average}(1/12) = 0$$

$$\gamma_{xy\ average} = 0.8\,C, \text{ where } C = 1.5\,Q/(GA)$$

$$\therefore\ \gamma_{xy\ average} = 1.2\,Q/(GA) = (Q/G)[1.2/A]$$

Figure 4.46(b) shows that if the depth of the beam is divided into a series of elements, then each of the individual elements will deform. When the deformed elements are assembled, then the cross-section deforms in the shape of an S. A plot of the variation of the 'exact' value of $u = Cd\{y/d - (4/3)(y/d)^3\}$ and the linear approximation to the variation of u given by $u = 0.8\,Cd\{y/d\}$ is also shown in the figure, to indicate the closeness of the approximation to the actual deformation.

In the case of I beams bending about the major axis parallel to the flange, because most of the shear is carried out by the web, A_s is approximately equal to web area/1.2. Therefore the area shear factor is given by $1.2A$/web area. If the beam is bending about an axis parallel to the web, then the shear is carried by the flanges. Therefore A_s is approximately the area of two flanges/1.2 and the area shear factor = $1.2A$/total flange area.

4.14 ● Stiffness matrix for a beam element – bending deformation only

In Section 4.1.3 the stiffness matrix for a bar element was derived. The stiffness matrix in general represents the relationship between the forces and the corresponding independent displacements, i.e. translations and rotations at the ends of an element.

In a beam element, the possible independent displacements at the ends are a translation and a rotation. Consider the beam element shown in Fig. 4.47(a). It carries a uniformly distributed load of q per unit length and has translations Δ_{n1}, Δ_{n2} and rotations θ_1, θ_2 at ends 1 and 2 respectively. Because the structure is linearly elastic, the principle of superposition is valid. Therefore the final state of the member can be arrived at in five stages as shown in Fig. 4.47(b). It should be noted that the results for translational and rotational stiffness for a propped cantilever derived in Section 4.5, Examples 3 and 4, have been made use of. Summing up the forces at the ends for the five cases we have to consider:

(a) translation Δ_{n1} at end 1;
(b) rotation θ_1 at end 1;
(c) translation Δ_{n2} at end 2;
(d) rotation θ_2 at end 2;
(e) fixed end condition;

and taking upward forces and translations as positive and clockwise couples and rotations also as positive, we have

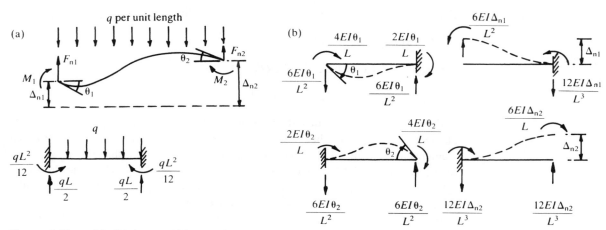

Figure 4.47 ● (a), (b) A general beam element with lateral load and support displacements

$$F_{n1} = (EI/L^2)\{12\Delta_{n1}/L - 6\theta_1 - 12\Delta_{n2}/L - 6\theta_2\} + 0.5qL$$

$$M_1 = (EI/L)\{-6\Delta_{n1}/L + 4\theta_1 + 6\Delta_{n2}/L + 2\theta_1\} - qL^2/12$$

$$F_{n2} = (EI/L^2)\{-12\Delta_{n1}/L + 6\theta_1 + 12\Delta_{n2}/L + 6\theta_2\} + 0.5qL$$

$$M_2 = (EI/L)\{-6\Delta_{n1}/L + 2\theta_1 + 6\Delta_{n2}/L + 4\theta_2\} + qL^2/12$$

The above four equations can be expressed in matrix form as

$$
\begin{bmatrix} F_{n1} \\ M_1 \\ F_{n2} \\ M_2 \end{bmatrix} = EI/L
\begin{bmatrix}
12/L^2 & -6/L & -12/L^2 & -6/L \\
-6/L & 4 & 6/L & 2 \\
-12/L^2 & 6/L & 12/L^2 & 6/L \\
-6/L & 2 & 6/L & 4
\end{bmatrix}
\times
\begin{bmatrix} \Delta_{n1} \\ \theta_1 \\ \Delta_{n2} \\ \theta_2 \end{bmatrix}
+
\begin{bmatrix} 0.5qL \\ -qL^2/12 \\ 0.5qL \\ qL^2/12 \end{bmatrix}
\tag{4.7}
$$

As can be seen, as in the case of a bar element, the stiffness relationship for a beam element consists of {Force vector}, {Stiffness matrix}, {Displacement vector} and {Fixed end force vector}. As already explained, the concept of the stiffness matrix forms the foundation of modern methods of structural analysis oriented to the use of digital computers. Further discussion of the stiffness matrix for a beam element is given in Chapter 6.

4.15 ● Stiffness matrix for a beam element including bending and shear deformation

In the previous section, the stiffness matrix for a beam element was derived where the deformations were caused by bending deformation, i.e. the extension and contraction of the fibres of the beam. As shown in Section 4.13, if the span/depth ratio of the beam is 'small', then it becomes important to include deformations due to both bending and shear in order to get a realistic idea of total deformation of the beam. The inclusion of shearing deformations will introduce corrections to the coefficients of the stiffness matrix given in the previous section. The corrected stiffness coefficients can be derived as follows.

Consider the cantilever shown in Fig. 4.48(a). It is subjected to a force F_{n2} and a couple M_2 at end 2. The total deflection Δ_{n2} and rotation θ_2 at end 2 can be calculated in two stages as follows.

1. Bending deformation. Using the information from Table 4.1,

$$\Delta_{n2bending} = F_{n2}L^3/(3EI) - M_2L^2/(2EI)$$

$$\theta_{2bending} = -F_{n2}L^2/(2EI) + M_2L/(EI)$$

Note that $\Delta_{n2bending}$ is considered positive upwards and $\theta_{2bending}$ which is the rotation of the cross-section is considered positive clockwise.

2. Shearing deformation. Because the shear force is constant, the deflection due to shear force is given by

$$\Delta_{n2shear} = F_{n2}L/(GA_s)$$

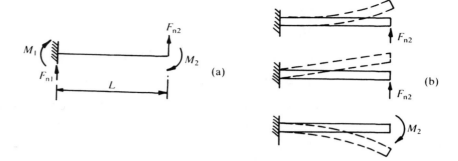

Figure 4.48 ● (a), (b) A cantilever under a concentrated load and couple at the tip

The deformation due to shear takes place as shown in Fig. 4.48(b), and so clearly $\theta_{2\text{shear}}$ is zero because the cross-section of the beam does not rotate.

3. Total displacements. Adding the deformations due to bending and shear, we have

$$\Delta_{n2} = \Delta_{n2\text{bending}} + \Delta_{n2\text{shear}}$$

$$= F_{n2}L^3/(3EI) - M_2L^2/(2EI) + F_{n2}L/(GA_s)$$

$$= F_{n2}L^3/(3EI)\{1 + \beta/4\} - M_2L^2/(2EI)$$

where $\beta = 12EI/(GA_sL^2)$. Similarly,

$$\theta_2 = \theta_{2\text{bending}} + \theta_{2\text{shear}} = -F_{n2}L^2/(2EI) + M_2L/(EI) + 0$$

Solving for F_{n2} and M_2

$$F_{n2} = [12EI/\{(1 + \beta)L^3\}]\Delta_{n2} + [6EI/\{(1 + \beta)L^2\}]\theta_2$$

$$M_2 = [6EI/\{(1 + \beta)L^2\}]\Delta_{n2} + [(4 + \beta)EI/\{(1 + \beta)L\}]\theta_2$$

From equilibrium, the force F_{n1} and the couple M_1 at end 1 are given by

$$F_{n1} = -F_{n2} = -[12EI/\{(1 + \beta)L^3\}]\Delta_{n2} - [6EI/\{(1 + \beta)L^2\}]\theta_2$$

$$M_1 + M_2 - F_{n2}L = 0$$

Therefore

$$M_1 = [6EI/\{(1 + \beta)L^2\}]\Delta_{n2} + [(2 - \beta)EI/\{(1 + \beta)L\}]\theta_2$$

Expressed in matrix form:

$$\begin{bmatrix} F_{n1} \\ M_1 \\ F_{n2} \\ M_2 \end{bmatrix} = EI/\{L(1 + \beta)\} \begin{bmatrix} -12/L^2 & -6/L \\ 6/L & 2 - \beta \\ 12/L^2 & 6/L \\ 6/L & 4 + \beta \end{bmatrix} \begin{bmatrix} \Delta_{n2} \\ \theta_2 \end{bmatrix}$$

The above equation gives all the necessary corrected stiffness coefficients. The correction factors are as follows.

In the beam stiffness matrix obtained by ignoring the shearing deformations, replace 12 by $12/(1 + \beta)$, 6 by $6/(1 + \beta)$, 4 by $(4 + \beta)/(1 + \beta)$ and 2 by $(2 - \beta)/(1 + \beta)$. The final corrected stiffness matrix is as follows.

$$\begin{bmatrix} F_{n1} \\ M_1 \\ F_{n2} \\ M_2 \end{bmatrix} = [EI/\{L(1+\beta)\}] \times \begin{bmatrix} 12/L^2 & -6/L & -12/L^2 & -6/L \\ -6/L & 4+\beta & 6/L & 2-\beta \\ -12/L^2 & 6/L & 12/L^2 & 6/L \\ -6/L & 2-\beta & 6/L & 4+\beta \end{bmatrix} \begin{bmatrix} \Delta_{n1} \\ \theta_1 \\ \Delta_{n2} \\ \theta_2 \end{bmatrix} \tag{4.8}$$

4.15.1 Effect of including shearing deformations on fixed end forces

As can be seen from Fig. 4.49, if the lateral load is symmetric with respect to the centre line, then shearing deformation does not disturb the boundary conditions. In the case when the loading is not symmetrical, then the procedure for calculating the fixed end forces is illustrated by considering the 'off-centre' loaded beam as shown in Fig. 4.50.

The various steps involved are as follows.

1. Fixed end forces ignoring shear deformation. As shown in Section 4.5.1, Example 2, ignoring the shear deformation, the fixed end forces are given by

(a) At left support: $V_1 = Wb^2(3a + b)/L^3$

$$M_1 = -Wab^2/L^2$$

(b) At right support: $V_2 = Wa^2(a + 3b)/L^3$

$$M_2 = +Wa^2b/L^2$$

Figure 4.49 ●
Deformation due to
shear of a clamped
beam under midspan
concentrated load

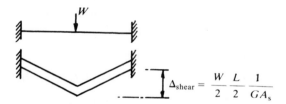

$$\Delta_{\text{shear}} = \frac{W}{2}\frac{L}{2}\frac{1}{GA_s}$$

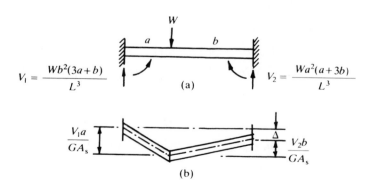

Figure 4.50 ● (a)–(c)
Deformation due to
shear of a fixed end
beam under off-centre
concentrated load

2. Calculate shear deformation due to forces in (1) above. In the region to the left of the load the shear force is $-V_1$ and in the region to the right of the load the shear force is V_2. As shown in Section 4.12, the shear deformation will be as shown in Fig. 4.50(b). Thus the net downward displacement Δ at the right support is given by

$$\Delta = (V_1a - V_2b)/(GA_s) = Wab(b^2 - a^2)/\{L^3GA_s\}$$

$$= Wab(b - a)/\{L^2GA_s\}$$

Note that for a symmetrical load, $a = b$ and therefore $\Delta = 0$. Therefore, when $\Delta \neq 0$ the fixed end forces calculated using bending deformation only lead to violation of the zero deflection condition at the right support.

3. Calculate the correction forces to restore the boundary conditions. From the stiffness matrix for the beam element including bending and shear deformation, the forces required to cause an upward deflection of Δ at the right support, without affecting other support displacements, is calculated by using the conditions $\Delta_{n1} = 0$, $\theta_1 = 0$, $\Delta_{n2} = \Delta$ and $\theta_2 = 0$. Therefore the necessary forces from the stiffness matrix are

$$\begin{bmatrix} F_{n1} = V_1 \\ M_1 \\ F_{n2} = V_2 \\ M_2 \end{bmatrix} = EI/\{L(1 + \beta)\} \begin{bmatrix} -12/L^2 \\ 6/L \\ 12/L^2 \\ 6/L \end{bmatrix} [\Delta]$$

where $\Delta = Wab(b - a)/(L^2GA_s)$.

4. Calculate the true fixed end forces. The true fixed end forces are the sum of the fixed end forces calculated by considering bending deformation only as in (1) + correction forces as in (3). Therefore

$$M_1 = -Wab^2/L^2 + 6EI\Delta/\{L^2(1 + \beta)\}$$

$$= -Wab^2/L^2 + Wab(b - a)/L^2[6EI/\{L^2GA_s(1 + \beta)\}]$$

Because $\beta = 12EI/(GA_sL^2)$, the above equation can be expressed as

$$M_1 = -Wab^2/L^2[1 - 0.5C]$$

where $C = \{1 - a/b\}\beta/(1 + \beta)$. Similarly,

$$M_2 = Wa^2b/L^2 + 6EI\,\Delta/\{L^2(1 + \beta)\}$$

$$= Wa^2b/L^2 + Wab(b - a)/L^2[6EI/\{L^2GA_s(1 + \beta)\}]$$

$$= Wa^2b/L^2[1 + 0.5C]$$

$$V_1 = Wb^2(3a + b)/L^3 - 12EI\Delta/\{L^3(1 + \beta)\}$$

$$= Wb^2(3a + b)/L^3 - Wab(b - a)\beta/\{L^3(1 + \beta)\}$$

$$= Wb^3/L^3[(3a/b + 1) - (a/b)C]$$

$$V_2 = Wa^2(3a + b)/L^3 + 12EI\,\Delta/\{L^3(1 + \beta)\}$$

$$= Wa^2(a + 3b)/L^3 + Wab(b - a)\beta/\{L^3(1 + \beta)\}$$

$$= Wa^3/L^3[(1 + 3b/a) + (a/b)C]$$

Summarizing,

$$M_1 = -Wab^2/L^2[1 - 0.5C]$$

$$M_2 = Wa^2b/L^2[1 + 0.5C]$$

$$V_1 = Wb^3/L^3[(3a/b + 1) - (a/b)C]$$

$$V_2 = Wa^3/L^3[(3b/a + 1) - (a/b)C]$$

As is to be expected, if $\beta = 0$ or $b = a$, then $C = 0$ and the expressions revert to the case of fixed end forces obtained by considering bending deformations only.

4.16 ● Deformation due to torsion

When a shaft is subjected to a twisting moment, two types of deformations occur.

1. A rigid body rotation of the cross-section about the shaft axis. This is called the angle of twist ψ.
2. A displacement in the axial direction. This is called the warping displacement.

If there is no restraint to warping displacement and the twisting moment T is constant along the shaft, then the relationship between torque T and the angle of twist ψ, as shown in works on the theory of elasticity, is given by

$$T = GJ \, d\psi/dx$$

where G = shear modulus and J = a section property known as the Saint Venant torsion constant. GJ is called torsional rigidity. The above equation can be integrated with respect to x to give

$$\psi = \{T/(GJ)\}x + C_1$$

where C_1 = constant of integration. If at $x = 0$, $\psi = \psi_1$, then $C_1 = \psi_1$. Therefore

$$\psi = \{T/(GJ)\}x + \psi_1$$

If at $x = L$, $\psi = \psi_2$, then

$$T = \{GJ/L\}(\psi_2 - \psi_1)$$

Using the notation in Fig. 4.51, if at $x = 0$, $T_1 = -T$ and at $x = L$, $T_2 = T$, then

$$T_1 = \{GJ/L\}(\psi_1 - \psi_2)$$

$$T_2 = \{GJ/L\}(\psi_2 - \psi_1)$$

Using the notation in Fig. 4.51, if at $x = 0$, $T_1 = -T$ and at $x = L$, $T_2 = T$, then

$$T_1 = \{GJ/L\}(\psi_1 - \psi_2)$$

$$T_2 = \{GJ/L\}(\psi_2 - \psi_1)$$

The above two equations can be expressed in matrix form as

$$\begin{bmatrix} T_1 \\ T_2 \end{bmatrix} = GJ/L \begin{bmatrix} 1 & -1 \\ -1 & 1 \end{bmatrix} \begin{bmatrix} \psi_1 \\ \psi_2 \end{bmatrix}$$

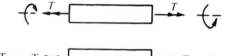

Figure 4.51 ● A shaft element under torsional forces at the ends

$T_1 = -T$ ← ← ← $T_2 = T$

This is known as the stiffness matrix for a shaft without any restraint to warping displacement.

4.16.1 Saint Venant's torsion constant J

The exact method of torsional stress analysis shows that the torsion constant J is given by

(a) Solid circular cylinder of radius R: $J = \pi R^4/2$.
(b) Hollow circular cylinder of internal radius R_i and outer radius R_o:

$$J = \pi(R_o^4 - R_i^4)/2.$$

(c) Solid rectangle $b \times d$:
Approximate value of J is given by $J \simeq 0.3b^3d^3/(b^2 + d^2)$ Exact value is given by

$$J = kbd^3, \, d > b$$

where the value of k is shown in Table 4.10.

Table 4.10 ● Torsion constant J for rectangular section

d/b	k	d/b	k
1.0	0.1406	3.0	0.263
1.2	0.166	4.0	0.281
1.5	0.196	5.0	0.291
2.0	0.229	10.0	0.312
2.5	0.249	∞	0.333

(d) Rolled steel sections such as I, L, T and $[$ sections: as discussed in Section 3.18.1, these sections behave, for all practical purposes, as if consisting of a series of individual thin rectangles. Therefore, although not exact, it can be assumed that

$$J = 0.333 \sum b_i t_i^3$$

where b_i and t_i are the length and thickness of individual rectangles into which the section can be divided, as shown in Fig. 3.55.

(e) Thin rectangular tube: $J = 4b^2d^2/\{(t_{ft} + t_{fb})/b + 2t_w/d\}$ where b = flange 'width', d = web 'depth', t_w = thickness of web, and t_{ft} and t_{fb} are the thickness of the top and bottom flanges respectively.
Note: The above expression for J for a thin-walled rectangular tube is known as Bredt's formula.

Figure 4.52 ●
An infinitesimal torsional
element subjected to
distributed torsional
moments

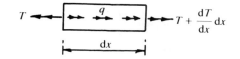

4.16.2 Differential equation for a shaft element – Saint Venant's torsion

The equation relating the 'force' which is the torque T and the 'displacement' which is the twist ψ for a shaft is given by

$$T = GJ \, d\psi/dx$$

This equation is similar to the equation relating axial force F and the axial displacement u:

$$F = AE \, du/dx$$

Similarly, considering the equilibrium of an infinitesimal element of a shaft with distributed torsional moment q per unit length along its length as shown in Fig. 4.52, the equation of equilibrium is given by

$$\{T + (dT/dx)\,dx\} - T + q\,dx = 0$$

Therefore

$$dT/dx + q = 0$$

In this case clearly torsion T is not constant along the length of the shaft and therefore Saint Venant's equation $T = GJ \, d\psi/dx$ is not strictly valid. However, *assuming* that the errors are likely to be small and substituting for $T = GJ \, d\psi/dx$, we have

$$d/dx\{GJ \, d\psi/dx\} + q = 0$$

If GJ is constant, then

$$GJ \, d^2\psi/dx^2 + q = 0$$

This equation is similar to the differential equation for an axially loaded bar:

$$AE \, d^2u/dx^2 + q = 0$$

Because the governing equations for a bar and a shaft are similar, all the results of Section 4.1 for a bar are applicable to a shaft if F, AE and u are replaced by T, GJ and ψ respectively.

The general stiffness relationship for a shaft is given by

$$\begin{bmatrix} T_1 \\ T_2 \end{bmatrix} = GJ/L \begin{bmatrix} 1 & -1 \\ -1 & 1 \end{bmatrix} \begin{bmatrix} \psi_1 \\ \psi_2 \end{bmatrix} + \begin{bmatrix} T_{f1} \\ T_{f2} \end{bmatrix}$$

The fixed end forces T_{f1} and T_{f2} for the two standard cases shown in Fig. 4.53 are given in Table 4.11.

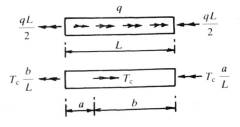

Figure 4.53 ● Fixed end twisting moments due to uniformly distributed torque and off-centre concentrated torque

Table 4.11 ● Fixed end torques

Loading	T_{f1}	T_{f2}
Uniformly distributed torque q	$-0.5qL$	$-0.5qL$
Concentrated torque T_c	$-T_c b/L$	$-T_c a/L$

4.16.3 Effect of restraining warping

In order to illustrate the effect of restraining warping, consider the I section cantilever shown in Fig. 4.54. When a twisting moment is applied to the tip of the cantilever, the section warps. If warping is not restrained, then warping is constant along the axis of the shaft. If warping is prevented, then the flanges behave as cantilever beams bending in their own plane. Thus the flanges can resist shear force

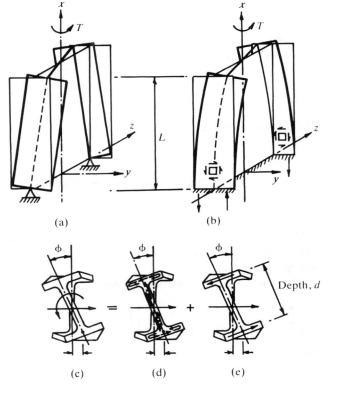

Figure 4.54 ●
Effect of restraining warping displacement due to torsion:
(a) warping displacement unrestrained,
(b) warping displacement restrained,
(c) twisting of the cross-section, (d) shear stresses due to Saint Venant's torsion,
(e) shear forces in the flange due to restraint to warping displacement

parallel to the flange. The shear forces in the flanges form a couple which can resist torsion. Thus when warping is restrained, the following two actions are involved in resisting torsion:

1. Stress distribution as in free warping torsion. This action is known as Saint Venant's torsional stress distribution. See Fig 4.54(d).
2. Bending action of flanges. This is known as torsion bending stress distribution. See Fig 4.54(e).

In general warping is 'small' in closed form sections and Saint Venant's torsion stresses are dominant. On the other hand, in the case of 'open' types of section, torsion bending stress distribution is the dominant stress distribution. Therefore, without restraint to warping, 'open' sections are very flexible indeed. If it is necessary to reduce deformation due to torsion then closed form sections should always be used.

4.17 ● References

Fraser D J 1981 *Conceptual Design and Preliminary Analysis of Structures* Pitman (Contains useful material on equilibrium and compatibility. Chapters 3, 4, 5 and 10 are worth studying.)

Gere J M, Timoshenko S P 1987 *Mechanics of Materials* (2nd edn) Van Nostrand Reinhold (Contains many examples for practice.)

Iwinski T 1958 *Theory of Beams* Pergamon Press (Use of Laplace transforms for beam problems.)

Lin T Y, Stotesbury S D 1981 *Structural Concepts and Systems for Architects and Engineers* Wiley (A 'general' book which can be studied for gaining useful 'feel' for the overall concept, design and performance of structures.)

Pilkey W D, Pilkey P H 1974 *Mechanics of Solids* Quantum (A good book.)

Popov E P 1978 *Mechanics of Materials* (2nd edn) Prentice-Hall

Thomson W T 1950 *Laplace Transformation* Wiley (Contains application to beam problems.)

Volterra E, Gaines J H 1971 *Advanced Strength of Materials* Prentice-Hall (Chapter 6 contains derivation of differential equations for the deflection and rotation of bow girders.)

4.18 ● Problems

1. A uniformly tapering bar of length L and of area A at one end and the area linearly varying to an area $(1 + \alpha)A$ at the other end is subjected to a tensile load of P. Determine the extension of the bar.

Answer: Area A_x at a distance x is given by

$$A_x = A(1 + \alpha x/L), \quad \Delta = \int P\,dx/(A_x E)$$

$$= [PL/(AE)] \int dx/(1 + \alpha x)$$

$$= [PL/(AE\alpha)] \ln(1 + \alpha)$$

(*Note*: Formula is not applicable for the case of a bar of constant cross-section for which $\alpha = 0$ because we get the indeterminate form 0/0. However, $\alpha^{-1} \ln(1 + \alpha) = 1/(1 + \alpha) = 1$ as $\alpha \to 0$.)

2. A bar of stepped cross-section and with stepped loading on the bar is shown in Fig. E4.1. Determine the extension of the bar.

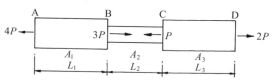

Figure E4.1

Answer: $F = 4P - 3P\langle x - L_1 \rangle^0 + P\langle x - L_1 - L_2 \rangle^0 - 2P\langle x - L_1 - L_2 - L_3 \rangle^0$, $\Delta_B = 4PL_1/(AE)$, $\Delta_C = \Delta_B + PL_2/(A_2 E)$, $\Delta_D = \Delta_C + 2PL_3/(A_3 E)$.

3. If the bar in Fig. E4.1 is fully fixed at both ends and the ambient temperature is increased by $T\,°\text{C}$,

calculate the resultant force P in the bar.
α = coefficient of linear thermal expansion.

Answer: $\alpha T(L_1 + L_2 + L_3) = PL_1/(A_1E) + PL_2/(A_2E) + PL_3/(A_3E)$.

4. A brass sleeve is fitted over the steel bolt as shown in Fig. E4.2 and the nut is tightened until it is snug. Determine the resulting force in the sleeve and the bolt due to an increase of the ambient temperature by $T°$ C. A_s, A_b = area of cross-section, α_s, α_b = coefficient of thermal expansion, E_s, E_b = Young's modulus of sleeve and bolt respectively. Assume $\alpha_b < \alpha_s$.

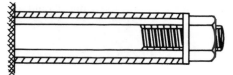

Figure E4.2

Answer: Due to temperature: bolt expands by $\Delta_{bt} = \alpha_b TL$, sleeve expands by $\Delta_{st} = \alpha_s TL$. Due to a *tensile* force P in the bolt it extends by $\Delta_{bp} = PL/(A_bE_b)$ and because, for equilibrium, the force in the sleeve is a compression of P, the sleeve *contracts* by $\Delta_{sp} = PL/(A_sE_s)$. For compatibility $\Delta_{bt} + \Delta_{bp} = \Delta_{st} - \Delta_{sp}$. Therefore

$$T(\alpha_s - \alpha_b) = P[1/(A_sE_s) + 1/(A_bE_b)]$$

(*Note*: If $\alpha_s = \alpha_b$, $P = 0$. Even if $\alpha_s < \alpha_b$, then $P = 0$ because from the equation P is compressive which is not valid for the system under consideration because the sleeve expands *less* than the bolt due to thermal change.)

5. Calculate the deflection profile for the beam with overhang shown in Fig. E4.3. Calculate the deflection at the centre and under the load and check the results using standard cases of beam loading shown in Tables 4.2 and 4.3.

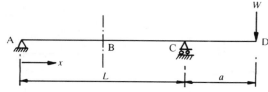

Figure E4.3

Answer: $V_A = -Wa/L$, $V_C = W(1 + a/L)$,
$6EIv = V_A(x^3 - L^2x) + V_C\langle x - L\rangle^3 - W\langle x - L - a\rangle^3$,
$v_{centre} = WaL^2/(16EI)$, $v_{load} = Wa^2(L + a)/(3EI)$.

6. Calculate the deflection profile for the beam shown in Fig. E4.4. What is the deflection under the loads? Check the answer using the results for standard cases of cantilever loading given in Table 4.2.

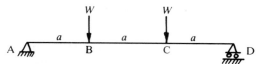

Figure E4.4

Answer: $6EIv = Wx^3 - W\langle x - a\rangle^3 - W\langle x - 2a\rangle^3 - Wa^2x$, $\Delta_W = 5Wa^3/(6EI)$, $\Delta_{centre} = 23Wa^3/(24EI)$.

7. Calculate the deflection profile for the simply supported beam in Fig. E4.5. Also calculate the deflection at the centre. Check the answer using the results for standard cases of cantilever loading in Table 4.2.

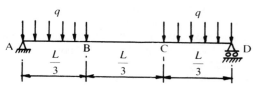

Figure E4.5

Answer: $24EIv = 4qax^3 - x^4 + \langle x - a\rangle^4 - \langle x - 2a\rangle^4 - 14qa^3x$, $v_{centre} = 25qa^4/(48EI)$, $a = L/3$.

8. Calculate the deflection profile for the simply supported beam shown in Fig. E4.6.

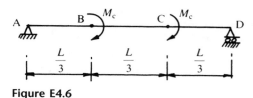

Figure E4.6

Answer: $18EIv = -2M_cx^3/a + 9M_c\langle x - a\rangle^2 + 9M_c\langle x - 2a\rangle^2 + 3M_cax$, $v_B = M_ca^2/(18EI)$, $v_{centre} = 0$, $a = L/3$.

9. Calculate the deflection under the loads in the propped cantilevers with an internal hinge shown in Fig. E4.7. Note that the flexural rigidity is *not* constant.

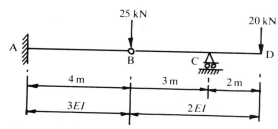

Figure E4.7

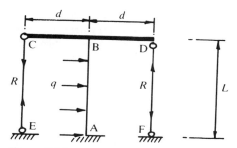

Figure E4.9

Answer: Take moments about B of the free body BCD, $V_c = 20 \times 5/3 = 33.33$. For equilibrium, V_B on BCD = 13.33 (down). Load at B on cantilever AB = $(25 - 13.33) = 11.67$ down. From cantilever AB with a load 11.33 at the tip: $v_B = (11.67)4^3/ \{3(3EI)\} = 83.0/(EI)$ (down). For BCD starting from B: $6(2EI)v = -13.33x^3 + 33.33\langle x - 3\rangle^3 - 20\langle x - 5\rangle^3 + C_1 x + C_2$, $v = 0$ at $x = 3$ and $v_B = -83.0/(EI)$ at $x = 0$, determine C_1 and C_2. $v_D = 11.4/(EI)$ down.

10. Calculate the deflection at B and under the loads of the beam on three supports with an internal hinge shown in Fig. E4.8. Use standard cases of simply supported beam and cantilever loadings to check the answer.

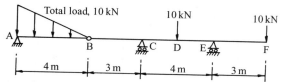

Figure E4.8

Answer: Take moment about B of the free body AB: $V_A = 6.67$, $V_B = 3.33$. Beam BCDEF loaded by $-V_B$ at B, 10 each at D and F. $V_E = 20.0$, $V_C = 3.33$, $6EIv = -V_B x^3 + V_C\langle x - 3\rangle^3 - 10\langle x - 5\rangle^3 + V_E\langle x - 7\rangle^3 - 10\langle x - 10\rangle^3 + C_1 x + C_2$. $v = 0$ at $x = 3$ at C and $x = 7$ at E. $C_1 = 230$, $C_2 = -600$, $v_C = -100/(EI)$, $v_D = 26.67/(EI)$, $v_F = -200/(EI)$.

11. Figure E4.9 shows a cantilever AB connected to rigid outriggers BC and BD. The outriggers are connected to the foundation through pin connected members CE and DF. Calculate the deflection at the tip of the cantilever under a uniformly distributed lateral load of q. Assume flexural rigidity of the cantilever AB is EI and the extensional rigidities of the pin-ended members to be AE.

Answer: Let the axial force in the members CE and DF be $\pm R$. The cantilever is subjected to a u.d.l. of q and an external anticlockwise moment at the top of $2Rd$. Using the standard results for cantilever loading, the deflection $\Delta_B = qL^4/(8EI) - 2RdL^2/(2EI)$. The slope at the top is $\theta_B = qL^3/(6EI) - 2RdL/(EI)$. Deflection at C and D is equal to $\Delta_C = \theta_B d$. Axial extension of the member CE is Δ_C. Therefore the axial force R in CE is given by $\Delta_C = RL/(AE)$; $RL/(AE) = [qL^3/(6EI) - 2RdL/(EI)]d$; $R = qL/C$, $C = 6(d/L)\{2 + EI/(AEd^2)\}$, $\Delta_B = qL^4 C_1/(8EI)$, $C_1 = 1 - 16d/(LC)$. As AE increases, C decreases. Therefore R increases and C_1 also decreases. Therefore the 'outriggers' stiffen the cantilever. (*Note*: This 'outrigger' concept of stiffening is used in many tall buildings. By having more outriggers along the height, considerable stiffening can be achieved.)

12. Figure E4.10 shows a 'stabilized' shelter for a grandstand. Calculate the axial force in AE and the deflection at C.

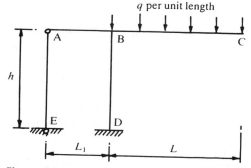

Figure E4.10

Answer: Let R be the force in the tie AE. Due to q, moment at B on cantilever BD is $0.5qL^2$. Rotation of B = $\theta_1 = 0.5qL^2h/(EI)$. Moment at B on BD due to R is RL_1 anticlockwise. Rotation at B of cantilever BD due to this moment is $\theta_2 = RL_1h/(EI)$ in the anticlockwise direction. Upward deflection of A is $\Delta_A = (\theta_1 - \theta_2)L_1 - R(L_1)^3/(3EI)$ where the second term is due to the bending of the cantilever BA loaded by R at A. Extension of AE is $\Delta_A = Rh/(AE)$. Equating the two expressions for Δ_A, $R = qL/C$, where

$$C = 2\{EI/(LL_1AE) + (L_1/L) + L_1^2/(3Lh)\}$$

$\Delta_C = qL^4/(8EI) + (\theta_1 - \theta_2)L$. It is worth noting that increasing AE decreases C and therefore increases R. Increasing R increases θ_2 and therefore decreases Δ_C.

13. To span an opening $2L \times L$, a grillage of four beams is used as shown in Fig. E4.11. The I for the longer beam is eight times the I for the shorter beam. Loads W are applied at P and Q. Calculate the deflection at the four points and construct the BMD and SFD. Assume that the beams are simply supported at the ends and the connection between the beams consists of a single vertical force only as in Example 7, Fig. 4.24 in Section 4.5.

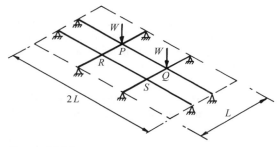

Figure E4.11

Answer: Let the load be applied to the long beams. Let the 'interactive' forces on the long beam at P and Q be R_1 acting upward and R_2 acting downwards at R and S. For compatibility, at the intersection of the beams, deflection of two sets of the beams must be same. It can be shown that for a simply supported beam of span L loaded by a single load at the left third point only, deflection under the load is $8Wa^3/(18EI)$ and at the unloaded third point the deflection is $7Wa^3(18EI)$ where $a = L/3$.

(a) Considering the 'loaded' long beam:

$$\Delta_p = \Delta_q = 8(W - R_1)a^3/\{18E(8I)\} + 7(W - R_1)a^3/\{18E(8I)\}, \ a = (2L)/3$$

(b) Considering the 'unloaded' long beam:

$$\Delta_r = \Delta_s = 8R_2a^3/\{18E(8I)\} + 7R_2a^3/\{18(8I)\}, \ a = (2L)/3.$$

(c) Short beam: It is loaded by an *upward* load R_2 at R and S and by *downward* load R_1 at P and Q. Therefore

$$\Delta_r = \Delta_s = 8(-R_2)b^3/\{18EI\} + 7(+R_1)b^3/\{18EI\}$$
$$\Delta_p = \Delta_q = 8(+R_1)b_3/\{18EI\} + 7(-R_2)b^3/\{18EI\}$$

where $b = L/3$. Equating the deflection at common intersection points, $15(W - R_1) = 8R_1 - 7R_2$, $15R_2 = 7R_1 - 8R_2$. Solving $R_1 = 0.719W$, $R_2 = 0.219W$.

14. For a uniform square box loaded as shown in Fig. E4.12(a), show that the bending moment at the corners is given by $M_c = WL/16$ and the vertical deflection at the load point $\Delta_v = 5WL^3/(384EI)$ and horizontal deflection at the centre of the vertical members $\Delta_h = 3WL/(384EI)$. Using the above result, determine the axial force in the rigid link connecting the two boxes as shown in Fig. E4.12(b).

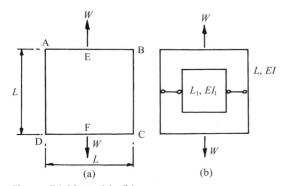

Figure E4.12 ● (a), (b)

Answer: Rotation at the ends of the beam is $= WL^2/(16EI) - M_cL^2/(2EI) =$ rotation at the ends of the columns $= M_cL^2/(2EI)$. Therefore $M_c = WL/16$.

Let the interactive force in the link be R, then for compatibility of horizontal displacement in the two boxes,

$$3WL^3/(384EI) - 5RL^3/(384EI) = 5RL_1^3/(384EI_1)$$

Therefore $R = 0.6W/(1 + k)$, $k = L_1^3I/(L^3I_1)$.

15. Determine the fixed end moments for the partially loaded beam shown in Fig. E4.13.

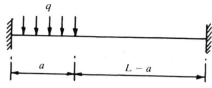

Figure E4.13

Answer: If a concentrated load W is applied at x, then $M_A = W(L - x)^2 x/L^2$, $M_B = Wx^2(L - x)/L^2$. Considering an element of load $q\,dx$ at x,

$$M_A = \int q\,dx(L - x)^2 x/L^2$$

$$M_B = \int q\,dx\, x^2(L - x)/L^2$$

$$a \geqslant x \geqslant 0$$

Therefore

$$M_A = qa^2\{6 - 8a/L + 3(a/L)^2\}/12$$

$$M_B = qa^2\{4a/L - 3(a/L)^2\}/12$$

Alternatively,

$$EI\,d^4v/dx^4 = -q + q\langle x - a\rangle^0$$

$$EI\,d^3v/dx^3 = -qx + q\langle x - a\rangle^1 + C_1$$

$$EI\,d^2v/dx^2 = -qx^2/2 + q\langle x - a\rangle^2/2 + C_1x + C_2$$

$$EI\,dv/dx = -qx^3/6 + q\langle x - a\rangle^3/6 + C_1x^2/2 + C_2x + C_3$$

$$EIv = -qx^4/24 + q\langle x - a\rangle^4/24 + C_1x^3/6 + C_2x^2/2 + C_3x + C_4$$

$v = 0$, $x = 0$, therefore $C_4 = 0$. $dv/dx = 0$ at $x = 0$, therefore $C_3 = 0$. $v = 0$ and $dv/dx = 0$ at $x = L$. Thus

$$-qL^4/24 + q\langle L - a\rangle^4/24 + C_1L^3/6 + C_2L^2/2 = 0$$

and

$$-qL^3/6 + q\langle L - a\rangle^3/6 + C_1L^2/2 + C_2L = 0$$

Solving

$$C_1 = 0.5qL\{1 - (L - a)^3(L + a)/L^4\}$$

$$C_2 = (-qL^2/12)\{1 - (L - a)^3(L + 3a)/L^4\}$$

Simplifying, $\quad C_2 = -\dfrac{qa^2}{12}\left\{6 - \dfrac{8a}{L} + 3\left(\dfrac{a}{L}\right)^2\right\}$

Fixed end moments from $EI\,d^2v/dx^2$ at $x = 0$ and $x = L$. The problem can also be approached by considering a cantilever as a basic structure. The first approach is the best.

16. Figure E4.14(a) shows a three-beam grillage. Beam BE is simply supported on the other two simply supported beams AC and DF. Assuming EI is the same for all the beams, determine the centre line deflections of all three beams.

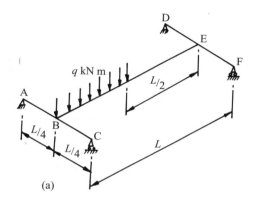

(a)

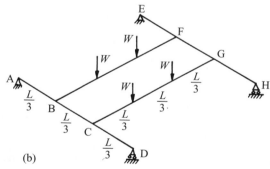

(b)

Figure E4.14 ● (a), (b)

Answer: AC = $qL^4/(1024\,EI)$, DF = $qL^4/(3072\,EI)$, BE = $11qL^4/(1536\,EI)$

17. Figure E14.4(b) shows a four-beam grillage. Beams BF and CG are simply supported on the other two simply supported beams AD and EH. Assuming EI is the same for all the beams, determine the centre line deflections of all four beams.

Answer: BF and CG = $43WL^3/(648\,EI)$, AD and EH = $23WL^3/(648\,EI)$

Chapter five

Deformation of structures

In Chapter 4, the deformation of bar, beam and shaft elements was considered and the equations for their calculation were derived. In this chapter the results of Chapter 4 will be applied to the calculation of deformation of simple structures. The object here is to develop manual calculation procedures with the *main aim of emphasizing the concepts of equilibrium and compatibility in the analysis of structures*. General procedures for the practical analysis of complex structures suitable for computers will be developed in Chapter 6.

5.1 ● Deflection of pin-jointed trusses

It was shown in Chapter 4, that for the bar shown in Fig. 5.1(a), the axial tensile force F in a bar is given by

$$F = (AE/L)\Delta$$

where Δ = extension of the bar, L = length of the bar, AE = axial rigidity. Therefore

$$\Delta = FL/(AE)$$

If the axial displacements are Δ_{a1} and Δ_{a2} at the ends 1 and 2 respectively of the bar, then

$$\Delta = \text{extension of the bar} = (\Delta_{a2} - \Delta_{a1})$$

If the components of Δ_{a1} and Δ_{a2} in the x- and y-directions at ends 1 and 2 of a bar are (u_1, v_1) and (u_2, v_2) respectively as shown in Fig. 5.1(b), then from geometry

$$\Delta_{a1} = u_1 l + v_1 m, \ \Delta_{a2} = u_2 l + v_2 m$$

Therefore

$$\Delta = \Delta_{a2} - \Delta_{a1} = (u_2 - u_1)l + (v_2 - v_1)m$$

where $l = \cos \alpha = (x_2 - x_1)/L$, $m = \sin \alpha = (y_2 - y_1)/L$, and α = inclination of the member to the x-axis. α is measured from the x-axis in the anticlockwise direction. l and m are called direction cosines of the vector 1–2. In order to avoid confusion in using the correct value of angle α, the direction cosines are best calculated from the co-ordinates of the joints.

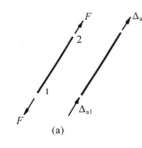

(a)

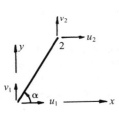

(b)

Figure 5.1 ● (a), (b) Forces and deformations of a bar element

Substituting for Δ in terms of the displacements at the ends of the member,

$$FL/(AE) = (u_2 - u_1)l + (v_2 - v_1)m$$

The above equation will be used for the computation of the deflections of pin-jointed trusses.

5.1.1 Examples

Example 1

Figure 5.2(a) shows a simple two bar pin-jointed structure. Determine the displacements of the joint 2. It is given that for bar 2–1, $A = 400$ mm², for bar 2–3, $A = 260$ mm² and $E = 210$ kN mm⁻² for both members.

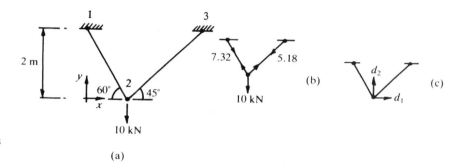

Figure 5.2 ● (a)–(c)
Forces and deformations
of a pin-jointed truss

Solution   The forces in the two bars are determined from statics and are shown in Fig. 5.2(b). The 'start' (end 1) and 'finish' (end 2) joints for defining the direction of the bar are chosen arbitrarily. Once the choice is made, from the dimensions given, the direction cosines etc. can be calculated. The notation used for bar designation is joint number at end 1 followed by joint number at end 2. Therefore Bar 2–1 indicates that end 1 is joint 2 and end 2 is joint 1. Since only joint 2 can displace, its displacements in the x- and y-directions are assumed to be d_1 and d_2 respectively as shown in Fig. 5.2(c).

In Table 5.1 all linear dimensions are in meters and force F and axial rigidity AE are in kN. However, $FL/(AE)$ is given in mm. Substituting in the equation

$$FL/(AE) = (u_2 - u_1)l + (v_2 - v_1)m$$

we have:

(a) Bar 2–1.

At end 1 which corresponds to joint 2, $u_1 = d_1$, $v_1 = d_2$. At end 2 which corresponds to joint 1, $u_2 = v_2 = 0$ because the joint is restrained. Therefore

$$0.201 = \{(0 - d_1)(-0.5) + (0 - d_2)0.866\}$$

Simplifying

$$0.5d_1 - 0.866d_2 = 0.201$$

Table 5.1

Member	x_1	x_2	y_1	y_2	L	l	m	$FL/(AE)$
2–1	1.155	0	0	2.0	2.31	–0.5	0.866	0.201
2–3	1.155	3.155	0	2.0	2.83	0.707	0.707	0.269

(b) Bar 2–3.

At end 1 which corresponds to joint 2, $u_1 = d_1$, $v_1 = d_2$. At end 2 which corresponds to joint 3, $u_2 = v_2 = 0$ because the joint is restrained. Therefore

$$0.269 = \{(0 - d_1)0.707 + (0 - d_2)0.707\}$$

Simplifying

$$0.707d_1 + 0.707d_2 = -0.269$$

Solving the two equations gives $d_1 = -0.094$ mm and $d_2 = -0.286$ mm. The negative sign indicates that the displacements are in the negative direction of the co-ordinate axes. The joint 2 moves horizontally to the left and vertically down.

Example 2

Determine the displacements of the pin-jointed truss shown in Fig. 5.3(a). Take $E = 210$ kN mm^{-2}. The areas of cross-section in square millimetres are shown against the members.

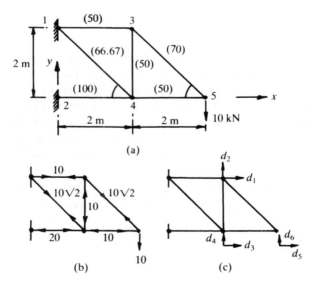

Figure 5.3 ● (a)–(c) Forces and deformations of a cantilever truss

Solution The forces in the members are calculated and are shown in Fig. 5.3(b). Figure 5.3(c) shows the numbering of unknown joint displacements. The data required for the calculations are set out in Table 5.2.

It is convenient to start with two members having known displacements at one end and intersecting at the same joint and solve for the displacements at the common joint.

Table 5.2

Member	x_1	x_2	y_1	y_2	L	l	m	$FL/(AE)$
2–4	0	2	0	0	2	1	0	−1.905
4–5	2	4	0	0	2	1	0	−1.905
1–3	0	2	2	2	2	1	0	1.905
4–3	2	2	0	2	2	0	1	−1.905
4–1	2	0	0	2	$2\sqrt{2}$	$-1/\sqrt{2}$	$1/\sqrt{2}$	2.857
5–3	4	2	0	2	$2\sqrt{2}$	$-1/\sqrt{2}$	$1/\sqrt{2}$	2.720

(a) Consider members 2–4 and 4–1 which have zero displacements at joints 1 and 2 and meet at joint 4.

(i) Member 2–4:

End 1 = Joint 2: $u_1 = v_1 = 0$

End 2 = Joint 4: $u_2 = d_3$, $v_2 = d_4$

$-1.905 = (d_3 - 0)1 + (d_4 - 0)0$,

therefore $d_3 = -1.905$ mm.

(ii) Member 4–1:

End 1 = Joint 4: $u_1 = d_3$, $v_1 = d_4$

End 2 = Joint 1: $u_2 = v_2 = 0$

$2.857 = (0 - d_3)(-1/\sqrt{2}) + (0 - d_4)(1/\sqrt{2})$

$d_3 = -1.905$, therefore $d_4 = -5.945$ mm.

(b) Now consider members 1–3 and 4–3 whose displacements at joints 1 and 4 are known and which meet at joint 3.

(i) Member 1–3:

End 1 = Joint 1: $u_1 = v_1 = 0$

End 2 = Joint 3: $u_2 = d_1$, $v_2 = d_2$

-1.905 mm $= (d_1 - 0)1 + (d_2 - 0)0$,

therefore $d_1 = 1.905$ mm.

(ii) Member 4–3:

End 1 = Joint 4: $u_1 = d_3$, $v_1 = d_4$

End 2 = Joint 3: $u_2 = d_1$, $v_2 = d_2$

$1.905 = (d_1 - d_3)0 + (d_2 - d_4)1$

$d_4 = -5.945$, therefore $d_2 = -7.850$ mm.

(c) Consider members 4–5 and 5–3 whose displacements at joints 3 and 4 are known and which meet at joint 5.

Table 5.3

Joint	x-displacement (mm)	y-displacement (mm)
1	0	0
2	0	0
3	1.905	−7.850
4	−1.905	−5.945
5	−3.810	−17.412

(i) Member 4–5:

End 1 = Joint 4: $u_1 = d_3$, $v_1 = d_4$

End 2 = Joint 5: $u_2 = d_5$, $v_2 = d_6$

$-1.905 = (d_5 - d_3)1 + (d_6 - d_4)0$

$d_3 = -1.905$, therefore $d_5 = -3.810$ mm.

(ii) Member 5–3:

End 1 = Joint 5: $u_1 = d_5$, $v_1 = d_6$

End 2 = Joint 3: $u_2 = d_1$, $v_2 = d_2$

$-2.720 = (d_1 - d_5)(-1/\sqrt{2}) + (d_6 - d_2)(1/\sqrt{2})$.

Solving for the unknown displacement, $d_6 = -17.412$ mm.

The results are summarized in Table 5.3.

Example 3

Calculate the displacements of all the joints of the Warren truss shown in Fig. 5.4(a). Take $E = 210$ kN mm^{-2}. The areas of cross-section in mm^2 are shown against the members in Fig. 5.4(a).

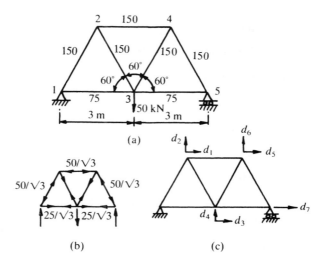

Figure 5.4 ● (a)–(c) Forces and deformations of pin-jointed Warren truss

Table 5.4

Member	x_1	x_2	y_1	y_2	L	l	m	$FL/(AE)$
1–3	0	3	0	0	3	1	0	2.749
3–5	3	6	0	0	3	1	0	2.749
2–4	1.5	4.5	2.598	2.598	3	1	0	−2.749
1–2	0	1.5	0	2.598	3	0.5	0.866	−2.749
3–4	3	4.5	0	2.598	3	0.5	0.866	2.749
3–2	3	1.5	0	2.598	3	−0.5	0.866	2.749
5–4	6	4.5	0	2.598	3	−0.5	0.866	−2.749

Solution   Figure 5.4(b) shows the forces in the members obtained from statics. Figure 5.4(c) shows the numbering of unknown joint displacements. Table 5.4 shows the data required for the calculation of displacements.

In this example, it is not possible to start from two members with one end fully restrained and meeting at a common joint. Therefore some thought is necessary to reduce numerical complexity.

Extension of member 1–3 = 2.749 = d_3

Extension of member 3–5 = $d_7 - d_3$ = 2.749. ∴ $d_7 = 5.498$

Extension of member 2–4 = $d_5 - d_1 = -2.749$ (i)

Extension of member 1–2 = $(0 + d_1)0.5 + (0 + d_2)0.866 = -2.749$

$$\therefore d_1 0.5 + d_2 0.866 = -2.749 \qquad (ii)$$

Extension of member 3–4 = $(d_5 - d_3)0.5 + (d_6 - d_4)0.866 = 2.749$. Because $d_3 = 2.749$ and $d_2 = d_6$ by symmetry, the above equation becomes

$$d_5 0.5 + (d_2 - d_4)0.866 = 4.1240 \qquad (iii)$$

Extension of member 3–2 = $(d_1 - d_3)(-0.5) + (d_2 - d_4)0.866 = 2.749$. Because $d_3 = 2.749$, the above equation becomes

$$-d_1 0.5 + (d_2 - d_4)0.866 = 1.375 \qquad (iv)$$

Subtracting (iv) from (iii)

$$0.5(d_5 + d_1) = 2.7493 \qquad (v)$$

Solve for d_1 and d_5 using (i) and (v). d_5 = 1.375 mm, d_1 = 4.124 mm. Using (ii) $d_2 = -5.555$ and using (iii), $d_4 = -9.523$. This is summarized in Table 5.5.

Table 5.5

Joint	x-displacement (mm)	y-displacement (mm)
1	0	0
2	4.124	−5.555
3	2.749	−9.523
4	1.375	−5.555
5	5.498	0

5.2 ● Effect of temperature on deflection of pin-jointed trusses

When there is a change in the temperature of a member, then the corresponding extension Δ is given by

$$\Delta = \alpha_t TL$$

where T = change in temperature and α_t = coefficient of linear thermal expansion. In the case of statically determinate structures, changes in the temperature of the members simply cause the deflection of the joints of the structures but do not induce any forces in the members. The resulting joint displacements can be calculated following the previous procedure except that the extension of members, instead of being caused by the loads, is caused by temperature changes.

Example

Calculate the deflection of the joints of the truss shown in Fig. 5.4, caused by a temperature change of 30 °C in member 2–4 only. Assume $\alpha_t = 10 \times 10^{-6}$ °C^{-1}.

Because there is no change in the temperature of members 1–3 and 3–5,

Extension of member 1–3 = 0 = d_3. Therefore $d_3 = 0$

Extension of member 3–5 = 0 = d_7. Therefore $d_7 = 0$

The equations (i) to (iv) in Example 3 of Section 5.1.1 become

Extension of member 2–4:

$$d_5 - d_1 = 10 \times 10^{-6} \times 30 \times (3000) = 0.9 \text{ mm} \tag{i}$$

Extension of member 1–2: $d_1 0.5 + d_2 0.866 = 0$ (ii)

Extension of member 3–4: $d_5 0.5 + (d_2 - d_4)0.866 = 0$ (iii)

Extension of member 3–2: $-d_1 0.5 + (d_2 - d_4)0.866 = 0$ (iv)

Subtracting (iv) from (iii), $0.5(d_5 + d_1) = 0$ (v)

Solve for d_1 and d_5 using (i) and (v). $d_1 = -0.45$ mm, $d_5 = 0.45$ mm. Using (ii), $d_2 = 0.260$ mm and using (iii), $d_4 = 0.520$ mm. This is summarized in Table 5.6.

Table 5.6

Joint	x-displacement (mm)	y-displacement (mm)
1	0	0
2	−0.45	0.260
3	0	0.520
4	0.45	0.260
5	0	0

5.3 ● Calculation of deformations of 2-D rigid-jointed structures

Many examples were given of the calculation of the displacements in rigid-jointed structures in Chapter 4. Some additional examples are given below.

Example 1

Calculate the horizontal displacement and rotation at the roller support of the rigid-jointed structure shown in Fig. 5.5(a). Assume that the members are axially rigid.

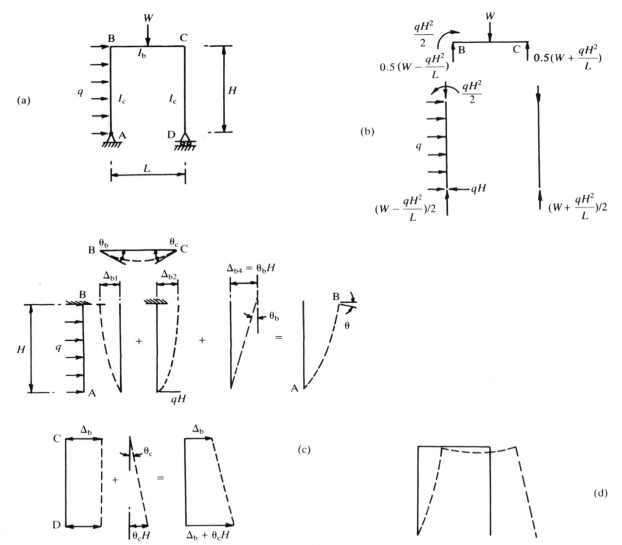

Figure 5.5 ● (a)–(d) Forces and deformations of a portal frame

Solution   Reactions: Because the support A is pinned, the reactions at A are a horizontal force H_A and a vertical reaction V_A. At D, because it is a roller support, only a vertical reaction V_D exists. From the equations of statics, $H_A = qH$, acting from right to left and $V_A + V_D = W$.

Taking moments about A,

$$qH(H/2) + W(L/2) = V_D L$$

$$V_D = W/2 + qH^2/(2L)$$

$$V_A = W - V_D = W/2 - qH^2/(2L)$$

Member CD is subjected to an axial force of V_D and zero bending moment. The bending moment in the beam BC at B is given, by taking moments about B, by

$$M_B = V_D L - W(L/2) = qH^2/2$$

The forces at the ends of the members obtained from statics are shown in Fig. 5.5(b). Using the data in Tables 4.2 and 4.3, the displacements at the ends of the various elements are calculated as follows.

(a) Beam BC. Considering it as a simply supported segment, the rotation at the ends of the beam BC are given by

   (i) Due to W (see Table 4.3):

$$\text{At joint } B, \ \theta_{b1} = WL^2/(16EI_b) \text{ clockwise}$$

$$\text{At joint } C, \ \theta_{c1} = WL^2/(16EI_b) \text{ anticlockwise}$$

   (ii) Due to moment $0.5qH^2$ at joint B (see Table 4.3):

$$\text{At joint } B, \ \theta_{b2} = 0.5qLH^2/(3EI_b) \text{ clockwise}$$

$$\text{At joint } C, \ \theta_{c2} = 0.5qLH^2/(6EI_b) \text{ anticlockwise}$$

   Therefore net rotations are

$$\theta_b = \{WL^2/(16EI_b) + qLH^2/(6EI_b)\} \text{ clockwise}$$

$$\theta_c = \{WL^2/(16EI_b) + qLH^2/(12EI_b)\} \text{ anticlockwise}$$

(b) Left-hand column AB. Consider it as a cantilever segment fixed at B and subjected to a u.d.l. q and a concentrated load at A due to the horizontal reaction of qH as shown in Fig. 5.5(c).

   With respect to the support A, the horizontal deflection at B is given by (see Table 4.2):

   (i) Due to q: $\Delta_{b1} = qH^4/(8EI_c)$ to the left

$$= -qH^4/(8EI) \text{ to the right}$$

   (ii) Due to concentrated load qH at A: $\Delta_{b2} = qH \cdot H^3/(3EI_c)$ to the right (see Table 4.2):

$$\text{Net deflection at B is } \Delta_{b3} = -qH^4/(8EI_c) + qH^4/(3EI_c)$$

$$= 5qH^4/(24EI_c) \text{ to the right}$$

However, for this deflection pattern the slope at B of column AB is zero while that of beam BC at B is $\theta_b = \{WL^2/(16EI_b) + qLH^2/(6EI_b)\}$ clockwise. Therefore to establish compatibility of rotations at B, the column AB must rotate clockwise as a rigid body by Δ_{b4} such that $\theta_b = \Delta_{b4}/H$.

Therefore the net horizontal translation to the right at B is given by

$\Delta_b =$ Translation Δ_{b3} due to loads acting on the cantilever AB fixed at B
    $+$ Translation Δ_{b4} to establish compatibility of rotations at B

$$= 5qH^4/(24EI_c) + \theta_b H$$

Therefore

$$\Delta_b = 5qH^4/(24EI_c) + \{WL^2/(16EI_b) + qLH^2/(6EI_b)\}H$$

(c) Right-hand column DC. The column CD is free from bending loads. Therefore it does not deform but translates and rotates as a rigid body to maintain compatibility of deformations at the joint C. The horizontal deflection Δ_C at C is the same as the horizontal deflection Δ_b at B. Therefore $\Delta_c = \Delta_b$. The horizontal translation of D is $\Delta_{d1} = \Delta_c$.

However, the slope of beam BC at C is θ_c in the anticlockwise direction. Therefore for compatibility of rotations at C, the column CD has to rotate in an anticlockwise direction such that the horizontal translation Δ_{d2} at D with respect to C is $\Delta_{d2} = \theta_c H$. The net horizontal translation of D is

$$\Delta_d = \Delta_{d1} + \Delta_{d2}$$

$$= [5qH^4/(24EI_c) + \{WL^2/(16EI_b) + qLH^2/(6EI_b)\}H] + \{WL^2/(16EI_b) + qLH^2/(12EI_b)\}H$$

Therefore

$$\Delta_d = 5qH^4/(24EI_c) + \{WL^2/(8EI_b) + qLH^2/(4EI_b)\}H$$

This example clearly shows how the deformations of individual elements are made up of deformations due to loads and rigid body displacements to maintain compatibility of displacements (i.e. translations and rotations) where the members intersect. Figure 5.5(d) shows the deformed shape of the structure.

Example 2

Calculate the deformations of the rigid-jointed structure shown in Fig. 5.6(a).

Solution   Reactions: Because the support A is pinned, the reactions at A are a horizontal force H_A and a vertical reaction V_A. At D, because it is a roller support, only a vertical reaction V_D exists. From the equations of statics, $H_A = W$, acting from right to left and $V_A + V_D = 0$.

Taking moments about A

$$W(2a) = V_D L$$

$$V_D = 2Wa/L$$

$$V_A = -V_D = -2Wa/L$$

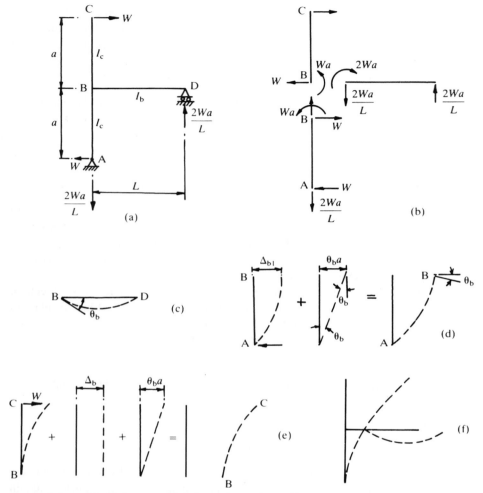

Figure 5.6 ● (a)–(f) Forces and deformations of a rigid-jointed frame

The bending moment in column BC at B is given, by taking moments about B, by

$$M_{BC} = Wa$$

The bending moment in the beam BD at B is given, by taking moments about B, by

$$M_{BD} = V_D L = 2Wa$$

The bending moment in column BA at B is given, by taking moments about B, by

$$M_{BA} = H_A a = Wa$$

The free-body diagram in Fig. 5.6(b) shows the forces acting on each element. The deformations of the individual elements are calculated as follows.

(a) Beam BD: Consider it as a simply supported segment. The vertical deflection is zero at B (because the column AB is axially rigid) and D. The deformations of interest are

(i) Rotation θ_b at B due to couple $2Wa$:

$$\theta_b = (2Wa)L/(3EI_b) = 2WaL/(3EI_b) \text{ clockwise}$$

(ii) The vertical deflection Δ_e at the mid-point of BD:

$$\Delta_e = (2Wa)L^2/(16EI_b) = WaL^2/(8EI_b)$$

(b) Column AB: Consider it as a cantilever segment fixed at B, Fig. 5.6(d).

(i) The horizontal deflection of B with respect to A due to the concentrated load W at A is $\Delta_{b1} = Wa^3/(3EI_c)$ to the right.

(ii) To maintain compatibility of rotations at B, the column has to rotate clockwise as a rigid body by $\Delta_{b2} = \theta_b a$. Therefore $\Delta_{b2} = \{2WaL/(3EI_b)\}a = 2Wa^2L/(3EI_b)$. The net horizontal translation at B is given by

$$\Delta_b = \Delta_{b1} + \Delta_{b2} = Wa^3/(3EI_c) + 2Wa^2L/(3EI_b)$$

$$= Wa^3/(3EI_c)[1 + 2(I_c/I_b)(L/a)]$$

(c) Column BC: Consider it as a cantilever segment fixed at B, Fig. 5.6(e). The horizontal deflection at C is given by

(i) Due to cantilever action with support B fixed: $\Delta_{c1} = Wa^3/(3EI_c)$

(ii) Due to translation of support B:

$$\Delta_{c2} = \Delta_b = Wa^3/(3EI_c)[1 + 2(I_c/I_b)(L/a)]$$

(iii) Due to rotation of support B:

$$\Delta_{c3} = a\theta_b = a[2WaL/(3EI_b)] = 2Wa^2L/(3EI_b)$$

Net deflection of C is

$$\Delta_c = \Delta_{c1} + \Delta_{c2} + \Delta_{c3}$$

$$= Wa^3/(3EI_c) + Wa^3/(3EI_c)[1 + 2(I_c/I_b)(L/a)] + 2Wa^2L/(3EI_b)$$

$$\Delta_c = \{2Wa^3/(3EI_c)\}[1 + 2(I_c/I_b)(L/a)]$$

The final deflected shape is shown in Fig. 5.6(f).

5.3.1 Deformations due to thermal changes

Consider the beam segment shown in Fig. 5.7. Let it be subjected to a uniform temperature change of T_t and T_b on the top and bottom faces respectively. It is further assumed that the temperature variation is linear with depth. This is equivalent to a constant temperature through the depth of $0.5(T_t + T_b)$ and a linear variation given by a change on the top face of $0.5(T_t - T_b)$ and on the bottom face of $-0.5(T_t - T_b)$ respectively. The constant temperature through the depth causes an axial extension of the member. On the other hand, the linear variation causes the extension of the fibres resulting in bending deformations.

Considering the bending deformations, the strain ε_x at any level is defined by

$$\varepsilon_x = \alpha_t T$$

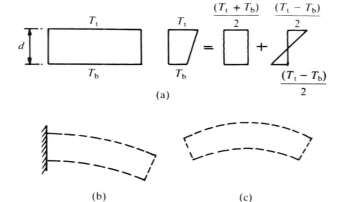

Figure 5.7 ● (a)–(c)
Deformations of a beam
element due to
temperature change

where α_t = coefficient of linear thermal expansion and T = change in temperature at the level where ε_x is being calculated. Assuming that the neutral axis is at mid-depth and because the temperature variation through the depth d is linear, then

$$T = [0.5(T_t - T_b)]\{y/(0.5d)\} = (T_t - T_b)y/d$$

Therefore

$$\varepsilon_x = \alpha_t T = \alpha_t(T_t - T_b)(y/d)$$

Assuming simple theory of bending, from Section 4.2.1, $\varepsilon_x = -y/R$, where $1/R$ is the curvature. Because $1/R \approx d^2v/dx^2$

$$\varepsilon_x = -(d^2v/dx^2)y$$

Equating the two expressions for ε_x we have

$$d^2v/dx^2 = -\alpha_t(T_t - T_b)/d$$

Integrating with respect to x

$$dv/dx = -\{\alpha_t(T_t - T_b)/d\}x + C_1$$

$$v = -\{\alpha_t(T_t - T_b)/d\}x^2/2 + C_1x + C_2$$

As shown in Fig. 5.7(b) and (c), two possible boundary conditions can be used as follows:

(a) Cantilever fixed at $x = 0$: In this case $v = dv/dx = 0$ at $x = 0$. Therefore $C_1 = C_2 = 0$. Thus

$$v = -\{\alpha_t(T_t - T_b)/d\}x^2/2$$

The deflection and slope at the tip are

$$dv/dx_{(x=L)} = \alpha_t(T_t - T_b)L/d \text{ in a clockwise direction}$$

$$v_{(x=L)} = \alpha_t(T_t - T_b)L^2/(2d) \text{ in the downward direction}$$

(b) Simply supported beam. The element is simply supported at $x = 0$ and $x = L$. Therefore $v = 0$ at $x = 0$ and $x = L$. This requires that $C_2 = 0$ and $C_1 = \alpha_tL(T_t - T_b)/(2d)$. The final expression for v is given by

$$v = \alpha_t(T_t - T_b)(Lx - x^2)/(2d)$$

The rotation at the supports and deflection at the midspan are given by

$$dv/dx_{\text{(supports)}} = \alpha_t(T_t - T_b)L/(2d)$$

which is anticlockwise at $x = 0$ and clockwise at $x = L$

$$v_{(x=0.5L)} = \alpha_t(T_t - T_b)L^2/(8d) \text{ upwards}$$

5.3.2   Example

Example

Calculate the horizontal displacement at the roller support and the vertical deflection at the centre of the beam shown in Fig. 5.8(a), when the temperature increases on the outsides of column AB and beam BC only and is equal to $T\,°\text{C}$. Ignore axial deformation of members.

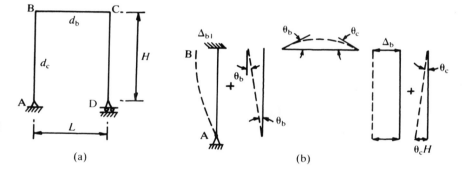

Figure 5.8 ● (a), (b)
Deformations of a portal
frame due to
temperature changes

Solution   Using the results from Section 5.3.1, the deformations of the column AB and the beam BC are calculated first.

(a) Beam BC. Consider it as a simply supported segment. The rotation at the ends B and C are

$$\theta_b = 0.5C_bL \text{ anticlockwise}$$

$$\theta_c = 0.5C_bL \text{ clockwise}$$

where $C_b = \alpha_t T/d_b$, and d_b = depth of the beam. The deflection at midspan is $v = C_bL^2/8$ upward.

(b) Column AB. Consider it as a cantilever segment fixed at B.

(i) Due to a change of temperature in the column, the horizontal deflection to the left at B is

$$\Delta_{b1} = 0.5C_cH^2$$

where $C_c = \alpha_t T/d_c$, and d_c = depth of column.

(ii) In order to maintain compatibility of rotation at B, the column has to rotate as a rigid body in an anticlockwise direction through an angle equal to θ_b. The deflection Δ_{b2} to the left at B would then be $\Delta_{b2} = \theta_b H$. This gives

$$\Delta_{b2} = 0.5C_bLH$$

The net horizontal deflection at B is

$$\Delta_b = \Delta_{b1} + \Delta_{b2} = 0.5H(C_bL + C_cH)$$

(c) Column CD. This column is free from thermal deformations. Therefore it translates and rotates as a rigid body to maintain compatibility.

   (i) To maintain compatibility of displacements, the joint C also moves to the left by Δ_b. This requires that the column CD moves as a rigid body to the left by Δ_b. The movement at the roller is therefore $\Delta_{d1} = \Delta_b$.

   (ii) To maintain compatibility of rotations at C, the column CD has to rotate in a clockwise direction such that the displacement Δ_{d2} to the left at D is $\Delta_{d2} = \theta_c H$. Therefore

$$\Delta_{d2} = 0.5C_bLH$$

The net translation to the left at the roller is

$$\Delta_d = \Delta_{d1} + \Delta_{d2} = 0.5H(C_bL + C_cH) + 0.5C_bLH$$
$$= 0.5H(2C_bL + C_cH)$$

5.4 ● Calculation of deformations of 'mixed' element structures

In the previous sections, the structures considered were made from either purely bar elements or bending elements. In this section, structures which have both bar and bending elements will be considered.

5.4.1 Examples

Example 1

Determine the tip deflection of the cable supported structure shown in Fig. 5.9(a).

Solution   The structure is statically determinate. The beam acts as a simply supported beam segment hinged at A and supported by the cable at B. The vertical reactions at the supports are $0.5qL$. Because the force in the cable provides the vertical reaction at B, the vertical components F_v of the force in the cable is equal to the reaction $0.5qL$. If the tensile force in the cable is F, then

$$F_v = F \sin \phi = 0.5qL$$

Therefore

$$F = 0.5qL/\sin \phi$$

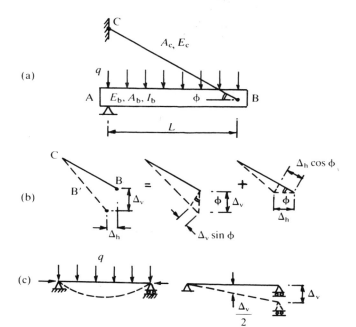

Figure 5.9 ● (a)–(c) Deformations of a cable stayed simply supported beam

The horizontal component F_h of the force in the cable is

$$F \cos \phi = 0.5qL \cot \phi$$

The horizontal component F_h causes a compressive force in the beam which shortens by

$$\Delta_h = F_h L/(A_b E_b) = 0.5qL^2 \cot \phi/(A_b E_b)$$

where E_b and A_b are respectively the Young's modulus and cross-sectional area of the beam. The extension Δ of the cable is

$$\Delta = FL \sec \phi/(A_c E_c) = 0.5qL^2/(A_c E_c \sin \phi \cos \phi)$$

where $L \sec \phi$ = length of the cable and A_c and E_c are respectively the area of cross-section and Young's modulus of the cable.

The extension Δ can be expressed as shown in Fig. 5.9(b) as

$$\Delta = -\Delta_h \cos \phi + \Delta_v \sin \phi$$

where Δ_v = vertical displacement of the cable at B. Therefore

$$\Delta_v = (\Delta + \Delta_h \cos \phi)/\sin \phi$$
$$= 0.5qL^2\{1/(A_c E_c \sin^2 \phi \cos \phi) + \cos^2 \phi/(A_b E_b \sin^2 \phi)\}$$

The deflection at midspan of the beam is given, as shown in Fig. 5.9(c), by two parts as follows.

Due to the deflection Δ_{v1} at the centre, considering the beam as simply supported with zero deflection at the ends A and B. Δ_{v1} is given by

$$\Delta_{v1} = 5qL^4/(384E_b I_b)$$

Due to the deflection Δ_v at B, the beam rotates as a rigid body about the hinge at A. The deflection Δ_{v2} at the centre of the beam is given by

$$\Delta_{v2} = 0.5\Delta_v$$

$$= 0.25qL^2\{1/(A_cE_c \sin^2 \phi \cos \phi) + \cos^2 \phi/(A_bE_b \sin^2 \phi)\}$$

The net deflection Δ_{vb} at the centre of the beam is

$$\Delta_{vb} = \Delta_{v1} + \Delta_{v2}$$

$$= 5qL^4/(384E_bI_b) + 0.25qL^2\{1/(A_cE_c \sin^2 \phi \cos \phi) + \cos^2 \phi/(A_bE_b \sin^2 \phi)\}$$

Example 2

Analyse the structure shown in Fig. 5.10(a).

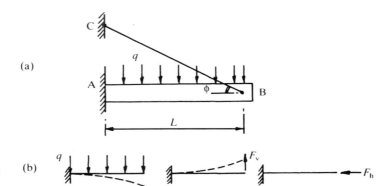

(a)

Figure 5.10 ● (a), (b)
Deformations of a cable
stayed cantilever

(b)

Solution This example differs from the previous example only because the beam AB is fixed at A. The structure is thus statically indeterminate. For the purposes of the solution it is convenient to treat the axial force F in the cable as statically indeterminate. The cable imposes at the end B of the beam a vertical force $F_v = F \sin \phi$ and a horizontal force $F_h = F \cos \phi$. The deformation of the beam and cable can be calculated as follows:

(a) Beam AB: As shown in Fig. 5.10(b) the deformations of the beam are given by:

    (i)   The horizontal translation Δ_h due to the compressive force F_h at the end B of the beam is given by

$$\Delta_h = F_hL/(A_bE_b) = FL \cos \phi/(A_bE_b)$$

    (ii) The vertical deflection Δ_{v1} at the end B of the beam due to the load q acting on the cantilever is $\Delta_{v1} = qL^4/(8E_bI_b)$ downward.

    (iii) The vertical deflection Δ_{v2} due to force F_v acting at the tip B of cantilever AB is given by

$$\Delta_{v2} = F_vL^3/(3E_bI_b) = FL^3 \sin \phi/(3E_bI_b)$$

The net downward deflection is given by

$$\Delta_v = \Delta_{v1} - \Delta_{v2} = qL^4/(8E_bI_b) - FL^3 \sin \phi/(3E_bI_b)$$

(b) Cable: As in Fig. 5.9(b), the extension Δ of the cable is

$$\Delta = FL \sec \phi/(A_cE_c)$$

where $L \sec \phi$ is the length of the cable. As shown in Example 1, Δ_v is given by $\Delta_v = (\Delta + \Delta_h \cos \phi)/\sin \phi$. Substituting for Δ and Δ_h we have

$$\Delta_v = \{FL \sec \phi/(A_cE_c) + FL \cos^2 \phi/(A_bE_b)\}/\sin \phi$$

Equating Δ_v calculated from the beam and cable deformations, we have

$$\Delta_v \text{ from beam} = qL^4/(8E_bI_b) - FL^3 \sin \phi/(3E_bI_b)$$

$$\Delta_v \text{ from cable} = \{FL \sec \phi/(A_cE_c) + FL \cos^2 \phi/(A_bE_b)\}/\sin \phi$$

Solving for F,

$$F = 3qL/(8C)$$

where

$$C = [\sin \phi + 3E_bI_b/(A_cE_cL^2 \sin \phi \cos \phi) + 3I_b \cos^2 \phi/(A_bL^2 \sin \phi)]$$

Note that, as is to be expected, as the axial rigidity A_cE_c of the cable increases, C decreases and the force F in the cable increases. The vertical deflection Δ_v of the beam at B is given by

$$\Delta_v = qL^4/(8E_bI_b) - FL^3 \sin \phi/(3E_bI_b)$$
$$= qL^4/(8E_bI_b) - qL^4 \sin \phi/(8E_bI_bC)$$
$$= \{qL^4/(8E_bI_b)\}(1 - 1/C)$$

Example 3

Analyse the trussed beam shown in Fig. 5.11(a). The properties of the various elements are as follows.

Beam: $E = 11.5$ kN mm^{-2}, $A = 6 \times 10^4$ mm^2,

$$I = 450 \times 10^6 \text{ mm}^4$$

Therefore

$$EI = 5175 \times 10^6 \text{ kN mm}^2, \quad AE = 69.0 \times 10^4 \text{ kN}$$

Strut: $E = 210$ kN mm^{-2}, $A = 2400$ mm^2

Ties: $E = 210$ kN mm^{-2}, $A = 706.9$ mm^2

Solution This structure combines both beam and truss action. The beam, together with the strut and ties, forms a truss and carries the external load resulting in axial tension and compression forces in the members. In addition, the beam resists the external load by bending and shearing actions. As shown in Fig. 5.11(b), the truss provides

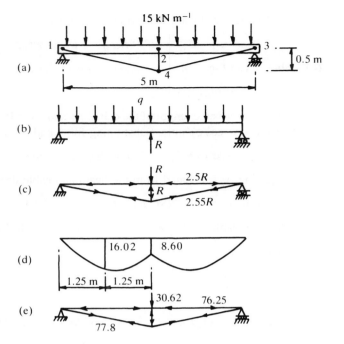

Figure 5.11 ●
(a)–(e) A trussed beam

Table 5.7 ●
Geometrical details

Member	x_1	y_1	x_2	y_2	L	I	m	$L/(AE)*$
1–2	0	0.5	2.5	0.5	2.50	1.0	0	3.623×10^{-3}
2–3	2.5	0.5	5.0	0.5	2.50	1.0	0	3.623×10^{-3}
1–4	0	0.5	2.5	0	2.55	0.981	−0.198	1.718×10^{-2}
4–3	2.5	0	5.0	0.5	2.55	0.981	0.198	1.718×10^{-2}
4–2	2.5	0	2.5	0.5	0.5	0	1.0	9.921×10^{-4}

*$L/(AE)$ is in mm per kN units

a reaction at the centre of the beam equal to R. The value of R is such that the vertical deflection of the truss and the beam are equal at the centre. Table 5.7 summarizes the basic geometric details of the structure and also the value of $L/(AE)$ for each member. Table 5.8 summarizes the extension of each member expressed in terms of force as well as in terms of displacement at its ends.

(a) Beam: As shown in Fig. 5.11(c), the beam is simply supported and subjected to a uniformly distributed load q of 15 kN m^{-1} and a central concentrated load of R kN, the reaction from the truss. Therefore the vertical deflection of the beam is

$$\Delta = 5qL^4/(384EI) - RL^3/(48EI)$$

$$= 5(15 \times 5) \times (5000)^3/(384 \times 5175 \times 10^6) - R(5000)^3/(48 \times 5175 \times 10^6)$$

$$= 23.59 - 0.5032R \text{ mm}$$

(b) Truss: Using the member forces as shown in Fig. 5.11(c), the extensions of the members can be calculated as follows. The displacements of the joints are indicated in Table 5.8.

Table 5.8

Joint	u	v
1	0	0
2	d_1	d_2
3	d_3	0
4	d_4	d_5

Table 5.9 ● Member force and axial extension

Member	F	$FL/(AE)$	$(u_2 - u_1)l + (v_2 - v_1)m$
1–2	$-2.5R$	$-9.0575R \times 10^{-3}$	d_1
2–3	$-2.5R$	$-9.0575R \times 10^{-3}$	$(d_3 - d_1)$
1–4	$2.55R$	$43.809R \times 10^{-3}$	$0.981d_4 - 0.198d_5$
4–3	$2.55R$	$43.809R \times 10^{-3}$	$0.981(d_3 - d_4) + 0.198(0 - d_5)$
4–2	$-1.0R$	$-0.9921R \times 10^{-3}$	$(d_2 - d_5)$

Because $FL/(AE) = (u_2 - u_1)l + (v_2 - v_1)m$, solving the simultaneous equations we have, from the first two equations, $d_1 = -9.0575R \times 10^{-3}$ mm, $d_3 = -18.115R \times 10^{-3}$ mm. From the third and fourth equations we have $d_5 = -266.1334R \times 10^{-3}$ mm, $d_4 = -9.0575R \times 10^{-3}$ mm. From the last equation $d_2 = -267.1255R \times 10^{-3}$ mm. Therefore for compatibility of deflections at the centre, we have, noting that Δ is a downward deflection and d_2 is an upward deflection,

$$\Delta = -d_2 \text{ i.e. } 23.59 - 0.5032R = 0.2671R$$

Therefore

$$R = 30.62 \text{ kN}$$

The bending moment distribution in the beam and the axial forces in the members of the truss are easily calculated and are shown in Fig. 5.11(d) and (e).

5.5 ● Calculation of deformations of 3-D rigid-jointed structures

The ideas of equilibrium and compatibility used in the previous sections for 2-D structures can be used for 3-D structures as well. This is illustrated by a simple example.

Example 1

Determine the displacements of the joints of the 3-D rigid-jointed structure shown in Fig. 5.12(a). Assume that L, EA, EI and GJ, which are respectively the length, the axial rigidity, the flexural rigidity and torsional rigidity, are constant.

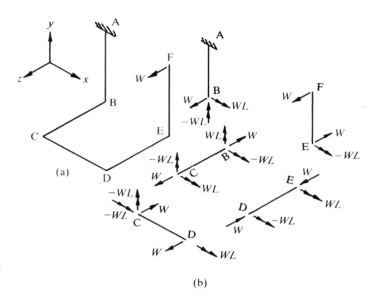

Figure 5.12 ● (a), (b)
Force distribution in a 3-
D rigid-jointed structure

Solution The forces at the ends of the elements are easily determined. These are shown in Fig. 5.12(b). The technique used for calculating the displacements is to consider each element as a cantilever fixed at the 'far end' and subjected to forces at the 'near end'. The final displacements at the 'near end' are modified by taking into consideration the displacements at the 'far end'. In the following, couples and rotations are represented by double-headed arrows as discussed in Chapter 1, Fig. 1.76.

For simplicity, the following notation is used:

$$C_1 = WL^3/(EI),\ C_2 = WL^2/(EI),\ C_3 = WL^3/(GJ),$$
$$C_4 = WL^2/(GJ),\ C_5 = WL/(EA)$$

(a) Element AB: It is a cantilever fixed at the 'far end' A and subjected to forces $F_z = W$, $M_x = WL$ and $M_y = -WL$ at the 'near end' B. The displacements at the near end B are as follows.

(i) Force $F_z = W$: This causes bending in the y–z plane. The deformations are

$$\Delta_{z1} = WL^3/(3EI),\ \theta_{x1} = -WL^2/(2EI)$$

(ii) Couple $M_x = WL$: This also causes bending in the y–z plane. The deformations are

$$\Delta_{z2} = -(WL)L^2/(2EI),\ \theta_{x2} = (WL)L/(EI)$$

(iii) Couple $M_y = -WL$: This is a torsional moment and causes twisting about the y-axis. The twist is calculated by $T = (GJ/L)\theta_y$. Therefore

$$-WL = (GJ/L)\theta_y \text{ and } \theta_y = -WL^2/(GJ)$$

Therefore the net displacements at B are

$$\Delta_z = WL^3/(3EI) - WL^3/(2EI) = -WL^3/(6EI)$$
$$\theta_x = -WL^2/(2EI) + WL^2/(EI) = WL^2/(2EI)$$
$$\theta_y = -WL^2/(GJ)$$

Table 5.10

Cause	Δ_x	Δ_y	Δ_z	θ_x	θ_y
Load at B	0	0	$-C_1/6$	$C_2/2$	$-C_4$
Support at A*	0	0	0	0	0
Rotation at A†	0	0	0	0	0
Net B	0	0	$-C_1/6$	$C_2/2$	$-C_4$

* Records the displacements at the 'far end'
† Records the translations at the 'near end' due to rotations at the 'far end'

Table 5.11

Cause	Δ_x	Δ_y	Δ_z	θ_x	θ_y
Load at C	$-C_1/2$	$-C_1/2$	C_5	C_2	$-C_2$
Support at B*	0	0	$-C_1/6$	$C_2/2$	$-C_4$
Rotation at B†	$-C_3$	$-C_1/2$	0	0	0
Net C	$-C_1/2 - C_3$	$-C_1$	$C_5 - C_1/6$	$1.5C_2$	$-C_2 - C_4$

* Records the displacements at B
† Note that a rotation of θ_x at B causes a translation of $\Delta_z = -\theta_x L$ at C. Similarly a rotation of θ_y at B causes a translation of $\Delta_x = \theta_y L$ at C

(b) Element BC: Considering B as the 'far end', the forces at the 'near end' C are $F_z = W$, $M_x = WL$ and $M_y = -WL$. The deformations at C due to the loads can be calculated as follows.

  (i) Load $F_z = W$. This is an axial force and causes an extension of $\Delta_z = C_5$, where $C_5 = WL/(EA)$. Table 5.10 summarizes how, using the basic displacement and rotation values for a cantilever, the displacement and rotation at the 'tip' corresponding to joint B can be calculated once the displacement at the 'support' corresponding to joint A and the load acting at the 'tip' B are known. In a similar manner, Tables 5.11 to 5.14 summarize the calculation of displacements at joints C to F respectively.

  (ii) Couple $M_x = WL$. This is a bending moment and causes pure bending in the y–z plane. The deformations at C are

$$\Delta_y = -(WL)L^2/(2EI) = -C_1/2$$

$$\theta_x = (WL)L/(EI) = C_2$$

  (iii) Couple $M_y = -WL$. This is a bending moment and causes pure bending in the x–z plane. The deformations are given by

$$\Delta_x = -(WL)L^2/(2EI) = -C_1/2$$

$$\theta_y = -(WL)L/(EI) = -C_2$$

(c) Member CD: This member is subjected at the 'near end' D to forces $F_z = W$ and $M_x = WL$. Considering the 'far end' C as fixed, the displacements at D are

  (i) Force $F_z = W$: This causes bending in the x–z plane and the deformations at D are

Table 5.12

Cause	Δ_x	Δ_y	Δ_z	θ_x	θ_y
Load at D	0	0	$C_1/3$	C_4	$-C_2/2$
Support C*	$-C_1/2 - C_3$	$-C_1$	$C_5 - C_1/6$	$1.5C_2$	$-C_2 - C_4$
Rotation C†	0	0	$C_1 + C_3$	0	0
Net D	$-C_1/2 - C_3$	$-C_1$	$C_5 + 7C_1/6 + C_3$	$1.5C_2 + C_4$	$-1.5C_2 - C_4$

* Records the displacements at C
† Note that a rotation of θ_y at C causes a translation of $\Delta_z = -\theta_y L$ at D

Table 5.13

Cause	Δ_x	Δ_y	Δ_z	θ_x	θ_y
Load at E	0	$C_1/2$	C_5	C_2	0
Support D*	$-C_1/2 - C_3$	$-C_1$	$C_5 + 7C_1/6 + C_3$	$1.5C_2 + C_4$	$-1.5C_2 - C_4$
Rotation D†	$1/5C_1 + C_3$	$1.5C_1 + C_3$	0	0	0
Net E	C_1	$C_1 + C_3$	$2C_5 + 7C_1/6 + C_3$	$2.5C_2 + C_4$	$-1.5C_2 - C_4$

* Records the displacements at D
† Note that a rotation of θ_x at D causes a translation of $\Delta_y = \theta_x L$ at E and a rotation of θ_y at D causes a translation of $\Delta_x = -\theta_y L$ at E

Table 5.14

Cause	Δ_x	Δ_y	Δ_z	θ_x	θ_y
Load at F	0	0	$C_1/3$	$C_2/2$	0
Support E*	C_1	$C_1 + C_3$	$2C_5 + 7C_1/6 + C_3$	$2.5C_2 + C_4$	$-1.5C_2 - C_4$
Rotation E†	0	0	$5C_1/2 + C_3$	0	0
Net F	C_1	$C_1 + C_3$	$2C_5 + 4C_1 + 2C_3$	$3C_2 + C_4$	$-1.5C_2 - C_4$

* Records the displacements at E
† Note that a rotation of θ_x at E causes a translation of $\Delta_z = \theta_x L$ at F

$$\Delta_z = WL^3/(3EI) = C_1/3$$

$$\theta_y = -WL^2/(2EI) = -C_2/2$$

(ii) Couple $M_x = WL$: This is a twisting moment and the twist at D is given by

$$\theta_x = (WL)L/(GJ) = C_4$$

(d) Element DE: This element is subjected at the 'near end' E to forces $F_z = W$ and $M_x = WL$. Assuming that the 'far end' D is fixed, the displacements at the 'near end' are

(i) Force $F_z = W$: This causes an axial compression. Therefore $\Delta_z = WL/(AE)$ = C_5.

(ii) Couple $M_x = WL$: This causes bending in the y–z plane and the deformations are

$$\Delta_y = (WL)L^2/(2EI) = C_1/2$$

$$\theta_x = (WL)L/(2EI) = C_2$$

(e) Segment EF: This is subjected at the 'near end' F to a force $F_z = W$. Assuming that the 'far end' E is fixed, the displacements at F are

$$\Delta_z = WL^3/(3EI) = C_1/3, \quad \theta_x = WL^2/(2EI) = C_2/2$$

Thus the displacements at all the joints are determined.

5.6 ● Use of stiffness matrices for deflection calculations

The above manual procedures used for the calculation of displacements give an insight into the behaviour of stuctures. However, they become unwieldy for practical structures. In Chapter 4, the stiffness matrices for bar, beam and shaft elements were derived and, as indicated, the stiffness matrices form the basis of computer-orientated methods of structural analysis for the calculation of forces and displacements in practical structures. This topic will be dealt with at length in Chapter 6.

5.7 ● Problems

1. Determine the displacements of the pin-jointed structure in Fig. E5.1. Assume that AE is the same for all members.

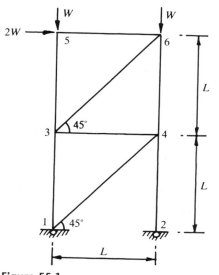

Figure E5.1

Table E5.1

Member	$FL/(AE)$	Δ
1–4	$4\alpha*$	$(d_3 + d_4)/\sqrt{2}$, knowing d_4, d_3 = $(5 + 4\sqrt{2})WL/(AE)$
1–3	α	d_2 ∴ $d_2 = WL/(AE)$
2–4	-5α	d_4 ∴ $d_4 = -5WL/(AE)$
3–4	-2α	$(d_3 - d_1)$ ∴ $d_1 = (7 + 4\sqrt{2})WL/(AE)$
3–6	4α	$(d_7 - d_1)/\sqrt{2} + (d_8 - d_2)/\sqrt{2}$ ∴ $d_7 = (16 + 8\sqrt{2})WL/(AE)$
3–5	$-\alpha$	$(d_6 - d_2)$ ∴ $d_6 = 0$
4–6	-3α	$(d_8 - d_4)$ ∴ $d_8 = -8WL/(AE)$
5–6	-2α	$(d_7 - d_5)$ ∴ $d_5 = (18 + 8\sqrt{2})WL/(AE)$

$*\alpha = WL/(AE)$

Answer: Let the joint deflections be (d_1, d_2) at 3, (d_3, d_4) at 4, (d_5, d_6) at 5 and (d_7, d_8) at 6. Answers are shown in Table E5.1.

2. Determine the forces in members and displacements of the joints of the structure shown in Fig. E5.2. Assume AE is the same for all members.

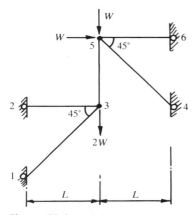

Figure E5.2

Table E5.2

Member	$FL/(AE)$	Δ
1–3	$(-4W + 2R)L/(AE)$	$(d_1 + d_2)/\sqrt{2}$
2–3	$(2W - R)L/(AE)$	d_1
3–5	$RL/(AE)$	$(d_4 - d_2)$
5–6	$RL/(AE)$	$-d_3$
5–4	$(-2W - 2R)L/(AE)$	$(-d_3 + d_4)/\sqrt{2}$

Answer: This is a statically indeterminate structure.
Let the force in the member 3–5 be unknown
tensile force R. Let the joint deflections be
(d_1, d_2) at 3 and (d_3, d_4) at 5.
Solving, $R = 0.56W$,
$(d_1, d_2, d_3, d_4) = (1.44, -5.52, -0.55, -4.96)WL/(AE)$

3. Determine the displacements of the pin-jointed
structure shown in Fig. E5.1 due to a temperature
change of T in members 1–3, 3–5 and 5–6. Assume
the coefficient of linear thermal expansion is
constant. The answer is shown in Table E5.2.

4. Figure E5.3 shows a rigid-jointed frame subjected
to a u.d.l. q normal to the cantilever. Calculate
the deformation of the structure.

Answer: Total normal load W_n in the cantilever
$= qL \sec \alpha$. Moment M_s at the top of the column
is $0.5qL^2 \sec^2 \alpha$,
vertical load $= W_n \cos \alpha = qL$.
Horizontal load $= W_n \sin \alpha = qL \tan \alpha$.

Table E5.3

Member	Thermal extension	Δ
1–4	0	$(d_3 + d_4)/\sqrt{2}$, knowing d_4, $d_3 = 0$
1–3	αLT	$d_2 \therefore d_2 = \alpha LT$
2–4	0	$d_4 \therefore d_4 = 0$
3–4	0	$(d_3 - d_1) \therefore d_1 = 0$
3–6	0	$(d_7 - d_1)/\sqrt{2} + (d_8 - d_2)/\sqrt{2}$ $\therefore d_7 = \alpha LT$
3–5	αLT	$(d_6 - d_2) \therefore d_6 = 2\alpha TL$
4–6	0	$(d_8 - d_4) \therefore d_8 = 0$
5–6	αLT	$(d_7 - d_5) \therefore d_5 = 0$

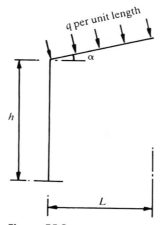

Figure E5.3

Horizontal deflection u_1 at the top of the column
$$= M_s h^2/(2EI) + (qL \tan \alpha)h^3/(3EI)$$

Rotation θ at the top of the column
$$= M_s h/(EI) + (qL \tan \alpha)h^3/(2EI)$$

The total vertical deflection of the tip of the
cantilever
$$= \theta L + qL^4/(8EI)\{\sec^3 \alpha\}$$

Total horizontal deflection of the cantilever tip
$$= u_1 + qL^4/(8EI)\{\sec^4 \alpha \sin \alpha\}$$

Chapter six

Stiffness method

In the previous chapters, manual methods of calculation were developed for the determination of forces and displacements of statically determinate and some simple statically indeterminate structures. These manual methods are only suitable for the analysis of 'small' structures. However, they enable the reader to understand how the necessary conditions of equilibrium, compatibility and material laws peculiar to each type of structure are satisfied. Computers have revolutionized the analysis of complex structures and, in practice, computer-orientated methods of analysis are extensively used. These methods are merely an adaptation, in a more systematic way, of the manual methods already discussed. Irrespective of what method of analysis is used, whether manual or computer-orientated, it is necessary to satisfy the following fundamental requirements.

1. Equilibrium as required by equations of statics.
2. Compatibility of displacements not only along an element but also at the joints where elements intersect.
3. Prescribed force or displacement constraints at supports.
4. Forces and deformations being compatible with the assumed element and material properties.

This chapter presents in detail one of the most popular computer-orientated general methods of structural analysis, called the stiffness method. Before the general method is presented, important relevant results from the previous chapters are summarized.

6.1 ● Behaviour of elements

In previous chapters, the emphasis was on the understanding of the behaviour of elements such as bars, beams and shafts from which all types of skeletal structures are assembled. The behaviour of the various types of elements can be summarized as follows.

Bar element

A bar element is subjected to axial forces only and suffers an axial extension in the process. At a cross-section only uniform axial stresses exist. The main parameters influencing the stresses, strains and axial extension of the bar are its area of cross-section A, length L and the Young's modulus of the material E.

The stiffness relationship for a bar element such as that shown in Fig. 4.5 is given, as shown in Section 4.1.3, by

$$\begin{bmatrix} F_{a1} \\ F_{a2} \end{bmatrix} = (AE/L) \begin{bmatrix} 1 & -1 \\ -1 & 1 \end{bmatrix} \begin{bmatrix} \Delta_{a1} \\ \Delta_{a2} \end{bmatrix} + \begin{bmatrix} F_{fa1} \\ F_{fa2} \end{bmatrix}$$

where F_{a1} and F_{a2} are the forces in the axial direction at the ends 1 and 2 respectively; Δ_{a1} and Δ_{a2} are the displacements in the axial direction at the ends 1 and 2 respectively; F_{fa1} and F_{fa2} are the fixed end forces at the ends 1 and 2 respectively due to axial load on the member or temperature change. It is useful to note that the stiffness matrix is symmetrical.

Beam element

A beam element resists bending moment and shear forces. The stresses at a cross-section consist of linearly varying bending stress and parabolically varying shear stress. The displacements of the beam at a section are characterized by a translation normal to the axis of the beam and a rotation about an axis normal to the plane of bending. The main parameters influencing the bending and shear stresses, translational and rotational displacements are the second moment of area I, span L and Young's modulus E of the material of the beam. It was also noticed that except in the case of very 'stocky' beams, i.e. beams with small span/depth ratio, flexural or bending deformations dominate and shear deformations can be safely neglected. The stiffness relationship for a beam element such as that shown in Fig. 4.47, taking into consideration only the flexural deformations, is given, as shown in Section 4.14, by

$$\begin{bmatrix} F_{n1} \\ M_1 \\ F_{n2} \\ M_2 \end{bmatrix} = (EI/L) \begin{bmatrix} 12/L^2 & -6/L & -12/L^2 & -6/L \\ -6/L & 4 & 6/L & 2 \\ -12/L^2 & 6/L & 12/L^2 & 6/L \\ -6/L & 2 & 6/L & 4 \end{bmatrix} \begin{bmatrix} \Delta_{n1} \\ \theta_1 \\ \Delta_{n2} \\ \theta_2 \end{bmatrix} + \begin{bmatrix} F_{fn1} \\ M_{f1} \\ F_{fn2} \\ M_{f2} \end{bmatrix}$$

where F_{n1} and F_{n2} are the forces normal to the axis of the beam at the ends 1 and 2 respectively. M_1 and M_2 are the clockwise couples at the ends 1 and 2 respectively. F_{fn1} and F_{fn2} are the fixed end forces normal to the axis of the beam at the ends 1 and 2 respectively. M_{f1} and M_{f2} are the fixed end clockwise couples at the ends 1 and 2 respectively. These fixed end actions are due to 'lateral' loads and temperature variations along the beam length. It is useful to note that the stiffness matrix is symmetrical.

If shear deformation effects are to be included, then as shown in Section 4.15, replace (12, 6, 4 and 2) by $\{12/(1 + \beta), 6/(1 + \beta), (4 + \beta)/(1 + \beta), (2 - \beta)/(1 + \beta)\}$ respectively, where $\beta = 12EI/(GA_sL^2)$ with the usual notation.

Shaft element

A shaft element is subjected to twisting moments and twisting deformation. The main parameters influencing the stresses (pure shear stresses), strains and twist of the shaft are the Saint Venant's torsional constant J, length L and shear modulus G of the material.

The stiffness relationship for a shaft element such as that shown in Fig. 4.50 is given, as shown in Section 4.16, by

$$\begin{bmatrix} T_1 \\ T_2 \end{bmatrix} = GJ/L \begin{bmatrix} 1 & -1 \\ -1 & 1 \end{bmatrix} \begin{bmatrix} \psi_1 \\ \psi_2 \end{bmatrix} + \begin{bmatrix} T_{f1} \\ T_{f2} \end{bmatrix}$$

where T_1 and T_2 are the twisting moments are the ends 1 and 2 respectively; ψ_1 and ψ_2 are the twists at the ends 1 and 2 respectively; T_{f1} and T_{f2} are the fixed end torques at the ends 1 and 2 respectively due to applied torque on the element.

It is useful to note that the stiffness matrix is symmetrical.

6.2 ● Equilibrium and compatibility requirements for structures assembled from various types of elements

The behaviour of various types of structures considered in the previous chapters can be summarized as follows.

2-D pin-jointed structures

Pin-jointed structures are assembled from bar elements. The basic conditions to be satisfied in the analysis are

(a) Equilibrium. As was shown in Section 1.7.2, at any joint i, the sum of the x-components of the axial forces in all the members meeting at the joint i together with the externally applied force in the x-direction at the joint i must be in equilibrium. Similar conditions apply to equilibrium in the y-direction.

$$-\sum F_{xij} + W_{xi} = 0, \text{ and } -\sum F_{yij} + W_{yi} = 0$$

where the summation Σ is over $j = 1$ to number of members meeting at the ith joint. W_{xi} and W_{yi} are respectively the external forces applied in the x- and y-directions at the ith joint. F_{xij} and F_{yij} are respectively the x- and y-components of the force in the jth member at the ith joint.

(b) Compatibility. All members meeting at a joint have at that joint the same displacements in the x- and y-directions. See examples in Section 5.1.1.

(c) Supports generally consist of pins or rollers. At a pin-support, all translations are suppressed. At a roller support only the displacement normal to the roller is suppressed.

(d) Bar elements are linearly elastic and the force–deformation relationship is expressed by the stiffness relationship for the bar element.

2-D rigid-jointed structures

2-D rigid-jointed structures like portal frames, arches, etc. are structures lying in the vertical x–y plane and subjected to loads acting in the same plane. At any section in a member, in general there exists an axial force, a shear force and a bending

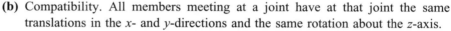

Figure 6.1 ●
Equilibrium with respect
to couples at a joint

moment. The element can thus be considered as being a combination of beam and
bar behaviour. The basic conditions to be satisfied in the analysis are

(a) Equilibrium. As in the case of trusses, at the ith joint,

$$-\sum F_{xij} + W_{xi} = 0, \quad -\sum F_{yi} + W_{yi} = 0$$

and in addition, as shown in Fig. 6.1,

$$-\sum M_{zij} + M_{zi} = 0$$

where the summation Σ is over $j = 1$ to number of members meeting at the ith
joint. W_{xi}, W_{yi} and M_{zi} are respectively the external forces applied in the x- and
y-directions and the couple about the z-axis at the ith joint. F_{xij} and F_{yij} are
respectively the x- and y-components of the axial and shear force in the jth
member at the ith joint. M_{zij} = the couple acting on the jth member at the ith
joint.

(b) Compatibility. All members meeting at a joint have at that joint the same
translations in the x- and y-directions and the same rotation about the z-axis.
  While this is generally true for translations, it may not be true for rotations.
For example, for the case shown in Fig. 6.2, members 1 and 2 have the same
rotation at the joint because they are 'rigidly' joined but member 3 is not
rigidly connected. Therefore the rotation of member 3 at the joint is different
from that of members 1 and 2.

Figure 6.2 ●
Compatibility with
respect to rotations
at a joint

(c) Supports generally consist of pins, rollers or fixed supports. At a fixed support
all translations and rotation are suppressed, at a pinned support, all translations
are suppressed and at a roller support only the displacement normal to the
roller is suppressed.

(d) The (bar + beam) elements are linearly elastic and the force–deformation relationship is expressed by the stiffness relationship which is a combination of the
stiffness relationships for the bar and beam elements. This is obtained by combining the stiffness relationship for the bar element and beam element as follows.

$$
\begin{bmatrix} F_{a1} \\ F_{n1} \\ M_1 \\ F_{a2} \\ F_{n2} \\ M_2 \end{bmatrix} = (EI/L)
\begin{bmatrix}
A/I & 0 & 0 & -A/I & 0 & 0 \\
0 & 12/L^2 & -6/L & 0 & -12/L^2 & -6/L \\
0 & -6/L & 4 & 0 & 6/L & 2 \\
-A/I & 0 & 0 & A/I & 0 & 0 \\
0 & -12/L^2 & 6/L & 0 & 12/L^2 & 6/L \\
0 & -6/L & 2 & 0 & 6/L & 4
\end{bmatrix}
$$

$$\times \begin{bmatrix} \Delta_{a1} \\ \Delta_{n1} \\ \theta_1 \\ \Delta_{a2} \\ \Delta_{n2} \\ \theta_2 \end{bmatrix} + \begin{bmatrix} F_{fa1} \\ F_{fn1} \\ M_{f1} \\ F_{fa2} \\ F_{fn2} \\ M_{f2} \end{bmatrix}$$

If the shear deformation effect is to be included, then as shown in Section 4.15, replace (12, 6, 4 and 2) by $\{12/(1 + \beta), 6/(1 + \beta), (4 + \beta)/(1 + \beta), (2 - \beta)/(1 + \beta)\}$ respectively, where $\beta = 12EI/(GA_sL^2)$ with the usual notation.

It is useful to note that the stiffness matrix is symmetrical.

Plane grid structure

Plane grid structures generally lie in the horizontal plane x–z and the loads are applied normal to the plane of the structure in the y-direction. At any cross-section there exist a shear force, a bending moment and a twisting moment. The element can thus be considered as having a combination of beam and shaft behaviour. The basic conditions to be satisfied in the analysis are

(a) Equilibrium. At any joint i,

$$-\sum F_{yij} + W_{yi} = 0$$

$$-\sum M_{xij} + M_{xi} = 0$$

and

$$-\sum M_{zij} + M_{zi} = 0$$

where the summation Σ is over $j = 1$ to number of members meeting at the ith joint. W_{yi}, M_{xi} and M_{zi} are respectively the external force applied in the y-direction and couples about the x-axis and z-axis respectively at the joint i. F_{yij} is the normal force in the jth member at the ith joint. M_{xij} and M_{zij} are respectively the x- and z-components of the bending and twisting moments in the jth member at the ith joint.

(b) Compatibility. All members meeting at a joint have at that joint the same translation in the y-direction and the same rotation about the x- and z-axes.

(c) Supports generally consist of pinned and fixed supports. At a fixed support translation in the y-direction and rotations about the x- and z-axes are suppressed; at a pinned support only translation in the y-direction is suppressed. It is also possible to have a pin support where the translation in the y-direction and only one rotation about an axis are suppressed.

(d) The (beam + shaft) elements are linearly elastic and the force–deformation relationship is expressed by the stiffness relationship which is a combination of the stiffness relationships for the shaft and beam elements. This is obtained by combining the stiffness relationships for the beam and shaft elements as follows.

$$
\begin{bmatrix} F_{n1} \\ M_1 \\ T_1 \\ F_{n2} \\ M_2 \\ T_2 \end{bmatrix} = (EI/L)
$$

$$
\times \begin{bmatrix}
12/L^2 & -6/L & 0 & -12/L^2 & -6/L & 0 \\
-6/L & 4 & 0 & 6/L & 2 & 0 \\
0 & 0 & GJ/(EI) & 0 & 0 & -GJ/(EI) \\
-12/L^2 & 6/L & 0 & 12/L^2 & 6/L & 0 \\
-6/L & 2 & 0 & 6/L & 4 & 0 \\
0 & 0 & -GJ/(EI) & 0 & 0 & GJ/(EI)
\end{bmatrix}
$$

$$
\times \begin{bmatrix} \Delta_{n1} \\ \theta_1 \\ \psi_1 \\ \Delta_{n2} \\ \theta_2 \\ \psi_2 \end{bmatrix} + \begin{bmatrix} F_{fn1} \\ M_{f1} \\ T_{f1} \\ F_{fn2} \\ M_{f2} \\ T_{f2} \end{bmatrix}
$$

It is useful to note that the stiffness matrix is symmetrical.

If the shear deformation effect is to be included, then, as shown in Section 4.15, replace (12, 6, 4 and 2) by $\{12/(1 + \beta), 6/(1 + \beta), (4 + \beta)/(1 + \beta), (2 - \beta)/(1 + \beta)\}$ respectively, where $\beta = 12EI/(GA_sL^2)$ with the usual notation.

6.3 ● Relationship between forces and displacements in general coordinate axes and member axes

It should be noted that from the point of view of satisfying the equilibrium and compatibility requirements, the forces and displacements at a joint are best specified along a convenient system of x–y–z axes. However, in the stiffness relationships given above, the displacements and forces at a joint are specified along and normal to the axis of the member. The relationship between the two sets is easily established by resolution of vectors as follows.

2-D pin-jointed structure element

Let the joint translations at joint 1 and joint 2 respectively in the x- and y-directions be (u_1, v_1) and (u_2, v_2) as shown in Fig. 6.3.

Their components along the member axis are given by

$$\Delta_{a1} = u_1 l + v_1 m, \quad \Delta_{a2} = u_2 l + v_2 m$$

where $l = \cos \phi$, $m = \sin \phi$ and $\phi =$ inclination of the member to the x-axis (see Section 5.1). Similarly, if the force in the member axis at the ith end is F_{ai}, then its components along the x- and y-axes are given by

$$F_{xi} = F_{ai} l, \quad F_{yi} = F_{ai} m$$

where $i = 1, 2$.

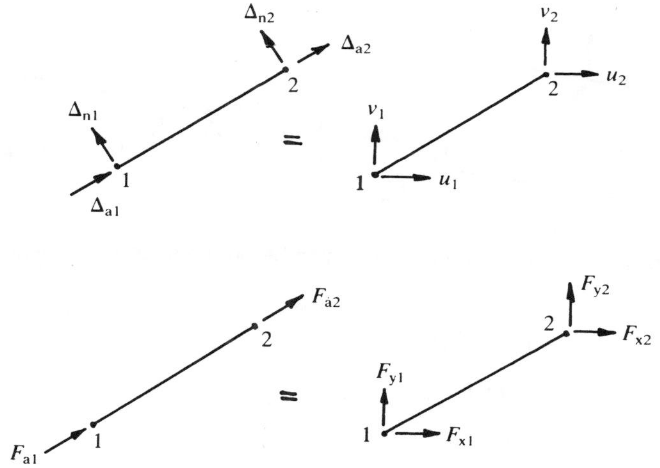

Figure 6.3 ● Forces and displacements in a truss member in member and general coordinate axes

2-D rigid-jointed structure element

Let the joint translations at the ith end be u_i and v_i respectively in the x- and y-directions as shown in Fig. 6.4. Their components along the normal to the member axis are given by

$$\Delta_{ai} = u_i l + v_i m$$

$$\Delta_{ni} = -u_i m + v_i l, \quad i = 1, 2$$

where l and m are the direction cosines of the vector 1–2 with respect to the x–y system.

The rotation θ_i about the z-axis remains unaltered because the z-axis is normal to the plane in which the member lies.

Similarly, if at the ith end the forces along and normal to the member axis are respectively F_{ai} and F_{ni}, then their components along the x- and y-axes are given by

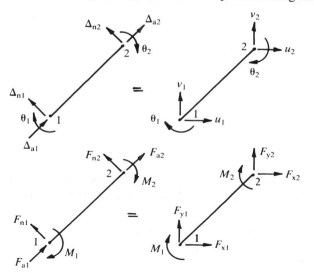

Figure 6.4 ● Forces and deformations in a 2-D rigid-jointed frame member in member and general coordinate axes

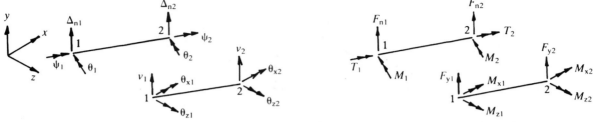

Figure 6.5 ● Forces and deformations in a plane grid member in member and general coordinate axes

$$F_{xi} = F_{ai}l - F_{ni}m$$

$$F_{yi} = F_{ai}m + F_{ni}l$$

and finally

$$M_{zi} = M_i, \quad i = 1, 2$$

F_{ni} and F_{ai} are the shear force and axial force at the ith end of the member.

Plane grid structure element

Let the joint rotations at the ith joint be θ_{xi} and θ_{zi} respectively about the x- and z-axes as shown in Fig. 6.5. Their components along and normal to the member axis are given by

$$\psi_i = \theta_{xi}l + \theta_{zi}n$$

$$\theta_i = -\theta_{xi}n + \theta_{zi}l, \quad i = 1, 2$$

where l and n are the direction cosines of the vector 1–2 with respect to the x–z system. ψ_i and θ_i are respectively the twist and bending rotation at the ith end of the member. The translation in the y-direction $v_i = \Delta_{n1}$.

Similarly, if the couples about the axis and normal to the axis are T_i and M_i respectively, then their components along the x- and z-axes are given by

$$M_{xi} = T_il - M_in$$

$$M_{yi} = T_in + M_il$$

and finally

$$F_{yi} = F_{ni}, \quad i = 1, 2$$

M_i and T_i are respectively the bending moment and twisting moment and F_{yi} is the shear force at the ith end of the member.

6.4 ● Stiffness matrix for a member with displacements and forces in general coordinate directions

In the equations for the stiffness matrices given in Section 6.2, the forces and displacements were expressed in terms of the forces and displacements given in member axes directions. These can be expressed in terms of forces and displacements along general directions using the expressions developed in Section 6.3. When this is done and the calculations are simplified, the stiffness matrices given in matrix equations [6.1] to [6.3] are obtained. It should be noted that if $l = 1$ and $m = 0$, then the member is orientated along the x-axis and the stiffness matrices in Tables 6.1 and 6.3 revert to those given in Section 6.2.

The detailed derivation of the stiffness relationships are given below.

2-D truss element

From the stiffness matrix for a bar element

$$F_{a1} = (AE/L)(\Delta_{a1} - \Delta_{a2}) + F_{fa1}$$

$$F_{a2} = (AE/L)(\Delta_{a2} - \Delta_{a1}) + F_{fa2}$$

Substituting for $\Delta_{a1} = (u_1 l + v_1 m)$ and $\Delta_{a2} = (u_2 l + v_2 m)$ we have

$$F_{a1} = (AE/L)\{(u_1 - u_2)l + (v_1 - v_2)m\} + F_{fa1}$$

$$F_{a2} = (AE/L)\{(u_2 - u_1)l + (v_2 - v_1)m\} + F_{fa2}$$

Expressing $F_{x1} = F_{a1}l$, $F_{y1} = F_{a1}m$ at end 1 and $F_{x2} = F_{a2}l$, $F_{y2} = F_{a2}m$ at end 2 we have

$$F_{x1} = (AE/L)\{(u_1 - u_2)l^2 + (v_1 - v_2)lm\} + F_{fa1}l$$

$$F_{y1} = (AE/L)\{(u_1 - u_2)lm + (v_1 - v_2)m^2\} + F_{fa1}m$$

$$F_{x2} = (AE/L)\{(u_2 - u_1)l^2 + (v_2 - v_1)lm\} + F_{fa2}l$$

$$F_{y2} = (AE/L)\{(u_2 - u_1)lm + (v_2 - v_1)m^2\} + F_{fa2}m$$

The above four equations can be expressed in matrix notation as

$$
\begin{bmatrix} F_{x1} \\ F_{y1} \\ F_{x2} \\ F_{y2} \end{bmatrix} = (AE/L)
\begin{bmatrix}
l^2 & lm & -l^2 & -lm \\
lm & m^2 & -lm & -m^2 \\
-l^2 & -lm & l^2 & lm \\
-lm & -m^2 & lm & m^2
\end{bmatrix}
\times
\begin{bmatrix} u_1 \\ v_1 \\ u_2 \\ v_2 \end{bmatrix}
+
\begin{bmatrix} F_{fa1}l \\ F_{fa1}m \\ F_{fa2}l \\ F_{fa2}m \end{bmatrix}
\tag{6.1}
$$

It is useful to note that the stiffness matrix is symmetrical.

2-D rigid-jointed frame element

Substituting

$$\Delta_{a1} = u_1 l + v_1 m \text{ and } \Delta_{n1} = -u_1 m + v_1 l \quad \text{at joint 1}$$

$$\Delta_{a2} = u_2 l + v_2 m \text{ and } \Delta_{n2} = -u_2 m + v_2 l \quad \text{at joint 2}$$

we have

$$F_{a1} = (AE/L)\{(u_1 - u_2)l + (v_1 - v_2)m\} + F_{fa1}$$

$$F_{n1} = (EI/L^2)[12\{(u_2 - u_1)m + (v_1 - v_2)l\}/L - 6\theta_1 - 6\theta_2] + F_{fn1}$$

$$M_1 = (EI/L)[6\{(u_1 - u_2)m + (v_2 - v_1)l\}/L + 4\theta_1 + 2\theta_2] + M_{f1}$$

$$F_{a2} = (AE/L)\{(u_2 - u_1)l + (v_2 - v_1)m\} + F_{fa2}$$

$$F_{n2} = (EI/L^2)[12\{(u_1 - u_2)m + (v_2 - v_1)l\}/L + 6\theta_1 + 6\theta_2] + F_{fn2}$$

$$M_2 = (EI/L)[6\{(u_1 - u_2)m + (v_2 - v_1)l\}/L + 2\theta_1 + 4\theta_2] + M_{f2}$$

Expressing

$$F_{x1} = F_{a1}l - F_{n1}m, \quad F_{y1} = F_{a1}m + F_{n1}l \quad \text{at joint 1}$$

$$F_{x2} = F_{a2}l - F_{n2}m, \quad F_{y2} = F_{a2}m + F_{n2}l \quad \text{at joint 2}$$

we have

$$F_{x1} = (EI/L^2)[12\{(u_1 - u_2)m^2 + (v_2 - v_1)lm\}/L + 6\theta_1 m + 6\theta_2 m] - F_{fn1}m \\ + (AE/L)\{(u_1 - u_2)l^2 + (v_1 - v_2)lm\} + F_{fa1}l$$

$$F_{y1} = (EI/L^2)[12\{(u_2 - u_1)lm + (v_1 - v_2)l^2\}/L - 6\theta_1 l - 6\theta_2 l] + F_{fn1}l \\ + (AE/L)\{(u_1 - u_2)lm + (v_1 - v_2)m^2\} + F_{fa1}m$$

$$M_1 = (EI/L)[6\{(u_1 - u_2)m + (v_2 - v_1)l\}/L + 4\theta_1 + 2\theta_2] + M_{f1}$$

$$F_{x2} = (EI/L^2)[12\{(u_2 - u_1)m^2 + (v_1 - v_2)lm\}/L - 6\theta_1 m - 6\theta_2 m] - F_{fn2}m \\ + (AE/L)\{(u_2 - u_1)l^2 + (v_2 - v_1)lm\} + F_{fa2}l$$

$$F_{y2} = (EI/L^2)[12\{(u_1 - u_2)lm + (v_2 - v_1)l^2\}/L + 6\theta_1 l + 6\theta_2 l] + F_{fn2}l \\ + (AE/L)\{(u_2 - u_1)lm + (v_2 - v_1)m^2\} + F_{fa2}m$$

$$M_2 = (EI/L)[6\{(u_1 - u_2)m + (v_2 - v_1)l\}/L + 2\theta_1 + 4\theta_2] + M_{f2}$$

If shear deformation effects are to be included, then, as shown in Section 4.15, replace (12, 6, 4 and 2) by $\{12/(1 + \beta), 6/(1 + \beta), (4 + \beta)/(1 + \beta), (2 - \beta)/(1 + \beta)\}$ respectively, where $\beta = 12EI/(GA_sL^2)$ with the usual notation.

When the above equations are expressed in matrix form, the matrix relationship shown in equation [6.2] is obtained. It is useful to note that the stiffness matrix is symmetrical.

$$\begin{bmatrix} F_{x1} \\ F_{y1} \\ M_1 \\ F_{x2} \\ F_{y2} \\ M_2 \end{bmatrix} = \frac{EI}{L(1 + \beta)} \times$$

$$\begin{bmatrix} \frac{12}{L^2}(m^2 + \alpha l^2) & \frac{-12}{L^2}(1 - \alpha)lm & \frac{6m}{L} & \frac{-12}{L^2}(m^2 + \alpha l^2) & \frac{12}{L^2}(1 - \alpha)lm & \frac{6m}{L} \\ & \frac{12}{L^2}(l^2 + \alpha m^2) & \frac{-6l}{L} & \frac{12}{L^2}(1 - \alpha)lm & \frac{-12}{L^2}(l^2 + \alpha m^2) & \frac{-6l}{L} \\ & & 4 + \beta & \frac{-6m}{L} & \frac{6l}{L} & 2 - \beta \\ & & & \frac{12}{L^2}(m^2 + \alpha l^2) & \frac{-12}{L^2}(1 - \alpha)lm & \frac{-6m}{L} \\ & \text{symmetrical} & & & \frac{12}{L^2}(l^2 + \alpha m^2) & \frac{6}{L}l \\ & & & & & 4 + \beta \end{bmatrix}$$

$$\times \begin{bmatrix} u_1 \\ v_1 \\ \theta_1 \\ u_2 \\ v_2 \\ \theta_2 \end{bmatrix} + \begin{bmatrix} F_{fa1}l - F_{fn1}m \\ F_{fa1}m + F_{fn1}l \\ M_{f1} \\ F_{fa2}l - F_{fn2}m \\ F_{fa2}m + F_{fn2}l \\ M_{f2} \end{bmatrix} \qquad (6.2)$$

$$\beta = \frac{12\,EI}{GA_sL^2}, \quad \alpha = \frac{AL^2}{12I}(1 + \beta)$$

Plane grid element

Substituting

$$\psi_1 = \theta_{x1}l + \theta_{z1}n \text{ and } \theta_1 = -\theta_{x1}n + \theta_{z1}l \quad \text{at joint 1}$$

$$\psi_2 = \theta_{x2}l + \theta_{z2}n \text{ and } \theta_2 = -\theta_{x2}n + \theta_{z2}l \quad \text{at joint 2}$$

we have

$$F_{n1} = F_{y1} = (EI/L^2)[12(v_1 - v_2)/L - 6\{-(\theta_{x1} + \theta_{x2})n + (\theta_{z1} + \theta_{z2})l\}] + F_{fn1}$$

$$T_1 = (GJ/L)\{(\theta_{x1} - \theta_{x2})l + (\theta_{z1} - \theta_{z2})n\} + T_{f1}$$

$$M_1 = (EI/L)[4(-\theta_{x1}n + \theta_{z1}l) + 2(-\theta_{x2}n + \theta_{z2}l) - (6/L)(v_1 - v_2)] + M_{f1}$$

$$F_{n2} = F_{y2} = (EI/L^2)[12(v_2 - v_1)/L + 6\{-(\theta_{x1} + \theta_{x2})n + (\theta_{z1} + \theta_{z2})l\}] + F_{fn2}$$

$$T_2 = (GJ/L)\{(\theta_{x2} - \theta_{x1})l + (\theta_{z2} - \theta_{z1})n\} + T_{f2}$$

$$M_2 = (EI/L)[2(-\theta_{x1}n + \theta_{z1}l) + 4(-\theta_{x2}n + \theta_{z2}l) - (6/L)(v_1 - v_2)] + M_{f2}$$

Expressing

$$M_{x1} = T_1l - M_1n, \quad M_{y1} = T_1n - M_1l \quad \text{at joint 1}$$

$$M_{x2} = T_2l - M_2n, \quad M_{y2} = T_2n - M_2l \quad \text{at joint 2}$$

we have

$$F_{n1} = F_{y1} = (EI/L^2)[12(v_1 - v_2)/L - 6\{-(\theta_{x1} + \theta_{x2})n + (\theta_{z1} + \theta_{z2})l\}] + F_{fn1}$$

$$M_{x1} = (GJ/L)\{(\theta_{x1} - \theta_{x2})l^2 + (\theta_{z1} - \theta_{z2})ln\} + T_{f1}l - \left(\frac{EI}{L}\right)[4(-\theta_{x1}n^2 + \theta_{z1}ln)$$
$$+ 2(-\theta_{x2}n^2 + \theta_{z2}ln) - (6/L)(v_1 - v_2)n] - M_{f1}n$$

$$M_{y1} = (GJ/L)\{(\theta_{x1} - \theta_{x2})ln + (\theta_{z1} - \theta_{z2})n^2\} + T_{f1}n + \left(\frac{EI}{L}\right)[4(-\theta_{x1}ln + \theta_{z1}l^2)$$
$$+ 2(-\theta_{x2}ln + \theta_{z2}l^2) - (6/L)(v_1 - v_2)l] + M_{f1}l$$

$$F_{n2} = F_{y2} = \left(\frac{EI}{L^2}\right)[12(v_2 - v_1)/L + 6\{-(\theta_{x1} + \theta_{x2})n + (\theta_{z1} + \theta_{z2})l\}] + F_{fn2}$$

$$M_{x2} = (GJ/L)\{(\theta_{x2} - \theta_{x1})l^2 + (\theta_{z2} - \theta_{z1})ln\} + T_{f2}l - \left(\frac{EI}{L}\right)[2(-\theta_{x1}n^2 + \theta_{z1}ln)$$
$$+ 4(-\theta_{x2}n^2 + \theta_{z2}ln) - (6/L)(v_1 - v_2)n] - M_{f2}n$$

$$M_{y2} = (GJ/L)\{(\theta_{x2} - \theta_{x1})ln + (\theta_{z2} - \theta_{z1})n^2\} + T_{f2}n + \left(\frac{EI}{L}\right)[2(-\theta_{x1}ln + \theta_{z1}l^2)$$
$$+ 4(-\theta_{x2}ln + \theta_{z2}l^2) - (6/L)(v_1 - v_2)l] + M_{f2}l$$

If shear deformation effects are to be included, then, as shown in Section 4.15, replace (12, 6, 4 and 2) by $\{12/(1 + \beta), 6/(1 + \beta), (4 + \beta)/(1 + \beta), (2 - \beta)/(1 + \beta)\}$ respectively, where $\beta = 12EI/(GA_sL^2)$ with the usual notation. When the above equations are expressed in matrix form, the matrix relationship shown in equation [6.3] is obtained. It is useful to note that the stiffness matrix is symmetrical.

$$\begin{bmatrix} F_{y1} \\ M_{x1} \\ M_{z1} \\ F_{y2} \\ M_{x1} \\ M_{z2} \end{bmatrix} = \frac{EI}{L(1 + \beta)}$$

$$\times \begin{bmatrix} \dfrac{12}{L^2} & \dfrac{6n}{L} & \dfrac{-6l}{L} & \dfrac{-12}{L^2} & \dfrac{6n}{L} & \dfrac{-6l}{L} \\[2mm] & (4+\beta)n^2 + \alpha l^2 & -(4+\beta-\alpha)ln & \dfrac{-6n}{L} & (2-\beta)n^2 - \alpha l^2 & -(2-\beta+\alpha)ln \\[2mm] & & (4+\beta)l^2 + \alpha n^2 & \dfrac{6l}{L} & -(2-\beta+\alpha)ln & (2-\beta)l^2 - \alpha n^2 \\[2mm] & & & \dfrac{12}{L^2} & \dfrac{-6n}{L} & \dfrac{6l}{L} \\[2mm] & \text{symmetrical} & & & (4+\beta)n^2 + \alpha l^2 & -(4+\beta-\alpha)ln \\[2mm] & & & & & (4+\beta)l^2 + \alpha n^2 \end{bmatrix}$$

$$\times \begin{bmatrix} v_1 \\ \theta_{x1} \\ \theta_{z1} \\ v_2 \\ \theta_{x2} \\ \theta_{z2} \end{bmatrix} + \begin{bmatrix} F_{n1} \\ -M_{f1}n + T_{f1}l \\ M_{f1}l + T_{f1}n \\ F_{n2} \\ -M_{f2}n + T_{f2}l \\ M_{f2}l + T_{f2}n \end{bmatrix} \qquad (6.3)$$

$$\beta = \frac{12EI}{GA_sL^2}, \ \alpha \times 10^3 = \frac{GJ}{EI}(1 + \beta)$$

6.5 ● Direction cosines

The direction cosines l and m of a member are best calculated in terms of the coordinates of the ends of the member. If (x_1, y_1, z_1) and (x_2, y_2, z_2) are the coordinates of end 1 and end 2 of the member as shown in Fig. 5.1, then

$$l = (x_2 - x_1)/L, \ m = (y_2 - y_1)/L, \text{ and } n = (z_2 - z_1)/L$$

where L = length of the member. It is perfectly arbitrary as to which end of a member is chosen as end 1 and which as end 2. However, once a choice is made we have to stick to it throughout the analysis.

6.6 ● Equations of equilibrium for the whole structure – structural stiffness matrix for a pin-jointed structure

As a simple example to develop the concept of the structural stiffness matrix, consider the 2-D pin-jointed truss shown in Fig. 6.6(a). Joints 1 and 3 are fully restrained. Joints 2 and 4 can displace both horizontally and vertically. Let the unknown joint displacements at joint 2 be $u = d_1$ and $v = d_2$. Similarly, at joint 4, $u = d_3$ and $v = d_4$. Thus there are four unknown joint displacements to be determined. This information is shown in Fig. 6.6(b) and in Table 6.1.

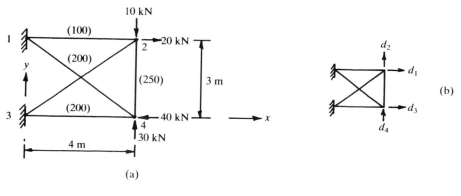

Figure 6.6 ● (a), (b) A statically indeterminate cantilever pin-jointed truss

Table 6.1

Joint	Joint displacements	
	u	v
1	0	0
2	d_1	d_2
3	0	0
4	d_3	d_4

The four equations of equilibrium which have to be satisfied are Σ horizontal forces = 0, and Σ vertical forces = 0 at joints 2 and 4. It should be noted that there is no need to consider equilibrium in the direction of reaction forces, because if each member is in equilibrium, then the supports will provide the necessary forces at the supported end of the members to ensure their equilibrium. Therefore, if the unknown displacements are determined, then the forces in each bar of the truss can be calculated.

The equations of equilibrium at each joint are written in terms of the forces in the bars meeting at the joint and the external load acting at that joint. The contributions from the members to the equations of equilibrium are best expressed in terms of displacements as given by the stiffness matrix for the member.

Table 6.2

Member	End 1	End 2	x_1	y_1	x_2	y_2	L	l	m
1–2	1	2	0	3	4	3	4	1.0	0
1–4	1	4	0	3	4	0	5	0.8	−0.6
2–3	2	3	4	3	0	0	5	−0.8	−0.6
3–4	3	4	0	0	4	0	4	1.0	0
4–2	4	2	4	0	4	3	3	0	1.0

The stiffness matrix for the member is a function of area A, length L, Young's modulus E and the direction cosines l and m. Table 6.2 can be used to calculate the direction cosines for the five members of the truss.

Table 6.3

Member	Joint number at		Displacements			
	End 1	End 2	u_1	v_1	u_2	v_2
1–2	1	2	0	0	d_1	d_2
1–4	1	4	0	0	d_3	d_4
2–3	2	3	d_1	d_2	0	0
3–4	3	4	0	0	d_3	d_4
4–2	4	2	d_3	d_4	d_1	d_2

Note: The designation of the members follows the convention joint number at end 1 followed by joint number at end 2. The designation is purely arbitrary but once a particular choice is made, it must be adhered to rigidly.

It is also necessary to identify for each member the end displacements (u_1, v_1) and (u_2, v_2) in terms of the known and unknown joint displacements (see Table 6.3).

Element stiffness calculation

The stiffness matrices for each element can now be calculated. Because there is no applied external load on the element, the fixed end force vector is zero and therefore it is not shown. The stiffness relationships for the various elements of the structure are as follows.

Member 1–2. $E = 210 \times 10^3$ N mm^{-2}, $L = 4000$ mm, $A = 100$ mm^2, $l = 1$, $m = 0$. Equation [6.1] becomes:

$$\begin{bmatrix} F_{x1} = \text{reaction at joint 1} \\ F_{y1} = \text{reaction at joint 1} \\ F_{x2} = F_1 \\ F_{y2} = F_2 \end{bmatrix} = \begin{bmatrix} a_{11} & a_{12} & a_{13} & a_{14} \\ a_{21} & a_{22} & a_{23} & a_{24} \\ a_{31} & a_{32} & a_{33} & a_{34} \\ a_{41} & a_{42} & a_{43} & a_{44} \end{bmatrix} \begin{bmatrix} u_1 = 0 \\ v_1 = 0 \\ u_2 = d_1 \\ v_2 = d_2 \end{bmatrix}$$

$$= 10^3 \begin{bmatrix} 5.25 & 0 & -5.25 & 0 \\ 0 & 0 & 0 & 0 \\ -5.25 & 0 & 5.25 & 0 \\ 0 & 0 & 0 & 0 \end{bmatrix} \begin{bmatrix} 0 \\ 0 \\ d_1 \\ d_2 \end{bmatrix}$$

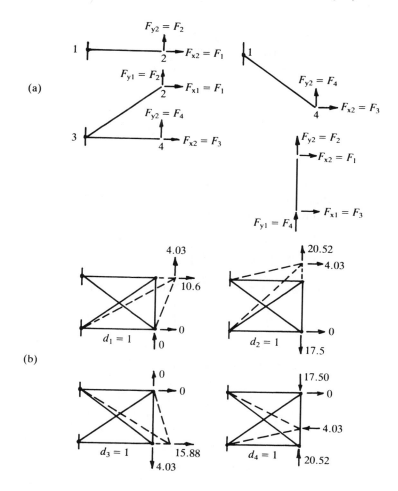

Figure 6.7 ● (a), (b) Interpretation of stiffness coefficients for the cantilever pin-jointed truss

Note that F_{x2} and F_{y2} act in the directions of displacements d_1 and d_2 respectively as shown in Fig. 6.7(a). Hence they are designated F_1 and F_2 respectively. The subscripts 1 and 2 simply indicate that these forces act in the direction of displacements d_1 and d_2 respectively.

The coefficients a_{ij} are numerical values depending on the values of A, L, E, l and m of the element. They are obtained by substituting the appropriate values in the stiffness matrix shown in equation [6.1].

From the above matrix, we have

$$\begin{bmatrix} F_1 \\ F_2 \end{bmatrix} = \begin{bmatrix} a_{33} & a_{34} \\ a_{43} & a_{44} \end{bmatrix} \begin{bmatrix} d_1 \\ d_2 \end{bmatrix} = 10^3 \begin{bmatrix} 5.25 & 0 \\ 0 & 0 \end{bmatrix} \begin{bmatrix} d_1 \\ d_2 \end{bmatrix}$$

Similarly, we can calculate the stiffness matrices for the other four members of the truss as follows.

Member 1–4. $E = 210 \times 10^3 \,\text{N mm}^{-2}$, $L = 5000 \,\text{mm}$, $A = 200 \,\text{mm}^2$, $l = 0.8$, $m = -0.6$

$$\begin{bmatrix} F_{x1} = \text{reaction at joint 1} \\ F_{y1} = \text{reaction at joint 1} \\ F_{x2} = F_3 \\ F_{y2} = F_4 \end{bmatrix} = \begin{bmatrix} b_{11} & b_{12} & b_{13} & b_{14} \\ b_{21} & b_{22} & b_{23} & b_{24} \\ b_{31} & b_{32} & b_{33} & b_{34} \\ b_{41} & b_{42} & b_{43} & b_{44} \end{bmatrix} \begin{bmatrix} u_1 = 0 \\ v_1 = 0 \\ u_2 = d_3 \\ v_2 = d_4 \end{bmatrix}$$

$$= 10^3 \begin{bmatrix} 5.38 & -4.03 & -5.38 & 4.03 \\ -4.03 & 3.02 & 4.03 & -3.02 \\ -5.38 & 4.03 & 5.38 & -4.03 \\ 4.03 & -3.02 & -4.03 & 3.02 \end{bmatrix} \begin{bmatrix} 0 \\ 0 \\ d_3 \\ d_4 \end{bmatrix}$$

Note the designation of forces as F_3 and F_4 to indicate that these forces act in the directions of displacement d_3 and d_4 respectively, as shown in Fig. 6.7(a).

From the above matrix we have

$$\begin{bmatrix} F_3 \\ F_4 \end{bmatrix} = \begin{bmatrix} b_{33} & b_{34} \\ b_{43} & b_{44} \end{bmatrix} \begin{bmatrix} d_3 \\ d_4 \end{bmatrix} = 10^3 \begin{bmatrix} 5.38 & -4.03 \\ -4.03 & 3.02 \end{bmatrix} \begin{bmatrix} d_3 \\ d_4 \end{bmatrix}$$

Member 2–3. $E = 210 \times 10^3$ N mm^{-2}, $L = 5000$ mm, $A = 200$ mm^2, $l = -0.8$, $m = -0.6$

$$\begin{bmatrix} F_{x1} = F_1 \\ F_{y1} = F_2 \\ F_{x2} = \text{reaction at joint 3} \\ F_{y2} = \text{reaction at joint 3} \end{bmatrix} = \begin{bmatrix} c_{11} & c_{12} & c_{13} & c_{14} \\ c_{21} & c_{22} & c_{23} & c_{24} \\ c_{31} & c_{32} & c_{33} & c_{34} \\ c_{41} & c_{42} & c_{43} & c_{44} \end{bmatrix} \begin{bmatrix} u_1 = d_1 \\ v_1 = d_2 \\ u_2 = 0 \\ v_2 = 0 \end{bmatrix}$$

$$= 10^3 \begin{bmatrix} 5.38 & 4.03 & -5.38 & -4.03 \\ 4.03 & 3.02 & -4.03 & -3.02 \\ -5.38 & -4.03 & 5.38 & 4.03 \\ -4.03 & -3.02 & 4.03 & 3.02 \end{bmatrix} \begin{bmatrix} d_1 \\ d_2 \\ 0 \\ 0 \end{bmatrix}$$

Note the designation of forces at F_1 and F_2 to indicate that these forces act in the directions of displacement d_1 and d_2 respectively, as shown in Fig. 6.7(a).

From the above matrix we have

$$\begin{bmatrix} F_1 \\ F_2 \end{bmatrix} = \begin{bmatrix} c_{11} & c_{12} \\ c_{21} & c_{22} \end{bmatrix} \begin{bmatrix} d_1 \\ d_2 \end{bmatrix} = 10^3 \begin{bmatrix} 5.38 & 4.03 \\ 4.03 & 3.02 \end{bmatrix} \begin{bmatrix} d_1 \\ d_2 \end{bmatrix}$$

Member 3–4. $E = 210 \times 10^3$ N mm^{-2}, $L = 4000$ mm, $A = 200$ mm^2, $l = 1$, $m = 0$

$$\begin{bmatrix} F_{x1} = \text{reaction at joint 3} \\ F_{y1} = \text{reaction at joint 3} \\ F_{x2} = F_3 \\ F_{y2} = F_4 \end{bmatrix} = \begin{bmatrix} e_{11} & e_{12} & e_{13} & e_{14} \\ e_{21} & e_{22} & e_{23} & e_{24} \\ e_{31} & e_{32} & e_{33} & e_{34} \\ e_{41} & e_{42} & e_{43} & e_{44} \end{bmatrix} \begin{bmatrix} u_1 = 0 \\ v_1 = 0 \\ u_2 = d_3 \\ v_2 = d_4 \end{bmatrix}$$

$$= 10^3 \begin{bmatrix} 10.5 & 0 & -10.5 & 0 \\ 0 & 0 & 0 & 0 \\ -10.5 & 0 & 10.5 & 0 \\ 0 & 0 & 0 & 0 \end{bmatrix} \begin{bmatrix} 0 \\ 0 \\ d_3 \\ d_4 \end{bmatrix}$$

From the above matrix we have

$$\begin{bmatrix} F_3 \\ F_4 \end{bmatrix} = \begin{bmatrix} e_{33} & e_{34} \\ e_{43} & e_{44} \end{bmatrix} \begin{bmatrix} d_3 \\ d_4 \end{bmatrix} = 10^3 \begin{bmatrix} 10.5 & 0 \\ 0 & 0 \end{bmatrix} \begin{bmatrix} d_3 \\ d_4 \end{bmatrix}$$

Member 4–2. $E = 210 \times 10^3$ N mm^{-2}, $L = 3000$ mm, $A = 250$ mm^2, $l = 0$, $m = 1$

$$
\begin{bmatrix} F_{x1} = F_3 \\ F_{y1} = F_4 \\ F_{x2} = F_1 \\ F_{y2} = F_2 \end{bmatrix} = \begin{bmatrix} f_{11} & f_{12} & f_{13} & f_{14} \\ f_{21} & f_{22} & f_{23} & f_{24} \\ f_{31} & f_{32} & f_{33} & f_{34} \\ f_{41} & f_{42} & f_{43} & f_{44} \end{bmatrix} \begin{bmatrix} u_1 = d_3 \\ v_1 = d_4 \\ u_2 = d_1 \\ v_2 = d_2 \end{bmatrix}
$$

$$
= 10^3 \begin{bmatrix} 0 & 0 & 0 & 0 \\ 0 & 17.5 & 0 & -17.5 \\ 0 & 0 & 0 & 0 \\ 0 & -17.5 & 0 & 17.5 \end{bmatrix} \begin{bmatrix} d_3 \\ d_4 \\ d_1 \\ d_2 \end{bmatrix}
$$

From the above equation we have

$$
\begin{bmatrix} F_3 \\ F_4 \\ F_1 \\ F_2 \end{bmatrix} = \begin{bmatrix} f_{11} & f_{12} & f_{13} & f_{14} \\ f_{21} & f_{22} & f_{23} & f_{24} \\ f_{31} & f_{32} & f_{33} & f_{34} \\ f_{41} & f_{42} & f_{43} & f_{44} \end{bmatrix} \begin{bmatrix} d_3 \\ d_4 \\ d_1 \\ d_2 \end{bmatrix} = 10^3 \begin{bmatrix} 0 & 0 & 0 & 0 \\ 0 & 17.5 & 0 & -17.5 \\ 0 & 0 & 0 & 0 \\ 0 & -17.5 & 0 & 17.5 \end{bmatrix} \begin{bmatrix} d_3 \\ d_4 \\ d_1 \\ d_2 \end{bmatrix}
$$

Formation of equations of equilibrium

The equations of equilibrium to be satisfied are

$$
-\sum F_{xji} + W_{xi} = 0 \quad \text{and} \quad -\sum F_{yji} + W_{yi} = 0
$$

at joint 2 and joint 4.

Consider the equilibrium equation in the x-direction at joint 2. $W_{xi} = 20$ and the members meeting at joint 2 are 1–2, 2–3 and 4–2. The x-direction at joint 2 is the same as the direction of displacement d_1. Therefore equation $-\Sigma F_{xji} + W_{xi} = 0$ can be expressed as

$$
W_{xi} = \sum F_{xji}
$$

Therefore

$$
20 = \sum F_1
$$

$$
= \{a_{33}d_1 + a_{34}d_2\} \text{ from member } 1\text{–}2 + \{c_{11}d_1 + c_{12}d_2\} \text{ from member } 2\text{–}3
$$
$$
+ \{f_{31}d_3 + f_{32}d_4 + f_{33}d_1 + f_{34}d_2\} \text{ from member } 4\text{–}2
$$

$$
= \{(a_{33} + c_{11} + f_{33})d_1 + (a_{34} + c_{12} + f_{34})d_2 + f_{31}d_3 + f_{32}d_4\}
$$

Similarly, since the y-direction at joint 2 is the same as the direction of displacement d_2, we have

$$
W_{yi} = \sum F_{yji}
$$

Therefore

$$
-10 = \sum F_2
$$

$$
= \{a_{43}d_1 + a_{44}d_2\} \text{ from member } 1\text{–}2 + \{c_{21}d_1 + c_{22}d_2\} \text{ from member } 2\text{–}3
$$
$$
+ \{f_{41}d_3 + f_{42}d_4 + f_{43}d_1 + f_{44}d_2\} \text{ from member } 4\text{–}2
$$

$$
= \{(a_{43} + c_{21} + f_{43})d_1 + (a_{44} + c_{22} + f_{44})d_2 + f_{41}d_3 + f_{42}d_4\}
$$

Note that $W_{yi} = -10$ because the load of 10 acts opposite to the positive direction of d_2.

Similarly, by considering the equation of equilibrium in the x- and y-directions at joint 4 we have

$$-40 = \sum F_3$$

$$= \{b_{33}d_3 + b_{34}d_4\} \text{ from member 1–4} + \{e_{33}d_3 + e_{34}d_4\} \text{ from member 3–4}$$
$$+ \{f_{11}d_3 + f_{12}d_4 + f_{13}d_1 + f_{14}d_2\} \text{ from member 4–2}$$

$$= \{f_{13}d_1 + f_{14}d_2 + (b_{33} + e_{33} + f_{11})d_3 + (b_{34} + e_{34} + f_{12})d_4\}$$

$$30 = \sum F_4$$

$$= \{b_{43}d_3 + b_{44}d_4\} \text{ from member 1–4} + \{e_{43}d_3 + e_{44}d_4\} \text{ from member 3–4}$$
$$+ \{f_{21}d_3 + f_{22}d_4 + f_{23}d_1 + f_{24}d_2\} \text{ from member 4–2}$$

$$= \{f_{23}d_1 + f_{24}d_2 + \{b_{43} + e_{43} + f_{21})d_3 + (b_{44} + e_{44} + f_{22})d_4\}$$

The above four equations can be expressed in matrix form as

$$
\begin{bmatrix} \sum F_1 \\ \sum F_2 \\ \sum F_3 \\ \sum F_4 \end{bmatrix} = \begin{bmatrix} 20 \\ -10 \\ -40 \\ 30 \end{bmatrix}
$$

$$
= \begin{bmatrix}
a_{33} + c_{11} + f_{33} & a_{34} + c_{12} + f_{34} & f_{31} & f_{32} \\
a_{43} + c_{21} + f_{43} & a_{44} + c_{22} + f_{44} & f_{41} & f_{42} \\
f_{13} & f_{14} & b_{33} + e_{33} + f_{11} & b_{34} + e_{34} + f_{12} \\
f_{23} & f_{24} & b_{43} + e_{43} + f_{21} & b_{44} + e_{44} + f_{22}
\end{bmatrix}
\begin{bmatrix} d_1 \\ d_2 \\ d_3 \\ d_4 \end{bmatrix}
$$

Therefore

$$
\begin{bmatrix} 20 \\ -10 \\ -40 \\ 30 \end{bmatrix} = 10^3 \begin{bmatrix}
5.25 + 5.38 & 4.03 & 0 & 0 \\
4.03 & 3.02 + 17.5 & 0 & -17.5 \\
0 & 0 & 5.38 + 10.5 & -4.03 \\
0 & -17.5 & -4.03 & 3.02 + 17.5
\end{bmatrix}
\begin{bmatrix} d_1 \\ d_2 \\ d_3 \\ d_4 \end{bmatrix}
$$

Therefore

$$
\begin{bmatrix} 20 \\ -10 \\ -40 \\ 30 \end{bmatrix} = 10^3 \begin{bmatrix}
10.63 & 4.03 & 0 & 0 \\
4.03 & 20.52 & 0 & -17.50 \\
0 & 0 & 15.88 & -4.03 \\
0 & -17.50 & -4.03 & 20.52
\end{bmatrix}
\begin{bmatrix} d_1 \\ d_2 \\ d_3 \\ d_4 \end{bmatrix}
$$

The matrix connecting the vector of applied forces at the joints and the vector of the corresponding joint displacements is known as the structural stiffness matrix. Like the element stiffness matrices from which it is assembled, the structural stiffness matrix is also symmetrical.

It is useful to have an understanding of what the elements of the structural stiffness matrix mean. Referring to Fig. 6.7(b), if joint 2 is given a unit displacement in the direction of displacement d_1 but keeping all other displacements equal to zero, then the forces developed in the direction of the displacements d_1 to d_4 represent the first column of the structural stiffness matrix. During the displacement pattern $d_1 = 1$ and $d_2 = d_3 = d_4 = 0$, members 1–4 and 3–4 do not deform and 4–2 does not suffer any extension (it is assumed that the displacements are 'small'). That is why the restraining forces developed in the direction of displacements d_3 and d_4 are zero. Members 1–2 and 2–3 extend and ΣF_1 is the sum of the horizontal components of the forces in these two members and ΣF_2 is the vertical component of the tensile force in member 2–3.

In a similar manner, by displacing joint 2 vertically by unity and calculating the forces developed in the direction of the displacements d_1 to d_4, we obtain the second column of the stiffness matrix.

The reader should carefully study the displacement patterns in Fig. 6.7(b) and understand the physical significance of the elements of the structural stiffness matrix.

6.7 ● Determination of displacements at the joints of the structure – Gaussian elimination procedure

The applied loads at the joints of the structure and the corresponding displacements are connected, as shown in the previous section, by the structural stiffness matrix. The equations of equilibrium as given by the structural stiffness relationship can be solved to determine the unknown joint displacements. The usual procedure adopted for solving displacements is the Gaussian elimination procedure. In this procedure, the given set of equations is modified in a systematic way such that, using the current example, the first equation has all four unknowns, the second equation has only three unknowns, d_2, d_3 and d_4, the third equation has only two unknowns, d_3 and d_4, and finally the last equation has only one unknown, d_4. This step is called forward elimination. Because the last equation has only one unknown d_4, it can be determined. Once d_4 is known, then d_3 is determined from the third equation. Once d_4 and d_3 are known d_2 can be determined from the second equation and finally d_1 is determined from the first equation. This step is called back substitution.

The method is applied to the present problem.

Forward elimination

(a) Given the set of equations:

$$10^3 \begin{bmatrix} 10.63 & 4.03 & 0 & 0 \\ 4.03 & 20.52 & 0 & -17.50 \\ 0 & 0 & 15.88 & -4.03 \\ 0 & -17.50 & -4.03 & 20.52 \end{bmatrix} \begin{bmatrix} d_1 \\ d_2 \\ d_3 \\ d_4 \end{bmatrix} = \begin{bmatrix} 20 \\ -10 \\ -40 \\ 10 \end{bmatrix}$$

(b) Eliminate d_1 from equations 2 to 4.

    (i) To eliminate d_1 from the second equation, multiply the first equation by $-(4.03/10.63) = -0.379$ and add it to the second equation. Therefore

$$
\begin{array}{llllll}
+ & -0.379 & \{10.63 & 4.03 & 0 & 0 & 20.0\} & \text{equation 1} \\
& & \{ 4.03 & 20.52 & 0 & -17.50 & -10.0\} & \text{equation 2} \\
\hline
= & & \{ 0 & 18.99 & 0 & -17.50 & -17.58\} & \text{modified equation 2}
\end{array}
$$

(ii) d_1 is not involved in equations 3 and 4 so they do not require altering. Therefore at the end of this stage the given set of equations has been modified to

$$10^3 \begin{bmatrix} 10.63 & 4.03 & 0 & 0 \\ 0 & 18.99 & 0 & -17.50 \\ 0 & 0 & 15.88 & -4.03 \\ 0 & -17.50 & -4.03 & 20.52 \end{bmatrix} \begin{bmatrix} d_1 \\ d_2 \\ d_3 \\ d_4 \end{bmatrix} = \begin{bmatrix} 20.0 \\ -17.58 \\ -40.0 \\ 30.0 \end{bmatrix}$$

(c) Eliminate d_2 from equations 3 to 4.

(i) Because equation 3 is free from d_2, it need not be altered.

(ii) Eliminate d_2 from equation 4. Multiply equation 2 by $-(17.50/18.99) = 0.922$ and add it to the fourth equation. Therefore

$$\begin{array}{cccccc} _+ & 0.922 & \{0 & 18.99 & 0 & -17.50 & -17.58\} & \text{equation 2} \\ & & \{0 & -17.50 & -4.03 & 20.52 & 30.0\} & \text{equation 4} \end{array}$$

$$= \quad \{0 \quad 0 \quad -4.03 \quad 4.39 \quad 13.80\} \text{ modified equation 4}$$

Therefore at the end of this stage the given set of equations has been modified to

$$10^3 \begin{bmatrix} 10.63 & 4.03 & 0 & 0 \\ 0 & 18.99 & 0 & -17.50 \\ 0 & 0 & 15.88 & -4.03 \\ 0 & 0 & -4.03 & 4.39 \end{bmatrix} \begin{bmatrix} d_1 \\ d_2 \\ d_3 \\ d_4 \end{bmatrix} = \begin{bmatrix} 20.0 \\ -17.58 \\ -40.0 \\ 13.80 \end{bmatrix}$$

(d) Eliminate d_3 from equation 4.

(i) Multiply equation 3 by $-(-4.03/15.88) = 0.254$ and add it to the fourth equation. Therefore

$$\begin{array}{cccccc} _+ & 0.254 & \{0 & 0 & 15.88 & -4.03 & -40.0\} & \text{equation 3} \\ & & \{0 & 0 & -4.03 & 4.39 & 13.80\} & \text{equation 4} \end{array}$$

$$= \quad \{0 \quad 0 \quad 0 \quad 3.37 \quad 3.65\} \text{ modified equation 4}$$

Therefore at the end of this stage the given set of equations has been modified to

$$10^3 \begin{bmatrix} 10.63 & 4.03 & 0 & 0 \\ 0 & 18.99 & 0 & -17.50 \\ 0 & 0 & 15.88 & -4.03 \\ 0 & 0 & 0 & 3.37 \end{bmatrix} \begin{bmatrix} d_1 \\ d_2 \\ d_3 \\ d_4 \end{bmatrix} = \begin{bmatrix} 20.0 \\ -17.58 \\ -40.0 \\ 3.65 \end{bmatrix}$$

This completes the forward elimination process.

Back substitution

The unknowns can now be determined by working backwards from the last equation. This stage is called back substitution.

Thus from the last equation

$$d_4 = 3.65/(3.37 \times 10^3) = 1.0828 \times 10^{-3}$$

From the third equation

$$d_3 = \{-40.0 + 4.03 \times 10^3 d_4\}/\{15.88 \times 10^3) = -2.2441 \times 10^{-3}$$

From the second equation

$$d_2 = \{-17.58 + 17.50 \times 10^3 d_4 + 0 d_3\}/(18.99 \times 10^3) = 0.0721 \times 10^{-3}$$

From the first equation

$$d_1 = \{20.0 + 0 d_4 \times 0 d_3 - 4.03 \times 10^3 d_2\}/(10.63 \times 10^3) = 1.8541 \times 10^{-3}$$

6.8 ● Determination of axial forces in the members from joint displacements

Once the displacements of all the joints are known, then the axial force F in each member can be determined from the equation

$$F = (AE/L)\{(u_2 - u_1)l + (v_2 - v_1)m\}$$

From Table 6.4, the values of u_1, v_1, u_2 and v_2 can be determined for each member.

Table 6.4

Member	u_1	v_1	u_2	v_2	l	m
1–2	0	0	1.8541×10^{-3}	0.0721×10^{-3}	1.0	0
1–4	0	0	-2.2441×10^{-3}	1.0828×10^{-3}	0.8	−0.6
2–3	1.8541×10^{-3}	0.0721×10^{-3}	0	0	−0.8	−0.6
3–4	0	0	-2.2441×10^{-3}	1.0828×10^{-3}	1.0	0
4–2	-2.2441×10^{-3}	1.0828×10^{-3}	1.8541×10^{-3}	0.0721×10^{-3}	0	1.0

Table 6.5 ● Forces in members

Members	F
1–2	9.73
1–4	−20.54
2–3	12.83
3–4	−23.56
4–2	−17.69

It is interesting to note that although this structure is one degree statically indeterminate, as can be checked from the fact that the number of joints $j = 4$, number of members $m = (5 + wall) = 6$, giving $2j - 3 = 5$, this fact was never required in the solution procedure. In other words, the solution procedure applies to both statically determinate and indeterminate structures and has the additional merit that both the forces in the members and the displacements of the joints are determined.

6.9 ● Automatic assembly of structural stiffness matrix

It is clear from the way the equations of equilibrium at the joints are established that the procedure consists simply of adding the appropriate contribution from each member to the particular equation of equilibrium. Because the method is meant to be programmed for a computer, it is useful to establish a systematic procedure for establishing the equations of equilibrium.

Consider the stiffness matrix for the member 4–2. The displacements at the ends are given by $u_1 = d_3$, $v_1 = d_4$, $u_2 = d_1$ and $v_2 = d_2$. Let the freedom number of the displacements at the ends be stored in the vector MEMDIS. The name of the vector MEMDIS stands for MEMber end DISplacements. For element 4–2 the elements of MEMDIS are given by

MEMDIS(1) = 3 because $u_1 = d_3$

MEMDIS(2) = 4 because $v_1 = d_4$

MEMDIS(3) = 1 because $u_2 = d_1$

MEMDIS(4) = 2 because $v_2 = d_2$

The stiffness relationship for the element 4–2 is given by

$$\begin{bmatrix} F_3 \\ F_4 \\ F_1 \\ F_2 \end{bmatrix} = \begin{bmatrix} f_{11} & f_{12} & f_{13} & f_{14} \\ f_{21} & f_{22} & f_{23} & f_{24} \\ f_{31} & f_{32} & f_{33} & f_{34} \\ f_{41} & f_{42} & f_{43} & f_{44} \end{bmatrix} \begin{bmatrix} d_3 = \text{MEMDIS}(1) \\ d_4 = \text{MEMDIS}(2) \\ d_1 = \text{MEMDIS}(3) \\ d_2 = \text{MEMDIS}(4) \end{bmatrix}$$

Let the *Element STIF*fness matrix be matrix ESTIF and the structural or *Global* stiffness matrix be GSTIF. Obviously element f_{11} contributes to the location (3, 3) of the structural stiffness matrix as can be verified from an examination of the structural stiffness matrix. Location (3, 3) is the same as location {MEMDIS(1), MEMDIS(1)}. Similarly, element f_{32} contributes to the location (1, 4) of the structural stiffness matrix. Location (1, 4) is the same as location {MEMDIS(3), MEMDIS(2)}. It can be easily verified by taking further elements that the rule

Element ESTIF(i, j) of the member stiffness matrix is added to the element GSTIF{MEMDIS(i), MEMDIS(j)} of the structural stiffness matrix GSTIF

is true for all valid values of i and j.

As another example of the use of the MEMDIS vector, consider the stiffness matrix for member 1–2. For this element $u_1 = 0$, $v_1 = 0$, $u_2 = d_1$ and $v_2 = d_2$. Therefore

MEMDIS(1) = 0 because $u_1 = 0$

MEMDIS(2) = 0 because $v_1 = 0$

MEMDIS(3) = 1 because $u_2 = d_1$

MEMDIS(4) = 2 because $v_2 = d_2$

The stiffness relationship for the element is given by

$$\begin{bmatrix} F_{x1} = \text{reaction at joint 1} \\ F_{y1} = \text{reaction at joint 1} \\ F_{x2} = F_1 \\ F_{y1} = F_2 \end{bmatrix} = \begin{bmatrix} a_{11} & a_{12} & a_{13} & a_{14} \\ a_{21} & a_{22} & a_{23} & a_{24} \\ a_{31} & a_{32} & a_{33} & a_{34} \\ a_{41} & a_{42} & a_{43} & a_{44} \end{bmatrix} \begin{bmatrix} u_1 = 0 = \text{MEMDIS}(1) \\ v_1 = 0 = \text{MEMDIS}(2) \\ u_2 = d_1 = \text{MEMDIS}(3) \\ v_2 = d_2 = \text{MEMDIS}(4) \end{bmatrix}$$

Consider element a_{31}. Because $F_1 = a_{31}(0) + a_{32}(0) + a_{33}d_1 + a_{34}d_2 = (a_{33}d_1 + a_{34}d_2)$, the element a_{31} does not figure in the structural stiffness matrix. According to the rule element a_{31} is added to the structural stiffness matrix at the location $\{\text{MEMDIS}(3), \text{MEMDIS}(1)\} = (1, 0)$. Since the location $(1, 0)$ has no meaning, element a_{31} is simply ignored.

6.10 ● Fixed end forces and joint load vector

It has already been pointed out in Section 6.1 that the stiffness relationship **for an element** is given by

$$\{\text{Force vector}\} = \{\text{Stiffness matrix}\}\{\text{Displacement vector}\}$$
$$+ \{\text{Fixed end force vector}\}$$

When considering the equilibrium of the joints in a 2-D pin-jointed truss, the equation to be satisfied is

$$W_{xi} - \sum F_{xji} = 0, \text{ and } W_{yi} - \sum F_{yji} = 0$$

where the summation sign Σ applies to all members j joining at the ith joint. However, F_{xji} and F_{yji} can be expressed as the sum of two parts. The first part is due to displacements and the second part is due to fixed end force. For example, expressing F_{xji} as

$$F_{xji} = F_{xji} \text{ due to displacement} + F_{xji} \text{ due to fixed end force}$$

the equation of equilibrium can be written as

$$W_{xi} - \sum \{F_{xji} \text{ due to displacement} + F_{xji} \text{ due to fixed end force}\} = 0$$

Simplifying,

$$W_{xi} - \sum F_{xji} \text{ due to fixed end force} = \sum F_{xji} \text{ due to displacement}$$

Similarly,

$$W_{yi} - \sum F_{yji} \text{ due to fixed end force} = \sum F_{yji} \text{ due to displacement}$$

Thus the force vector in the stiffness relationship **for the structure** consists of contributions from two sources – the external forces applied at the joints and the fixed forces arising from external load on the element or due to restraining thermal deformations. For a degree of freedom at a joint, the contribution to the load vector from the loads applied at the joint consists of a single term. However, the contribution to the load vector from the fixed end forces is contributed by all the elements meeting at that joint.

6.10.1 Analysis of pin-jointed truss for thermal loads

As was pointed out in Chapter 1, Section 1.9, in the case of statically determinate trusses, temperature changes will cause only displacements of the joints. On the other hand, if the truss is statically indeterminate, then both forces and joint displacements are caused. The analysis of a pin-jointed truss subjected to ambient temperature changes also shows how to calculate the joint load vector from fixed end forces. These ideas are illustrated by a simple example.

Example 1

Analyse the structure shown in Fig. 6.6 when the members of the truss are subjected to a temperature rise of 20 °C. Assume that the coefficient of linear thermal expansion $\alpha_t = 10 \times 10^{-6} \, °C^{-1}$.

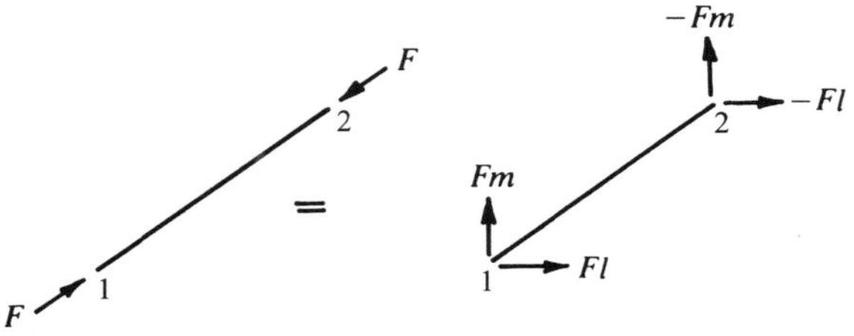

Figure 6.8 ● Forces at the joints for preventing thermal expansion

Solution As shown in Fig. 6.8, the compressive force F required to restrain thermal expansion of $\alpha_t TL$ is

$$F = (AE/L)\alpha_t TL = AE\alpha_t T$$

where L = length of the member, T = rise in temperature. Therefore the fixed end forces in the axial directions are

$$F_{fa1} = AE\alpha_t T \text{ and } F_{fa2} = -AE\alpha_t T$$

Resolving these forces in the x- and y-directions, the fixed end force vector is given by

$$\begin{bmatrix} F_{x1} \\ F_{y1} \\ F_{x2} \\ F_{y2} \end{bmatrix} = AE\alpha_t T \begin{bmatrix} l \\ m \\ -l \\ -m \end{bmatrix}$$

The fixed end force vector for all the members of the truss and the directions in which they act can be calculated (see Fig. 6.9(a)) as shown in Table 6.6.

Note: In Table 6.6 the directions in which the fixed end forces act are shown next to the value of the force in brackets. An asterisk indicates that the force acts in the direction of a reaction.

Figure 6.9(a) shows the fixed end forces in the individual members. Summing up the forces ΣF_i in the direction of displacements d_i, we have

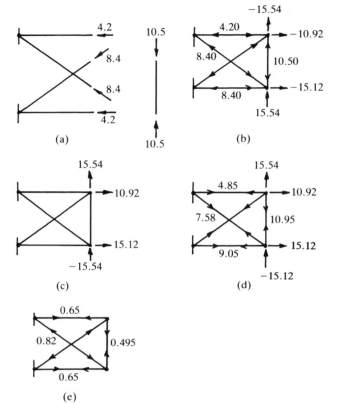

Figure 6.9 ● (a), (b) Forces in a pin-jointed statically indeterminate truss due to temperature changes

Table 6.6

	Member				
	1–2	1–4	2–3	3–4	4–2
F	−4.2	−8.40	−8.40	−8.40	−10.5
F_{x1}	4.2(*)	6.72(*)	−6.72(d_1)	8.40(*)	0(d_3)
F_{y1}	0(*)	−5.04(*)	−5.04(d_2)	0(*)	10.5(d_4)
F_{x2}	−4.2(d_1)	−6.72(d_3)	6.72(*)	−8.40(d_3)	0(d_1)
F_{y2}	0(d_2)	5.04(d_4)	5.04(*)	0(d_4)	−10.5(d_2)

$$\sum F_1 = -4.20 \text{ from } 1\text{--}2 + (-6.72) \text{ from } 2\text{--}3 + 0 \text{ from } 4\text{--}2 = -10.92$$

$$\sum F_2 = 0 \text{ from } 1\text{--}2 + (-5.04) \text{ from } 2\text{--}3 + (-10.5) \text{ from } 4\text{--}2 = -15.54$$

$$\sum F_3 = -6.72 \text{ from } 1\text{--}4 + (-8.40) \text{ from } 3\text{--}4 + 0 \text{ from } 4\text{--}2 = -15.12$$

$$\sum F_4 = 5.04 \text{ from } 1\text{--}4 + 0 \text{ from } 3\text{--}4 + 10.5 \text{ from } 4\text{--}2 = 15.54$$

These fixed end forces, shown in Fig. 6.9(b), will prevent extension of the members and hence the displacements of joints.

Because in the actual structures the restraint forces do not exist, they have to be removed. When they are removed, joints displace. The displacement of the joints can be calculated for the forces removed which are simply opposite to the resultant fixed end forces at the joints calculated before. The forces causing displacements are as shown in Fig. 6.9(c):

$$\sum F_1 = -(-10.92) = 10.92, \quad \sum F_2 = -(-15.54) = 15.54$$

$$\sum F_3 = -(-15.12) = 15.12, \quad \sum F_4 = -(15.54) = -15.54$$

The stiffness matrix of the structure is a property of the structure and does not depend on the loads acting on it and therefore it remains the same as before. The equations to be solved for determining the displacements are

$$10^3 \begin{bmatrix} 10.63 & 4.03 & 0 & 0 \\ 4.03 & 20.52 & 0 & -17.50 \\ 0 & 0 & 15.88 & -4.03 \\ 0 & -17.50 & -4.03 & 20.52 \end{bmatrix} \begin{bmatrix} d_1 \\ d_2 \\ d_3 \\ d_4 \end{bmatrix} = \begin{bmatrix} 10.92 \\ 15.54 \\ 15.12 \\ -15.54 \end{bmatrix}$$

When Gaussian elimination is carried out on the above set of equations, the equation becomes

$$10^3 \begin{bmatrix} 10.63 & 4.03 & 0 & 0 \\ 0 & 18.99 & 0 & -17.50 \\ 0 & 0 & 15.88 & -4.03 \\ 0 & 0 & 0 & 3.37 \end{bmatrix} \begin{bmatrix} d_1 \\ d_2 \\ d_3 \\ d_4 \end{bmatrix} = \begin{bmatrix} 10.92 \\ 11.40 \\ 15.12 \\ -1.20 \end{bmatrix}$$

Note: The structural stiffness matrix goes through the same changes as before. The load vector changes as follows:

$$\begin{bmatrix} 10.92 \\ 15.54 \\ 15.12 \\ -15.54 \end{bmatrix} \rightarrow \begin{bmatrix} 10.92 \\ 11.40 \\ 15.12 \\ -15.54 \end{bmatrix} \rightarrow \begin{bmatrix} 10.92 \\ 11.40 \\ 15.12 \\ -5.04 \end{bmatrix} \rightarrow \begin{bmatrix} 10.92 \\ 11.40 \\ 15.12 \\ -1.20 \end{bmatrix}$$

Note: The major effort involved in the solution of a set of equilibrium equations is the forward elimination of the structural stiffness matrix. When analysing a structure under many sets of loads, the forward elimination of the structural stiffness matrix needs to be done once only. On the other hand, the forward elimination of the load vector has to be done as many times as there are new load cases to analyse. When writing computer programs it is useful to keep the forward elimination of the structural stiffness matrix and the load vector as two separate and independent operations.

The displacements are determined by backward substitution as follows:

$$d_4 = -1.20/(3.37 \times 10^3) = -0.3555 \times 10^{-3}$$

$$d_3 = \{15.12 + 4.03 \times 10^3 d_4\}/(15.88 \times 10^3) = 0.8619 \times 10^{-3}$$

$$d_2 = \{11.40 + 17.5 \times 10^3 d_4 + 0d_3\}/(18.99 \times 10^3) = 0.2728 \times 10^{-3}$$

$$d_1 = \{10.92 + 0d_4 + 0d_3 - 4.03 \times 10^3 d_2\}/(10.63 \times 10^3) = 0.9239 \times 10^{-3}$$

The forces due to displacement are determined using the formula

$$F = (AE/L)\{(u_2 - u_1)l + (v_2 - v_1)m\}$$

and the information in Table 6.7.

Table 6.7

Member	u_1	v_1	u_2	v_2	l	m
1–2	0	0	0.9239×10^{-3}	0.2728×10^{-3}	1.0	0
1–4	0	0	0.8619×10^{-3}	-0.3555×10^{-3}	0.8	−0.6
2–3	0.9239×10^{-3}	0.2728×10^{-3}	0	0	−0.8	−0.6
3–4	0	0	0.8619×10^{-3}	-0.3555×10^{-3}	1.0	0
4–2	0.8619×10^{-3}	-0.3555×10^{-3}	0.9239×10^{-3}	0.2728×10^{-3}	0	1.0

The forces in the members arising from the displacements of the joints are shown in Fig. 6.9(d). The final force in the member shown in Fig. 6.9(e) is the sum of that due to the fixed end force shown in Fig. 6.9(b) and the forces due to displacements which are shown in Fig. 6.9(d). The reason why the force in a member due to fixed end force and that due to displacement are different is that the fixed end force is dependent on the properties of individual members. On the other hand, when the restraint to expansion is removed, allowing the joints to displace, then the forces at the joints are shared by the members. The final forces are shown in Table 6.8.

Table 6.8 ● Forces in members

Members	F displacement	F fixed end	F final
1–2	4.85	−4.2	0.65
1–4	7.58	−8.4	−0.82
2–3	7.58	−8.4	−0.82
3–4	9.05	−8.4	0.65
4–2	11.00	−10.5	0.50

6.11 ● Automatic assembly of the load vector

It was shown in the previous example that the fixed end forces from many members contribute towards the load vector. The ideas developed for the automatic assembly of the structural stiffness matrix can be used for the assembly of the load vector as well. For any member the ith component of the fixed end load vector is added to the MEMDIS(i)th component of the structural or Global LOAD vector GLOAD.

6.12 ● Analysis of 2-D rigid-jointed structures

The analysis of rigid-jointed structures follows the same steps as the analysis of pin-jointed structures. In fact this is one of the great advantages of the stiffness method because computer programs written for the analysis of one type of structure can be easily adapted for the analysis of another type of structure. The analysis of rigid-jointed structures is demonstrated by a simple example.

Example 1

Analyse the 2-D rigid-jointed structure shown in Fig. 6.10. The cross-sectional properties of the beams and columns are as follows:

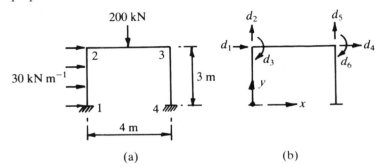

Figure 6.10 ● (a), (b)
Forces and deformations
of a portal frame

Beam: $I = 116.86 \times 10^{-6}$ m⁴, $A = 6830 \times 10^{-6}$ m², area shear factor = 4.056.
Columns: $I = 55.44 \times 10^{-6}$ m⁴, $A = 4740 \times 10^{-6}$ m², area shear factor = 4.116,
$E = 210 \times 10^{6}$ kN m⁻², $G = 80.77 \times 10^{6}$ kN m⁻², $\alpha_t =$ coefficient of linear thermal
expansion = 11×10^{-6} °C⁻¹.
There are two loading cases.
External loads as shown in Fig. 6.10(a).
Temperature rise of 20°C on the outside of members 1–2 and 2–3 only.

Solution **(a)** At any joint in a 2-D rigid-jointed structure there are three possible independent displacements: a translation in the x-direction, a translation in the y-direc-
tion and a rotation about the z-axis. Because joints 1 and 4 are fully fixed, all
three possible displacements are zero. At joints 2 and 3, the three displacements
are respectively (d_1, d_2, d_3) and (d_4, d_5, d_6) as shown in Fig. 6.10(b). Note that
the word 'displacement' is used in a general sense to include both translations
and rotations. Thus the structure has three degrees of freedom at each joint as
shown in Table 6.9.

Table 6.9

Node	u	v	θ
1	0	0	0
2	d_1	d_2	d_3
3	d_4	d_5	d_6
4	0	0	0

The direction cosines and the displacements at the ends of the members are
shown in Table 6.10.

Table 6.10

Member	Joint at End 1	End 2	x_1	y_1	x_2	y_2	L	l	m	u_1	v_1	θ_1	u_2	v_2	θ_2
1–2	1	2	0	0	0	3	3	0	1	0	0	0	d_1	d_2	d_3
2–3	2	3	0	3	4	3	4	1	0	d_1	d_2	d_3	d_4	d_5	d_6
4–3	4	3	4	0	4	3	3	0	1	0	0	0	d_4	d_5	d_6

(b) Calculate the element stiffness matrices

Element 1–2. Column: $I = 55.44 \times 10^{-6}$, $A = 4740 \times 10^{-6}$, area shear factor $= 4.116$, therefore $A_s = 1152 \times 10^{-6}$, $\beta = 12EI/(GA_s^2) = 0.167$, $L = 3$, $l = 0$, $m = 1$. Substituting these values in the matrix in equation [6.2] the element stiffness matrix is given by

$$
\begin{bmatrix} F_{x1} = * \\ F_{y1} = * \\ M_1 = * \\ F_{x2} = F_1 \\ F_{y2} = F_2 \\ M_2 = F_3 \end{bmatrix} = 10^3 \begin{bmatrix} 4.43 & 0 & 6.65 & -4.43 & 0 & 6.65 \\ 0 & 331.80 & 0 & 0 & -331.80 & 0 \\ 6.65 & 0 & 13.86 & -6.65 & 0 & 6.10 \\ -4.43 & 0 & -6.65 & 4.43 & 0 & -6.65 \\ 0 & -331.80 & 0 & 0 & 331.80 & 0 \\ 6.65 & 0 & 6.10 & -6.65 & 0 & 13.86 \end{bmatrix}
$$

$$
\times \begin{bmatrix} u_1 = 0 \\ v_1 = 0 \\ \theta_1 = 0 \\ u_2 = d_1 \\ v_2 = d_2 \\ \theta_2 = d_3 \end{bmatrix}
$$

(*Note:* * indicates that the force acts in the direction of a reaction.) From which

$$
\begin{bmatrix} F_1 \\ F_2 \\ F_3 \end{bmatrix} = 10^3 \begin{bmatrix} 4.43 & 0 & -6.65 \\ 0 & 331.80 & 0 \\ -6.65 & 0 & 13.86 \end{bmatrix} \begin{bmatrix} d_1 \\ d_2 \\ d_3 \end{bmatrix}
$$

Element 2–3. Beam: $I = 116.86 \times 10^{-6}$, $A = 6830 \times 10^{-6}$, area shear factor $= 4.056$. Therefore $A_s = 1684 \times 10^{-6}$, $\beta = 12EI/(GA_s^2) = 0.135$, $L = 4$, $l = 1$, $m = 0$. Substituting these values in the matrix in equation [6.2], the element stiffness matrix is given by

$$
\begin{bmatrix} F_{x1} = F_1 \\ F_{y1} = F_2 \\ M_1 = F_3 \\ F_{x2} = F_4 \\ F_{y2} = F_5 \\ M_2 = F_6 \end{bmatrix} = 10^3 \begin{bmatrix} 358.58 & 0 & 0 & -358.58 & 0 & 0 \\ 0 & 4.05 & -8.11 & 0 & -4.05 & -8.11 \\ 0 & -8.11 & 22.35 & 0 & 8.11 & 10.08 \\ -358.58 & 0 & 0 & 358.58 & 0 & 0 \\ 0 & -4.05 & 8.11 & 0 & 4.05 & 8.11 \\ 0 & -8.11 & 10.08 & 0 & 8.11 & 22.35 \end{bmatrix}
$$

$$
\times \begin{bmatrix} u_1 = d_1 \\ v_1 = d_2 \\ \theta_1 = d_3 \\ u_2 = d_4 \\ v_2 = d_5 \\ \theta_2 = d_6 \end{bmatrix}
$$

From which

$$
\begin{bmatrix} F_1 \\ F_2 \\ F_3 \\ F_4 \\ F_5 \\ F_6 \end{bmatrix} = 10^3 \begin{bmatrix} 358.58 & 0 & 0 & -358.58 & 0 & 0 \\ 0 & 4.05 & -8.11 & 0 & -4.05 & -8.11 \\ 0 & -8.11 & 22.35 & 0 & 8.11 & 10.08 \\ -358.58 & 0 & 0 & 358.58 & 0 & 0 \\ 0 & -4.05 & 8.11 & 0 & 4.05 & 8.11 \\ 0 & -8.11 & 10.08 & 0 & 8.11 & 22.35 \end{bmatrix} \begin{bmatrix} d_1 \\ d_2 \\ d_3 \\ d_4 \\ d_5 \\ d_6 \end{bmatrix}
$$

Element 4–3. Properties as for member 1–2.

$$
\begin{bmatrix} F_{x1} = * \\ F_{y1} = * \\ M_1 = * \\ F_{x2} = F_4 \\ F_{y2} = F_5 \\ M_2 = F_6 \end{bmatrix} = 10^3 \begin{bmatrix} 4.43 & 0 & 6.65 & -4.43 & 0 & 6.65 \\ 0 & 331.80 & 0 & 0 & -331.80 & 0 \\ 6.65 & 0 & 13.86 & -6.65 & 0 & 6.10 \\ -4.43 & 0 & -6.65 & 4.43 & 0 & -6.65 \\ 0 & -331.80 & 0 & 0 & 331.80 & 0 \\ 6.65 & 0 & 6.10 & -6.65 & 0 & 13.86 \end{bmatrix} \begin{bmatrix} 0 \\ 0 \\ 0 \\ d_4 \\ d_5 \\ d_6 \end{bmatrix}
$$

(*Note:* * indicates that the force acts in the direction of a reaction.) From which

$$
\begin{bmatrix} F_4 \\ F_5 \\ F_6 \end{bmatrix} = 10^3 \begin{bmatrix} 4.43 & 0 & -6.65 \\ 0 & 331.80 & 0 \\ -6.65 & 0 & 13.86 \end{bmatrix} \begin{bmatrix} d_4 \\ d_5 \\ d_6 \end{bmatrix}
$$

(c) Assemble the structural stiffness matrix

$$
\begin{bmatrix} \Sigma F_1 \\ \Sigma F_2 \\ \Sigma F_3 \\ \Sigma F_4 \\ \Sigma F_5 \\ \Sigma F_6 \end{bmatrix} = 10^3
\begin{bmatrix}
\begin{array}{c} 358.58 \\ \underline{4.43} \\ 363.01 \end{array} &
\begin{array}{c} 0 \\ - \\ 0 \end{array} &
\begin{array}{c} 0 \\ \underline{-6.65} \\ -6.65 \end{array} &
358.58 & 0 & 0 \\
0 &
\begin{array}{c} 4.05 \\ \underline{331.80} \\ 335.85 \end{array} &
\begin{array}{c} -8.11 \\ 0 \\ -8.11 \end{array} &
0 & -4.05 & -8.11 \\
0 &
\begin{array}{c} -8.11 \\ \underline{0} \\ -8.11 \end{array} &
\begin{array}{c} 22.35 \\ \underline{13.86} \\ 36.21 \end{array} &
0 & 8.11 & 10.08 \\
& & \text{(above with } -6.65) & & & \\
\end{bmatrix}
\begin{bmatrix} d_1 \\ d_2 \\ d_3 \\ d_4 \\ d_5 \\ d_6 \end{bmatrix}
$$

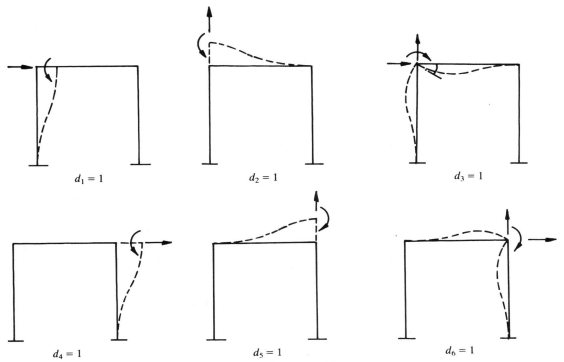

Figure 6.11 ● Independent deformation patterns for a portal frame

Figure 6.11 shows the deformation patterns of the six independent displacements which give rise to the forces recorded in each column of the structural stiffness matrix.

(d) Fixed end forces due to external loading on elements. Consider a member as shown in Fig. 6.12, loaded by a u.d.l. of q_x and q_y per unit length in the x- and y-directions respectively. The fixed end reactions in the x- and y-directions at ends 1 and 2 are

$$F_{x1} = F_{x2} = -0.5q_x L_y, \qquad F_{y1} = F_{y2} = -0.5q_y L_x$$

Figure 6.12 ● Fixed end forces due to uniformly distributed load in the x- and y-directions

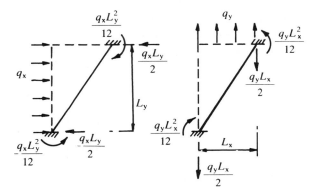

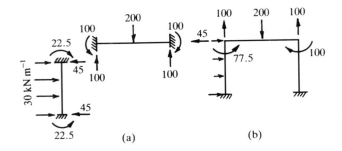

Figure 6.13 ● (a), (b)
Restraining forces on a
portal frame

Table 6.11

Forces	Members		
	1–2	2–3	4–3
F_{x1}	−45.0(*)	0(d_1)	0(*)
F_{y1}	0(*)	100.0(d_2)	0(*)
M_1	−22.5(*)	−100.0(d_3)	0(*)
F_{x2}	−45.0(d_1)	0(d_4)	0(d_4)
F_{y2}	0(d_2)	100.0(d_5)	0(d_5)
M_2	22.5(d_3)	100.0(d_6)	0(d_6)

The fixed end moments are

$$M_1 = -q_x L_y^2/12 + q_y L_x^2/12$$
$$M_2 = q_x L_y^2/12 - q_y L_x^2/12$$

If instead of a u.d.l. we have midspan concentrated loads of W_x and W_y in the x- and y-directions respectively, then the fixed end forces are

$$F_{x1} = F_{x2} = -0.5W_x, \quad F_{y1} = F_{y2} = -0.5W_y$$

The fixed end moments are

$$M_1 = -W_x L_y/8 + W_y L_x/8, \quad M_2 = W_x L_y/8 - W_y L_x/8$$

For the present problem, the fixed end forces as shown in Fig. 6.13(a) are given in Table 6.11. In the table next to the value of the forces, the direction of the force is shown in brackets by the direction of the displacement. * indicates it is the direction of a reaction.

Summing up the forces in the directions of displacements, the net fixed end forces are

$$\sum F_1 = -45.0 \text{ from member } 1 + 0 \text{ from member } 2 = -45.0$$

$$\sum F_2 = 0 \text{ from member } 1 + 100.0 \text{ from member } 2 = 100.0$$

$$\sum F_3 = 22.5 \text{ from member } 1 - 100.0 \text{ from member } 2 = -77.5$$

$$\sum F_4 = 0 \text{ from member } 2 + 0 \text{ from member } 3 = 0$$

$$\sum F_5 = 100.0 \text{ from member } 2 + 0 \text{ from member } 3 = 100.0$$

$$\sum F_6 = 100.0 \text{ from member } 2 + 0 \text{ from member } 3 = 100.0$$

The joint forces are shown in Fig. 6.13(b).

(e) Solve the equations of equilibrium to determine the displacements of the joints caused by *removing* the restraining forces at the joints.

The equations to be solved in the case of external loads are

$$
10^3
\begin{bmatrix}
363.01 & 0 & -6.65 & -358.58 & 0 & 0 \\
0 & 335.85 & -8.11 & 0 & -4.05 & -8.11 \\
-6.65 & -8.11 & 36.21 & 0 & 8.11 & 10.08 \\
-358.58 & 0 & 0 & 363.01 & 0 & -6.65 \\
0 & -4.05 & 8.11 & 0 & 335.85 & 8.11 \\
0 & -8.11 & 10.08 & -6.65 & 8.11 & 36.21
\end{bmatrix}
\times
\begin{bmatrix}
d_1 \\ d_2 \\ d_3 \\ d_4 \\ d_5 \\ d_6
\end{bmatrix}
$$

$$
=
\begin{bmatrix}
45.0 \\ -100.0 \\ 77.5 \\ 0 \\ -100.0 \\ -100.0
\end{bmatrix}
$$

After Gaussian forward elimination, the equations become

$$
10^3
\begin{bmatrix}
363.01 & 0 & -6.65 & -358.58 & 0 & 0 \\
0 & 335.85 & -8.11 & 0 & -4.05 & -8.11 \\
0 & 0 & 35.89 & -6.57 & 8.01 & 9.88 \\
0 & 0 & 0 & 7.61 & 1.47 & -4.84 \\
0 & 0 & 0 & 0 & 333.73 & 6.74 \\
0 & 0 & 0 & 0 & 0 & 30.08
\end{bmatrix}
\times
\begin{bmatrix}
d_1 \\ d_2 \\ d_3 \\ d_4 \\ d_5 \\ d_6
\end{bmatrix}
$$

$$
=
\begin{bmatrix}
45.0 \\ -100.0 \\ 75.91 \\ 58.35 \\ -129.40 \\ -83.58
\end{bmatrix}
$$

The displacements are: $d_1 = 6.0907 \times 10^{-3}$, $d_2 = -0.2712 \times 10^{-3}$, $d_3 = 4.0459 \times 10^{-3}$, $d_4 = 5.9654 \times 10^{-3}$, $d_5 = -0.3316 \times 10^{-3}$, $d_6 = -2.7789 \times 10^{-3}$.

Note: The net axial extensions Δ of the members are

Member 1–2: $\Delta = d_2 = -0.2712 \times 10^{-3}$

Member 2–3: $\Delta = d_4 - d_1 = -0.1253 \times 10^{-3}$

Member 4–3: $\Delta = d_5 = -0.3316 \times 10^{-3}$

These extensions are an order of magnitude less than other displacements. Therefore, in manual calculation procedures, it is usually assumed that the members of a rigid-jointed structure are axially inextensible.

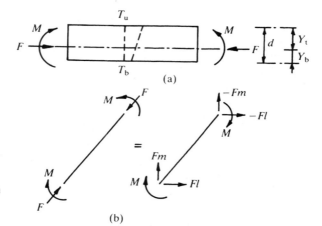

Figure 6.14 ● (a), (b)
Fixed end forces due to
temperature change in a
beam element

(f) Fixed end forces due to thermal restraint.

As shown in Fig. 6.14(a), let a uniform member be subjected to a temperature rise of T_t and T_b at the top and bottom surfaces respectively. Let the temperature variation be linear through the depth. Let the forces required to prevent expansion be an axial force F and a bending moment M. The total strain ε_x at the top and bottom faces must be zero for zero expansion. Therefore

At top face: $\varepsilon_x = \alpha_t T_t - F/(AE) - My_t/(EI) = 0$

At bottom face: $\varepsilon_x = \alpha_t T_b - F/(AE) + My_b/(EI) = 0$

where y_t and y_b are the distance to the top and bottom faces from the neutral axis. Solving

$$M = EI\,\alpha_t(T_t - T_b)/d$$

$$F = AE\,\alpha_t(T_t y_b + T_b y_t)/d$$

where d = depth of beam = $y_t + y_b$. The fixed end forces as shown in Fig. 6.14(b) are therefore given by $F_{x1} = Fl$, $F_{y1} = Fm$, $F_{x2} = -Fl$, $F_{y2} = Fm$, $M_1 = M$, $M_2 = -M$.

In the present example,

Member 1–2: $T_t = 10$, $T_b = 0$, $d = 0.254$, $y_t = y_b = 0.5d$, $F = 54.75$, $M = 5.04$.

Member 2–3: $T_t = 10$, $T_b = 0$, $d = 0.305$, $y_t = y_b = 0.5d$, $F = 78.99$, $M = 8.85$.

Member 4–3: No change in temperature. Therefore $F = M = 0$. (See Table 6.12.) Summing up the forces in the directions of displacements, we have

$$\sum F_1 = 0 \text{ from member 1} + 78.99 \text{ from member 2} = 78.99$$

$$\sum F_2 = -54.75 \text{ from member 1} + 0 \text{ from member 2} = -54.75$$

$$\sum F_3 = -5.04 \text{ from member 1} + 8.85 \text{ from member 2} = 3.81$$

$$\sum F_4 = -78.99 \text{ from member 2} + 0 \text{ from member 3} = -78.99$$

Table 6.12

Forces	Members		
	1–2	2–3	4–3
F_{x1}	0(*)	78.99(d_1)	0(*)
F_{y1}	54.75(*)	0(d_2)	0(*)
M_1	5.04(*)	8.85(d_3)	0(*)
F_{x2}	0(d_1)	−78.99(d_4)	0(d_4)
F_{y2}	−54.75(d_2)	0(d_5)	0(d_5)
M_2	−5.04(d_3)	−8.85(d_6)	0(d_6)

$$\sum F_5 = 0 \text{ from member } 2 + 0 \text{ from member } 3 = 0$$

$$\sum F_6 = -8.85 \text{ from member } 2 + 0 \text{ from member } 3 = -8.85$$

(g) Solve the equations of equilibrium to determine the displacements of the joints caused by *removing* the restraining forces at the joints.

The equations to be solved in the case of temperature changes are

$$10^3 \begin{bmatrix} 363.01 & 0 & -6.65 & -358.58 & 0 & 0 \\ 0 & 335.85 & -8.11 & 0 & -4.05 & -8.11 \\ -6.65 & -8.11 & 36.21 & 0 & 8.11 & 10.08 \\ -358.58 & 0 & 0 & 363.01 & 0 & -6.65 \\ 0 & -4.05 & 8.11 & 0 & 335.85 & 8.11 \\ 0 & -8.11 & 10.08 & -6.65 & 8.11 & 36.21 \end{bmatrix} \times \begin{bmatrix} d_1 \\ d_2 \\ d_3 \\ d_4 \\ d_5 \\ d_6 \end{bmatrix}$$

$$= \begin{bmatrix} -78.99 \\ 54.75 \\ -3.81 \\ 78.99 \\ 0 \\ 8.85 \end{bmatrix}$$

After Gaussian forward elimination, they become

$$10^3 \begin{bmatrix} 363.01 & 0 & -6.65 & -358.58 & 0 & 0 \\ 0 & 335.85 & -8.11 & 0 & -4.05 & -8.11 \\ 0 & 0 & 35.89 & -6.57 & 8.01 & 9.88 \\ 0 & 0 & 0 & 7.61 & 1.47 & -4.84 \\ 0 & 0 & 0 & 0 & 333.73 & 6.74 \\ 0 & 0 & 0 & 0 & 0 & 30.08 \end{bmatrix} \times \begin{bmatrix} d_1 \\ d_2 \\ d_3 \\ d_4 \\ d_5 \\ d_6 \end{bmatrix} = \begin{bmatrix} -78.99 \\ 54.75 \\ -3.94 \\ 0.24 \\ 1.49 \\ 11.38 \end{bmatrix}$$

The displacements are: $d_1 = 4.9574 \times 10^{-5}$, $d_2 = 0.1682 \times 10^{-3}$, $d_3 = -0.1633 \times 10^{-3}$, $d_4 = 0.2735 \times 10^{-3}$, $d_5 = -3.1712 \times 10^{-6}$, $d_6 = 0.3786 \times 10^{-3}$.

(h) Forces induced due to displacements. The axial force F, the shear force Q and bending moments M_1 and M_2 caused by the displacements are

$$F = (AE/L)\{(u_2 - u_1)l + (v_2 - v_1)m\}$$

From the stiffness matrix

$$M_1 = \{EI/(1 + \beta)\}[(4 + \beta)\theta_1 + (2 - \beta)\theta_2 + (6/L)\{(v_2 - v_1)l - (u_2 - u_1)m\}]$$

$$M_2 = \{EI/(1 + \beta)\}[(2 - \beta)\theta_1 + (4 + \beta)\theta_2 + (6/L)\{(v_2 - v_1)l - (u_2 - u_1)m\}]$$

From equilibrium

$$F_{n1} = -(M_1 + M_2)/L, \quad Q = F_{n2} = -F_{n1}$$

For the two cases of loading considered, the forces due to displacements are:

(i) External loading case. Using the displacements calculated in (e) above, the forces due to displacements and the fixed end forces calculated in (d) are shown in Table 6.13.

Table 6.13 ● External loading case

Forces	Members					
	1–2		2–3		4–3	
	Disp.	Fixed	Disp.	Fixed	Disp.	Fixed
F	−89.98	0	−44.93	0	−110.03	0
M_1	−15.85	−22.5	61.92	−100.0	−56.61	0
M_2	15.55	22.5	−21.81	100.0	−78.17	0
F_{n1}	0.10	45.0	−10.03	100.0	44.93	0
F_{n2}	−0.10	45.0	10.03	100.0	−44.93	0

(j) Temperature change. Using the displacements calculated in (g) above, the forces due to displacements and the thermal fixed end forces calculated in (f) are shown in Table 6.14.

Table 6.14 ● Thermal change case

Forces	Members					
	1–2		2–3		4–3	
	Disp.	Fixed	Disp.	Fixed	Disp.	Fixed
F	55.81	−54.75	80.29	−78.99	−1.05	0
M_1	−1.33	5.04	−1.22	8.85	0.49	0
M_2	−2.59	−5.04	5.43	−8.85	3.43	0
F_{n1}	1.30	0	−1.05	0	−1.31	0
F_{n2}	−1.30	0	1.05	0	1.31	0

(k) Final forces in the members. This is simply the sum of fixed end forces and the forces due to displacement caused by the removal of restrains. Table 6.15 gives the final forces.

(l) The final bending moment and shear force diagrams can be drawn as shown in Fig. 6.15.

Table 6.15 ●
Final forces

Forces	Members					
	1–2		2–3		4–3	
	External	Temp.	External	Temp.	External	Temp.
F	−89.98	1.06	−44.93	1.30	−110.03	−1.05
M_1	−38.35	3.71	−38.08	7.63	−56.61	0.49
M_2	38.05	−7.63	78.19	−3.42	−78.17	3.43
F_{n1}	45.10	1.30	89.97	−1.05	44.93	−1.31
F_{n2}	44.90	−1.30	110.03	1.05	−44.93	1.31

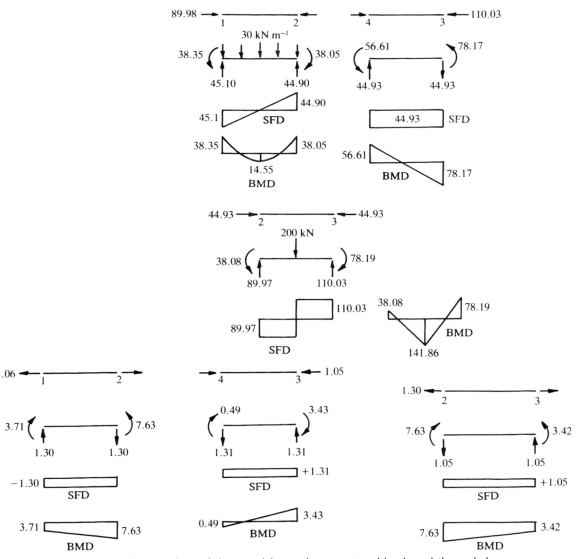

Figure 6.15 ● Forces in the members of the portal frame due to external loads and thermal changes

6.13 ● Analysis of plane grids

The analysis of plane grids using the stiffness method is demonstrated by a simple example.

Example 1

Analyse the plane grid structure shown in Fig. 6.16(a), where $E = 14 \times 10^6$ kN m^{-2}, $G = 6.36 \times 10^6$ kN m^{-2}, $I = 66.67 \times 10^{-6}$ m^4, $J = 46.67 \times 10^{-6}$ m^4, $A = 20 \times 10^{-3}$ m^2, area shear factor = 1.2, $L = 3$ m, $\beta = 12EI/(GA_sL^2) = 0.0117$.

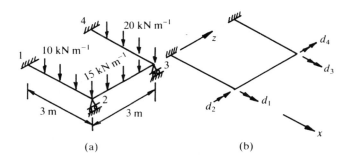

Figure 6.16 ● (a), (b)
A plane grid

(a)   (b)

Solution Joints 1 and 4 are fully restrained and at joints 2 and 3 vertical displacement is restrained. Therefore rotations about the x- and z-axes are possible at joints 2 and 3. These are designated d_1 to d_4 as shown in Fig. 6.16(b).

Table 6.16 shows the nodal displacements.

Table 6.16

Node	v	θ_x	θ_z
1	0	0	0
2	0	d_1	d_2
3	0	d_3	d_4
4	0	0	0

Table 6.17 shows the information required for establishing the element stiffness matrices.

Table 6.17

Element	x_1	z_1	x_2	z_2	L	l	n	v_1	θ_{x1}	θ_{z1}	v_2	θ_{x2}	θ_{z2}
1–2	0	0	3	0	3	1	0	0	0	0	0	d_1	d_2
2–3	3	0	3	3	3	0	1	0	d_1	d_2	0	d_3	d_4
4–3	0	3	3	3	3	1	0	0	0	0	0	d_3	d_4

(a) Calculate element stiffness matrices. Using the member stiffness matrix shown in equation [6.3], we have

Element 1–2:

$$
\begin{bmatrix} F_{y1} = * \\ M_{x1} = * \\ M_{z1} = * \\ F_{y2} = * \\ M_{x2} = F_1 \\ M_{z2} = F_2 \end{bmatrix} = 10^2 \begin{bmatrix} 4.10 & 0 & -6.15 & -4.10 & 0 & -6.15 \\ 0 & 0.99 & 0 & 0 & -0.99 & 0 \\ -6.15 & 0 & 12.34 & 6.15 & 0 & 6.12 \\ -4.10 & 0 & 6.15 & 4.10 & 0 & 6.15 \\ 0 & -0.99 & 0 & 0 & 0.99 & 0 \\ -6.15 & 0 & 6.12 & 6.15 & 0 & 12.34 \end{bmatrix}
$$

$$
\times \begin{bmatrix} v_1 = 0 \\ \theta_{x1} = 0 \\ \theta_{z2} = 0 \\ v_2 = 0 \\ \theta_{x2} = d_1 \\ \theta_{z2} = d_2 \end{bmatrix}
$$

From which

$$
\begin{bmatrix} F_1 \\ F_2 \end{bmatrix} = 10^2 \begin{bmatrix} 0.99 & 0 \\ 0 & 12.34 \end{bmatrix} \begin{bmatrix} d_1 \\ d_2 \end{bmatrix}
$$

Element 2–3:

$$
\begin{bmatrix} F_{y1} = * \\ M_{x1} = F_1 \\ M_{z1} = F_2 \\ F_{y2} = * \\ M_{x2} = F_3 \\ M_{z2} = F_4 \end{bmatrix} = 10^2 \begin{bmatrix} 4.10 & 6.15 & 0 & -4.10 & 6.15 & 0 \\ 6.15 & 12.34 & 0 & -6.15 & 6.12 & 0 \\ 0 & 0 & 0.99 & 0 & 0 & -0.99 \\ -4.10 & -6.15 & 0 & 4.10 & -6.15 & 0 \\ 6.15 & 6.12 & 0 & -6.15 & 12.34 & 0 \\ 0 & 0 & -0.99 & 0 & 0 & 0.99 \end{bmatrix}
$$

$$
\times \begin{bmatrix} v_1 = 0 \\ \theta_{x1} = d_1 \\ \theta_{z2} = d_2 \\ v_2 = 0 \\ \theta_{x2} = d_3 \\ \theta_{z2} = d_4 \end{bmatrix}
$$

From which

$$
\begin{bmatrix} F_1 \\ F_2 \\ F_3 \\ F_4 \end{bmatrix} = 10^2 \begin{bmatrix} 12.34 & 0 & 6.12 & 0 \\ 0 & 0.99 & 0 & -0.99 \\ 6.12 & 0 & 12.34 & 0 \\ 0 & -0.99 & 0 & 0.99 \end{bmatrix} \begin{bmatrix} d_1 \\ d_2 \\ d_3 \\ d_4 \end{bmatrix}
$$

Element 4–3:

$$\begin{bmatrix} F_{y1} = * \\ M_{x1} = * \\ M_{z1} = * \\ F_{y2} = * \\ M_{x2} = F_3 \\ M_{z2} = F_4 \end{bmatrix} = 10^2 \begin{bmatrix} 4.10 & 0 & -6.15 & -4.10 & 0 & -6.15 \\ 0 & 0.99 & 0 & 0 & -0.99 & 0 \\ -6.15 & 0 & 12.34 & 6.15 & 0 & 6.12 \\ -4.10 & 0 & 6.15 & 4.10 & 0 & 6.15 \\ 0 & -0.99 & 0 & 0 & 0.99 & 0 \\ -6.15 & 0 & 6.12 & 6.15 & 0 & 12.34 \end{bmatrix}$$

$$\times \begin{bmatrix} v_1 = 0 \\ \theta_{x1} = 0 \\ \theta_{z2} = 0 \\ v_2 = 0 \\ \theta_{x2} = d_3 \\ \theta_{z2} = d_4 \end{bmatrix}$$

From which

$$\begin{bmatrix} F_3 \\ F_4 \end{bmatrix} = 10^2 \begin{bmatrix} 0.99 & 0 \\ 0 & 12.34 \end{bmatrix} \begin{bmatrix} d_3 \\ d_4 \end{bmatrix}$$

(b) Assemble the structural stiffness matrix:

$$\begin{bmatrix} \Sigma F_1 \\ \\ \\ \Sigma F_2 \\ \\ \\ \Sigma F_3 \\ \\ \\ \Sigma F_4 \end{bmatrix} = 10^2 \begin{bmatrix} 12.34 & 0 & 6.12 & 0 \\ \underline{0.99} & 0 & & \\ 13.33 & 0 & 6.12 & 0 \\ 0 & 0.99 & 0 & -0.99 \\ \underline{0} & \underline{12.34} & & \\ 0 & 13.33 & 0 & -0.99 \\ 6.12 & 0 & 12.34 & 0 \\ & & \underline{0.99} & 0 \\ \overline{6.12} & 0 & 13.33 & 0 \\ 0 & -0.99 & 0 & 0.99 \\ & & \underline{0} & \underline{12.34} \\ \overline{0} & -0.99 & 0 & 13.33 \end{bmatrix} \begin{bmatrix} d_1 \\ \\ \\ d_2 \\ \\ \\ d_3 \\ \\ \\ d_4 \end{bmatrix}$$

(c) Calculate the fixed end forces as follows. If a u.d.l. of q_y acts in the positive y-direction, then the fixed end moments at the ends are $\pm q_y L^2/12$ as shown in Fig. 6.17(a). As shown in Fig. 6.17(b), the components of the fixed end moments in the x- and z-directions are

$$M_{x1} = -(q_y L^2/12)n, \ M_{z1} = (q_y L^2/12)l, \ F_{y1} = -0.5q_y L$$

$$M_{x2} = (q_y L^2/12)n, \ M_{z2} = -(q_y L^2/12)l, \ F_{y2} = -0.5q_y L$$

Similarly, if a concentrated load W_y acts at midspan, then the fixed end reactions are

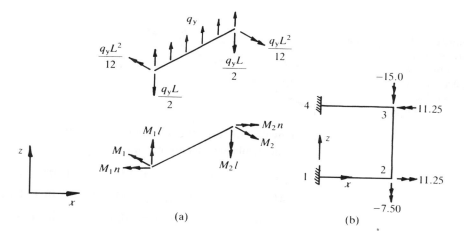

Figure 6.17 ● (a), (b)
Fixed end forces due
to external loads on
a plane grid

(a)   (b)

Table 6.18 ●
Fixed end reactions

Forces	Members		
	1–2	2–3	4–3
F_{y1}	15.0(*)	22.5(*)	30.0(*)
M_{x1}	0(*)	11.25(d_1)	0(*)
M_{z1}	−7.5(*)	0(d_2)	−15.0(*)
F_{y2}	15.0(*)	22.5(*)	30.0(*)
M_{x2}	0(d_1)	−11.25(d_3)	0(d_3)
M_{z2}	7.5(d_2)	0(d_4)	15.0(d_4)

$$M_{x1} = -(W_yL/8)n, \quad M_{z1} = (W_yL/8)l, \quad F_{y1} = -0.5W_y$$

$$M_{x2} = (W_yL/8)n, \quad M_{z2} = -(W_yL/8)l, \quad F_{y2} = -0.5W_y$$

The fixed end reactions for the present loading for each element are shown in Table 6.18, from which $\Sigma F_1 = 11.25$, $\Sigma F_2 = 7.5$, $\Sigma F_3 = -11.25$, $\Sigma F_4 = 15.0$.

(d) The equations to be solved for calculating the displacements caused by the *removal* of restraints are

$$10^2 \begin{bmatrix} 13.33 & 0 & 6.12 & 0 \\ 0 & 13.33 & 0 & -0.99 \\ 6.12 & 0 & 13.33 & 0 \\ 0 & -0.99 & 0 & 13.33 \end{bmatrix} \begin{bmatrix} d_1 \\ d_2 \\ d_3 \\ d_4 \end{bmatrix} = \begin{bmatrix} -11.25 \\ -7.50 \\ 11.25 \\ -15.00 \end{bmatrix}$$

After the forward Gaussian elimination the above equations become

$$10^2 \begin{bmatrix} 13.33 & 0 & 6.12 & 0 \\ 0 & 13.33 & 0 & -0.99 \\ 0 & 0 & 10.52 & 0 \\ 0 & 0 & 0 & 13.26 \end{bmatrix} \begin{bmatrix} d_1 \\ d_2 \\ d_3 \\ d_4 \end{bmatrix} = \begin{bmatrix} -11.25 \\ -7.50 \\ 16.42 \\ -15.56 \end{bmatrix}$$

The displacements can now be determined. They are $d_1 = -1.5606 \times 10^{-2}$, $d_2 = -0.6498 \times 10^{-2}$, $d_3 = 1.5608 \times 10^{-2}$, $d_4 = -1.1737 \times 10^{-2}$.

(e) Determine the forces at the ends of the element caused by displacements. From the stiffness matrix for the member in equation [6.3] and for the orientation

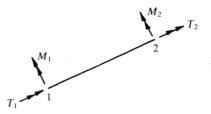

Figure 6.18 ● General orientation of a plane grid member for calculating member end forces due to displacements

shown in Fig. 6.18 (i.e. viewing end 1 at left and end 2 at right), the forces due to the displacements are given by

$$M_1 = \{EI/(L(1 + \beta))\}[(4 + \beta)\theta_1 + (2 - \beta)\theta_2 + (6/L)(v_2 - v_1)]$$
$$M_2 = \{EI/(L(1 + \beta))\}[(2 - \beta)\theta_1 + (4 + \beta)\theta_2 + (6/L)(v_2 - v_1)]$$

where $\theta_1 = -\theta_{x1}n + \theta_{z1}l$, $\theta_2 = -\theta_{x2}n + \theta_{z2}l$. The torque T is given by

$$-T_1 = T_2 = \{GJ/L\}(\psi_2 - \psi_1)$$

where $\psi_1 = \theta_{x1}l + \theta_{z1}n$, $\psi_2 = \theta_{x2}l + \theta_{z2}n$.
From equilibrium

$$F_{y1} = -(M_1 + M_2)/L, \quad F_{y2} = -F_{y1}$$

Table 6.19 shows the forces due to the displacements and the fixed end reactions.

Table 6.19 ● Fixed end and displacement forces

Forces	Members					
	1–2		2–3		4–3	
	Fixed	Disp.	Fixed	Disp.	Fixed	Disp.
F_{y1}	15.0	4.0	22.5	0	30.0	7.22
M_1	−7.5	−3.98	−11.25	9.71	−15.0	−7.18
T_1	0	1.55	0	0.52	0	−1.56
F_{y2}	15.0	−4.0	22.5	0	30.0	−7.22
M_2	7.5	−8.02	11.25	−9.71	15.0	−14.48
T_2	0	−1.55	0	−0.52	0	1.56

(f) The final forces are the sum of the fixed end reactions and forces due to displacements caused by the removal of restraints. They are shown in Table 6.20 and also in Fig. 6.19.

Table 6.20 ● Final forces

Forces	Members		
	1–2	2–3	4–3
F_{y1}	19.0	22.5	37.22
M_1	−11.48	−1.54	−22.18
T_1	1.55	0.52	−1.56
F_{y2}	11.0	22.5	22.88
M_2	−0.52	1.54	0.52
T_2	−1.55	−0.52	1.56

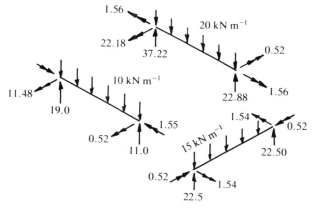

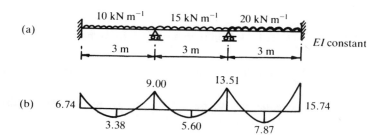

Figure 6.19 ● Final forces in the members of the plane grid due to external loads

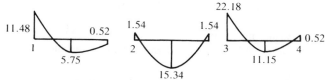

(a)

10 kN m⁻¹   15 kN m⁻¹   20 kN m⁻¹

3 m   3 m   3 m   *EI* constant

Figure 6.20 ● (a), (b) Bending moment distribution in a continuous beam

(b)   9.00   13.51

6.74   15.74

3.38   5.60   7.87

The bending moment diagrams can now be drawn. They are shown in Fig. 6.19.

It is interesting to note that if the loading on the grid had been applied to a continuous beam obtained by 'straightening out' the grid as shown in Fig. 6.20(a), then the bending moment distribution can be shown to be as depicted in Fig. 6.20(b). As can be seen, the support moments over supports 2 and 3 are much larger than those in the grid analysed. The reason for this is quite simple. In the case of the plane grid, the restraint to bending rotation is provided by the torsional stiffness $GJ/L = 0.99 \times 10^2$ of the adjacent beams. On the other hand, in the case of the continuous beam it is provided by the rotational stiffness which (assuming that $\beta = 0$ and the adjacent beam is hinged at the far end) will not be less than $3EI/L = 9.33 \times 10^2$ which is an order of magnitude larger than GJ/L. Thus the elements in the continuous beam are restrained to a much greater extent at their ends than the elements of the plane grid.

6.13.1 Analysis of grids for temperature change

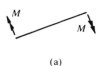

(a)

(b)

Figure 6.21 ● (a), (b) Fixed end couples in a plane grid member due to thermal restraint

The fixed end reactions due to restraining thermal expansion in the case of plane grid members consist of fixed end moment only as shown in Fig. 6.21(a). The magnitude of the fixed end moment is given by

$$M = EI\,\alpha_t(T_t - T_b)/d$$

where EI = flexural rigidity, α_t = coefficient of linear thermal expansion, d = depth of beam, T_t and T_b are the temperature changes at the top and bottom faces respectively.

The components of the fixed end moments in the x- and z-directions as shown in Fig. 6.21(b) are given by

$$F_{y1} = 0, \quad M_{x1} = -Mn, \quad M_{z1} = Ml, \quad F_{y2} = 0, \quad M_{x2} = Mn, \quad M_{z2} = -Ml$$

Assuming in the present example $\alpha_t = 11 \times 10^{-6}\,°C^{-1}$ and $T_t = 20\,°C$, $T_b = 0$, depth $d = 0.2$ m, the fixed end forces are as shown in Table 6.21. From which

$$\sum F_1 = 0 \text{ from member } 1 - 1.03 \text{ from member } 2 = -1.03$$

$$\sum F_2 = -1.03 \text{ from member } 1 + 0 \text{ from member } 2 = -1.03$$

$$\sum F_3 = 1.03 \text{ from member } 2 + 0 \text{ from member } 3 = 1.03$$

$$\sum F_4 = 0 \text{ from member } 2 - 1.03 \text{ from member } 3 = -1.03$$

The equations to be solved for calculating the displacements caused by the *removal* of restraints are

$$10^2 \begin{bmatrix} 13.33 & 0 & 6.12 & 0 \\ 0 & 13.33 & 0 & -0.99 \\ 6.12 & 0 & 13.33 & 0 \\ 0 & -0.99 & 0 & 13.33 \end{bmatrix} \begin{bmatrix} d_1 \\ d_2 \\ d_3 \\ d_4 \end{bmatrix} = \begin{bmatrix} 1.03 \\ 1.03 \\ -1.03 \\ 1.03 \end{bmatrix}$$

After the forward Gaussian elimination, the above equations become

Table 6.21 ● Fixed end thermal restraint forces

Forces	Members		
	1–2	2–3	4–3
F_{y1}	0(*)	0(*)	0(*)
M_{x1}	0(*)	$-1.03(d_1)$	0(*)
M_{z1}	1.03(*)	0(d_2)	1.03(*)
F_{y2}	0(*)	0(*)	0(*)
M_{x2}	0(d_1)	1.03(d_3)	0(d_3)
M_{z2}	$-1.03(d_2)$	0(d_4)	$-1.03(d_4)$

$$10^2 \begin{bmatrix} 13.33 & 0 & 6.12 & 0 \\ 0 & 13.33 & 0 & -0.99 \\ 0 & 0 & 10.52 & 0 \\ 0 & 0 & 0 & 13.26 \end{bmatrix} \begin{bmatrix} d_1 \\ d_2 \\ d_3 \\ d_4 \end{bmatrix} = \begin{bmatrix} 1.03 \\ 1.03 \\ -1.50 \\ 1.11 \end{bmatrix}$$

The displacements can now be determined. They are $d_1 = 0.1427 \times 10^{-2}$, $d_2 = 8.3468 \times 10^{-4}$, $d_3 = -0.1426 \times 10^{-2}$, $d_4 = 8.3465 \times 10^{-4}$.

Determine the forces at the ends of the element caused by displacements. From Section 6.15 the forces due to displacements are given by

$$M_1 = \{EI/(L(1 + \beta))\}[(4 + \beta)\theta_1 + (2 - \beta)\theta_2 + (6/L)(v_2 - v_1)]$$

$$M_2 = \{EI/(L(1 + \beta))\}[(2 - \beta)\theta_1 + (4 + \beta)\theta_2 + (6/L)(v_2 - v_1)]$$

where $\theta_1 = -\theta_{x1}n + \theta_{z1}l$, $\theta_2 = -\theta_{x2}n + \theta_{z2}l$. The torque T is given by

$$-T_1 = T_2 = \{GJ/L\}(\psi_2 - \psi_1)$$

where $\psi_1 = \theta_{x1}l + \theta_{z1}n$, $\psi_2 = \theta_{x2}l + \theta_{z2}n$. From equilibrium

$$F_{y1} = -(M_1 + M_2)/L, \quad F_{y2} = -F_{y1}$$

Table 6.22 shows the forces due to displacements and fixed end reactions.

Table 6.22 ● Fixed end and displacement forces

Forces	Members					
	1–2		2–3		4–3	
	Fixed	Disp.	Fixed	Disp.	Fixed	Disp.
F_{y1}	0	−0.51	0	0	0	−0.51
M_1	1.03	0.51	1.03	−0.89	1.03	0.51
T_1	0	−0.14	0	0	0	0.14
F_{y2}	0	0.51	0	0	0	0.51
M_2	−1.03	1.03	−1.03	0.89	−1.03	1.03
T_2	0	0.14	0	0	0	−0.14

The final forces are the sum of the fixed end forces and forces due to displacements caused by the removal of restraints. They are shown in Table 6.23. The BMD can now be drawn as shown in Fig. 6.22.

Table 6.23 ● Final forces

Forces	Members		
	1–2	2–3	4–3
F_{y1}	−0.51	0	−0.51
M_1	1.54	0.14	1.54
T_1	−0.14	0	0.14
F_{y2}	0.51	0	0.51
M_2	0	−0.14	0
T_2	0.14	0	−0.14

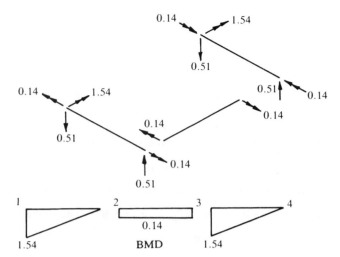

Figure 6.22 ●
Force distribution in the
plane grid due to
thermal loads

BMD

6.14 ● Use of symmetry and skew symmetry

The examples in the previous sections have shown that the major work involved is in the solution of the simultaneous equations to determine the displacements. In large-scale structures an effort should be made to reduce the number of equations to be solved. In the case of structures having one or more axes of symmetry parallel to the co-ordinate axes and subjected to symmetrical or skew symmetrical loads, it is possible to reduce the number of equations to be solved to approximately half the number that would need to be solved if the whole structures is solved.

6.14.1 Symmetrical structures under symmetrical loading

2-D rigid-jointed structure

Consider the rigid-jointed structure shown in Fig. 6.23(a) which is symmetrical about the centre line. If the loads are also symmetrical about the centre line, then so are the displacements. Therefore we only need to analyse one half of the structure. The points on the line of symmetry do not translate horizontally and do not rotate. However, there is a vertical translation. Therefore the members on the line of symmetry carry only an axial force. When analysing a symmetrical half of the structure, the geometric properties for members whose axes coincide with the axis of symmetry should be only one half of the actual values. Similarly, loads on the line of symmetry should be only one half of the actual values because the load is shared between the two symmetrical halves of the structure.

2-D pin-jointed structure

As in the case of a rigid-jointed structure, the members lying on the axis of symmetry suffer only vertical translation. Therefore what was said about rigid-jointed structures also applies to pin-jointed structures. However, there is an important

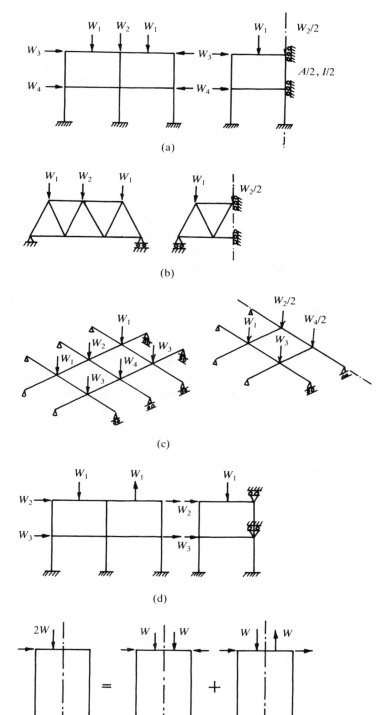

Figure 6.23 ●
Symmetrical structure under symmetrical and skew symmetrical loading: (a) 2-D rigid-jointed structure under symmetrical loading, (b) 2-D pin-jointed structure under symmetrical loading, (c) plane grid structure under symmetrical loading, (d) 2-D rigid-jointed structure under skew symmetrical loading, (e) a general load and its symmetrical and skew symmetrical components

point about the boundary conditions to be used at the supports. If one end is pin ended and the other end is on rollers, the structure is strictly not symmetrical. The object of providing a pinned support is to prevent rigid body translation. Even so, we can ignore this fact when analysing a structure under *symmetrical vertical loads only* provided that when analysing a symmetrical half, the pinned support is treated as a roller support because the zero horizontal displacement at the centre line prevents the possibility of horizontal rigid body translation of the whole structure. This is shown in Fig. 6.23(b).

Grid structures

If the structure is symmetrical about the x- or z-axes, then joints along the line of symmetry do not rotate about the x- or z-axis respectively. Again the geometric properties for members on the axis of symmetry should be only one half of the actual values. Similarly, loads on the line of symmetry should be only one half of the actual values because the load is shared between the two symmetrical halves of the structure. This is shown in Fig. 6.23(c).

6.14.2 Symmetrical structures under skew symmetrical loading

2-D rigid-jointed structure

Consider the rigid-jointed structure shown in Fig. 6.23(a) which is symmetrical about the centre line. If the loads are skew symmetrical about the centre line, then so also are the displacements. Therefore we only need to analyse one half of the structure. The points on the line of symmetry do not translate vertically but there is a horizontal translation and rotation. Because the deformations are skew symmetrical with respect to the centre line, the points on the line of symmetry are also points of contraflexure and there is no restraint to horizontal force. When analysing a symmetrical half of the structure, the geometric properties for members lying on the axis of symmetrical should be only one half of the actual values. This is shown in Fig. 6.23(d).

2-D pin-jointed structure

All the remarks made in connection with the rigid-jointed structures are applicable to pin-jointed structures. However, there is an important point about the boundary conditions to be used at the supports if one end is pin ended and the other end is on rollers. When analysing a symmetric half, the roller support is treated as a pinned support because at the centre line there is no resistance to horizontal displacement.

Grid structure

If the structure is symmetrical about the x- or z-axes, then points along the line of symmetry do not translate in the vertical direction and there is no resistance to rotation about the x- or z-axis respectively. Again the geometric properties for members on the axis of symmetry should be only one half of the actual values.

6.14.3 Symmetrical and skew symmetrical components of a general load

A general load can always be split into symmetrical and skew symmetrical components (Fig. 6.23(e)) and each load case analysed separately and the results combined. This is where we have to think carefully about the *total* numerical effort involved. It can be said that in general we should analyse the structure for the given loading without bothering to split into symmetrical and skew symmetrical analysis because this saves considerable manual effort. We should resort to symmetrical and skew symmetrical analysis *only when it is not possible to analyse the whole structure in one step because of storage limitations of the computer being used.*

6.15 ● Checking the output from computer programs

Complex and sophisticated computer programs are extensively used in practice. Unfortunately the quality of software available varies enormously. The user should never accept the output without sufficient random checks to ensure its accuracy. The output from computer programs is subject to two main sources of errors. They are

1. Errors arising from input of wrong data. Under this category come
   (a) Errors due to wrong member connections, missing members, wrong boundary conditions, etc. With the widespread use of graphical facilities to draw the structure from the input data, the error can be easily spotted and corrected. Therefore this should not be a source of serious error.
   (b) Errors due to wrong properties assigned to members. This is not as easy to eliminate as the previous error. However, because the number of different types of members in a structure is necessarily small, it is possible to use colour graphics to draw each type of member in a different colour so that the wrong type of members can be fairly easily identified.
   (c) Errors due to wrong data about loads. This is the hardest to eliminate and the only solution is thorough checking.
2. Programming errors. One advantage of doing the calculations using a computer program is that it eliminates random errors that are inevitable in manual calculations. The errors that arise in calculations done using a computer program are always systematic errors. They could arise for example because of accumulation of round-off errors when solving a very large number of simultaneous equations. Good programs should have in-built checks to at least warn the user when he or she might expect 'unreliable' answers. Unfortunately this is not always done. Again, although every responsible programmer thoroughly checks the programs to ensure that the output from the program is 'correct', it is always possible that checking is not 'perfect', with the result that unusual coincidence of circumstances not envisaged by the programmer might produce 'wrong' answers for the 'right' data. It is therefore important that users never accept the output without running random checks to satisfy themselves that the output is correct. When checking the output, it is useful to remember that the 'correct' result of any structural analysis of any structure by whatever method should satisfy the following conditions:

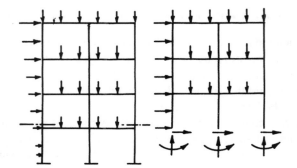

Figure 6.24 ● Checks for equilibrium of a 2-D rigid-jointed structure

(a) equilibrium between internal 'stresses' and external loads;
(b) compatibility of displacements;
(c) stress–strain law for the material;
(d) prescribed boundary conditions.

As stated before, because errors are systematic rather than random, the user should check at random that the above conditions are not violated by the output.

Prescribed boundary conditions in terms of connections to 'foundation' are easy to check from the output of displacements. Compatibility of displacements is not so easy to check from the output and can only be inferred when checking the member forces from the output of displacements of joints. The simplest to check is equilibrium. In the case of a truss, it is useful to check equilibrium at a random selection of joints using the output of member forces. The same thing can be done in the case of rigid-jointed structures. In multistorey frames, it is useful to check, from the output of shear and axial force in members, the horizontal and vertical equilibrium at storey levels as shown in Fig. 6.24. Another useful check is to see if the total vertical and horizontal reactions add up to the applied load. This might identify wrong load data input. Random checks should be made to see if the output for member forces agrees with the loads acting on the member and the output of displacements at its ends. When doing this both horizontal and vertical members should be included, and if there are any inclined members they should be checked first. Finally, a graphical output of the deformed shape of the structure is very useful for checking the overall response of the structure to applied loads.

6.16 ● References

Bhatt P 1981 *Problems in Matrix Analysis of Skeletal Structures* Construction Press (Reprinted by A H Wheeler, Allahabad, India, 1988) (Contains a large number of numerical examples. It also discusses many related problems such as analysis of structures with settlement of supports, structures with many different types of elements, etc.)

Bhatt P 1986 *Programming the Matrix Analysis of Skeletal Structures* Ellis Horwood (Contains 'sophisticated',

readable and well documented programs in FORTRAN for the linear elastic analysis of pin-jointed, rigid-jointed and plane grid structures. Also additional material on semi-rigid connections, beams on elastic foundations, etc.)

Catchick B K 1978 Prestress analysis for continuous beams: some developments in the equivalent load method *Structural Engineer* **56B** (2): 29–36 (Contains procedure for the determination of 'fixed

end forces' due to prestress in statically indeterminate structures.)
McGuire W, Gallagher R H 1979 *Matrix Structural Analysis* Wiley (A useful text.)

Weaver W, Gere J M 1980 *Matrix Analysis of Framed Structures* (2nd edn) Van Nostrand (Another usable text.)

6.17 ● Problems

1. Analyse the pin-jointed structure shown in Fig. E5.2. Assume that the extensional rigidity AE is constant.

Answer:

$$(AE/L)\begin{bmatrix} 1.35 & 0.35 & 0 & 0 \\ 0.35 & 1.35 & 0 & -1.0 \\ 0 & 0 & 1.35 & -0.35 \\ 0 & -1.0 & -0.35 & 1.35 \end{bmatrix}\begin{bmatrix} d_1 \\ d_2 \\ d_3 \\ d_4 \end{bmatrix} = \begin{bmatrix} 0 \\ -2W \\ W \\ -W \end{bmatrix}$$

After Gaussian elimination

$$(AE/L)\begin{bmatrix} 1.35 & 0.35 & 0 & 0 \\ 0 & 1.26 & 0 & -1.0 \\ 0 & 0 & 1.35 & -0.35 \\ 0 & 0 & 0 & 0.47 \end{bmatrix}\begin{bmatrix} d_1 \\ d_2 \\ d_3 \\ d_4 \end{bmatrix} = \begin{bmatrix} 0.0 \\ -2.0W \\ 1.0W \\ -2.33W \end{bmatrix}$$

Solving
$(d_4, d_3, d_2, d_1) = \{WL/(AE)\}[-4.96, -0.55, -5.52, 1.43]$.
Forces in members
$(1–3, 2–3, 3–5, 5–6, 5–4) = W\{-2.04, 1.44, 0.56, 0.56, -2.21\}$

2. Analyse the pin-jointed structure shown in Fig. E5.2, when the temperature in members 3–5, 5–6 and 5–4 increases by $T\,°C$. Assume constant value of coefficient of thermal expansion α.

Answer: Fixed end axial forces in members 3–5, 5–6 and 5–4 are $-AE\alpha T$. Restraint load vector due to fixed end forces: $(AE\alpha T)[0, 1.0, (1 + 1/\sqrt{2}), -(1 + 1/\sqrt{2}]$. Displacements are caused by *removing* the restraint load vector. After Gaussian elimination, the force vector: $(AE\alpha T)[0, -1.0, -1.707, 0.468]$. Displacements $(d_4, d_3, d_2, d_1) = (\alpha TL)[1.0, -1.0, 0, 0, 0]$.
Forces due to displacements:
$(1–3, 2–3, 3–5, 5–6, 5–4) = (AE\alpha T)[0, 0, 1.0, 1.0, 1.0]$.
Final forces $(1–3, 2–3, 3–5, 5–6, 5–4)$
$= (AE\alpha T)[0, 0, 0, 0]$.

3. Determine the degrees of freedom for the rigid-jointed structures shown in Fig. E6.1. Consider cases with and without axial deformations. (*Note*: Including or ignoring shear deformations does not affect the degrees of freedom.)

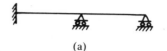

(a)

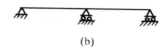

(b)

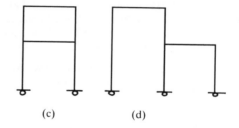

(c)   (d)

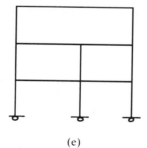

(e)

Figure E6.1 ● (a)–(e)

Answer: (a) (4, 2); (b) (5, 3); (c) (14, 8); (d) (15, 9); (e) (27, 13) where the first number indicates the total number of degrees of freedom when axial deformations are included.

4. Analyse the continuous beam shown in Fig. E6.2. Ignore axial and shearing deformations.

Answer: Let $d_1 = \theta_B$, $d_2 = \theta_C$, $d_3 = \theta_D$.
Fixed end couples at the ends of members:
$(\mp 0.333$ for AB, ∓ 0.375 for BC, ∓ 0.1667 for CD$)qL^2$

$$(EI/L)\begin{bmatrix} 5.667 & 1.333 & 0.0 \\ 1.333 & 10.667 & 4.0 \\ 0 & 4.0 & 8.0 \end{bmatrix}\begin{bmatrix} d_1 \\ d_2 \\ d_3 \end{bmatrix} = (qL^2)\begin{bmatrix} 0.0417 \\ -0.2083 \\ -0.1667 \end{bmatrix}$$

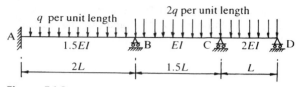

Figure E6.2

After Gaussian elimination:

$$(EI/L) \begin{bmatrix} 5.667 & 1.333 & 0.0 \\ 0 & 10.353 & 4.0 \\ 0 & 0 & 6.455 \end{bmatrix} \begin{bmatrix} d_1 \\ d_2 \\ d_3 \end{bmatrix} = (qL^2) \begin{bmatrix} 0.0417 \\ -0.2181 \\ -0.0824 \end{bmatrix}$$

Solving:
$(EI/L)\{d_3, d_2, d_1\} = (qL^2)[-1.2772, -1.6132, 1.1154] 10^{-2}$.
Couples at the ends of members due to rotations:
$(qL^2)[(0.0167, 0.0335)$ for AB, $(0.0082, -0.0282)$
for BC, $(-0.1801, -0.1667)$ for CD]. Final couples
at the ends of members: $(qL^2)[(-0.317, 0.367)$ for
AB, $(-0.367, 0.347)$ for BC, $(-0.347, 0)$ for CD].

5. Analyse the 2-D rigid-jointed structure shown in
Fig. E6.3. Ignore axial and shear deformations.

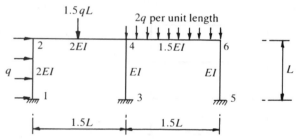

Figure E6.3

Answer: Let $d_1 = \theta_2$, $d_2 = \theta_4$, $d_3 = \theta_6$, $d_4 =$ sway
displacement Fixed end moments: column 1–2:
$\mp 0.0833 qL^2$, beams: $(\mp 0.2813$ for 2–4, ∓ 0.3750
for 4–6)qL^2, reaction at the top of the column
$= 0.5 qL$ to the left.

$$(EI/L) \begin{bmatrix} 13.33 & 2.67 & 0 & -12.0 \\ 2.67 & 13.33 & 2.0 & -6.0 \\ 0 & 2.0 & 8.0 & -6.0 \\ -12.0 & -6.0 & -6.0 & 48.0 \end{bmatrix} \begin{bmatrix} d_1 \\ d_2 \\ d_3 \\ d_4/L \end{bmatrix}$$

$$= (qL^2) \begin{bmatrix} 0.1980 \\ 0.0937 \\ -0.3750 \\ 0.5000 \end{bmatrix}$$

After Gaussian elimination:

$$(EI/L) \begin{bmatrix} -13.33 & 2.67 & 0 & -12.0 \\ 0 & 12.80 & 2.0 & -3.60 \\ 0 & 0 & 7.69 & -5.44 \\ 0 & 0 & 0 & 32.34 \end{bmatrix} \begin{bmatrix} d_1 \\ d_2 \\ d_3 \\ d_4/L \end{bmatrix}$$

$$= (qL^2) \begin{bmatrix} 0.1980 \\ 0.0541 \\ -0.3835 \\ 0.4221 \end{bmatrix}$$

Solving:
$(EI/L)[d_4/L, d_3, d_2, d_1] = (qL^2)[1.3053, -4.0654,$
$1.4250, 2.3748] 10^{-2}$.
Couples at the ends of members due to
displacements: Columns: $(qL^2)[(-0.0616, 0.0334)$
for 1–2, $(-0.0496, -0.0209)$ for 3–4, $(-0.1596,$
$-0.2409)$ for 5–6]. Beams: $(qL^2)[(-0.1649, 0.1399)$
for 2–4, $(-0.0239, -0.1339)$ for 4–6]. Final
couples at the ends of members: Columns:
$(qL^2)[(-0.1449, 0.1167)$ for 1–2, $(-0.0496,$
$-0.0209)$ for 3–4, $(-0.1596, -0.2409)$ for 5–6].
Beams: $(qL^2)[(-0.1164, 0.4213)$ for 2–4,
$(-0.3989, 0.2411)$ for 4–6].

6. Analyse the rigid-jointed structure shown in
Fig. E6.3 due to a rise in temperature on the
outside of member 1–2 only by T °C. Ignore axial
and shear deformations.

Answer: Fixed end moment in column 1–2 =
$\mp 2EI \alpha T/d$, $d =$ depth of member. Load vector:
$(EI \alpha T/d) [2.0, 0, 0, 0]$. After Gaussian elimination,
load vector: $(EI \alpha T/d)[2.0, -0.40, 0.063, 1.732]$.
Displacements: $(EI/L)[d_4/L, d_3, d_2, d_1]$
$= (EI \alpha T/d)[5.355, 4.602, -2.338, 20.287]10^{-2}$.
Couples due to displacement:
Columns: $EI \alpha T/d)[(0.169, 0.980)$ for 1–2, $(-0.368,$
$-0.415)$ for 3–4, $(-0.229, -0.137)$ for 5–6].
Beams: $(EI \alpha T/d)[(1.02, 0.416)$ for 2–4, $(-0.002,$
$0.137)$ for 4–6].
Final couples:
Columns: $(EI \alpha T/d)[(2.169, -1.02)$ for 1–2,
$(-0.368, -0.415)$ for 3–4, $(-0.229, -0.137)$
for 5–6]. Beams: $(EI \alpha T/d) [(1.02, 0.416)$ for 2–4,
$(-0.002, 0.137)$ for 4–6].

7. Determine the degrees of freedom for the plane
grids shown in Fig. E6.4. The hatched edges are
clamped.

Answers: (a) 18 (b) 17

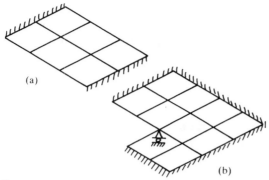

(a)

(b)

Figure E6.4 ● (a), (b)

8. Analyse the plane grid shown in Fig. E6.5. Ignore shearing deformations. Assume $GJ = 0.2EI$ for all the members.

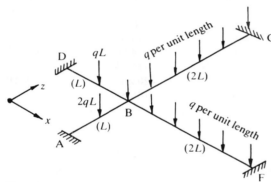

Figure E6.5

Answer: Let $d_1 = \theta_{xB}$, $d_2 = \theta_{zB}$, $d_3 = \Delta_B$.
Fixed end forces:
x-direction beams:

(i) Couples about z-axis, i.e. bending moments: (qL^2) [∓0.250 for AB, ∓0.333 for BC]
(ii) Force in the normal y-direction: $0.5[2qL + q(2L)] = 2qL$

z-direction beams:
(i) Couples about x-axis, i.e. bending moments: (qL^2) [∓0.125 for DB, ∓0.333 for BF]
(ii) Force in the normal y-direction: $0.5[qL + q(2L)] = 1.5qL$

$$(EI/L)\begin{bmatrix} 6.30 & 0 & -4.50 \\ 0 & 6.30 & 4.50 \\ 4.50 & 4.50 & 27.0 \end{bmatrix}\begin{bmatrix} d_1 \\ d_2 \\ d_3/L \end{bmatrix} = (qL^2)\begin{bmatrix} -0.0833 \\ 0.2083 \\ -3.50 \end{bmatrix}$$

After Gaussian elimination:

$$(EI/L)\begin{bmatrix} 6.30 & 0 & -4.50 \\ 0 & 6.30 & 4.50 \\ 0 & 0 & 20.57 \end{bmatrix}\begin{bmatrix} d_1 \\ d_2 \\ d_3/L \end{bmatrix} = (qL^2)\begin{bmatrix} -0.0833 \\ 0.2083 \\ -3.7083 \end{bmatrix}$$

Solving: $(EI/L)[d_3/L, d_2, d_1] = [-0.1803, 0.1618, -0.1420]qL^2$
Couples at the ends due to displacements.

(i) z-direction beams:
Couples about the x-axis, i.e. bending moments: $(qL^2)[(-0.7978, -0.5138)$ for AB, $(0.5545, 0.4125)$ for BC]. Couples about the z-axis, i.e. twisting moments: (qL^2) [∓0.0322 for AB, ∓0.0161 for BC]

(ii) x-direction beams:
Couples about the z-axis, i.e. bending moments: $(qL^2)[(-0.7582, -0.4346)$ for DB, $(0.5941, 0.4323)$ for BF].
Couples about the x-axis, i.e. twisting moments: (qL^2) ±0.0284 for DB, ±0.0142 for BF].

Final couples:

(i) z-direction beams:
Couples about the x-axis, i.e. bending moments: $(qL^2)[(-1.0478, -0.2638)$ for AB, $(0.2212, 0.7458)$ for BC].
Couples about the z-axis, i.e. twisting moments: (qL^2) ∓0.0322 for AB, ±0.0161 for BC].

(ii) x-direction beams:
Couples about the z-axis, i.e. bending moments: $(qL^2)[(-0.8832, -0.3096)$ for DB, $(0.2608, 0.7656)$ for BF].
Couples about the z-axis, i.e. twisting moments: (qL^2)[∓0.0284 for DB, ±0.0142 for BF].

9. The computer output from the analysis of the 2-D pin-jointed structure shown in Fig. E6.6 is as follows. Check if the results are correct. Assume $E = 210$ kN mm^{-2}, length = 3000 mm, $A = 450$ mm^2 for all the members.

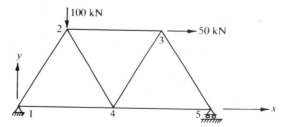

Figure E6.6

Table E6.1 ● Joint displacements (mm)

Joint	u	v
1	0	0
2	3.79	−4.44
3	3.66	−1.83
4	2.57	−3.18
5	3.42	0

Table E6.2 ● Forces in members (kN)

Member	Axial force
1–2	−61.6
1–4	80.8
2–3	−3.87
2–4	−53.87
3–4	53.87
3–5	−53.87
4–5	26.93

Table E6.3

Joint	u (mm) positive →	θ (radians) clockwise
1	0	9.7×10^{-3}
2	21.6	-1.4×10^{-3}
3	0	0
4	5.2	4.2×10^{-3}
5	21.6	3.3×10^{-3}
6	0	0
7	5.2	-0.2×10^{-3}

Table E6.4 ● Couples at the ends of members in kN m and positive clockwise

Member	M at end 1	M at end 2
1–2	0	53.81
2–5	−53.81	212.01
4–5	−174.90	−212.00
3–4	−11.55	39.01
6–7	−203.87	−210.50
4–7	135.89	210.50

10. The computer output from the analysis of the 2-D rigid-jointed structure shown in Fig. E6.7 is as follows. Check if the results are correct. Ignore axial and shear deformations. Assume $E = 210$ kN mm^{-2}, $I = 240 \times 10^{-6}$ m^4 for member 1–2, $I = 85 \times 10^{-6}$ m^4 for member 3–4 and $I = 270 \times 10^{-6}$ m^4 for the remainder.

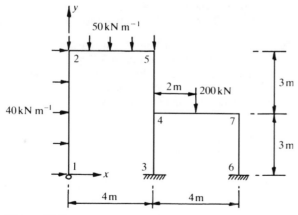

Figure E6.7

Chapter seven

Moment distribution method

In Chapter 6, the computer orientated stiffness method of analysis was presented. This is the most common method of analysis used in practice. However, in some rare cases where access to a computer is unobtainable, there may be a need to use a simple manual method for the analysis of continuous beams and single storey portal frames with one or two bays. The object of this chapter is to present one such method. It is called the moment distribution method. It was developed in the pre-computer era, and is based on the stiffness method except that the unknown displacements are solved iteratively. The numerical effort involved is not heavy provided that the structure needs to be analysed for one or two loading cases only because, unlike the stiffness method, each load case has to be analysed separately.

In the following sections the method is presented through a series of examples. Alongside the iterative approach of moment distribution is presented a 'simplified' stiffness approach adopted for 'manual' calculations. The object of this is to enable the reader to compare the numerical effort involved in the two procedures so that the one which is less onerous can be selected.

In order to explain the concepts involved, consider the continuous beam with constant flexural rigidity EI shown in Fig. 7.1. The equations of equilibrium to be solved are, from the procedure explained in Chapter 6,

$$(EI/L)\begin{bmatrix} 4+4 & 2 \\ 2 & 4+4 \end{bmatrix}\begin{bmatrix} \theta_b \\ \theta_c \end{bmatrix} = (qL^2/12)\begin{bmatrix} 1 \\ -1 \end{bmatrix}$$

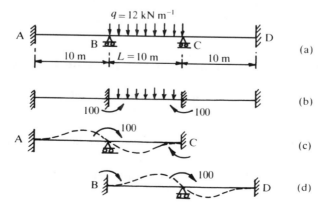

Figure 7.1 ● (a)–(d) Analysis of a continuous beam by the moment distribution method

where $qL^2/12 = 100$. The displacements θ_b and θ_c are determined by solving the above two equations. The values are $(EI/L)\,\theta_b = 16.67$, $(EI/L)\,\theta_c = -16.67$.

The moment distribution method is equivalent to solving these equations iteratively as follows.

(a) Assume that $\theta_c = 0$ and calculate θ_b from the first equation. Therefore $(EI/L)\,\theta_b = 12.5$. This is physically equivalent to clamping all joints and allowing only joint B to rotate as shown in Fig. 7.1(c).

(b) Similar to the first step, assume $\theta_b = 0$ and calculate θ_c from the second equation. Therefore $(EI/L)\,\theta_c = -12.5$. This is physically equivalent to clamping all the joints and allowing only joint C to rotate as shown in Fig. 7.1(d).

Thus at this stage we have approximations of 12.5 and -12.5 respectively for the values of $(EI/L)\,\theta_b$ and $(EI/L)\,\theta_c$.

(c) If the above approximate values are substituted into the equations of equilibrium, then the right-hand sides become 75 and -75 respectively instead of 100 and -100. Therefore the *corrections* to the values of joint rotations are obtained by solving the equations

$$(EI/L)\begin{bmatrix} 8 & 2 \\ 2 & 8 \end{bmatrix}\begin{bmatrix} \theta_b \\ \theta_c \end{bmatrix} = \begin{bmatrix} 25 \\ -25 \end{bmatrix}$$

In moment distribution terms this is viewed as removing the external restraints applied at B and C while joints C and B rotate under the action of the moments -100 and 100.

(d) Repeat step (a). This gives $(EI/L)\,\theta_b = 3.13$.

(e) Repeat step (b). This gives $(EI/L)\,\theta_c = -3.13$.

Thus at the end of the second iteration $(EI/L)\,\theta_b = 12.5 + 3.13 = 15.63$ and $(EI/L)\,\theta_c = -12.5 - 3.13 = -15.63$ compared with the exact values of 16.67 and -16.67 respectively.

Obviously the steps could be repeated until the corrections to the values of unknowns become negligibly small.

The moment distribution method does not attempt to determine the values of unknown displacements. Rather it determines the corrections to moments in members due to successive rotations of joints. In practice, if we have access to a program to carry out the stiffness method of analysis, then the choice is obvious. Even when such a program is not available, in many cases it may be quicker to set up and solve the equations of equilibrium manually.

7.1 ● Concept of rotational stiffness and carry-over factors

It was shown in Example 3, Section 4.5, that in a propped cantilever having uniform flexural rigidity EI the couple needed to cause unit rotation at the propped end is $4EI/L$ and the fixed end moment at the support is $2EI/L$. This is shown in Fig. 7.2(a). The value $4EI/L$ is called the rotational stiffness factor and the ratio of the couple at the fixed end to the couple causing unit rotation is called the carry-over factor. In the case of a propped cantilever the carry-over factor is $(2EI/L)/(4EI/L) = 0.5$.

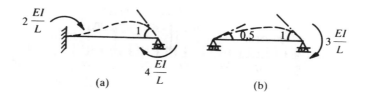

Figure 7.2 ● (a), (b)
Stiffness and carry-over
factors for a beam
element

(a)   (b)

Similarly, in the case of a simply supported beam having uniform flexural rigidity EI, as was shown in Example 4, Section 4.4, the rotational stiffness factor is $3EI/L$ and the carry-over factor is naturally zero. This is shown in Fig. 7.2(b).

7.2 ● Concept of distribution factors

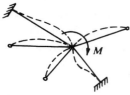

Figure 7.3 ●
Distribution of a couple
applied at a joint to
members meeting
at the joint

Consider a set of members meeting at a rigid joint as shown in Fig. 7.3. If a couple is applied to the common joint, then because the joint is 'rigid' all the members undergo the same rotation. Therefore for equilibrium,

$$\sum S_i \theta = M$$

where S_i is the rotational stiffness factor of the ith member meeting at the joint. $\sum S_i$ is called the total stiffness at the joint. Therefore

$$\theta = M / \sum S_i$$

The moment induced in the jth member is

$$M_j = S_j \theta = \left\{ S_j / \sum S_i \right\} M$$

The ratio $S_j / \sum S_i$ is called the distribution factor of member j because during common rotation, the moment induced in the jth member at the common joint is the distribution factor of the jth member multiplied by the couple M.

7.3 ● Examples of the analysis of structures with only rotation at the joints

Example 1

Analyse the continuous beam shown in Fig. 7.4. The relative values of second moment of area are shown against each member.

Solution   The solution consists of several steps as follows.

(a) Calculate the distribution factors at each joint. The analysis assumes that shear deformations are neglected.

Joints B, C, and D have more than one member meeting there. Therefore distribution factors are required at these joints.

Joint B. The members meeting are BA and BC. A is clamped. Therefore the stiffness factor S_{ba} for BA is $4E(4I)/5 = 3.2EI$. Member BC is continuous at the

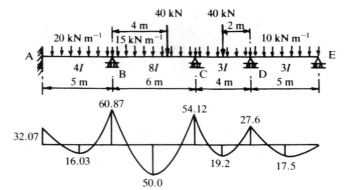

Figure 7.4 ● Loading and distribution of bending moments in a continuous beam

end C. However, when calculating the approximation to the rotation at B, we assume, as already explained, that the joint C is fixed. Therefore the stiffness factor S_{bc} for BC is $4E(8I)/6 = 5.33EI$; the total stiffness ΣS at B is equal to the sum of the stiffnesses of BA and BC; $\Sigma S = 3.20EI + 5.33EI = 8.53EI$; at B the distribution factors are

For BA the distribution factor $= 3.2EI/(8.53EI) = 0.375$

For BC the distribution factor $= 5.33EI/(8.53EI) = 0.625$

Check that the sum of the distribution factors $= 1.00$.

Joint C. Members CB and CD are both continuous at the ends B and D. So, for the reason explained before, i.e. calculation of the approximation to the rotation of the joints by considering the continuous end as fixed – the stiffness and distribution factors for members meeting at C are:

Stiffness factor for CB $= 4E(8I)/6 = 5.33EI$ (same as BC)

Stiffness factor for CD $= 4E(3I)/4 = 3.00EI$

$$\sum S = 5.33EI + 3.00EI = 8.33EI$$

For CB the distribution factor $= 5.33EI/(8.33EI) = 0.640$

For CD the distribution factor $= 3.00EI/(8.33EI) = 0.360$

Check that the sum of the distribution factors $= 1.00$.

Joint D. Joint C is continuous. So the stiffness factor for DC $= 4E(3I)/4 = 3.0EI$, which is the same as the stiffness factor for CD. Joint E is simply supported. Thus the stiffness factor for DE is $3E(3I)/5 = 1.8EI$. Note the fact that if the far end is simply supported, then the stiffness factor is $3EI/L$. Therefore

$$\sum S = 3.0EI + 1.8EI = 4.8EI$$

For DC the distribution factor $= 3.0EI/(4.8EI) = 0.625$

For DE the distribution factor $= 1.8EI/(4.8EI) = 0.375$

Check that the sum of the distribution factors $= 1.00$.

(b) Calculation of carry-over factors.

Joint B. When joint B is rotated, because A is fixed and C is assumed to be fixed, the carry-over factors from B to A and from B to C are 0.5 each.

Joint C. When joint C is rotated, because B and D are assumed to be fixed, the carry-over factors from C to B and from C to D are 0.5 each.

Joint D. When joint D is rotated, because C is assumed to be fixed, the carry-over factor from D to C is 0.5. On the other hand, because E is simply supported, the carry-over factor from D to E is zero.

(c) Calculation of fixed end moments.

Assuming that clockwise couples are positive, the fixed end couples at the ends of the members are calculated using the fixed end moments in Tables 4.6 and 4.7 as follows.

Member
AB.
$$M_{ab} = -20 \times 5^2/12 = -41.67$$
$$M_{ba} = +20 \times 5^2/12 = 41.67$$

Member
BC.
$$M_{bc} = -\{15 \times 6^2/12 + 40 \times 2^2 \times 4/6^2\} = -71.67$$
$$M_{cb} = \{15 \times 6^2/12 + 40 \times 2 \times 4^2/6^2\} = 80.56$$

Member
CD.
$$M_{cd} = -\{10 \times 4^2/12 + 40 \times 4/8\} = -33.33$$
$$M_{dc} = 33.33$$

Member
DE.
$$M_{de} = -10 \times 5^2/8 = -31.25$$
$$M_{ed} = 0$$

The above information is set out in a tabular form in Table 7.1. The DF row shows the distribution factors, the COF row shows the carry-over factors and the FEM row shows the fixed end moments. Note that as joint A is fixed, there is no carry-over moment *from* A because the *fixed* joint is never allowed to rotate. Similarly, there is no carry-over *to the pinned joint* E because it is never prevented from rotating.

(d) Rotate joints one at a time while keeping the far ends fixed *except* when the far end is *simply supported*.

The fixed end couples are external couples applied at the joints to prevent them from rotating. When the restraint is removed, then the joint rotates as explained in Chapter 6.

Joint B. The net fixed end couple at the joint B is $41.67 - 71.67 = -30.0$. This restraint is *removed* by applying a couple of $+30.0$ to the joint B. This couple is shared by the members meeting at the joint in proportion to their distribution factors at the joint. Therefore in this case moment $M_{ba} = 30 \times 0.375 = 11.25$ and $M_{bc} = 30 \times 0.625 = 18.75$. These values are recorded at joint B in the row called RJ, i.e. 'rotate joints'.

Joint C. The net restraint couple at C is equal to $80.56 - 33.33 = 47.23$. When this restraint is *removed* by applying a couple of -47.23, the moments developed in the members at the end C are $M_{cb} = -47.23 \times 0.640 = -30.23$ and $M_{cd} = -47.23 \times 0.360 = -17.00$. These values are entered at joint C in the RJ row.

Table 7.1

A		B		C		D		E	Remarks
	0.375	0.625	0.640	0.360	0.625	0.375			DF
→0	0.5←	→0.5	0.5←	→0.5	−0.5←	→0			COF
−41.67	41.67	−71.67	80.56	−33.33	33.33	−31.25		0	FEM
0	11.25	18.75	−30.23	−17.00	−1.30	−0.78		0	RJ
5.63	0	−15.12	9.38	−0.65	−8.50	0		0	CO
	5.67	9.45	−5.59	−3.14	5.31	3.19		0	RJ
2.84	0	−2.80	4.73	2.66	−1.57	0		0	CO
0	1.05	1.75	−4.73	−2.66	0.98	0.59		0	RJ
0.53	0	−2.37	0.88	0.49	−1.33	0		0	CO
0	0.89	1.48	−0.88	−0.49	0.83	0.50		0	RJ
0.45	0	−0.44	0.74	0.42	−0.25	0		0	CO
0	0.17	0.28	−0.74	−0.42	0.16	0.09		0	RJ
0.08	0	−0.37	0.14	0.08	−0.21	0		0	CO
0	0.14	0.23	−0.14	−0.08	0.13	0.08		0	RJ
0.07	0	−0.07	0.13	0.07	−0.04	0		0	CO
0	0.03	0.04	−0.13	−0.07	0.03	0.01		0	RJ
−32.07	60.87	−60.86	54.12	−54.12	27.57	−27.57		0	Final

Joint D. The net restraint couple at D is equal to $33.33 - 31.25 = 2.08$. When this restraint is *removed* by applying a couple of -2.08, the moments developed are $M_{dc} = -2.08 \times 0.625 = -1.30$ and $M_{de} = -2.08 \times 0.375 = -0.78$. These values are entered at joint C in the RJ row.

When this set of calculations is completed, a first approximation to the joint rotations has been obtained.

A discontinuous line is drawn to indicate the end of an iteration.

(e) Carry-over moments to the far end.

Joint B. When joint B is rotated, it was *assumed* in calculating the stiffness factors and distribution factors at B that joints A and C are fixed. Therefore due to the rotation of joint B, $M_{ba} = 11.25$ and $M_{bc} = 18.75$. The corresponding carry-over moments are $M_{ab} = 0.5M_{ba} = 0.5 \times 11.25 = 5.63$ and $M_{cb} = 0.5M_{bc} = 0.5 \times 18.75 = 9.38$. These are entered at the appropriate places in the row headed CO, i.e. 'carry-over the moments'. It has to be remembered that the carry-over moments are restraint moments required to prevent rotation of the joint at the far end of the member which is or is *assumed* to be fixed.

Joint C. From rotation of joint C, the carry-over moments are

$$M_{bc} = 0.5 \times (-30.23) = -15.12$$

$$M_{dc} = 0.5 \times (-17.0) = -8.50$$

Joint D. From the rotation of joint D, $M_{cd} = 0.5 \times (-1.30) = -0.65$, $M_{ed} = 0$ because it is simply supported.

(f) Because when rotating joints in step (d) the far end, if it was continuous, is *assumed* to be fixed, external restraints have been applied here equal to the carry-over moments. Therefore step (d) will be repeated to remove these restraints.

For example, at joint B the net restraint moment of −15.12 is due to carry-over moment of −15.12 from C. *Removing* the restraint causes rotation of joint B and causes moments $M_{ba} = 15.12 \times 0.375 = 5.67$ and $M_{bc} = 15.12 \times 0.625 = 9.45$.

Similarly, at joint C, there are carry-over moments resulting from rotation of joints B and D. The total restraint moment at C is equal to $(9.38 - 0.65) = 8.73$. This restraint is removed by applying a moment of −8.73 at C. This moment is distributed to members CB and CD according to their distribution factors. Therefore

$$M_{cb} = 0.64(-8.73) = -5.59$$

$$M_{cd} = 0.36(-8.73) = -3.14$$

Similarly, at joint D the restraint moment at D results from the rotation of joint C and is equal to −8.50. This restraint is removed by applying a moment of 8.50 which results in $M_{dc} = 0.625(8.50) = 5.31$ and $M_{de} = 0.375(8.50) = 3.19$. This step is entered in the RJ row.

(g) Carry-over moments from step (f).

(h) Steps RJ and CO are repeated until the changes in moments are acceptably small.

(i) The final moment at the end of a member is obtained by summing up the moments at that end due to fixed end moments, the moments resulting from the rotation of that end and the carry-over moments resulting from the rotation of the far end of the member.

(j) Rotation of joints.

The rotations at the joints can be easily determined as follows.

Joint B. Summing up the moments in <u>RJ rows</u> for the moment M_{ba} we have

$$M_{ba} = 11.25 + 5.67 + 1.05 + 0.89 + 0.17 + 0.14 + 0.03 = 19.2$$

$$= \text{Rotational stiffness factor} \times \theta_b = \{4E(4I)/5\}\,\theta_b$$

Therefore

$$EI\,\theta_b = 6.0$$

Similarly, as a check, considering M_{bc} we have

$$M_{bc} = 18.75 + 9.45 + 1.75 + 1.48 + 0.28 + 0.23 + 0.04 = 31.98$$

$$= \{4E(8I)/6\}\,\theta_b$$

Therefore

$$EI\,\theta_b = 5.996$$

Joint C. Summing up the moments in <u>RJ rows</u> we have for M_{cb} and M_{cd}

$$M_{cd} = -30.23 - 5.59 - 4.73 - 0.88 - 0.74 - 0.14 - 0.13 = -42.44$$

$$= \{4E(8I)/6\}\, \theta_c$$

Therefore

$$EI\, \theta_c = -7.96$$

As a check,

$$M_{cd} = -17.00 - 3.14 - 2.66 - 0.49 - 0.42 - 0.08 - 0.07 = -23.86$$

$$= \{4E(3I)/4\}\, \theta_c$$

Therefore

$$EI\, \theta_c = -7.953$$

Joint D. Summing up the moments in <u>RJ rows</u> for M_{dc} and M_{de}, we have

$$M_{dc} = -1.30 + 5.31 + 0.98 + 0.83 + 0.16 + 0.13 + 0.03 = 6.14$$

$$= \{4E(3I)/4\}\, \theta_d$$

Therefore

$$EI\, \theta_d = 2.05$$

$$M_{de} = -0.78 + 3.19 + 0.59 + 0.50 + 0.09 + 0.08 + 0.01 = 3.68$$

$$= \{3E(3I)/5\}\, \theta_d$$

Therefore

$$EI\, \theta_d = 2.04$$

Note that because joint E is simply supported, the rotational stiffness of the beam DE at D is 3 *not* 4 times $\{E(3I)/5\}$.

Joint E. The rotation of joint E arises from two causes. These are shown in Fig. 7.5.

(i) Due to the rotation of joint D by $EI\, \theta_d = 2.04$, the rotation at E is $EI\, \theta_{e1}$ $= -0.5(EI\, \theta_d) = -1.02$ as explained in Chapter 4, Section 4.4, Example 4.

(ii) Due to the rotation at the propped end of the cantilever under applied load of 10 kN m. One simple procedure for calculating this rotation under any loading is shown in Fig. 7.5(b). Let the fixed end moments in the corresponding fixed beam be $-M_{s1}$ at the left support and M_{s2} at the right support. In the propped cantilever with the prop at the right-hand support, the fixed end moment must be zero. Therefore, if a couple of $-M_{s2}$ is applied at the right support, then the carry-over moment at the left support is $-0.5M_{s2}$. The corresponding joint rotation is given by

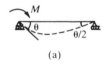

(a)

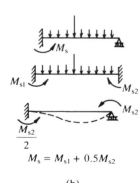

(b)

Figure 7.5 ● (a), (b) Joint rotations of a propped cantilever due to external loads and couple at the support

$$\{4EI/L\}\, \theta_e = -M_{s2}$$

The fixed end moment at the left support is $-(M_{s1} + 0.5M_{s2})$.

In the present example $-M_{s1} = M_{s2} = 10 \times 5^2/12 = 20.83$. The fixed end moment in the propped cantilever is $-(20.83 + 0.5 \times 20.83) = -31.25$.

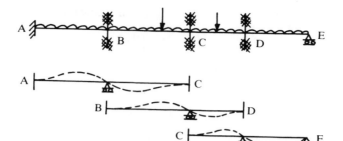

Figure 7.6 ●
Basic deformations
of a continuous
beam for simplified
stiffness analysis

$$\{4E(3I)/5\}\ \theta_{e2} = -20.83$$

Therefore

$$EI\ \theta_{e2} = -8.68$$

(iii) The total rotation at E is given by

$$EI\ \theta_e = -1.02 - 8.68 = -9.70$$

Figure 7.4(b) shows the BMD for the structure.

(k) Solution using the manual adaptation of the stiffness method. It is useful to contrast the numerical effort involved if the 'simplified' stiffness method had been adopted to solve the problem.

Structural stiffness matrix formulation. Using the deformation patterns shown in Fig. 7.6 and ignoring the shearing deformations, the structural stiffness matrix can be easily shown to be

$$\begin{bmatrix} M_B \\ M_C \\ M_D \end{bmatrix} = EI \begin{bmatrix} 4 \times \frac{4}{5} + 4 \times \frac{8}{6} & 2 \times \frac{8}{6} & 0 \\ 2 \times \frac{8}{6} & 4 \times \frac{8}{6} + 4 \times \frac{3}{4} & 2 \times \frac{3}{4} \\ 0 & 2 \times \frac{3}{4} & 4 \times \frac{3}{4} + 3 \times \frac{3}{5} \end{bmatrix} \begin{bmatrix} \theta_b \\ \theta_c \\ \theta_d \end{bmatrix}$$

Using the fixed end moments calculated before, the load vector is

$$\begin{bmatrix} -(41.67 - 71.67) \\ -(80.56 - 33.33) \\ -(33.33 - 31.25) \end{bmatrix}$$

Simplifying,

$$EI \begin{bmatrix} 8.533 & 2.667 & 0 \\ 2.667 & 8.333 & 1.50 \\ 0 & 1.50 & 4.80 \end{bmatrix} \begin{bmatrix} \theta_b \\ \theta_c \\ \theta_d \end{bmatrix} = \begin{bmatrix} 30.00 \\ -47.23 \\ -2.08 \end{bmatrix}$$

Gaussian elimination. Carrying out the Gaussian elimination we have:
Step 1. Eliminate θ_b from second and third equations:

$$EI \begin{bmatrix} 8.533 & 2.667 & 0 \\ 0 & 7.499 & 1.50 \\ 0 & 1.50 & 4.80 \end{bmatrix} \begin{bmatrix} \theta_b \\ \theta_c \\ \theta_d \end{bmatrix} = \begin{bmatrix} 30.00 \\ -56.607 \\ -2.08 \end{bmatrix}$$

Step 2. Eliminate θ_c from the third equation:

$$EI\begin{bmatrix} 8.533 & 2.667 & 0 \\ 0 & 7.499 & 1.50 \\ 0 & 0 & 4.50 \end{bmatrix}\begin{bmatrix} \theta_b \\ \theta_c \\ \theta_d \end{bmatrix} = \begin{bmatrix} 30.00 \\ -56.607 \\ 9.243 \end{bmatrix}$$

Solve for the unknowns:

$$EI\,\theta_d = 2.054, \quad EI\,\theta_c = -7.96, \quad EI\,\theta_b = 6.00$$

Solve for member forces. General equations used are – because translations are zero, and shearing deformations are ignored – given by (see Section 6.1)

$$M_1 = M_{f1} + (EI/L)\{4\theta_1 + 2\theta_2\}$$

$$M_2 = M_{f2} + (EI/L)\{2\theta_1 + 4\theta_2\}$$

where the subscript f indicates fixed end moment.

(i) $M_{ab} = -41.67 + (4/5)\{0 + 2 \times 6.00\} = -32.07$

      $M_{ba} = 41.67 + (4/5)\{0 + 4 \times 6.00\} = 60.87$

(ii) $M_{bc} = -71.67 + (8/6)\{4 \times 6.00 + 2(-7.96)\} = -60.90$

      $M_{cb} = 80.56 + (8/6)\{2 \times 6.00 + 4(-7.96)\} = 54.11$

(iii) $M_{cd} = -33.33 + (3/4)\{4 \times (-7.96) + 2 \times 2.054\} = -54.13$

      $M_{dc} = 33.33 + (3/4)\{2 \times (-7.96) + 4 \times 2.054\} = 27.55$

(iv) $M_{de} = -31.25 + (3/5)\{3 \times (2.054)\} = -27.55$

Note the use of the equation $M_{de} = M_{def} + \{3EI/L\}\,\theta_d$ to allow for the fact that the end E is simply supported.

The reader should contrast the numerical work involved in the two approaches to see which is less onerous.

Example 2

Analyse the continuous beam shown in Fig. 7.7.

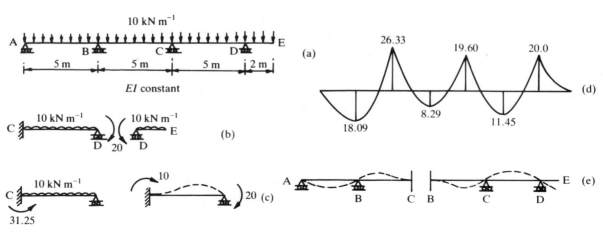

Figure 7.7 ● (a)–(e) Analysis of a continuous beam with a cantilever overhang

Solution The object of this example is to show how to handle cantilever elements.

(a) Fixed end moments.

The first step is to 'eliminate' the cantilever part and determine the fixed end moments as shown in Fig. 7.7.

(i) $M_{abf} = 0$, $M_{baf} = 10 \times 5^2/8 = 31.25$ because BA is a propped cantilever.

(ii) $M_{bcf} = -10 \times 5^2/12 = -20.83$, $M_{cbf} = 10 \times 5^2/12 = 20.83$.

(iii) $M_{cdf} = -10 \times 5^2/8 = -10 \times 5^2/8 = -31.25$ due to a u.d.l. of 10 kN m^{-1} because CD is considered as a propped cantilever.

The effect of the cantilever is to exert a clockwise moment of $10 \times 2^2/2$ = 20 kN m at the support D. This induces a *clockwise* carry-over moment of 10 at C. Therefore the *net* fixed end moment $M_{cdf} = -31.25$ due to u.d.l. + 10 due to the effect of cantilever $= -21.25$. It should be appreciated that the moment at D has to remain at 20 to maintain equilibrium with the cantilever moment.

(b) Distribution factors.

The rotational stiffness and distribution factors are easily determined as follows.

Joint B. A is simply supported and C is continuous. Therefore the stiffness factors are

$$S_{ba} = 3(EI)/5 = 0.6EI, \; S_{bc} = 4(EI)/5 = 0.8EI$$

$$\sum S = 0.6EI + 0.8EI = 1.4EI$$

Therefore the distribution factors are for BA = $0.6EI/(1.4EI) = 0.429$, and for BC = $0.8EI/(1.4EI) = 0.571$.

Joint C. B is continuous and because the moment at D has to remain constant at 20 from equilibrium considerations, D should be considered as simply supported. Therefore the stiffness factors are

$$S_{cb} = 4(EI)/5 = 0.8EI, \quad S_{cd} = 3(EI)/5 = 0.6EI$$

$$\sum S = 0.8EI + 0.6EI = 1.4EI$$

This leads to distribution factors for CB of $0.8EI/(1.4EI) = 0.571$ and for CD of $0.6EI/(1.4EI) = 0.429$.

(c) Carry-over factors.

Because in the analysis both A and D will be considered as simply supported, carry-over factors of 0.5 apply only from B to C and from C to B.

(d) The iterative calculations are set out in Table 7.2. Figure 7.7(d) shows the bending moment diagram.

(e) Calculation of rotation of joints.

The joint rotations are calculated by summing up the moment at the end of the member resulting from the rotation of the joint given in RJ rows and dividing it by the rotational stiffness.

Joint B. $M_{ba} = -4.47 - 0.05 - 0.37 - 0.03$

$$= -4.92 = 3(EI/5)\,\theta_b$$

Table 7.2

A		B		C		D	Remarks
	0.429	0.571	0.571	0.429			DF
$\rightarrow$0	0$\leftarrow$	$\rightarrow$0.5	0.5$\leftarrow$	$\rightarrow$0			COF
0	31.25	−20.83	20.83	−21.25		20.0	FEM
	−4.47	−5.95	0.24	0.18			RJ
0	0	0.12	−2.98	0			CO
	−0.05	−0.07	1.70	1.28			RJ
	0	0.85	−0.04	0			CO
	−0.37	−0.49	0.02	0.02			RJ
	0	0.01	−0.25	0			CO
	0	−0.01	0.14	0.11			RJ
	0	0.07	0	0			CO
	−0.03	−0.04	0	0			RJ
0	26.33	−26.34	19.66	−19.66		20.0	Final

Therefore

$$EI\ \theta_b = -8.2$$
$$M_{bc} = -5.95 - 0.07 - 0.49 - 0.01 - 0.04$$
$$= -6.56 = 4(EI/5)\ \theta_b$$

Therefore

$$EI\ \theta_b = -8.2 \text{ as a check on calculations.}$$

Joint C. $\quad M_{cb} = 0.24 + 1.70 + 0.02 + 0.14$
$$= 2.1 = 4(EI/5)\ \theta_c$$

Therefore

$$EI\ \theta_c = 2.63$$
$$M_{cd} = 0.18 + 1.28 + 0.02 + 0.11 = 1.59 = 3(EI/5)\ \theta_c$$

therefore

$$EI\ \theta_c = 2.65$$

Joint A. There are two components to the rotation. The first component arises when joint B is rotated considering joint A as simply supported. As shown in Fig. 7.7(b), then the rotation of joint A is $-0.5\theta_b$. Therefore $EI\ \theta_{a1} = 4.1$. The second component arises due to the rotation of joint A due to the u.d.l. acting on the propped cantilever. This is given, as explained in the previous example, by $10 \times 5^2/12 = 4(EI/5)\ \theta_{a2}$. Therefore $EI\ \theta_{a2} = 26.04$.

$$EI\ \theta_a = 4.1 + 26.04 = 30.14$$

Joint D. The rotation at joint D consists of three components as follows.

(i)   Due to rotation of joint C: $EI\ \theta_{d1} = -0.5EI\ \theta_c = -1.32$.

(ii) Rotation of joint due to u.d.l.: $-10 \times 5^2/12 = 4(EI/5)\ \theta_{d2}$, therefore $EI\ \theta_{d2} = -26.04$.

(iii) Rotation of joint due to moment from the cantilever: $20 = 4(EI/5)\ \theta_{d3}$; therefore $EI\ \theta_{d3} = 25$ and

$$EI\ \theta_d = -1.32 - 26.04 + 25 = -2.36$$

Joint E. The deflection and rotation of joint E is due to the u.d.l. acting on the cantilever DE plus the rotation of 'support' D. Rotation $\theta_e = 10 \times 2^3/(6EI)$ due to u.d.l. $+ \theta_d$, therefore $EI\ \theta_e = 10.697$. Deflection of $E = 10 \times 2^4/(8EI)$ due to u.d.l. $+ \theta_d\ 2$, therefore $EI\ \Delta_e = 15.28$ downwards.

(f) Structural stiffness matrix.

Using the displacement patterns shown in Fig. 7.7(e), the structural stiffness matrix relationship is easily shown to be

$$EI\begin{bmatrix} 3/5 + 4/5 & 2/5 \\ 2/5 & 4/5 + 3/5 \end{bmatrix}\begin{bmatrix} \theta_b \\ \theta_c \end{bmatrix} = \begin{bmatrix} -(31.25 - 20.83) \\ -(20.83 - 31.25 + 10) \end{bmatrix} = \begin{bmatrix} -10.42 \\ 0.42 \end{bmatrix}$$

$$EI\begin{bmatrix} 1.4 & 0.4 \\ 0.4 & 1.4 \end{bmatrix}\begin{bmatrix} \theta_b \\ \theta_c \end{bmatrix} = \begin{bmatrix} -10.42 \\ 0.42 \end{bmatrix}$$

Carrying out the Gaussian elimination we have

$$EI\begin{bmatrix} 1.4 & 0.4 \\ 0 & 1.286 \end{bmatrix}\begin{bmatrix} \theta_b \\ \theta_c \end{bmatrix} = \begin{bmatrix} -10.420 \\ 3.397 \end{bmatrix}$$

Therefore

$$EI\ \theta_c = 2.642, \quad EI\ \theta_b = -8.198$$

$$M_{ba} = M_{baf} + (3EI/5)\ \theta_b = 31.25 + (3/5)(-8.198) = 26.33$$

$$M_{bc} = M_{bcf} + (EI/5)\{4\theta_b + 2\theta_c\}$$

$$= -20.83 + (1/5)\{4(-8.198) + 2(2.642)\} = -26.33$$

$$M_{cb} = M_{cbf} + (EI/5)\{2\theta_b + 4\theta_c\}$$

$$= 20.83 + (1/5)\{2(-8.198) + 4(2.642)\} = 19.66$$

$$M_{cd} = M_{cdf} + (3EI/5)\ \theta_c = -21.25 + (3/5)(2.642) = -19.67$$

Once again the reader has to decide which procedure is numerically less onerous.

Example 3

Analyse the rigid-jointed structure shown in Fig. 7.8.

Solution   For simplicity, it is assumed that the members are axially inextensible with the result that the joints undergo only rotations.

In addition the effects of shearing deformations are ignored. Because the joints cannot translate but can only rotate, the steps to be followed are as for the analysis of continuous beams in the previous examples.

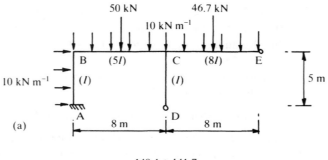

(a)

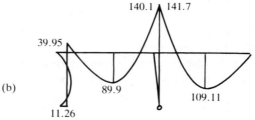

(b)

Figure 7.8 ● (a)–(c)
Analysis of a 'no-sway'
rigid-jointed frame

(c)

(a) Stiffness and distribution factors.

Joint B. A is fixed and C is continuous. Therefore

$$S_{ba} = 4E(I)/5 = 0.8EI, \quad S_{bc} = 4E(5I)/8 = 2.5EI$$

$$\sum S = 0.8EI + 2.5EI = 3.3EI$$

Distribution factors are for BA = 0.8/3.3 = 0.242, and for BC = 2.5/3.3 = 0.758.

Joint C. B is continuous but joints D and E are pinned. Therefore S_{cb} = 4E(5I)/8 = 2.5EI, S_{cd} = 3E(I)/5 = 0.6EI because joint D is pinned and S_{ce} = 3E(8I)/8 = 3.0EI again because joint E is pinned. Therefore $\sum S$ = 2.5EI + 0.6EI + 3.0EI = 6.1EI.

The distribution factors are for CB = 2.5/6.1 = 0.410, for CD = 0.6/6.1 = 0.098, for CE = 3.0/6.1 = 0.492.

(b) Carry-over factors.

Joint B. Because A is fixed and C is continuous, the carry-over factors from B to both A and C are 0.5.

Joint C. Because B is continuous, the carry-over factor from C to B is 0.5. Because both D and E are pinned, the carry-over factors from B to both D and E are zero.

(c) Fixed end moments.

AB: $M_{ab} = -10 \times 5^2/12 = -20.83$, $M_{ba} = 20.83$

BC: $M_{bc} = -\{10 \times 8^2/12 + 50 \times 8/8\} = -103.33$,

$\quad M_{cb} = 103.33$

CE: $M_{ce} = -\{10 \times 8^2/8 + 3 \times 46.67 \times 8/16\} = -150.0$

$\quad M_{ec} = 0$

CD: $M_{cd} = M_{dc} = 0$ because of zero 'lateral' load on the member.

(d) Rotation and carry-over of moments.

It is convenient to carry out the calculations on an outline of the 'structure itself' as shown in Table 7.3. The student must study carefully the arrangement in Table 7.3. All moment values for the beams are above the COF for beam rows. All moment values for columns are below the COF for column rows. The left-hand side of a vertical line is for the moment values at the top of the column and the right-hand side is for the moment values at the bottom of the column. The first few steps are explained below.

(i) Cycle 1 of joint rotation and carry-over

Rotation of joints

Joint B. The sum of the fixed end moments from BA and BC is (20.83 − 103.33) = −82.50. This is the net restraint present at B. Therefore *releasing* the restraint by allowing the joint to rotate develops

$$M_{ba} = -(-82.50) \times 0.242 = 19.97$$

$$M_{bc} = -(-82.50) \times 0.758 = 62.54$$

This is entered under RJ rows.

Joint C. The sum of the fixed end moments from CB and CE is (103.33 − 150.00) = −46.67. This is the net restraint present at C. Therefore *releasing* the restraint by allowing the joint to rotate develops

$$M_{bc} = -(-46.67) \times 0.410 = 19.14$$

$$M_{ce} = -(-46.67) \times 0.492 = 22.96$$

$$M_{cd} = -(-46.67) \times 0.098 = 4.57$$

In contrast to the continuous beam, where only two members meet at a joint, in this example, because three members meet at C, the *negative* of restraint moment is distributed among three members according to their distribution factors. After the moments due to rotation of joints are calculated, a short horizontal line is drawn to indicate that the iteration is complete.

Carry-over moments

This is carried out according to the appropriate carry-over factors. Thus

Table 7.3

(BC)	(CB)	(CE)	
−39.94	140.16	−141.79	Final
−0.05	−0.08	−0.09	RJ
0.06	0.19	0.0	CO
0.38	0.12	0.14	RJ
−0.50	−0.28	0.0	CO
−0.56	−1.00	−1.20	RJ
0.75	2.43	0.0	CO
4.86	1.49	1.79	RJ
−6.41	−3.63	0.0	CO
−7.25	−12.82	−15.39	RJ
9.57	31.27	0.0	CO
62.54	19.14	22.96	RJ
−103.33	103.33	−150.0	FEM

B		C	
(0.758)	(0.410)	(0.492)	DF (beams)
→0.5	0.5←		COF (beams)
(0.242) for BA		(0.098) for CD	DF (columns)
→0.5 to AB			COF (columns)

20.83	−20.83	0	FEM
19.97		4.57	RJ
	9.99 (CO)		
−2.32		−3.07	RJ
	−1.16 (CO)		
1.55		0.36	RJ
	0.78 (CO)		
−0.18		−0.24	RJ
	−0.09 (CO)		
0.12		0.03	RJ
	0.06 (CO)		
−0.02		−0.02	RJ
39.95	−11.26	1.63	Final
(BA)	(AB)	(CD)	

Joint A. $M_{ab} = 0.5M_{ba} = 0.5(19.97) = 9.99$

Note that because joint A is fixed, there is no carry-over moment from A to B because joint A is never rotated.

Joint B. $M_{bc} = 0.5M_{cb} = 0.5(19.14) = 9.57$

Joint C. $M_{cb} = 0.5M_{bc} = 0.5(62.54) = 31.27$

(ii) Cycle 2 of joint rotation and carry-over
Rotation of joints

Joint B. The sum of the fixed end moments from BA and BC is $(0 + 9.57) = 9.57$. This is the net restraint present at B. Therefore *removing* the restraint and allowing the joint to rotate develops at B,

$$M_{ba} = -(9.57) \times 0.242 = -2.32$$

$$M_{bc} = -(9.57) \times 0.758 = -7.25$$

Joint C. The sum of the fixed end moments from CB and CE is $31.27 + 0.0 = 31.27$. This is the net restraint present at C. Therefore *removing* the restraint and allowing the joint C to rotate develops

$$M_{cb} = -(31.27) \times 0.410 = -12.82$$

$$M_{ce} = -(31.27) \times 0.492 = -15.39$$

$$M_{cd} = -(31.27) \times 0.098 = -3.07$$

Carry-over moments
This is done according to the appropriate carry-over factors. Thus

Joint A. $M_{ab} = 0.5M_{ba} = 0.5(-2.32) = -1.16$

Joint B. $M_{bc} = 0.5M_{cb} = 0.5(-12.82) = -6.41$

Joint C. $M_{cb} = 0.5M_{bc} = 0.5(-7.25) = -3.63$

Further cycles of iterations are continued until the changes to moments are acceptably small (Table 7.3).

The final bending moment diagram is shown in Fig. 7.8.

(e) Rotation of joints
These are determined in the usual way by considering the moment developed due to the rotation of joints and equating it to the rotational stiffness multiplied by the joint rotation.

Joint B. $M_{ba} = 19.97 - 2.32 + 1.55 - 0.18 + 0.12 - 0.02 = 19.12 = \{4(EI)/5\}\theta_b$. Therefore $EI\,\theta_b = 23.90$.
Check using $M_{bc} = 62.54 - 7.25 + 4.86 - 0.56 + 0.38 - 0.05 = 59.92 = \{4E(5I)/8.00\}\theta_b$. Therefore $EI\,\theta_b = 23.97$.

Joint C. $M_{cb} = 19.14 - 12.82 + 1.49 - 1.00 + 0.12 - 0.08 = 6.85 = \{4E(5I)/8\}\theta_c$. Therefore $EI\,\theta_c = 2.74$. Check using $M_{ce} = 22.96 - 15.39 + 1.79 - 1.20 + 0.14 - 0.09 = 8.21 = \{3E(8I)/8\}\theta_c$. Therefore $EI\,\theta_c = 2.74$.

Note the use of the factor 3 instead of 4 in calculating the stiffness factor because joint E is pinned. Similarly

$$M_{cd} = 4.57 - 3.07 + 0.36 - 0.24 + 0.03 - 0.02$$

$$= 1.63 = \{3E(I)5\} \, \theta_c$$

$$EI \, \theta_c = 2.72.$$

Joint D. $EI \, \theta_d = -0.5(EI \, \theta_c) = -1.36$. Because there are no lateral loads on the member this is the net rotation.

Joint E. $EI \, \theta_{e1} = -0.5(EI \, \theta_c) = -1.37$ due to the rotation of the joint C. $4\{E(8I)/8\} \, \theta_{e2} = -\{10 \times 8^2/12 + 50 \times 8/8\}$ due to the lateral loads on the element when the beam CE is considered as a propped cantilever. Therefore

$$EI \, \theta_{e2} = -25.83$$

Therefore the net rotation at E is

$$EI \, \theta_e = -1.37 - 25.83 = -27.20$$

(f) Stiffness formulation

Using the deformation patterns shown in Fig. 7.8(c), the 'simplified' structural stiffness matrix is easily formulated and is given by

$$EI \begin{bmatrix} 4I/5 + 4(5I)/8 & 2(5I)/8 \\ 2(5I)/8 & 4(5I)/8 + 3(8I)/8 + 3I/5 \end{bmatrix} \begin{bmatrix} \theta_b \\ \theta_c \end{bmatrix} = \begin{bmatrix} -(20.83 - 103.33) \\ -(103.33 - 150.0) \end{bmatrix}$$

Simplifying, we have

$$EI \begin{bmatrix} 3.30 & 1.25 \\ 1.25 & 6.10 \end{bmatrix} \begin{bmatrix} \theta_b \\ \theta_c \end{bmatrix} = \begin{bmatrix} 82.50 \\ 46.67 \end{bmatrix}$$

Carrying out the Gaussian elimination, the above equation becomes

$$EI \begin{bmatrix} 3.30 & 1.25 \\ 0 & 5.627 \end{bmatrix} \begin{bmatrix} \theta_b \\ \theta_c \end{bmatrix} = \begin{bmatrix} 82.50 \\ 15.42 \end{bmatrix}$$

Solving $EI \, \theta_c = 2.740$, $EI \, \theta_b = 23.962$

$$M_{ab} = M_{abf} + (2EI/5) \, \theta_b = -20.83 + (2/5)23.962 = -11.25$$

$$M_{ba} = M_{baf} + (4EI/5) \, \theta_b = 20.83 + (4/5)23.962 = 40.00$$

$$M_{bc} = M_{bcf} + \{E(5I)/8\}[4\theta_b + 2\theta_c]$$

$$= -103.33 + (5/8)[4 \times 23.962 + 2 \times 2.740] = -40.00$$

$$M_{cd} = M_{cdf} + (3EI/5) \, \theta_c = 0 + (3/5)2.740 = 1.64$$

$$M_{ce} = M_{cef} + \{3E(8I)/8\} \, \theta_c = -150.0 + (3)2.740 = -141.78$$

Once again the reader should compare the numerical work involved in the two methods before embarking on one procedure or the other.

7.4 ● Examples of the analysis of structures with rotation and translation at the joints

In the previous examples the joints of the structure underwent rotations only. In general, joints undergo both translations and rotations. This section will discuss, using examples, the analysis of structures where the joints translate as well as rotate.

Example 1

Analyse the structure shown in Fig. 7.9.

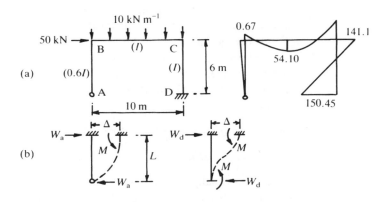

Figure 7.9 ● (a)–(c) Analysis of a single bay portal frame

Solution   The main difference between this structure and the ones analysed previously is that the structure can sway, i.e. joints B and C undergo a horizontal translation. The steps to be followed are similar to the ones followed in the previous examples, except for the consideration of joint translation.

(a) Fixed end moments due to lateral loads
   Element BC: The lateral load is 10 kN m^{-1}. Therefore $M_{bc} = -10 \times 10^2/12$ $= -83.33$, $M_{cb} = 83.33$.
(b) Stiffness and distribution factors

   Joint B. Joint A is pinned and joint C is continuous. Therefore the (rotational) stiffness factors for
   BA $= 3E(0.6I)/6 = 0.3EI$ and for BC $= 4E(I)/10 = 0.4EI$.
   Therefore $\Sigma S = 0.3EI + 0.4EI = 0.7EI$.
   The distribution factors are for BA $= 0.3EI/(0.7EI) = 0.429$ and for BC $= 0.4EI/$ $(0.7EI) = 0.571$.

Joint C. Joint B is continuous and joint D is fixed. Therefore the (rotational) stiffness factors for CB = $4E(I)/10 = 0.4EI$ and for CD = $4E(I)/6 = 0.67EI$. ΣS = $0.4EI + 0.67EI = 1.07EI$. The distribution factors are for CB = $0.4EI/(1.07EI)$ = 0.375 and for CD = $0.67EI/(1.07EI) = 0.625$.

(c) Carry-over factors

Joint B. Because A is pinned and C is continuous, the carry-over factor from B to A is zero but from B to C is 0.5.

Joint C. Because B is continuous and D is fixed, the carry-over factors from C to B and D are 0.5.

(d) Consideration of sway

The distribution factors were calculated on the assumption that joints do not translate. It is therefore important to ensure that *during translation joints do not rotate and during rotation, joints do not translate.* It is assumed that because the members are assumed to be axially rigid, all the joints at any horizontal level undergo the same translation.

Column pinned at the base: Consider the cantilever shown in Fig. 7.9(b). The relationship between the load W causing deflection Δ and the support moment is given by

$$\Delta = WL^3/(3EI) \text{ or } W = 3EI\Delta/L^3$$

$$M = -WL = -3EI\Delta/L^2$$

Column rotationally fixed at both top and bottom: As was shown in Chapter 4, Section 4.4, Example 5, for the column shown in Fig. 7.9(b), the load W causing a relative displacement of Δ between the ends of the column is given by

$$W = 12EI\Delta/L^3$$

From equilibrium the moments at the supports $M = -6EI\Delta/L^2$.

From the above considerations, the moments induced at the ends of the columns when sway is permitted but not rotation of joints B, C and D, are given by

Column AB. Horizontal reaction

$$W_a = 3E(0.6I)\Delta/6^3 = 1.8(EI\Delta/6^3)$$

$$M_{ab} = 0, M_{ba} = -3E(0.6I)\Delta/6^2 = -1.8(EI\Delta/6^2)$$

Column DC. Horizontal reaction

$$W_d = 12E(I)\Delta/6^3 = 12(EI\Delta/6^3)$$

$$M_{cd} = M_{dc} = -6E(I)\Delta/6^2 = -6(EI\Delta/6^2)$$

Therefore net horizontal reaction

$$= W_a + W_d = (EI\Delta/6^3)(1.8 + 12.0)$$

$$= (EI\Delta/6^3)13.8$$

(e) Fixed end moments due to sway with no joint rotation allowed. In the example a horizontal force of 50 kN has to be balanced. Therefore

$$50 = W_a + W_d = 13.8(EI\Delta/6^3)$$

Therefore

$$EI\Delta/6^2 = 21.74$$

The corresponding moments at the ends of the column are

$$M_{ab} = 0, \quad M_{ba} = -1.8(EI\Delta/6^2)$$

$$= -1.8 \times 21.74 = -39.13$$

$$M_{cd} = M_{dc} = -6(EI\Delta/6^2) = -6 \times 21.74 = -130.44$$

(f) Correction to sway to restore horizontal equilibrium.

During rotation of joints and corresponding carry-over moments, inevitably moments in the columns change from the above fixed end values, leading to a lack of horizontal equilibrium. If the change of moments is given by dM_{ba} in column AB and dM_{cd} and dM_{dc} in column DC, then allowing joints B and C to sway an additional amount $d\Delta$, so as to restore equilibrium, we have $dM_{ba} + dM_{cd} + dM_{dc} - 1.8EI\, d\Delta/6^2$ due to sway in column AB $- \{6EI/d\Delta/6^2 + 6EI/d\Delta/6^2\}$ due to sway in column CD $= 0$. Therefore

$$EI\, d\Delta/6^2 = \{dM_{ba} + dM_{cd} + dM_{dc}\}/13.8$$

or moment M_{ba} in column AB

$$= -1.8EI\, d\Delta/6^2 = -(1.8/13.8)\{dM_{ba} + dM_{cd} + dM_{dc}\}$$

$$M_{ba} = -0.130\{dM_{ba} + dM_{cd} + dM_{dc}\}$$

Similarly in column CD

$$M_{dc} = M_{cd} = -6EI\, d\Delta/6^2$$

$$= -(6/13.8)\{dM_{ba} + dM_{cd} + dM_{dc}\}$$

$$= 0.435\{dM_{ba} + dM_{cd} + dM_{dc}\}$$

These factors, i.e. -0.130 and 0.435 are entered in Table 7.4 as sway factors.

(g) Iteration cycles.

In the previous examples, the iteration cycle consisted of RJ (rotate joints) and CO (carry-over) moments. In this example there is an additional step, SJ (sway joints), to restore horizontal equilibrium. Apart from this additional step, the analysis procedure is as for previous examples. The steps are shown in Table 7.4.

The following notes are added by way of explanation.

Cycle 1. The FEM at the joints consist of two parts. The first part is due to the lateral loads on the members and the second part is due to the sway of the columns so as to restore the horizontal equilibrium with respect to the horizontal forces acting at the joints.

Step RJ is carried out as normal. At B the restraining moment = (-39.13 from BA from sway and -83.33 from BC from applied loads) = -122.46. The

Table 7.4

	B		C		D	
0.429	0.571	0.375	0.625			DF
0←	→0.5	0.5←	→0.5			CF
−0.13			−0.435	−0.435		sway factors
0.0	−83.33	83.33	0.0	0.0		FEM lateral load
−39.13	0	0.0	−130.44	−130.44		FEM due to sway
52.54*	69.93	17.67	29.44*	0		RJ
0	8.84	34.97	0	14.72*		CO
−12.57	0	0	−42.07	−42.07		SJ
1.60*	2.13	2.66	4.44*	0		RJ
0	1.33	1.07	0	2.22*		CO
−1.07	0	0	−3.59	−3.59		SJ
−0.11*	−0.15	0.95	1.58*	0		RJ
0	0.48	−0.08	0	0.79*		CO
−0.29	0	0	−0.98	−0.98		SJ
−0.08*	−0.11	0.40	0.66*	0		RJ
0	0.20	−0.06	0	0.33*		CO
−0.12	0	0	−0.40	−0.40		SJ
−0.03*	−0.05	0.17	0.29*	0		RJ
0	0.09	−0.03	0	0.15*		CO
−0.05	0	0	−0.18	−0.18		SJ
−0.02*	−0.02	0.08	0.13*	0		RJ
0.67	−0.66	141.13	−141.12	−159.45		Final

negative of this is distributed to BA and BC according to distribution factors. Thus $M_{ba} = -(-122.46) \times 0.429 = 52.44$ and $M_{bc} = -(-122.46) \times 0.571 = 69.93$.

Similarly, at C, the restraining moment = (83.33 from applied loads − 130.44 from sway) = −47.11, the negative of which is distributed according to the distribution factors. Therefore $M_{cb} = -(-47.11) \times 0.375 = 17.67$, $M_{dc} = -(47.11) \times 0.625 = 29.44$. This completes the first cycle. This is indicated by a horizontal line.

Cycle 2. The moments are carried over according to the appropriate carry-over moments.

Calculate the additional sway to restore horizontal equilibrium: The moments in the columns have changed as shown by *. Therefore $dM_{ba} = 52.54$, $dM_{cd} = 29.44$ and $dM_{dc} = 14.72$. Therefore $\{dM_{ba} + dM_{cd} + dM_{dc}\} = 96.70$. To restore horizontal equilibrium, we need $M_{ba} = -0.13 \times 96.70 = -12.57$, $M_{cd} = M_{dc} = -0.435 \times 96.70 = -42.07$. These figures are entered in sway joints, i.e. the SJ row.

Rotate joints as normal. The restraint moments at B arise from two sources: carry-over moment of 8.84 from C and sway moment of −12.57. Therefore the net restraint moment is (8.84 − 12.57) = −3.73, the negative of which is distributed according to the distribution factors at B. Similarly, the total restraint moment at C is made up of 34.97 from carry-over from B and −42.07 due to sway. Therefore the negative of the net restraint moment −(34.97 − 42.07) = −7.10 is distributed at C according to distribution factors at C.

This completes the second cycle which is indicated by a discontinuous horizontal line.

Cycle 3. Carry-over the moments as normal.
$dM_{ba} = 1.60$, $dM_{cd} = 4.44$, $dM_{dc} = 2.22$. Therefore $\{dM_{ba} + dM_{cd} + dM_{dc}\} = 8.26$. Therefore sway moments are $M_{ba} = -0.13 \times 8.26 = -1.07$, $M_{cd} = M_{dc} = -0.435 \times 8.26 = -3.59$. These figures are entered in the SJ row.
Rotate joints in the usual way.
Further cycles are repeated until additional moments are negligibly small.

(h) Determination of displacements.

Joint B. M_{ba} due to the rotation of the joint = $52.54 + 1.60 - 0.11 - 0.08 - 0.03 - 0.02 = 53.90 = \{3E(0.6I)/6\} \, \theta_b$ because A is pinned. Therefore $EI \, \theta_b = 179.67$. As a check, due to the rotation of the joint

$$M_{bc} = 69.93 + 2.13 - 0.15 - 0.11 - 0.05 - 0.02 = 71.73 = \{4EI/10\} \, \theta_b$$

$$EI \, \theta_b = 179.33$$

Joint C. M_{cb} due to the rotation of the joint

$$= 17.67 + 2.66 + 0.95 + 0.40 + 0.17 + 0.08 = 21.93$$

$$= \{4EI/10\} \, \theta_b$$

$$EI \, \theta_c = 54.83$$

As a check $M_{dc} = 29.44 + 4.44 + 1.58 + 0.66 + 0.29 + 0.13$

$$= 36.54 = \{4EI/6\} \, \theta_c$$

Therefore

$$EI \, \theta_c = 54.81$$

Sway of CD. Summing up the moments due to the sway *including* the fixed end moment due to the sway of column CD at C we have

$$M_{cd} = -130.44 - 42.07 - 3.59 - 0.98 - 0.40 - 0.18 = -177.66$$

$$= -6EI \, \Delta/6^2$$

$$EI \, \Delta = 1065.96$$

As a check, doing a similar calculation on column AB we have

$$M_{ba} = -39.13 - 12.57 - 1.07 - 0.29 - 0.12 - 0.05 = -53.23$$

$$= -3E(0.6I)\Delta/6^2$$

$$EI \, \Delta = 1064.60$$

Joint A. There are two components to the rotation at A.

(i) Due to the rotation of θ_b: the rotation $EI\ \theta_{a1} = -0.5(EI\ \theta_b) = -89.84$ as explained in Chapter 4, Section 4.4, Example 4.

(ii) As explained in Chapter 4, Section 4.3, Example 2, in a cantilever subjected to an end load, the tip deflection $\Delta = WL^3/(3EI)$ and rotation at the tip is $\theta = WL^2/(2EI)$. Therefore $\theta = 1.5\Delta/L$. In a similar way, the rotation at A due to sway is given by

$$EI\ \theta_{a2} = 1.5EI\ \Delta/6 = 1.5 \times 1064.60/6 = 266.15$$

Therefore

$$EI\ \theta_a = -89.84 + 266.15 = 176.31$$

(i) Simplified structural stiffness matrix.

As shown in Fig. 7.9, considering the three independent displacement patterns, the simplified structural stiffness matrix relationship is given by

$$EI \begin{bmatrix} \dfrac{3(0.6)}{6} + \dfrac{4(1)}{10} & \dfrac{2(1)}{10} & \dfrac{-3(0.6)}{6^2} \\[2mm] \dfrac{2(1)}{10} & \dfrac{4(1)}{10} + \dfrac{4(1)}{6} & \dfrac{-6(1)}{6^2} \\[2mm] \dfrac{-3(0.6)}{6^2} & \dfrac{-6(1)}{6^2} & \dfrac{3(0.6)}{6^3} + \dfrac{12(1)}{6^3} \end{bmatrix} \begin{bmatrix} \theta_b \\[2mm] \theta_c \\[2mm] \Delta \end{bmatrix}$$

Simplifying,

$$EI \begin{bmatrix} 0.70 & 0.20 & -0.05 \\ 0.20 & 1.067 & -0.167 \\ -0.05 & -0.167 & 0.0639 \end{bmatrix} \begin{bmatrix} \theta_b \\ \theta_c \\ \Delta \end{bmatrix}$$

The load vector is given by

$$\begin{bmatrix} -(-83.33) \\ -(83.33) \\ 50.0 \end{bmatrix}$$

Carrying out Gaussian elimination we have:
Eliminating θ_b from the second and third equations,

$$EI \begin{bmatrix} 0.70 & 0.20 & -0.05 \\ 0 & 1.01 & -0.152 \\ 0 & -0.153 & 0.0603 \end{bmatrix} \begin{bmatrix} \theta_b \\ \theta_c \\ \Delta \end{bmatrix} = \begin{bmatrix} 83.33 \\ -107.14 \\ 55.95 \end{bmatrix}$$

Eliminating θ_c from the third equation,

$$EI \begin{bmatrix} 0.70 & 0.20 & -0.05 \\ 0 & 1.01 & -0.152 \\ 0 & 0 & 0.0373 \end{bmatrix} \begin{bmatrix} \theta_b \\ \theta_c \\ \Delta \end{bmatrix} = \begin{bmatrix} 83.33 \\ -107.14 \\ 39.78 \end{bmatrix}$$

Solving: $EI\ \Delta = 1066.46$, $EI\ \theta_c = 54.86$, $EI\ \theta_b = 179.55$

$$M_{ba} = M_{baf} + \{E(0.6I)/6\}[3\theta_b - 3\Delta/6]$$
$$= 0 + (0.6)/6[3 \times 179.55 - 3 \times 1066.46/6] = 0.54$$

$$M_{bc} = M_{bcf} + (EI/10)[4\theta_b + 2\theta_c]$$
$$= -83.33 + (1/10)[4 \times 179.55 + 2 \times 54.86] = -0.54$$

$$M_{cb} = M_{cbf} + (EI/10)[2\theta_b + 4\theta_c]$$
$$= 83.33 + (1/10)[2 \times 179.55 + 4 \times 54.86] = 141.19$$

$$M_{cd} = M_{cdf} + \{E(I)/6\}[4\theta_c - 6\Delta/6]$$
$$= 0 + (1/6)[4 \times 54.86 - 6 \times 1066.46/6] = -141.17$$

$$M_{dc} = M_{dcf} + \{E(I)/6\}[2\theta_c - 6\Delta/6]$$
$$= 0 + (1/6)[2 \times 54.86 - 6 \times 1066.46/6] = -159.46$$

Once again the reader should compare the numerical effort involved in the two methods.

Example 2

Analyse the structure shown in Fig. 7.10.

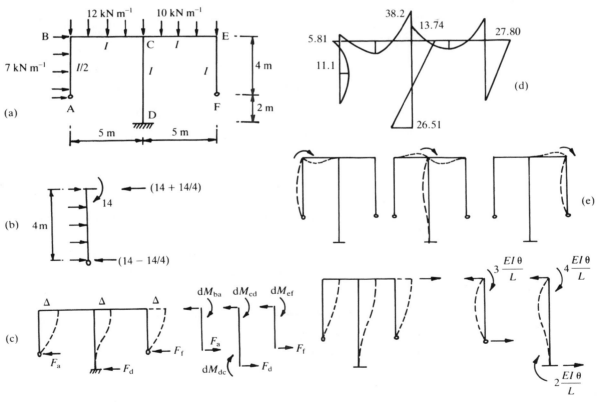

Figure 7.10 ● (a)–(e) Analysis of a portal frame

Solution   **(a)** Distribution factors.

Joint B. Because the joint A is pinned, $S_{ba} = 3E(0.5I)/4 = 0.375EI$ and because the joint C is continuous, $S_{bc} = 4EI/5 = 0.8EI$. Therefore $\Sigma S = 0.375EI + 0.8EI = 1.175EI$. Thus the distribution factor for BA $= S_{ba}/\Sigma S = 0.319$, and for BC $= S_{bc}/\Sigma S = 0.681$.

Joint C. Because the joint B is continuous, $S_{cb} = 4EI/5 = 0.8EI$. Similarly, because E is continuous, $S_{ce} = 4EI/5 = 0.8EI$. Finally, because D is fixed, $S_{cd} = 4EI/6 = 0.667EI$. Summing up stiffness factors, $\Sigma S = EI(0.8 + 0.8 + 0.667) = 2.267EI$. The distribution factors are for CB and CE $= 0.8EI/(2.267EI) = 0.353$ and for CD $= 0.667EI/(2.267EI) = 0.294$.

Joint E . Because C is continuous $S_{ec} = 4EI/5 = 0.8EI$ and because F is pinned, $S_{ef} = 3EI/4 = 0.75EI$. $\Sigma S = EI(0.80 + 0.75) = 1.55EI$. Therefore distribution factors are for EC $= S_{ec}/\Sigma S = 0.516$ and for EF $= S_{ef}/\Sigma S = 0.484$.

(b) Carry-over factors.

Joint B. From B to C is 0.5 and from B to A is zero because A is pinned.

Joint C. It is 0.5 from C to B, E and D.

Joint E. From E to C is 0.5 but from E to F is zero because F is pinned.

(c) Fixed end forces.

Member AB. Treating it as a propped cantilever, $M_{baf} = 7 \times 4^2/8 = 14.0$.

Horizontal reaction at B

$= 0.5(7 \times 4)$ from the u.d.l. $- M_{baf}/4$

$= 14.0 + 3.5 = 17.5$ to the left as shown in Fig. 7.10(b)

Member BC. $M_{bcf} = -12 \times 5^2/12 = -25.0$, $M_{cbf} = 25.0$

Member CE. $M_{cef} = -10 \times 5^2/12 = -20.83$, $M_{ecf} = 20.83$

(d) Sway factors.

Due to a horizontal displacement of Δ to the right without any rotation of the joints (see Fig. 7.10(c)), the horizontal forces developed at the base of the columns and the fixed end moments at the ends of the columns which develop are given by

Member AB.	$F_a = 3E(0.5I)\Delta/4^3 = 2.3438 \times 10^{-2}EI\,\Delta$ $M_{ba} = -3E(0.5I)\Delta/4^2 = -9.375 \times 10^{-2}EI\,\Delta$
Member CD.	$F_d = 12EI\,\Delta/6^3 = 5.5556 \times 10^{-2}EI\,\Delta$ $M_{cd} = M_{dc} = -6EI\,\Delta/6^2 = -16.6667 \times 10^{-2}EI\,\Delta$
Member EF.	$F_f = 3EI\,\Delta/4^3 = 4.6875 \times 10^{-2}EI\,\Delta$ $M_{ef} = -3EI\,\Delta/4^2 = -18.75 \times 10^{-2}EI\,\Delta$

(e) Sway moments to remove horizontal restraint.

To remove the horizontal restraint force of 17.5 to the left calculated in (c) above, we need

$$17.5 = F_a + F_d + F_f = (2.3438 + 5.5556 + 4.6875) \times 10^{-2}EI\,\Delta$$

Therefore

$$EI\,\Delta = 139.0335$$

The corresponding moments in the columns are

$$M_{ba} = -9.375 \times 10^{-2}EI\,\Delta = -13.03$$

$$M_{cd} = M_{dc} = -16.6667 \times 10^{-2}EI\,\Delta = -23.17$$

$$M_{ef} = -18.75 \times 10^{-2}EI\,\Delta = -26.07$$

(f) Sway factors.

If at any stage the changes in the moments in the columns from the values calculated above are dM_{ba}, dM_{cd}, dM_{dc} and dM_{ef}, then the horizontal equilibrium is disturbed. This can be restored as follows.

As shown in Fig. 7.10(c), the horizontal reaction to the left at the tops of the column due to the change in moments become $F_a = dM_{ba}/4$, $F_d = (dM_{cd} + dM_{dc})/6$, $F_f = dM_{ef}/4$. Therefore to restore equilibrium we need to sway the columns to the right by $d\Delta$ such that

$$F_a + F_d + F_f = (2.3438 + 5.5556 + 4.6875) \times 10^{-2}EI\,d\Delta$$

$$= dM_{ba}/4 + (dM_{cd} + dM_{dc})/6 + dM_{ef}/4$$

Therefore

$$EI\,d\Delta = 1.9862\{dM_{ba} + 2(dM_{cd} + dM_{dc})/3 + dM_{ef}\}$$

Thus the induced moments at the tops of the columns are

$$M_{ba} = -3E(0.5I)d\Delta/4^2$$

$$= -0.1862\{dM_{ba} + 2(dM_{cd} + dM_{dc})/3 + dM_{ef}\}$$

$$M_{cd} = M_{dc} = -6EI\,d\Delta/6^2$$

$$= -0.3310\{dM_{ba} + 2(dM_{cd} + dM_{dc})/3 + dM_{ef}\}$$

$$M_{ef} = -3EI\,d\Delta/4^2$$

$$= -0.3724\{dM_{ba} + 2(dM_{cd} + dM_{dc})/3 + dM_{ef}\}$$

(g) All the factors needed for joint rotation, carry-over and sway of the structure are available. The moment distribution process can now proceed. The details are shown in Table 7.5.

The following notes are added by way of explanation.

(i) Fixed end moments. In the beams this is due to lateral loads only. In the column BA, +14.00 is due to the lateral loads and −13.04 is due to the sway to balance the external restraint of 17.5. However, in the column CD at both the top and the bottom and in the column EF at the top only, the fixed end moment is due to the sway caused by removing the horizontal restraint.

Table 7.5

(BC)	(CB)	(CE)	(EC)	
−5.81	38.19	−13.74	27.80	Final
0.05	−0.04	−0.04	0.05	RJ
−0.05	0.06	0.07	−0.05	CO
0.12	−0.10	−0.10	0.13	RJ
−0.10	0.25	0.18	−0.10	CO
0.50	−0.20	−0.20	0.36	RJ
−0.77	−0.15	0.65	−0.77	CO
−0.29	−1.53	−1.53	1.30	RJ
3.36	8.19	1.35	1.36	CO
16.37	6.71	6.71	2.69	RJ
−25.00	25.00	−20.83	20.83	FEM

	B		C		E	
	(0.681)	(0.353)	(0.353)		(0.516)	DF (beams)
	→0.5	←0.5	→0.5		←0.5	COF (beams)
		(0.319)	(0.294)		(0.484)	DF (columns)
	[SF = 0.1862		SF = 0.3310		SF = 0.3723]	
	BA only		CD and DC		EF only	

B (BA)	C (CD)	D (DC)	E (EF)	Label
14.00	0.0	0.0	0.0	FEM
−13.04	−23.17	−23.17	−26.04	SWAY
7.67*	5.59*		2.52*	RJ
		2.80*(CO)		
−2.94	−5.22	−5.22	−5.88	SJ
−0.13*	−1.27*		1.22*	RJ
		−0.64*(CO)		
0.03	0.06	0.06	0.07	SJ
0.24*	−0.17*		0.34*	RJ
		−0.09*(CO)		
−0.08	−0.14	−0.14	−0.15	SJ
0.06*	−0.08*		0.12*	RJ
		−0.04*(CO)		
−0.02	−0.03	−0.03	−0.04	SJ
0.02*	−0.03*		0.04*	RJ
		−26.47		
5.81 (BA)	−24.46 (CD)		−27.80 (EF)	Final

A   D   F

*Indicates the change in moment in columns which will disturb the horizontal equilibrium

(ii) The moments at the joints are balanced in the usual way by removing the rotational restraint and allowing the joints to rotate.

(iii) Carry over the moments according to the carry-over factors.

(iv) Calculate $\{dM_{ba} + 2(dM_{cd} + dM_{dc})/3 + dM_{ef}\}$ resulting from the rotation of the joints and the carry-over moments. These moments are shown with an asterisk in Table 7.5. For example, after the first cycle of joint rotations and carry-over of moments we have

$$\{dM_{ba} + 2(dM_{cd} + dM_{dc})/3 + dM_{ef}\}$$
$$= 7.67 + 2(5.59 + 2.80)/3 + 2.52 = 15.78$$

The negative of this is distributed to the columns as determined by sway factors. Thus

$M_{ba} = -15.78 \times 0.1862 = -2.94$, $M_{cd} = M_{dc} = -15.78 \times 0.3310 = -5.22$ and $M_{ef} = -15.78 \times 0.3723 = -5.88$.

(v) Steps 1–3 are repeated.

(vi) Repeat step 4. In the second cycle we have

$$\{dM_{ba} + 2(dM_{cd} + dM_{dc})/3 + dM_{ef}\}$$
$$= \{-0.13 + 2(-1.27 - 0.64)/3 + 1.22\} = -0.18$$

The process is repeated until the changes in moments are negligibly small. The BMD is shown in Fig. 7.10(d).

(h) Calculation of displacements.

(i) *Rotation of joints.* Summing up the moments due to the rotation of joints we have

Joint B.

$$M_{ba} = 7.67 - 0.13 + 0.24 + 0.06 + 0.02 = 7.86 = \{3E(0.5I)/4\}\,\theta_b$$

Therefore

$$EI\,\theta_b = 20.96$$

$$M_{bc} = 16.37 - 0.29 + 0.50 + 0.12 + 0.05 = 16.75 = \{4EI/5\}\,\theta_b$$

Therefore

$$EI\,\theta_b = 20.94$$

Joint C.

$$M_{cb} = M_{ce} = 6.71 - 1.53 - 0.20 - 0.10 - 0.04 = 4.84 = \{4EI/5\}\,\theta_c$$

Therefore

$$EI\,\theta_c = 6.05$$

$$M_{cd} = 5.59 - 1.27 - 0.17 - 0.08 - 0.03 = 4.04 = \{4EI/6\}\,\theta_c$$

Therefore

$$EI\,\theta_c = 6.06$$

Joint E.

$$M_{ec} = 2.69 + 1.30 + 0.36 + 0.13 + 0.05 = 4.53 = \{4EI/5\}\,\theta_e$$

Therefore

$$EI\,\theta_e = 5.66$$

$$M_{ef} = 2.52 + 1.22 + 0.34 + 0.12 + 0.04 = 4.24 = \{3EI/4\}\,\theta_e$$

Therefore

$$EI\,\theta_e = 5.65$$

(ii) *Sway translation.* Summing up the moments induced by sway we have

Joint B.

$$M_{ba} = -13.04 - 2.94 + 0.03 - 0.08 - 0.02$$
$$= -16.05 = -\{3E(0.5I)/4^2\}\,\Delta$$

Therefore

$$EI\,\Delta = 171.2$$

Joint C.

$$M_{cd} = M_{dc} = -23.17 - 5.22 + 0.06 - 0.14 - 0.03$$
$$= -28.5 = -\{6EI/6^2\}\,\Delta$$

Therefore

$$EI\,\Delta = 171.0$$

Joint E.

$$M_{ef} = -26.04 - 5.88 + 0.07 - 0.15 - 0.04$$
$$= -32.04 = -\{3EI/4^2\}\,\Delta$$

Therefore

$$EI\,\Delta = 170.88$$

(j) Structural stiffness matrix. Using the displacement patterns shown in Fig. 7.10(e), the simplified structural stiffness matrix is given by

$$
\begin{bmatrix} M_B \\ M_C \\ M_D \\ F \end{bmatrix} = EI
\begin{bmatrix}
\dfrac{3(0.5)}{4}+\dfrac{4}{5} & \dfrac{2}{5} & 0 & \dfrac{-3(0.5)}{4^2} \\[2mm]
\dfrac{2}{5} & \dfrac{4}{5}+\dfrac{4}{5}+\dfrac{4}{6} & \dfrac{2}{5} & \dfrac{-6}{6^2} \\[2mm]
0 & \dfrac{2}{5} & \dfrac{4}{5}+\dfrac{3}{4} & \dfrac{-3}{4^2} \\[2mm]
\dfrac{-3(0.5)}{4^2} & \dfrac{-6}{6^2} & \dfrac{-3}{4^2} & \dfrac{3(0.5)}{4^3}+\dfrac{12}{6^3}+\dfrac{3}{4^3}
\end{bmatrix}
\begin{bmatrix} \theta_b \\ \theta_c \\ \theta_e \\ \Delta \end{bmatrix}
$$

Simplifying,

$$EI \begin{bmatrix} 1.175 & 0.40 & 0 & -0.0938 \\ 0.40 & 2.267 & 0.40 & -0.1667 \\ 0 & 0.40 & 1.55 & -0.1875 \\ -0.0938 & -0.1667 & -0.1875 & 0.1259 \end{bmatrix}$$

The load vector is given by

$$\begin{bmatrix} -(14.00 - 25.00) \\ -(25.00 - 20.83) \\ -(20.83) \\ -(-17.5) \end{bmatrix} = \begin{bmatrix} 11.0 \\ -4.17 \\ -20.83 \\ 17.50 \end{bmatrix}$$

Carrying out the Gaussian elimination we have

(i) Original equations:

$$EI \begin{bmatrix} 1.175 & 0.40 & 0 & -0.0938 \\ 0.40 & 2.267 & 0.40 & -0.1667 \\ 0 & 0.40 & 1.55 & -0.1875 \\ -0.0938 & -0.1667 & -0.1875 & 0.1259 \end{bmatrix} \begin{bmatrix} \theta_b \\ \theta_c \\ \theta_e \\ \Delta \end{bmatrix} = \begin{bmatrix} 11.00 \\ -4.17 \\ -20.83 \\ 17.50 \end{bmatrix}$$

(ii) Eliminating θ_b from equations 2, 3 and 4:

$$EI \begin{bmatrix} 1.175 & 0.40 & 0 & -0.0938 \\ 0 & 2.131 & 0.40 & -0.1348 \\ 0 & 0.40 & 1.55 & -0.1875 \\ 0 & -0.1348 & -0.1875 & 0.1183 \end{bmatrix} \begin{bmatrix} \theta_b \\ \theta_c \\ \theta_e \\ \Delta \end{bmatrix} = \begin{bmatrix} 11.00 \\ -7.92 \\ -20.83 \\ 18.38 \end{bmatrix}$$

(iii) Eliminating θ_c from equations 3 and 4:

$$EI \begin{bmatrix} 1.175 & 0.40 & 0 & -0.0938 \\ 0 & 2.131 & 0.40 & -0.1348 \\ 0 & 0 & 1.475 & -0.1622 \\ 0 & 0 & -0.1622 & 0.1098 \end{bmatrix} \begin{bmatrix} \theta_b \\ \theta_c \\ \theta_e \\ \Delta \end{bmatrix} = \begin{bmatrix} 11.00 \\ -7.92 \\ -19.34 \\ 17.88 \end{bmatrix}$$

(iv) Eliminating θ_e from equation 4:

$$EI \begin{bmatrix} 1.175 & 0.40 & 0 & -0.0938 \\ 0 & 2.131 & 0.40 & -0.1348 \\ 0 & 0 & 1.475 & -0.1622 \\ 0 & 0 & 0 & 0.0920 \end{bmatrix} \begin{bmatrix} \theta_b \\ \theta_c \\ \theta_e \\ \Delta \end{bmatrix} = \begin{bmatrix} 11.00 \\ -7.92 \\ -19.34 \\ 15.75 \end{bmatrix}$$

Solving:

$$EI \Delta = 171.14, \ EI \theta_e = 5.71, \ EI \theta_c = 6.04, \ EI \theta_b = 20.97$$

Forces in the members:

Member AB.

$$M_{ba} = M_{baf} + \{E(0.5I)/4\}[3\theta_b - 3\Delta/4]$$

$$= 14.00 + (0.5/4)[3 \times 20.97 - 3 \times 171.14/4] = 5.82$$

Member BC.

$$M_{bc} = M_{bcf} + (EI/5)\{4\theta_b + 2\theta_c\}$$

$$= -25.00 + (1/5)\{4 \times 20.97 + 2 \times 6.04\} = -5.81$$

$$M_{cb} = M_{cbf} + (EI/5)\{2\theta_b + 4\theta_c\}$$

$$= 25.00 + (1/5)\{2 \times 20.97 + 4 \times 6.04\} = 38.22$$

Member CE.

$$M_{ce} = M_{cef} + (EI/5)\{4\theta_c + 2\theta_e\}$$

$$= -20.83 + (1/5)\{4 \times 6.04 + 2 \times 5.71\} = -13.71$$

$$M_{ec} = M_{ecf} + (EI/5)\{2\theta_c + 4\theta_e\}$$

$$= 20.83 + (1/5)\{2 \times 6.04 + 4 \times 5.71\} = 27.81$$

Member CD.

$$M_{cd} = M_{cdf} + (EI/6)\{4\theta_c - 6\Delta/6\}$$

$$= 0 + (1/6)\{4 \times 6.04 - 6 \times 171.14/6\} = -24.50$$

$$M_{dc} = M_{dcf} + (EI/6)\{2\theta_c - 6\Delta/6\}$$

$$= 0 + (1/6)\{2 \times 6.04 - 6 \times 171.14/6\} = -26.51$$

Member EF.

$$M_{ef} = M_{eff} + (EI/4)\{3\theta_e - 3\Delta/4\}$$

$$= 0 + (1/4)\{3 \times 5.71 - 3 \times 171.14/4\} = -27.81$$

The reader should once again compare the numerical effort involved in the two methods.

7.5 ● Special techniques in the moment distribution method

When moment distribution was the only practicable method for the analysis of rigid-jointed structures, many special techniques were developed to simplify and reduce the number of iteration cycles. In particular special stiffness and carry-over factors to simplify the analysis of symmetrical structures were derived. These are discussed in detail in the monographs on the moment distribution method by Lightfoot (1961) and Kani (1957). However, the advent of digital computers has radically altered the scene and, as remarked in the introduction to this chapter, nowadays the moment distribution method is used only when access to a computer is unobtainable. It is for this reason that special techniques will not be discussed. Interested readers should consult references in Section 7.6.

7.6 ● References

Gere J M 1963 *Moment Distribution Method* Van Nostrand

Kani G N J 1957 *Analysis of Multistorey Frames* Crosby-Lockwood

Lightfoot E 1961 *Moment Distribution Method* Spon

7.7 ● Problems

1. Analyse the continuous beam shown in Fig. E7.1. Check the answers using the 'simplified' stiffness method.

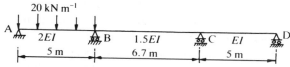

Figure E7.1

Answer: Distribution factors: BA : BC :: 0.57 : 0.43. CB : CD :: 0.6 : 0.4. Carry-over factors = 0.5 from B to C and C to B. Elsewhere zero. Fixed end moment at BA $= 20 \times 5^2/8 = 62.50$. Final moments: $M_{BA} = 24.42$, $M_{CD} = 5.75$. $EI(\theta_A, \theta_B, \theta_C, \theta_D)$ $= (41.9, -32.08, 9.60, -4.80)$.

2. Analyse the rigid-jointed frame shown in Fig. E7.2. Check the answers using the 'simplified' stiffness method.

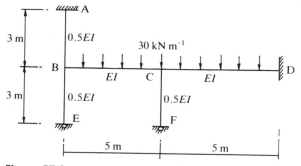

Figure E7.2

Answer: Distribution factors: BA : BC : BE :: 0.34 : 0.41 : 0.25, CB : CD : CF :: 0.38 : 0.38 : 0.24. Carry-over factors: 0.5 from B to A and C. 0.5 from C to B and D. FEM: $M_{BC} = -62.5$, $M_{CB} = 62.5$, $M_{CD} = -62.5$, $M_{DC} = 62.5$. Final moments: $M_{AB} = 11.06$, $M_{BA} = 22.11$, $M_{BE} = 16.26$, $M_{BC} = -38.37$, $M_{CB} = 70.76$, $M_{CD} = -67.56$, $M_{CF} = -3.20$, $M_{DC} = 59.96$. $EI(\theta_B, \theta_C, \theta_E, \theta_F) = (33.06, -6.30, -16.67, 3.16)$.

3. Analyse the rigid-jointed frame shown in Fig. E7.3. Check the answers using the 'simplified' stiffness method.

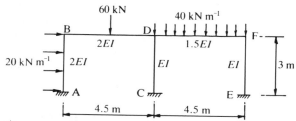

Figure E7.3

Answer: Distribution factors: BA: BD :: 0.6 : 0.4. DB : DF : DC :: 0.4 : 0.3 : 0.3, FD: FE :: 0.5 : 0.5. Carry-over factors are 0.5 except from fixed ends A, C and E. Fixed end moments: $-M_{BD} = M_{DB} = 33.75$, $-M_{DF} = M_{FD} = 67.5$, $-M_{AB} = M_{BA} = 15.0$. FEM at tops of columns due to sway: $M_{AB} = M_{BA} = -22.50$, $M_{DC} = M_{CD} = M_{FE} = M_{EF} = -11.25$. Sway factors: Let $K = dM_{BA} + dM_{AB} + dM_{DC} + dM_{CD} + dM_{FE} + dM_{EF}$. $M_{AB} = M_{BA} = -0.25\ K$, $M_{DC} = M_{CD} = M_{FE} = M_{EF} = -0.125\ K$. Final moments:

$M_{AB} = -29.38$, $M_{BA} = 10.07$, $M_{BD} = -10.07$
$M_{DB} = 62.19$, $M_{DF} = -66.87$, $M_{DC} = 4.68$
$M_{CD} = -3.62$, $M_{FD} = 43.86$, $M_{FE} = -43.86$
$M_{EF} = -27.89$

$$EI\begin{bmatrix} 4.44 & 0.89 & 0 & -1.33 \\ 0.89 & 4.44 & 0.67 & -0.67 \\ 0 & 0.67 & 2.67 & -0.67 \\ -1.33 & -0.67 & -0.67 & 1.78 \end{bmatrix}\begin{bmatrix} \theta_B \\ \theta_D \\ \theta_F \\ \Delta \end{bmatrix} = \begin{bmatrix} 18.75 \\ 33.75 \\ -67.50 \\ 30.0 \end{bmatrix}$$

After Gaussian elimination:

$$EI\begin{bmatrix} 4.44 & 0.89 & 0 & -1.33 \\ 0 & 4.27 & 0.67 & -0.40 \\ 0 & 0 & 2.56 & -0.60 \\ 0 & 0 & 0 & 1.20 \end{bmatrix}\begin{bmatrix} \theta_B \\ \theta_D \\ \theta_F \\ \Delta \end{bmatrix} = \begin{bmatrix} 18.7 \\ 30.0 \\ -72.19 \\ 21.42 \end{bmatrix}$$

$EI(\theta_B, \theta_D, \theta_F, \Delta) = (7.09, 12.45, -23.96, 17.88)$

Chapter eight

Elastic stability analysis

Elastic stress calculation for members subjected to axial and shear forces, bending and twisting moments was discussed in detail in previous chapters. A structural member which carries lateral load only is called a beam. Similarly, a member which resists axial loads only is called a column. Therefore a member which carries both lateral and axial loads is generally called a beam-column. Consider a beam-column shown in Fig. 8.1 subjected to a lateral load which causes bending moment and shear force at a section. Simultaneous with the lateral load, let an axial force P also act at the ends of the member. In the treatment adopted in the previous chapters it was assumed that the axial force causes only axial stress and that the bending moment at a section is caused by the lateral loads only. This is an obvious simplification as can be seen by noticing, as shown in Fig. 8.2, that if the beam deforms under the action of the lateral load, then the axial load will be acting eccentrically with respect to the *deformed* member and the axial forces cause an additional bending moment equal to axial load times the eccentricity. The object of this chapter is to study in detail the effect of this additional bending moment caused by the axial loads as this has serious implications for the overall behaviour of members subjected to axial and lateral loads.

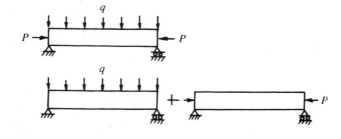

Figure 8.1 ● A beam-column

Figure 8.2 ● A beam-column

8.1 ● Axially loaded beam-column with initial deformation

As an introduction to the problem, consider a simply supported beam-column as shown in Fig. 8.3. Let us assume that, even in the absence of external loads, the

Figure 8.3 ●
An initially deformed
beam-column under
axial force P

member is initially bent. This is a common situation in practice because it is practically impossible to produce perfectly straight beam-columns. Let the initial lateral displacement at any point be v_i and the axial compressive force be P. Under the action of the axial load, the member deforms laterally by v. Therefore the total displacement at any point is $v + v_i$. The bending moment caused at a section is then $-P(v + v_i)$. The minus sign is used because the bending moment causes tension at the top face. Therefore, using the moment-curvature relationship established in Section 4.2, we have

$$EI\ d^2v/dx^2 = -P(v + v_i)$$

Therefore

$$d^2v/dx^2 + \alpha^2 v = -\alpha^2 v_i$$

where $\alpha^2 = P/(EI)$

The solution to the above second-order non-homogeneous equation is given by

$$v = A \cos \alpha x + B \sin \alpha x + \text{particular integral}$$

where A and B are constants of integration.

In order to be able to determine the particular integral, it is necessary to know v_i as a function of x. For simplicity, it is assumed that

$$v_i = v_0 \sin (\pi x/L)$$

where v_0 is the initial displacement at midspan. It should be noted that the assumed shape of initial displacement satisfies the initial conditions of zero displacement and curvature at the ends of the beam-column. For this assumption, the particular integral is given by

$$\text{Particular integral} = \frac{\alpha^2}{(\pi^2/L^2 - \alpha^2)} v_0 \sin (\pi x/L)$$

$$= \frac{P}{(P_e - P)} v_0 \sin (\pi x/L)$$

where $P_e = \pi^2 EI/L^2$ has the same dimensions as the axial load P and is called the Euler load in honour of the Swiss mathematician Euler, who was the first to investigate the buckling of columns.

The constants A and B can be determined from the boundary conditions as follows.

Because $v = 0$ at $x = 0$ and $x = L$, it can be verified that $A = B = 0$. Therefore the final displacement is equal to the value given by the particular integral only. Therefore

$$v = \{P/(P_e - P)\} v_i$$

$$v_{total} = v + v_i = P_e v_i/(P_e - P)$$

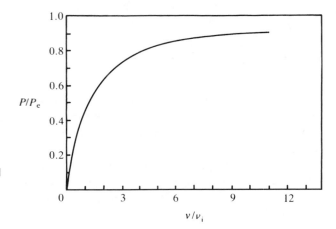

Figure 8.4 ● Axial load versus lateral deflection relationship for a beam-column

$$M = -P(v + v_i) = -Pv_i[P_e/(P_e - P)]$$

$$= M_i[P_e/\{P_e - P\}]$$

where M_i = initial bending moment = $-Pv_i$.

Figure 8.4 shows a plot of v/v_i vs. P/P_e. For values of $P/P_e \leqslant 0.4$, the response is almost linear. However, for values of $P/P_e > 0.4$, the response is decidedly non-linear. This is a simple case of an elastic structure exhibiting non-linear behaviour. It can be seen that v_{total} is the initial displacement v_i magnified by a factor $P_e/(P_e - P)$ and the bending moment M is the initial bending moment M_i magnified by the same factor. The magnification factor $P_e/(P_e - P)$ is non-linearly related to the axial load P. It is interesting to note that as P tends to P_e, then the displacement v and the bending moment both tend to infinity irrespective of however small the initial displacement v_i is. Of course at this large displacement, the material is unlikely to remain elastic as was assumed in the derivation of the above equation. When the displacement tends to infinity the structure is said to be unstable. The axial load $P = P_e$ at which the simply supported beam-column becomes unstable is known as the elastic critical or buckling load P_{cr} of the member. In this particular case $P_{cr} = P_e$. It is fair to say that in the majority of practical cases, because of the many simplifying assumptions made in the derivation of P_{cr}, such as the material remaining elastic even at very large displacements, the concept of elastic buckling load is a mathematical rather than a real entity. However, in the absence of anything more simple and convenient, the mathematical entity of elastic critical load is commonly used by engineers as a measure of the ability of columns to resist buckling.

The elastic critical load P_{cr} for a simply supported beam-column is given by $P_e = \pi^2 EI/L^2$. Expressing the second moment of area $I = Ar^2$, where A = cross-sectional area and r = radius of gyration, then $P_e/A = \sigma_{cr}$. σ_{cr} can be expressed as

$$\sigma_{cr} = \frac{\pi^2 EI}{AL^2} = \frac{\pi^2 Er^2}{L^2} = \frac{\pi^2 E}{(L/r)^2}$$

where σ_{cr} = axial stress to cause elastic buckling. In other words, for a given material, elastic critical stress σ_{cr} depends on the L/r ratio, generally known as the slenderness ratio. For a given material and cross-section, the critical load is governed entirely by the length of the beam-column. The longer the member becomes, the lower the elastic critical load becomes, which agrees with common experience.

A beam-column will, if it is not restrained in any particular plane, buckle about an axis with the least value of second moment of area. As was shown in Section 3.12.5, the maximum or minimum second moment of area will occur about a principal axis. For example, a rectangular column will buckle about an axis parallel to its depth and an I section will buckle about its minor axis which is parallel to the web. Similarly, an angle section will buckle about its principal minor axis.

8.2 ● Experimental determination of P_{cr} – Southwell plot

The concept of elastic critical load is a purely mathematical concept because near P_{cr} the displacements tend to infinity and the material therefore does not remain elastic. It appears at first sight that P_{cr} cannot be determined experimentally. However, the equation derived above,

$$v = \{P/(P_e - P)\} v_i$$

i.e. $(P_e - P)v = Pv_i$

and dividing throughout by PP_e,

$$(v/P) = (1/P_e)v + v_i/P_e$$

can be used for the experimental determination of P_{cr}, which is a purely mathematical entity. This is done by noting that the equation is of the form $y = mx + c$, an equation for a straight line where the ordinate $y = v/P$, the abscissa $x = v$, the slope of the line $m = 1/P_e$ and the intercept on the y-axis $C = v_i/P_e$. Therefore by conducting experiments on an imperfect column under loads well *below* the critical load $P_{cr} = P_e$, the critical load is determined from the slope of the best straight line fit to the experimental points. This observation was first made by Southwell and is generally known as the Southwell plot.

8.3 ● Derivation of the beam-column differential equation

In studying the elastic buckling load of a column, it is necessary to determine the load at which the structure remains in equilibrium in the deformed position. In order to derive the necessary equations, consider an element of a beam-column in the deformed position with the forces acting as shown in Fig. 8.5. It is assumed that during the deformation the axial load remains in its original direction. The vertical and moment equilibrium considerations require that

$$dQ/dx = q$$

$$\{M + (dM/dx)dx\} - M + Q\,dx + P(dv/dx)dx - q\,dx\,dx/2 = 0$$

Ignoring second-order terms we have

$$dM/dx + Q + P\,dv/dx = 0$$

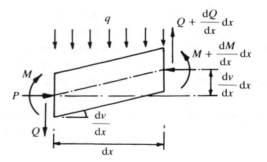

Figure 8.5 ● Forces acting on a beam-column in the deformed state

Therefore

$$Q = -\mathrm{d}M/\mathrm{d}x - P\,\mathrm{d}v/\mathrm{d}x$$

and

$$\mathrm{d}Q/\mathrm{d}x = -\mathrm{d}^2M/\mathrm{d}x^2 - P\,\mathrm{d}^2v/\mathrm{d}x^2 = q$$

Because $EI\,\mathrm{d}^2v/\mathrm{d}x^2 = M$, the above equation can be written as

$$\mathrm{d}^2\{EI\,\mathrm{d}^2v/\mathrm{d}x^2\}/\mathrm{d}x^2 + P\,\mathrm{d}^2v/\mathrm{d}x^2 = -q$$

If it is assumed that EI is constant, then the above equation can be written as

$$EI\,\mathrm{d}^4v/\mathrm{d}x^4 + P\,\mathrm{d}^2v/\mathrm{d}x^2 = -q$$

This equation is generally known as the beam-column equation.

Note that in a beam-column shear force Q is given by

$$Q = -\mathrm{d}M/\mathrm{d}x - P\,\mathrm{d}v/\mathrm{d}x$$

$$= -(EI\,\mathrm{d}^3v/\mathrm{d}x^3 + P\,\mathrm{d}v/\mathrm{d}x)$$

As shown in Fig. 8.5, the second term denotes the component of P normal to the neutral axis which is equal to $P \sin(\mathrm{d}v/\mathrm{d}x) = P\,\mathrm{d}v/\mathrm{d}x$ because $(\mathrm{d}v/\mathrm{d}x)$ is very small.

Note that if $P = 0$, then the above equations are identical to the equations derived in Section 4.2.

The solution to the beam-column equation, which is a fourth-order ordinary differential equation with constant coefficients, is given by

$$v = A \cos \alpha x + B \sin \alpha x + C\alpha x + D + \text{particular integral}$$

where $\alpha^2 = P/(EI)$ and A to D are constants of integration. Once the value of q is known, then the particular integral can be determined.

8.3.1 Solution of the beam-column equation for various boundary conditions

It is useful to calculate the elastic buckling loads for columns with commonly assumed ideal boundary conditions. This is done by calculating the axial load $P = P_{cr}$ at which the column remains in equilibrium in a deformed position even when $q = 0$. The procedure is illustrated by some simple examples. It is useful, before any

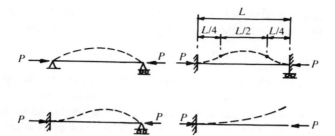

Figure 8.6 ● Mode of buckling of beam-columns with various boundary conditions

calculations are performed, to have the following expressions for the various derivatives of v:

$$v = A \cos \alpha x + B \sin \alpha x + C\alpha x + D$$

$$dv/dx = \alpha\{-A \sin \alpha x + B \cos \alpha x + C\}$$

$$d^2v/dx^2 = \alpha^2\{-A \cos \alpha x - B \sin \alpha x\}$$

$$d^3v/dx^3 = \alpha^3\{A \sin \alpha x - B \cos \alpha x\}$$

Because $M = EI\ d^2v/dx^2$
and $Q = -(EI\ d^3v/dx^3 + P\ dv/dx) = -EI\ (d^3v/dx^3 + \alpha^2\ dv/dx)$, then

$$M = EI\ \alpha^2\{-A \cos \alpha x - B \sin \alpha x\}$$

$$Q = -EI\ \alpha^3\{C\}$$

Note that there are four constants of integration to be determined. This requires four boundary conditions. The four cases shown in Fig. 8.6 will be considered.

1 Pin-ended column. In this case both ends are simply supported. Therefore the boundary conditions are: at the ends $x = 0$ and $x = L$, $v = 0$ and bending moment $M = EI\ d^2v/dx^2 = 0$. Therefore $v = 0$ at $x = 0$ requires that $A + D = 0$. Similarly, $d^2v/dx^2 = 0$ at $x = 0$ requires that $A = 0$. Therefore $A = D = 0$. Since at $x = L$, $v = 0$, and since $A = D = 0$,

$$B \sin \alpha L + C\alpha L = 0$$

Similarly, since at $x = L$, $d^2v/dx^2 = 0$, and since $A = 0$,

$$\therefore\ B \sin \alpha L = 0$$

$$\therefore\ C = 0$$

$$\therefore\ v = B \sin \alpha x$$

where $B \sin \alpha L = 0$.

If $B = 0$, then $v = 0$ and the column remains undeformed. This is a 'trivial' solution and is of no practical importance because, as remarked before, inevitable lack of straightness of the column will make the column deform in bending under axial load P. Therefore, in order for $v \neq 0$, then B cannot be zero. $B \sin \alpha L = 0$, therefore $\sin \alpha L = 0$. This is possible when $\alpha L = n\pi$, where $n = 1, 2, 3, \ldots$

Because $\alpha^2 = P/(EI)$, $P_{cr} = n^2\pi^2EI/L^2$. The smallest critical load occurs when $n = 1$. In other words, the elastic critical load for a pin-ended column is

$$P_{cr} = \pi^2EI/L^2 = P_e$$

a result obtained in Section 8.1 by considering the axial load at which the displacements of an initially deformed member tend to infinity.

It should be noted that although $v = B \sin \alpha x$ the value of the constant B has remained indeterminate. The solution has succeeded in determining only the *shape* of the deformed column, which at the critical load is a half sine curve. The shape of the deformed structure at buckling is known as the mode of buckling.

2 Fixed-ended or clamped column. In this case, because both the ends are clamped, the boundary conditions at $x = 0$ and $x = L$ are $v = 0$ and $dv/dx = 0$.
$v = 0$ at $x = 0$ requires that $A + D = 0$. Therefore

$$D = -A$$

$dv/dx = 0$ at $x = 0$ requires that $B + C = 0$. Therefore

$$C = -B$$

$v = 0$ at $x = L$ requires that

$$A \cos \alpha L + B \sin \alpha L + C\alpha L + D = 0$$

Substituting $D = -A$ and $C = -B$,

$$A(\cos \alpha L - 1) + B(\sin \alpha L - \alpha L) = 0$$

Therefore

$$B = -A(\cos \alpha L - 1)/(\sin \alpha L - \alpha L)$$

Using $C = -B$, $D = -A$ and $B = -A(\cos \alpha L - 1)/(\sin \alpha L - \alpha L)$, v can be expressed as

$$v = A[(\cos \alpha x - 1) - (\cos \alpha L - 1)(\sin \alpha x - \alpha x)/(\sin \alpha L - \alpha L)]$$

Therefore

$$dv/dx = -A\alpha[\sin \alpha x + (\cos \alpha L - 1)(\cos \alpha x - 1)/(\sin \alpha L - \alpha L)]$$

$$d^2v/dx^2 = -A\alpha^2[\cos \alpha x - (\cos \alpha L - 1)\sin \alpha x/(\sin \alpha L - \alpha L)]$$

Obviously if $A = 0$, then $v = 0$ and the column is undeformed. For $v \neq 0$, then $A \neq 0$. Therefore for $dv/dx = 0$ at $x = L$,

$$[\sin \alpha L + (\cos \alpha L - 1)(\cos \alpha L - 1)/(\sin \alpha L - \alpha L)] = 0$$

Simplifying,

$$\sin^2 \alpha L - \alpha L \sin \alpha L + \cos^2 \alpha L + 1 - 2\cos \alpha L = 0$$

Because $\sin^2 \alpha L + \cos^2 \alpha L = 1$, the above equation can be simplified to

$$2(1 - \cos \alpha L) = \alpha L \sin \alpha L$$

Because $\cos 2\alpha = 1 - 2\sin^2\alpha$ and $\sin 2\alpha = 2\sin\alpha\cos\alpha$, the above equation can be simplified to

$$\sin\alpha L/2\,\{\sin\alpha L/2 - (\alpha L/2)\,\cos\alpha L/2\} = 0$$

Therefore either $\sin\alpha L/2 = 0$ for which the smallest root is $\alpha L/2 = \pi$ or $\{\sin\alpha L/2 - (\alpha L/2)\cos\alpha L/2\} = 0$, i.e. $\alpha L/2 = \tan\alpha L/2$ for which the smallest root is $\alpha L/2 = 1.43\,\pi$. The lower value of αL is obtained when $\sin\alpha L/2 = 0$. Therefore $\alpha L/2 = \pi$ or $\alpha^2 = 4\pi^2/L^2$. Therefore, because $\alpha^2 = P/(EI)$

$$P_{cr} = 4\pi^2 EI/L^2 = 4P_e$$

The critical load for a fixed-ended beam-column is four times that of a pin-ended column. As in the case of a pin-ended column, the constant A has remained indeterminate. The deformed shape has zero deflection and slope at the ends and in addition, as in the case of fixed beams subjected to lateral load, there are two symmetrically positioned points of contraflexure. Their location is determined by setting $d^2v/dx^2 = 0$ when $\alpha L = 2\pi$:

$$d^2v/dx^2 = -A\alpha^2[\cos\alpha x - (\cos\alpha L - 1)\sin\alpha x/(\sin\alpha L - \alpha L)]$$

Because $\cos\alpha L - 1 = 0$, when $\alpha L = 2\pi$, the above equation simplifies to

$$d^2v/dx^2 = -A\alpha^2[\cos\alpha x] = -A\alpha^2[\cos 2\pi x/L]$$

Therefore

$$d^2v/dx^2 = 0 \text{ when } \cos 2\pi(x/L) = 0. \;\therefore\; 2\pi\left(\frac{x}{L}\right) = \frac{\pi}{2} \text{ or } \frac{3\pi}{2}$$

Thus $x/L = 0.25$ or 0.75 when $0 \leqslant x/L \leqslant 1$. Therefore the points of contraflexure lie at $0.25L$ from each end. The column between the points of contraflexure behaves, as in the case of laterally loaded beams, as a pin-ended column of length $0.5L$ with a buckling load equal to $\pi^2 EI/(0.5L)^2 = 4\pi^2 EI/L^2$ which is the same as the buckling load for the clamped column.

3 Propped cantilever. In this case the boundary conditions are at the fixed end at $x = 0$, $v = dv/dx = 0$ and at the simply supported end at $x = L$, $v = 0$ and $M = EId^2v/dx^2 = 0$.
Therefore $v = 0$ at $x = 0$ requires that $A + D = 0$. Thus

$$D = -A$$

$dv/dx = 0$ at $x = 0$ requires that $B + C = 0$. Therefore

$$C = -B$$

$v = 0$ at $x = L$ requires that

$$A\cos\alpha L + B\sin\alpha L + C\alpha L + D = 0$$

Substituting $D = -A$ and $C = -B$, the above equation simplifies to

$$A(\cos\alpha L - 1) + B(\sin\alpha L - \alpha L) = 0$$

$d^2v/dx^2 = 0$ at $x = L$ requires that

$$A\cos\alpha L + B\sin\alpha L = 0$$

Therefore

$$B = -A \cos \alpha L / \sin \alpha L$$

provided that $\sin \alpha L \neq 0$. Substituting for $B = -A \cos \alpha L / \sin \alpha L$ in $A(\cos \alpha L - 1) + B(\sin \alpha L - \alpha L) = 0$, the equation becomes

$$A[(\cos \alpha L - 1) - (\sin \alpha L - \alpha L) \cos \alpha L / \sin \alpha L] = 0$$

Therefore

$$A[\cos \alpha L \sin \alpha L - \sin \alpha L - \sin \alpha L \cos \alpha L + \alpha L \cos \alpha L] = 0$$

i.e. $A[-\sin \alpha L + \alpha L \cos \alpha L] = 0$

To ensure that the beam-column is in a deformed state, $\sin \alpha L = \alpha L \cos \alpha L$, $A \neq 0$. Therefore $\tan \alpha L = \alpha L$. The solution of this transcendental equation has the smallest root at $\alpha L = 1.43\pi$ giving

$$P_{cr} = 2.04 \pi^2 EI/L^2 = 2.04 P_e$$

4 Cantilever. The boundary conditions in this case are at the fixed end at $x = 0$, $v = dv/dx = 0$ and at the 'free' end at $x = L$, bending moment $M = 0$ and shear force $Q = 0$. Therefore, as in the case of a propped cantilever, the boundary conditions at the fixed end require that $D = -A$ and $C = -B$. The boundary conditions at $x = L$ require that

$$M = EI\, \alpha^2[A \cos \alpha L + B \sin \alpha L] = 0$$

$$Q = EI\, \alpha^3 C = 0$$

Therefore $A \cos \alpha L + B \sin \alpha L = 0$, $C = 0$.
Because $C = -B$, $B = 0$. Thus

$$A \cos \alpha L + B \sin \alpha L = A \cos \alpha L = 0$$

As $A \neq 0$, $\cos \alpha L = 0$.
The smallest root is given by $\alpha L = \pi/2$. Therefore $\alpha^2 = 0.25\pi^2/L^2$, leading to

$$P_{cr} = 0.25\pi^2 EI/L^2 = 0.25 P_e$$

8.3.2 Concept of effective length

In the previous section, the critical elastic buckling loads for columns with idealized boundary conditions were calculated, which can be expressed as follows:

(a) Cantilever: $P_{cr} = 0.25\pi^2 EI/L^2 = \pi^2 EI/(2L)^2$

$$= \pi^2 EI/L_{eff}^2, \qquad L_{eff} = 2L$$

(b) Pin-ended column: $P_{cr} = \pi^2 EI/L^2 = \pi^2 EI/L^2$

$$= \pi^2 EI/L_{eff}^2, \qquad L_{eff} = L$$

(c) Propped cantilever: $P_{cr} = 2.04\pi^2 EI/L^2 = \pi^2 EI/(0.7L)^2$

$$= \pi^2 EI/L_{eff}^2, \qquad L_{eff} = 0.7L$$

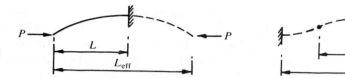

Figure 8.7 ● Effective lengths of cantilever and fixed-ended columns

(d) Clamped column: $P_{cr} = 4\pi^2 EI/L^2 = \pi^2 EI/(0.5L)^2$

$$= \pi^2 EI/L_{eff}^2, \qquad L_{eff} = 0.5L$$

Similarly, in general for a column with any boundary conditions,

$$P_{cr} = \pi^2 EI/L_{eff}^2, \quad \sigma_{cr} = P/A = \pi^2 E/(L_{eff}/r)^2$$

where r = radius of gyration about the axis of buckling. In other words, elastic buckling of any beam-column can be viewed as essentially the problem of buckling of a simply supported beam-column with a length equal not to the *actual* length L but equal to the appropriate *effective* length L_{eff}. Figure 8.7 shows the 'effective' simply supported beam-columns for the fixed-ended and cantilever beam-columns. It should be noted that using effective length for calculating the buckling load gives no additional information except that this concept is commonly used by structural engineers. The main advantage of this concept is that the formulae developed for a simply supported beam-column for which $L = L_{eff}$ can be used for columns with any other boundary conditions by using the appropriate L_{eff} instead of L, provided the effective length for the particular set of boundary conditions can either be calculated or more likely, in many cases at least, estimated to a reasonable accuracy.

It should be noted that the maximum buckling load is attained when both the ends are clamped. In this case $P_{cr} = 4P_e$ and $L_{eff} = 0.5L$. Similarly, if one can imagine a column free at the top and simply supported at the bottom, a patently unstable structure, then for this case $P_{cr} = 0$ and therefore $L_{eff} = \infty$. Thus in practice the buckling loads and the corresponding effective length for columns lie between the two limiting cases considered; therefore

$$0 \leqslant P_{cr} \leqslant 4P_e, \quad \infty \geqslant L_{eff} \geqslant 0.5L$$

8.4 ● Elastic stability stiffness matrix for a beam-column element

In the previous sections, the elastic critical load of an isolated beam-column with idealized boundary conditions was discussed. In practice, compression members occur as elements connected to other elements, as in the case of rigid-jointed frames and trusses as shown in Fig. 8.8. The boundary conditions at the ends of the element are difficult to specify without taking into consideration the interaction of other elements meeting at the ends. The most convenient procedure for analysing the whole structure is to use the stiffness method of analysis discussed in Chapter 6. The first step in using the stiffness method is to develop the element stiffness matrix. This is done in the following sections.

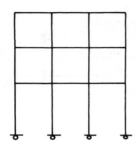

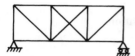

Figure 8.8 ● Rigid-jointed and pin-jointed structures

Consider the beam element shown in Fig. 8.9. Let the displacements at end 1 be (θ_1, Δ_{n1}) and at end 2 be (θ_2, Δ_{n2}) where at the ith end, θ_2 is the clockwise rotation and Δ_{ni} the normal translation positive upwards. Let the corresponding forces be (M_1, F_{n1}) at end 1 and (M_2, F_{n2}) at end 2 where, at the ith end, M_i is a clockwise couple and F_{ni} is a force positive upwards. Because there are no lateral loads on the element, bending deflection v and its derivatives are given, as shown in Section 8.3.1, by

$$v = A \cos \alpha x + B \sin \alpha x + C\alpha x + D$$

$$dv/dx = \alpha\{-A \sin \alpha x + B \cos \alpha x + C\}$$

$$M = EI\,\alpha^2\{-A \cos \alpha x - B \sin \alpha x\}$$

$$Q = -EI\,\alpha^3\{C\}$$

The boundary conditions are

(a) $v = \Delta_{n1}$ at $x = 0$. Therefore

$$A + D = \Delta_{n1} \tag{i}$$

(b) $dv/dx = -\theta_1$ at $x = 0$

Note the use of a negative sign because, for the co-ordinate axes adopted, (dv/dx) is positive anticlockwise but θ is positive clockwise. Therefore

$$\alpha(B + C) = -\theta_1 \tag{ii}$$

(c) $v = \Delta_{n2}$ at $x = L$. Therefore

$$A \cos \alpha L + B \sin \alpha L + C\alpha L + D = \Delta_{n2} \tag{iii}$$

(d) $dv/dx = -\theta_2$ at $x = L$. Therefore

$$\alpha\{-A \sin \alpha L + B \cos \alpha L + C\} = -\theta_2 \tag{iv}$$

Solving the four equations (i)–(iv) for the four constants of integration in terms of the displacements at the ends of the element, we have

$$A = [(-\Delta_{n1} + \Delta_{n2})(\cos \alpha L - 1) + \theta_1 L\{\cos \alpha L - \sin \alpha L/(\alpha L)\} + \theta_2 L\{\sin \alpha L/(\alpha L) - 1\}]/\{2(1 - \cos \alpha L) - \alpha L \sin \alpha L\}$$

$$B = [(-\Delta_{n1} + \Delta_{n2})\sin \alpha L + \theta_1 L\{\sin \alpha L - (\cos \alpha L - 1)/(\alpha L)\} + \theta_2 L\{(1 - \cos \alpha L)/(\alpha L)\}]/\{2(1 - \cos \alpha L) - \alpha L \sin \alpha L\}$$

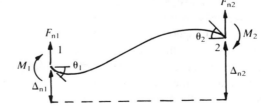

Figure 8.9 ● Forces and displacements at the ends of a beam-column element

$$C = -B - \theta_1 L/(\alpha L)$$

$$C = [(\Delta_{n1} - \Delta_{n2})\sin \alpha L - \theta_1 L \{(1 - \cos \alpha L)/(\alpha L)\}$$
$$-\theta_2 L \{(1 - \cos \alpha L)/(\alpha L)\}]/\{2(1 - \cos \alpha L) - \alpha L \sin \alpha L\}$$

$$D = \Delta_{n1} - A$$

The bending moment and shear force at any section are given by

$$M = EI \alpha^2 \{-A \cos \alpha x - B \sin \alpha x\}$$

$$Q = -EI \alpha^3 \{C\}$$

The forces at the ends of the element are given by

At $x = 0$, $M_1 = M = EI \alpha^2 \{-A\}$, $F_{n1} = -Q = EI \alpha^2 \{C\}$

At $x = L$, $M_2 = -M = EI \alpha^2 \{A \cos \alpha L + B \sin \alpha L\}$

$$F_{n2} = Q = -EI \alpha^3 \{C\}$$

Substituting for the constants of integration, we have

$$
\begin{bmatrix} F_{n1} \\ M_1 \\ F_{n2} \\ M_2 \end{bmatrix} = (EI/L) \times
\begin{bmatrix}
T_1/L^2 & -Q_1/L & -T_1/L^2 & -Q_1/L \\
 & S_1 & Q_1/L & S_2 \\
 & & T_1/L^2 & Q_1/L \\
 & \text{symmetrical} & & S_1
\end{bmatrix}
\begin{bmatrix} \Delta_{n1} \\ \theta_1 \\ \Delta_{n2} \\ \theta_2 \end{bmatrix}
$$

where

$$S_1 = (\alpha L)^2 \{\sin \alpha L/(\alpha L) - \cos \alpha L\}/\{2(1 - \cos \alpha L) - \alpha L \sin \alpha L\}$$

$$S_2 = (\alpha L)^2 \{1 - \sin \alpha L/(\alpha L)\}/\{2(1 - \cos \alpha L) - \alpha L \sin \alpha L\}$$

$$Q_1 = S_1 + S_2, \quad T_1 = 2Q_1 - (\alpha L)^2$$

Figure 8.10 shows the physical significance of the above stiffness coefficients. $S_1 (EI)/L$ is the rotational stiffness and $T_1 (EI)/L^3$ is the translational stiffness of a

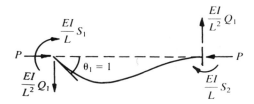

Figure 8.10 ●
Interpretation of
coefficients of the
stability stiffness matrix
for a beam-column
element

Table 8.1 ● (a)
Coefficients of the
stability matrix for a
beam element carrying
axial compressive
loading

ρ	S_1	S_2	Q_1	T_1
0.00	4.0	2.0	6.0	12.0
0.05	3.93	2.02	5.95	11.41
0.10	3.87	2.03	5.90	10.81
0.15	3.80	2.05	5.85	10.22
0.20	3.73	2.07	5.80	9.63
0.25	3.66	2.09	5.75	9.03
0.30	3.59	2.11	5.70	8.43
0.35	3.52	2.13	5.65	7.84
0.40	3.44	2.15	5.59	7.24
0.45	3.37	2.17	5.54	6.64
0.50	3.30	2.19	5.49	6.04
0.55	3.22	2.22	5.44	5.44
0.60	3.14	2.24	5.38	4.84
0.65	3.06	2.27	5.33	4.24
0.70	2.98	2.29	5.27	3.64
0.75	2.90	2.32	5.22	3.03
0.80	2.82	2.35	5.16	2.43
0.85	2.73	2.37	5.11	1.82
0.90	2.65	2.40	5.05	1.22
0.95	2.56	2.44	4.99	0.61
1.00	2.47	2.47	4.94	0.00
1.10	2.28	2.54	4.82	−1.22
1.20	2.09	2.61	4.70	−2.44
1.30	1.89	2.69	4.58	−3.67
1.40	1.68	2.78	4.46	−4.90
1.50	1.46	2.88	4.33	−6.14
1.60	1.22	2.98	4.20	−7.38
1.70	0.98	3.10	4.07	−8.63
1.80	0.72	3.22	3.94	−9.88
1.90	0.44	3.37	3.81	−11.14
2.00	0.14	3.53	3.67	−12.40
2.10	−0.18	3.70	3.53	−13.67
2.20	−0.52	3.90	3.38	−14.95
2.30	−0.89	4.13	3.23	−16.23
2.40	−1.30	4.38	3.08	−17.52
2.50	−1.75	4.68	2.93	−18.82
2.60	−2.25	5.02	2.77	−20.12
2.70	−2.81	5.42	2.61	−21.44
2.80	−3.45	5.88	2.44	−22.76
2.90	−4.18	6.44	2.27	−24.09
3.00	−5.03	7.12	2.09	−25.43
3.10	−6.05	7.96	1.91	−26.77
3.20	−7.30	9.02	1.72	−28.14
3.30	−8.86	10.40	1.53	−29.51
3.40	−10.91	12.24	1.33	−30.89
3.50	−13.72	14.85	1.13	−32.28
3.60	−17.87	18.79	0.92	−33.69
3.70	−24.68	25.39	0.70	−35.12
3.80	−38.17	38.65	0.48	−36.55
3.90	−78.34	78.58	0.24	−38.01
4.00	−Infinity	+Infinity	0.00	−39.48

Table 8.1 ● (b)
Coefficients of the
stability stiffness matrix
for a beam element
carrying axial tensile
loading

$-\rho$	S_1	S_2	Q_1	T_1
0	4.00	2.00	6.00	12.00
0.20	4.26	1.94	6.20	14.36
0.40	4.50	1.88	6.38	16.72
0.60	4.74	1.83	6.57	19.06
0.80	4.96	1.79	6.75	21.39
1.00	5.18	1.75	6.92	23.72
1.20	5.38	1.71	7.10	26.03
1.40	5.58	1.68	7.26	28.34
1.60	5.78	1.65	7.43	30.65
1.80	5.97	1.62	7.59	32.94
2.00	6.15	1.60	7.75	35.23
2.50	6.58	1.54	8.13	40.92
3.00	6.99	1.50	8.49	46.58
4.00	7.34	1.43	9.17	57.82
5.00	8.42	1.38	9.80	68.94
6.00	9.04	1.34	10.39	79.98
7.00	9.63	1.31	10.94	90.97
8.00	10.18	1.29	11.46	101.89
9.00	10.69	1.27	11.96	112.73
10.00	11.19	1.25	12.44	123.57
11.00	11.66	1.24	12.89	134.38
12.00	12.11	1.22	13.32	145.07
13.00	12.54	1.21	13.76	155.79
14.00	12.96	1.21	14.17	166.55
15.00	13.36	1.20	14.56	177.14
16.00	13.76	1.19	14.95	187.75
17.00	14.14	1.18	15.32	198.42
18.00	14.51	1.18	15.68	208.95
19.00	14.87	1.17	16.04	219.52
20.00	15.22	1.17	16.38	230.08

propped cantilever beam-column. Table 8.1(a) and (b) gives the numerical values of the stiffness coefficients S_1, S_2, Q_1 and T_1 for different values of the convenient non-dimensional factor $\rho = P/P_e$, where P_e = Euler load = $\pi^2 EI/L^2$. The variables ρ and αL are related by $\alpha L = \pi \sqrt{\rho}$.

As shown in Fig. 8.10, S_1 is the rotational stiffness factor of a beam-column with the far end fixed and the ratio S_2/S_1 is the moment carry-over factor. In the early literature on elastic stability of rigid-jointed frames, the factors $S = S_1$ and $C = S_2/S_1$ were used extensively for the moment distribution method of manual calculation of elastic stability and are called stability functions S and C respectively.

It should be noted that when the axial load $P = 0$, then the stiffness factors become $S_1 = 4$, $S_2 = 2$, $Q_1 = 6$ and $T_1 = 12$. These correspond to the orthodox beam stiffness factors.

In the above derivation, the shear deformation effect has been ignored. This is reasonable because only 'slender' members are liable to buckling for which shear deformation is generally unimportant.

8.4.1 Elastic stability matrix for a member in a 2-D rigid-jointed frame

The elastic stability matrix for the beam-column derived above is used as the basis for deriving the elastic stability stiffness matrix for a member of a 2-D rigid-jointed structure. The steps to be followed are identical to the steps in Section 6.4. The resulting matrix is shown in equation [8.1]. As can be seen, it is obtained from the matrix in equation [6.2] by replacing (12, 6, 4, 2) respectively by (T_1, Q_1, S_1, S_2). The stability stiffness relationship for an element in a 2-D rigid-jointed structure is given by

$$
\begin{bmatrix} F_{x1} \\ F_{y1} \\ M_1 \\ F_{x2} \\ F_{y2} \\ M_2 \end{bmatrix}
= \frac{EI}{L}
\begin{bmatrix}
\dfrac{T_1 m^2 + \alpha l^2}{L^2} & \dfrac{-(T_1 - \alpha)lm}{L^2} & \dfrac{Q_1}{L}m & \dfrac{-(T_1 m^2 + \alpha l^2)}{L^2} & \dfrac{(T_1 - \alpha)lm}{L^2} & \dfrac{Q_1}{L}m \\[2ex]
 & \dfrac{T_1 l^2 + \alpha m^2}{L^2} & \dfrac{-Q_1 l}{L} & \dfrac{(T_1 - \alpha)lm}{L^2} & \dfrac{-(T_1 l^2 + \alpha m^2)}{L^2} & \dfrac{-Q_1 l}{L} \\[2ex]
 & & S_1 & \dfrac{-Q_1 m}{L} & \dfrac{Q_1 l}{L} & S_2 \\[2ex]
 & & & \dfrac{T_1 m + \alpha l^2}{L^2} & \dfrac{-(T_1 - \alpha)lm}{L^2} & \dfrac{-Q_1 m}{L} \\[2ex]
 & & & & \dfrac{T_1 l^2 + \alpha m^2}{L^2} & \dfrac{Q_1 l}{L} \\[2ex]
 & \text{symmetrical} & & & & S_1
\end{bmatrix}
$$

$$
\times
\begin{bmatrix} u_1 \\ v_1 \\ \theta_1 \\ u_2 \\ v_2 \\ \theta_2 \end{bmatrix}
\tag{8.1}
$$

$$
\alpha = \frac{AL^2}{I}
$$

It should be noted that the stiffness matrix is a non-linear function of the axial load P. Knowing (or more likely assuming) an approximate value of P, the element stiffness matrix can be calculated and assembled on to the structural stiffness matrix as explained in Chapter 6.

8.5 ● Criterion for determining limit of elastic stability of 2-D rigid-jointed structures

In order to demonstrate the criterion used to determine the limit of elastic stability of structures, consider the propped cantilever shown in Fig. 8.11. The compressive

Figure 8.11 ●
A propped cantilever
beam-column

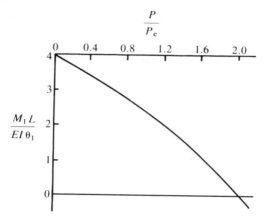

Figure 8.12 ●
Variation of the beam
stiffness with P

axial load is P. Let a moment M_1 be applied at end 1 to 'disturb' the structure from its initial straight position. The relationship between M_1 and the corresponding rotation θ_1 is given by

$$M_1 = (EI/L)\, S_1 \theta_1$$

Therefore

$$\{M_1 L/(EI)\}/\theta_1 = S_1$$

The curve relating stiffness $\{M_1 L/(EI)\}/\theta_1$ to $\rho = P/P_e$ shown in Fig. 8.12 shows that, as P/P_e approaches 2.04, the stiffness M_1/θ_1 of the structure tends to zero. As was shown previously, the elastic stability load for the propped cantilever is $P_{cr} = 2.04 P_e$. What this example has shown is that if M_1 is treated as a disturbing moment, then because at $P = P_{cr}$ the stiffness coefficient S_1 becomes zero, then however small the disturbing moment M_1, the resulting displacement θ_1 will be uncontrollably large. In other words, the structure becomes unstable. This is what was observed in Section 8.1. Therefore a convenient way of examining the stability of a structure is to examine its stiffness under increasing axial loads and the load at which the structure loses its stiffness altogether gives the elastic stability load.

In the case of structures with a single degree of freedom, the displacement tends to infinity when the corresponding stiffness factor tends to zero. However, in the case of structures with multiple degrees of freedom, the displacements tend to infinity when the *determinant* of the stiffness matrix rather than a single stiffness coefficient tends to zero. This is the criterion used for determining the load or load factor at which the structure becomes elastically unstable. This concept is illustrated by some simple examples in the next section.

8.5.1 Examples

Example 1

Calculate P_{cr} for the beam-column shown in Fig. 8.13. The left end is fixed and the right end is on a sliding support, i.e. the rotation and shear force are both zero.

Figure 8.13 ●
A beam-column element with fixed and 'guided' end conditions

Solution   There is only one degree of freedom which is the translation at the sliding support. The stiffness relationship between the disturbing force F_{n2} and the corresponding displacement Δ_{n2} is

$$F_{n2} = [EI/L^3] T_1 \Delta_{n2}$$

The displacement tends to infinity when the stiffness coefficient T_1 tends to zero. From Table 8.1, this happens when $\rho = 1$. Therefore $P_{cr} = P_e$.

Example 2

Calculate the necessary equation for determining the critical load for the structure shown in Fig. 8.14.

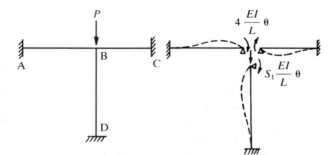

Figure 8.14 ●
Buckling load calculation for a rigid-jointed structure

Solution   In this example also the structure has only one degree of freedom which is the rotation at B. Assuming that the axial forces in the beams AB and BC are negligible and the axial load in the column is P, then the rotational stiffness factor S_1 for these beams becomes equal to 4, but for the column, because it carries an axial compressive load P, the rotational stiffness factor is $S_1 < 4$. The stiffness relationship is therefore given by

$$\sum M_B = \sum [4(EI/L)_{BA} + 4(EI/L)_{BC} + S_1(EI/L)_{BD}]\theta_B$$

The stiffness of the structure tends to zero when the terms inside the square brackets tend to zero. Therefore the structure becomes unstable when

$$S_1 = -4[(EI/L)_{AB} + (EI/L)_{BC}]/\{(EI/L)_{BD}$$

If EI/L = constant for all the three members, then $S_1 = -8$. From Table 8.1, using linear interpolation, this occurs when $\rho = 3.25$. Because the axial load in the column is P, $P_{cr} = 3.25 P_e$.

Example 3

Calculate the necessary equation for determining the critical load for the structure shown in Fig. 8.15.

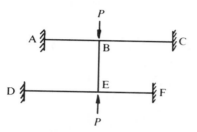

Figure 8.15 ● Elastic stability analysis of a 'no-sway' substructure

Solution   This structure has two degrees of freedom: rotation at joints B and E. Again assuming that the axial forces in the beams can be neglected and the axial load in the column is P, the stiffness matrix is given by

$$\begin{bmatrix} \Sigma M_B \\ \Sigma M_E \end{bmatrix} = \begin{bmatrix} \Sigma 4(EI/L)_{\text{Top}} + S_1(EI/L)_{\text{BE}} & S_2(EI/L)_{\text{BE}} \\ S_2(EI/L)_{\text{BE}} & \Sigma 4(EI/L)_{\text{Bottom}} + S_1(EI/L)_{\text{BE}} \end{bmatrix} \times \begin{bmatrix} \theta_B \\ \theta_E \end{bmatrix}$$

where the subscripts Top and Bottom refer respectively to beams (AB and BC) at the top of the column and beams (DE and EF) at the bottom of the column.
Let $\alpha_1 = [4\Sigma(EI/L)_{\text{Top}}]/(EI/L)_{\text{BE}}$, $\alpha_2 = [4\Sigma(EI/L)_{\text{Bottom}}]/(EI/L)_{\text{BE}}$, then

$$\begin{bmatrix} \Sigma M_B \\ \Sigma M_E \end{bmatrix} = (EI/L)_{\text{BE}} \begin{bmatrix} \alpha_1 + S_1 & S_2 \\ S_2 & \alpha_2 + S_1 \end{bmatrix} \begin{bmatrix} \theta_B \\ \theta_E \end{bmatrix}$$

The displacements tend to infinity when the determinant of the structural stiffness matrix becomes zero, i.e. when

$$(S_1 + \alpha_1)(S_1 + \alpha_2) - S_2^2 = 0$$

As a simple example, if EI/L is constant for all the elements, then $\alpha_1 = \alpha_2 = 8$, and the determinant becomes zero when $(S_1 + 8) = \pm S_2$. From the numerical values of the coefficients in Table 8.1, this occurs at $2.6 < \rho < 2.7$ and the specific value is given by $\rho_{cr} = 2.68$ or $P_{cr} = 2.68 P_e$.

It is interesting to note that in Example 2, where the base D of the column is fixed, $\rho_{cr} = 3.25$, but in Example 3, where the base of the column is not fixed but is elastically restrained by beams, $\rho_{cr} = 2.68$. The decrease in the value of ρ_{cr} is due to the decrease in the 'fixity' at the bottom D of the column BD. In practice, using the equation $(S_1 + \alpha_1)(S_1 + \alpha_2) - S_2^2 = 0$, charts such as that shown in Fig. 8.16 can be developed. Such charts can be used to estimate the buckling load of columns of a 'no-sway' multistorey structure such as that shown in Fig. 8.17. Note that if the

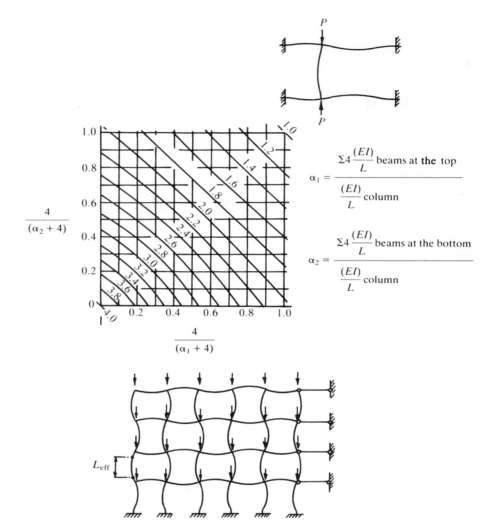

Figure 8.16 ● Critical buckling load for a column with rigidly jointed restraining beams and prevented from sway. (Numbers against curves indicate P_{cr}/P_e)

$$\alpha_1 = \frac{\Sigma 4 \frac{(EI)}{L} \text{ beams at the top}}{\frac{(EI)}{L} \text{ column}}$$

$$\alpha_2 = \frac{\Sigma 4 \frac{(EI)}{L} \text{ beams at the bottom}}{\frac{(EI)}{L} \text{ column}}$$

Figure 8.17 ●
Critical buckling mode for a 'no-sway' rigid-jointed frame

beams at both the top and the bottom are flexible, then the column BE behaves like a simply supported column with $\rho_{cr} = 1$. On the other hand, if the beams at the top and bottom are both very stiff, then the column behaves like a fixed column with $\rho_{cr} = 4$. Therefore for the no-sway column $P_e \leqslant P_{cr} \leqslant 4P_e$.

This example provides an important lesson in considering column design in the case of no-sway multistorey columns with fixed base. Although ground storey columns carry more load than the first storey columns, because of the fixed base, the ground storey columns are better restrained at the base than the first storey columns which are restrained by beams only. Therefore it is quite possible that the first storey columns may be weaker than the ground storey columns.

Example 4

Calculate the critical elastic buckling load for the structure shown in Fig. 8.18.

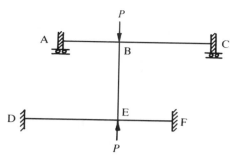

Figure 8.18 ● Elastic stability analysis of a 'sway' substructure

Solution   This structure is similar to the one in Fig. 8.15, except that the column is allowed to sway. There are therefore three degrees of freedom and the stiffness matrix for the structure is established as

$$\begin{bmatrix} \Sigma M_B \\ \Sigma M_E \\ \Sigma F_{xB} \end{bmatrix} = (EI/L)_{BE} \begin{bmatrix} \alpha_1 + S_1 & S_2 & -Q_1/L \\ S_2 & \alpha_2 + S_1 & -Q_1/L \\ -Q_1/L & -Q_1/L & T_1/L^2 \end{bmatrix} \begin{bmatrix} \theta_B \\ \theta_E \\ \Delta_B \end{bmatrix}$$

where $\alpha_1 = [4\Sigma(EI/L)_{Top}]/(EI/L)_{BE}$ and $\alpha_2 = [4\Sigma(EI/L)_{Bottom}]/(EI/L)_{BE}$.

The structure becomes unstable when the determinant of the structural stiffness matrix becomes zero. Because this is a simple 3×3 matrix, the determinant is easily calculated and charts such as the one shown in Fig. 8.19 can be prepared which are

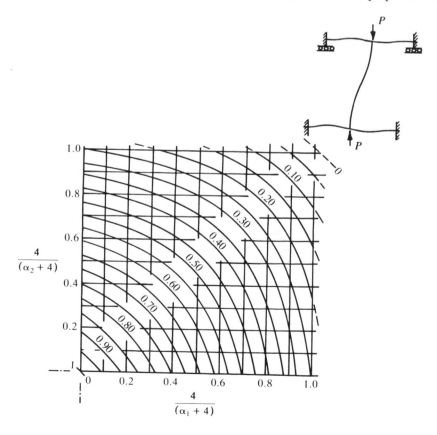

Figure 8.19 ● Critical buckling loads for a column rigidly connected to beams at top and bottom and allowed to sway freely. (Figures against curves indicate P_{cr}/P_e)

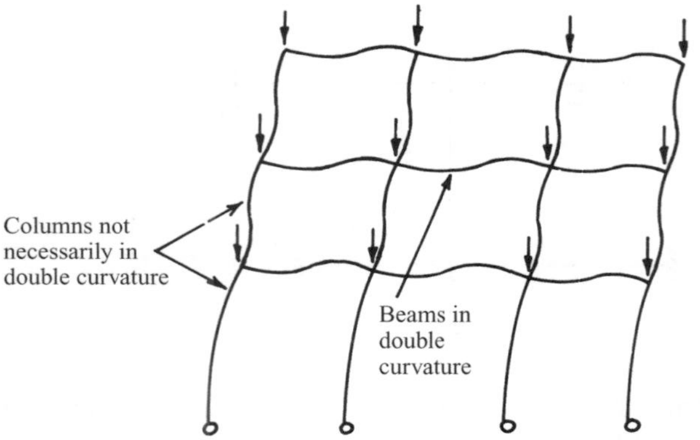

Columns not
necessarily in
double curvature

Beams in
double
curvature

Figure 8.20 ● Critical
buckling mode for a
sway frame

useful for estimating the critical load of columns in multistorey structures which are
free to sway as shown in Fig. 8.20.

Note that if the beams at both the top and the bottom are flexible, then the
column BE behaves like a column pinned at one end and free at the other end with
$\rho_{cr} = 0$. On the other hand, if the beams at top and bottom are both very stiff, then
the column behaves like a column fixed at one end and with a sliding support at the
other end with $\rho_{cr} = 1$. Therefore for a sway column $0 \leq P_{cr} \leq P_e$.

8.5.2 Determination of elastic critical load of 2-D rigid-jointed frames using computer programs

The numerical effort involved in determining the elastic critical load of a general
2-D rigid-jointed structure is onerous. In the examples of the previous section, at
any applied load level, the axial force in all the members could be estimated quite
accurately. This is not always possible. It should be remembered that the stiffness
coefficients of a member depend non-linearly on the level of axial load in the member. Therefore stiffness coefficients of the members of a structure change throughout the loading history. This in turn affects the axial load acting on the member.
The problem is highly non-linear. The procedure followed in automatic computation of the elastic stability load is as follows.

1. Assume that the axial loads in members are zero and set up the structural stiffness matrix.
2. Apply the external load and analyse the structure.
3. Determine the axial forces in the members from the analysis of the structure.
4. Assuming that these are the correct axial loads in the members, form the new
   element and structural stiffness matrices.
5. Repeat steps (1)–(4) until there is no 'appreciable' difference between the axial
   loads assumed and those calculated.
6. Having obtained the 'correct' axial loads in the members, calculate the determinant of the structural stiffness matrix. The determinant is simply the continued
   product of the diagonal elements of the structural stiffness matrix after forward
   Gaussian elimination.

7. Increase the load level and repeat steps 3–6. The process is repeated until, for two consecutive load levels, the determinant has a positive and negative value. The 'correct' elastic stability load level can be obtained by linear interpolation as was done in the examples.

The calculations are onerous and a good estimate of the elastic stability load using data in Figs 8.17 and 8.20 can drastically cut down the total amount of computations. Further information is given in Bhatt (1981) and Bhatt (1986).

8.6 ● Practical design of axially loaded struts

In the previous sections, methods were developed for the calculation of the elastic buckling loads of single columns or structures. As was pointed out, the elastic buckling load is a mathematical entity which represents the load at which an elastic structure will suffer infinitely large displacements when very small forces are applied to disturb the structure from its undeformed equilibrium position. While the elastic buckling load is certainly a measure of the sensitivity of the structure to become unstable for a given axial load, it is not by itself very useful for designing structural elements. Some of the important reasons are discussed below.

(a) Consider the simply supported beam-column shown in Fig. 8.3. Let the column be made of steel for which $E = 210 \times 10^3$ N mm^{-2} and the stress at yield $\sigma_0 = 275$ N mm^{-2}. The critical elastic buckling stress is given by

$$\sigma_{cr} = \pi^2 E (r/L_{eff})^2$$

A plot of this equation is shown in Fig. 8.21. Obviously at approximately $(L_{eff}/r) = 87$, $\sigma_{cr} = \sigma_0$, the yield stress of the material. Therefore for values of $(L_{eff}/r) < 87$, $\sigma_{cr} > \sigma_0$. In other words, even without considering other factors discussed below, the material ceases to be elastic at loads well below P_{cr} and

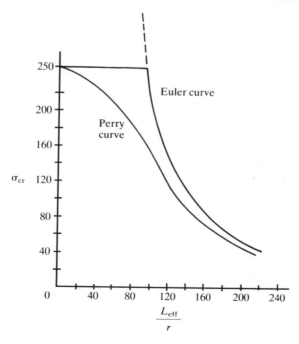

Figure 8.21 ●
Variation of critical stress
with slenderness ratio

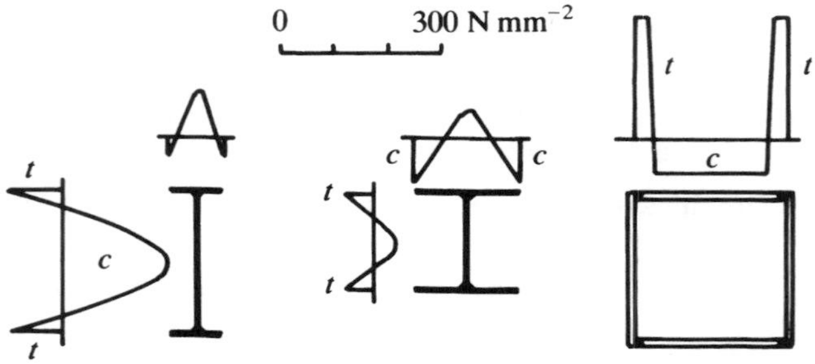

Figure 8.22 ● Typical
residual stress patterns
in rolled steel sections
and welded box section.
c = compression, t =
tension

the fundamental assumption used in deriving P_{cr}, i.e. that the material is elastic, ceases to be valid. So the yield stress σ_0 can be used as an upper limit for the validity of theoretical σ_{cr}.

(b) As was pointed out in Section 8.1, practical struts are never perfectly straight but have an initial 'crookedness' given by v_i. In other words, at a given value of the axial load P, the bending moment M accompanying it is given by

$$M = P(v + v_i)$$

where v_i = initial displacement and v = displacement caused by the axial load P. This means that even for columns for which $(L_{eff}/r) > 87$, on the compression face of the column, the yield stress is reached at a load lower than that calculated by considering the axial stress only and ignoring the accompanying bending moment.

(c) The processes involved in the manufacture of columns such as rolling or welding set up 'thermal' stresses due to unequal rates at which different parts of the column heat or cool. Typical thermal stress patterns in a rolled and welded steel column are shown in Fig. 8.22. What matters from the point of view of the buckling of the column is the severity and location of thermal compressive stress. For example, in a typical universal column section, thermal compressive stress as high as 100 N mm^{-2} occurs at the tips of the flanges. If the yield stress σ_0 is taken as, say, 275 N mm^{-2}, then the tips yield when the axial applied stress is only 175 N mm^{-2}. This is more serious for buckling about the minor axis, because the thermal stresses are at the extreme fibres. In practical column design these stresses have to be taken into account otherwise the axial load calculated to cause buckling will be in error on the unsafe side.

(d) The load P_{cr} is the load at which, in an elastic column, the deflections tend to infinity. In practical columns severe restrictions have to be put on allowable deflections.

It is for these major reasons that the value of P_{cr} cannot by itself be used for the design of columns. Figure 8.23 shows a plot of experimental failure load for columns of various slenderness ratio L_{eff}/r. As can be seen, the Euler curve $\sigma_{cr} = \pi^2 E \, (r/L_{eff})^2$ and the σ_{cr} = yield stress cut-off act as upper bounds to the experimental failure loads. In general 'semi-empirical formulae' with parameters to fit experimental data are used for the design of columns. One such formula used in British practice for example is called the Perry–Robertson formula after the names of its developers.

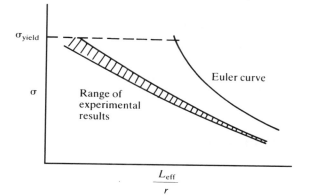

Figure 8.23 ●
Theoretical and actual
ultimate strength of
columns

8.6.1 Perry–Robertson formula

It was shown in Section 8.1 that in an initially deformed pin-ended beam-column, the displacement v caused by the axial load is given by

$$v = \{P/(P_e - P)\}\, v_i$$

where $v_i = v_0 \sin \pi x/L$. The total displacement v_t

$$v_t = v + v_i = P_e v_i/(P_e - P)$$

and the maximum bending moment M_{max} occurs at the centre at $x = L/2$ and is equal to

$$M_{max} = P v_t = P P_e v_0/(P_e - P)$$

The maximum compressive stress σ_{max} in the extreme fibre on the concave face caused by the axial load P and bending moment M_{max} is given by

$$\sigma_{max} = P/A + M_{max}\, y/I$$

where $y = $ distance to the concave face from the neutral axis. Letting $P/A = \sigma$ and $v_0 y/r^2 = \eta$, the above equation can be expressed as

$$\sigma_{max} = \sigma[1 + \eta\, \sigma_{cr}/(\sigma_{cr} - \sigma)]$$

$$(\sigma_{max} - \sigma)(\sigma_{cr} - \sigma) = \eta\, \sigma_{cr}\sigma$$

$$\sigma^2 - \sigma\{\sigma_{cr}(1 + \eta) + \sigma_{max}\} + \sigma_{cr}\sigma_{max} = 0$$

For a given value of the 'permissible' stress σ_{max}, solving the quadratic equation, the value of the axial stress σ is given by

$$\sigma = 0.5\{\sigma_{max} + \sigma_{cr}(1 + \eta)\} - 0.5\sqrt{[\{\sigma_{cr}(1 + \eta) + \sigma_{max}\}^2 - 4\sigma_{max}\sigma_{cr}]}$$

For universal beams or columns buckling about the minor axis, y/r is approximately 2.0. Assuming that the initial out of line displacement $v_0 = L/1000$, the parameter $\eta = 0.002L/r$. If the maximum stress $\sigma_{max} = \sigma_0 = 250$, say, and Young's modulus $E = 210 \times 10^3$ N mm^{-2}, then the stress σ is a function of (L_{eff}/r) only. In Fig. 8.21, the relationship between σ and (L_{eff}/r) is plotted as a Perry curve. As can be seen, the curve has a strong resemblance to the trend of experimental results for different values of L/r. Therefore, although the Perry–Robertson formula was obtained on the basis of a *stress* criterion, by suitable adjustment of the parameter η and setting $\sigma_{max} = \sigma_0$, it is possible to make the equation fit the experimental results for specific classes of columns so that $P = A\sigma$ gives the *ultimate* axial force of the column. This is the method adopted in the British steel code BS 5950.

8.7 ● Lateral instability of beams

In previous sections, the axial load required to cause elastic instability of a compression member was investigated. When a member is subjected to bending moment, compressive stresses are set up in the area on one side of the neutral axis and tensile stresses are set up on the opposite side. If we imagine that the beam is made up of a series of parallel strips then obviously the strips on the compressive side can buckle. In this case, although the bending moment is applied in the vertical x–y plane, the major deformation is in the horizontal or x–z plane. Obviously continuity of material prevents the imaginary individual strips from buckling on their own but this simple 'model' makes us aware of the fact that beams subjected to bending moment only, without any axial forces being applied, are also subjected to instability effects. In this section, the value of the critical pure bending moment required to cause the elastic instability of a beam of rectangular section will be derived. This type of instability is called lateral or lateral torsional buckling of beams.

8.7.1 Lateral torsional buckling of a rectangular beam under pure bending moment

Consider the rectangular beam shown in Fig. 8.24. Let the applied moments at the ends be M_0. It is assumed that with respect to bending in the vertical plane, the beam is simply supported at the ends. In addition, it is assumed that the beam is prevented from twisting about the x-axis at the ends. In order to determine the elastic critical moment, it is necessary to examine the equilibrium of the structure in the deformed position. As shown in Fig. 8.24, under the action of the bending moment M_0, the beam cross-section twists by an angle ϕ about the x-axis and the beam deflects by v in the y-direction and w in the z-direction. Consider an orthogonal set of axes centred on the deformed element such that the set of axes (x', y', z') coincide with the (x, y, z) axes when the section is undeformed. Examining the lateral bending of the beam about the z'-axis, we have from the displaced position in the cross-section:

$$EI_y d^2w'/dx'^2 = \text{component of } M_0 \text{ in the } y'\text{-direction}$$

$$= -M_0 \sin \phi$$

$$= -M_0\phi \text{ if } \phi \text{ is small.}$$

Assuming $d^2w'/dx'^2 = d^2w/dx^2$ for small values of ϕ, we have

$$EI_y d^2w/dx^2 = -M_0 \phi \tag{i}$$

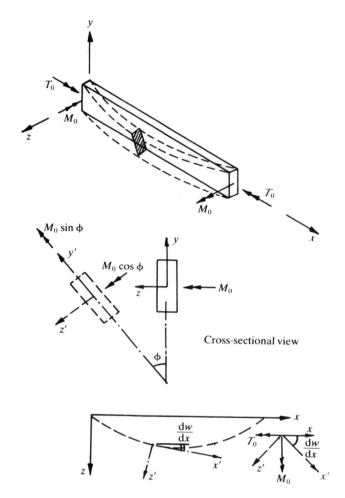

Figure 8.24 ● Lateral stability of beams

Because this equation consists of two variables, ϕ and w, we need to have one more equation involving these two variables in order to solve for ϕ and w. Examining the torque–twist relationship we have, from the deformed position shown in plan,

$$GJ \, d\phi/dx' = \text{torque} = -T \cos(dw/dx) + M_0 \sin(dw/dx)$$

$$= -T_0 + M_0 \, dw/dx \text{ for small values of } dw/dx$$

where GJ = torsional rigidity and T_0 = restraining torque at the ends of the beam. Assuming $d\phi/dx' = d\phi/dx$ for small deformations,

$$GJ \, d\phi/dx = -T_0 + M_0 \, dw/dx$$

Differentiating with respect to x we have

$$GJ \, d^2\phi/dx^2 = M_0 d^2w/dx^2 \qquad \text{(ii)}$$

Substituting for d^2w/dx^2 from equation (ii) in equation (i)

$$(EI_y GJ) \, d^2\phi/dx^2 + M_0^2 \phi = 0$$

Setting $M_0^2/(EI_y GJ) = \alpha^2$, the solution to the equation is given by

$$\phi = A \cos \alpha x + B \sin \alpha x$$

The boundary conditions are $\phi = 0$ at $x = 0$ and $x = L$ because the beam is prevented from twisting at the ends. This requires that $A = 0$ and $B \sin \alpha L = 0$. Since $B \neq 0$ to ensure that the beam is deformed, $\sin \alpha L = 0$ for which the smallest root is $\alpha L = \pi$. Therefore $\alpha^2 = \pi^2/L^2 = M_0^2/(EI_y GJ)$, and

$$M_{cr} = M_0 = (\pi/L)\sqrt{\{EI_y GJ\}}$$

It is interesting to note that while the buckling of a column was governed solely by the section property I_y for buckling about the yy-axis, the lateral buckling of the beam is governed by two properties, i.e. I_y and torsional inertia J.

For a rectangular section, $I_y = b^3 d/12$ and for a 'deep' rectangular section, $J = db^3/3$ as shown in Section 4.16.1. Assuming that the material is isotropic, $G = 0.5E/(1 + v)$, then M_{cr} is equal to

$$M_{cr} = (\pi/L)(b^3 d/6)E/\sqrt{\{2(1 + v)\}}$$

The maximum bending stress σ_{cr} is given by $M_{cr}/(bd^2/6)$ so that

$$\sigma_{cr} = \pi[E/\sqrt{\{2(1 + v)\}}](b/L)(b/d)$$

It is clear from the above expression that deep rectangles, i.e. sections having large values of d/b, have low values of σ_{cr}. This means that such 'slender' beams are likely to fail by lateral buckling rather than by bending.

In the case of symmetrical I beams, because an I section is an 'open' section, when considering the torsional resistance, we have to take account of resistance to torsion provided by the bending of flanges in their own plane as discussed in Section 4.16.3. When this is done, it can be shown that

$$M_{cr} = (\pi/L)\sqrt{\{EI_y GJ\}}[1 + \{\pi^2 EI_y/(GJ)\}(h^2/L^2)]$$

where h = depth between the flanges.

8.7.2 Practical design of beams liable to lateral torsional buckling

As in the case of buckling of columns under axial compression, the elastic critical bending moment serves to give an indication of the factors which influence the lateral torsional buckling load. In practice, apart from imperfections, residual stresses, etc. which necessarily influence the lateral torsional buckling strength, the bending moment is rarely constant in beams. In order to cope with the many complicated cases of loading, codes of practice use empirical formulae designed to give acceptably conservative values of critical moment for design purposes.

8.8 ● References

Bhatt P 1981 *Problems in Matrix Analysis of Skeletal Structures* Construction Press (Reprinted by A H Wheeler, Allahabad, India, 1988.) (Contains many numerical examples and additional theory.)

Bhatt P 1986 *Programming the Matrix Analysis of Skeletal Structures* Ellis Horwood (Contains a 'good' program in FORTRAN for the elastic stability analysis of 2-D rigid-jointed frames.)

Horne M R 1975 An approximate method for calculating the critical loads of multistorey plane frame. *Structural Engineer* **53** (6): 242–48 (Presents a method for calculating critical elastic buckling load using linear elastic analysis of plane frame. This method is included in the British steel code 5950, 1985, under deflection method.)

Horne M R, Merchant W 1965 *The Stability of Frames* Pergamon (A good book for fundamental concepts.)

Wood R H 1974 Effective length of columns in multistorey buildings, Parts 1, 2 and 3. *Structural Engineer* **52** (7, 8, 9): 235–44, 295–302 and 341–6. (Much thoughtful material.)

(Note: The following contain additional material which may be a little 'advanced' but still capable of being understood by the reader after mastering the present chapter.)

Chen W F, Lui E M 1982 *Structural Stability* Elsevier

Dubas P 1982 Ultimate capacity of compression members with intermittent lateral supports. In

Narayanan R (ed) 1982 *Axially Compressed Structures: Stability and Strength* Elsevier Applied Science Ch. 9

Medland I C, Segedin C M 1985 Interbraced columns and beams. In Narayanan R (ed) *Steel Framed Structures: Stability and Strength* Elsevier Applied Science pp 81–114

Nethercot D A 1983 Elastic lateral buckling of beams, Ch. 1, pp. 1–33; Inelastic lateral buckling of beams, Ch. 2, pp. 35–69; Design of laterally unsupported beams, Ch. 3, pp. 71–94. In Narayanan, R (ed) *Beams and Beam Columns: Stability and Strength* Elsevier Applied Science

Tall L 1982 Centrally compressed members. In Narayanan R (ed) *Axially Compressed Structures: Stability and Strength* Elsevier Applied Science Ch. 1.

8.9 ● Problems

1. In Fig. E8.1, the angle section shown is used as a pin-ended column and is free to buckle about any axis. Determine the elastic critical load. $E = 210 \text{ kN mm}^{-2}$, $L = 3$ m.

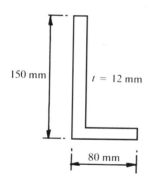

150 mm $t = 12$ mm

80 mm

Figure E8.1

Answer: As in problem 13, Fig. E3.8, in Chapter 3, $I_{zz} = 6.06 \times 10^6 \text{ mm}^4$, $I_{yy} = 1.23 \times 10^6 \text{ mm}^4$, $I_{yz} = 1.55 \times 10^6 \text{ mm}^4$. The principal second moments of area are I_{11} and I_{22} given by

$$I_{11} = 0.5(I_{zz} + I_{yy}) + \sqrt{[\{(I_{zz} - I_{yy})/2\}^2 + I_{yz}^2]}$$
$$= 6.52 \times 10^6 \text{ mm}^4$$

$$I_{22} = 0.5(I_{zz} + I_{yy}) - \sqrt{[\{(I_{zz} - I_{yy})/2\}^2 + I_{yz}^2]}$$
$$= 0.78 \times 10^6 \text{ mm}^4$$

Therefore

$$P_{cr} = \pi^2 E I_{22}/L^2 = 179.6 \text{ kN}$$

2. Determine the fixed end moments in a beam-column due to a uniformly distributed load of q.

Answer: The differential equation is given by

$$EI \, d^4v/dx^4 + P \, d^2v/dx^2 = -q$$

Solution is $v = A \cos \alpha x + B \sin \alpha x + Cx + D - qx^2/(2\alpha^2 EI)$, $v = dv/dx = 0$ at $x = 0$. Therefore $A + D = 0$, $B\alpha + C = 0$.

$$v = \frac{dv}{dx} = 0 \text{ at } x = L. \text{ Therefore}$$

$$v = A(\cos \alpha x - 1) + B(\sin \alpha x - \alpha x) - qx^2/(2\alpha^2 EI)$$

$$dv/dx = -\alpha A \sin \alpha x + \alpha B(\cos \alpha x - 1) - qx/(\alpha^2 EI)$$

$$d^2v/dx^2 = -\alpha^2 A \cos \alpha x - \alpha^2 B \sin \alpha x - q/(\alpha^2 EI)$$

$v = 0$ at $x = L$, therefore $A(\cos \alpha L - 1) + B(\sin \alpha L - \alpha L) = qL^2/(2a^2 EI)$. $dv/dx = 0$ at $x = L$, therefore $A\alpha L \sin \alpha L - B\alpha L(\cos \alpha L - 1) = -qL^2/(\alpha^2 EI)$. Solving for A:

$$EI \, \alpha^2 A = -\{qL^2/(2\alpha L)\} \cot(\alpha L/2)$$

At $x = 0$,

$$EI \, d^2v/dx^2 = -EI \, \alpha^2 A - qL^2/(\alpha L)^2 = -\mu qL^2/12$$

where $\mu = (12 - 6\alpha L \cot(\alpha L/2)/(\alpha L)^2$.

3. Determine the elastic critical load for the structure shown in Fig. E8.2.

Answer: P_{Euler} for column CD is twice that of AB. At any load level, the ratio ρ = axial load/Euler load, for both columns remains the same. The axial loads in the beams are approximately zero. The structural stability stiffness matrix is given by

$$(EI/L) \begin{bmatrix} S_1 + 4 & S_2 \\ S_2 & 2S_1 + 4 + 4 \end{bmatrix} \begin{bmatrix} \theta_B \\ \theta_D \end{bmatrix}$$

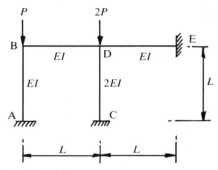

Figure E8.2

Table E8.1

ρ	S_1	S_2	Determinant
2.3	−0.89	4.13	2.29
2.4	−1.30	4.38	−4.60

Determinant $= (S_1 + 4)(2S_1 + 8) - S_2^2 = 0$
Therefore at $\rho = 2.3$, Det $= 2.29$ and at $\rho = 2.4$, Det $= -4.60$.
Therefore by linear interpolation, Det $= 0$, when

$$\rho_{cr} = 2.3 + (2.4 - 2.3) \times 2.29/\{2.29 - (-4.60)\}$$

$$= 2.33$$

$$P_{cr} = \rho_{cr}\pi^2 EI/L^2$$

4. Determine the elastic critical load for the portal frame shown in Fig. E8.3. Ignore axial and shear deformations.

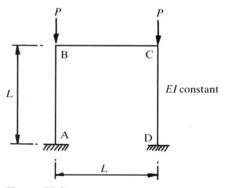

Figure E8.3

Answer: The elastic stability structural stiffness matrix is given by

$$(EI/L)\begin{bmatrix} S_1 + 4 & S_2 & -Q_1 \\ S_2 & S_1 + 4 & -Q_1 \\ -Q_1 & -Q_1 & T_1 \end{bmatrix}\begin{bmatrix} \theta_B \\ \theta_C \\ \Delta/L \end{bmatrix}$$

where Δ = sway displacement. The determinant is best determined by carrying out Gaussian elimination on the matrix and calculating the continued product of the diagonal terms.

$\rho = 0.45$, $S_1 = 3.37$, $S_2 = 2.17$, $Q_1 = 5.54$, $T_1 = 6.64$.

The stiffness matrix becomes:

$$(EI/L)\begin{bmatrix} 7.37 & 2.17 & -5.54 \\ 2.17 & 7.37 & -5.54 \\ -5.54 & -5.54 & 6.64 \end{bmatrix}$$

After Gaussian elimination:

$$(EI/L)\begin{bmatrix} 7.37 & 2.17 & -5.54 \\ 0 & 6.73 & -3.91 \\ 0 & 0 & 0.21 \end{bmatrix}$$

Determinant $= 7.37 \times 6.73 \times 0.21 = 10.42 > 0$, therefore $\rho_{cr} > 0.45$ $\rho = 0.50$, $S_1 = 3.30$, $S_2 = 2.19$, $Q_1 = 5.49$, $T_1 = 6.04$.
The stiffness matrix becomes:

$$(EI/L)\begin{bmatrix} 7.30 & 2.19 & -5.49 \\ 2.19 & 7.30 & -5.49 \\ -5.49 & -5.49 & 6.04 \end{bmatrix}$$

After Gaussian elimination:

$$(EI/L)\begin{bmatrix} 7.30 & 2.19 & -5.49 \\ 0 & 6.64 & -3.84 \\ 0 & 0 & -0.31 \end{bmatrix}$$

Determinant $= 7.30 \times 6.64 \times (-0.31) = -15.03 < 0$, therefore $\rho_{cr} < 0.50$
Interpolating,

$$\rho_{cr} = 0.45 + (0.5 - 0.45) \times (10.42)/\{10.42 - (-15.05)\} = 0.47$$

5. Figure E8.4 shows a rigid-jointed frame. Determine the elastic critical load P. Assume EI and L are constant.

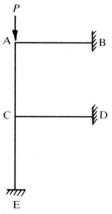

Figure E8.4

Answer: Assume axial loads in beams = 0, axial loads in columns = P. Stiffness matrix is

$$\begin{bmatrix} M_A \\ M_c \end{bmatrix} = (EI/L)\begin{bmatrix} S_1 + 4 & S_2 \\ S_2 & S_1 + S_1 + 4 \end{bmatrix}\begin{bmatrix} \theta_a \\ \theta_c \end{bmatrix}$$

Determinant = $(S_1 + 4)(2S_1 + 4) - S_2^2$

ρ	S_1	S_2	Determinant
2.10	−0.18	3.70	0.22
2.20	−0.52	3.90	−4.91

$P_{cr} = 2.10 + (2.2 - 2.1) \times 0.22/(0.22 + 4.91)$
$= 2.104$, $P_{cr} = 2.104\pi^2 EI/L^2$

6. What will be the value of P_{cr} in Exercise 5, if the flexural rigidity of column CE is 1.5 times that of other members?

Answer: $\rho_{ac} = P/(\pi^2 EI/L^2)$, $\rho_{ce} = P/(\pi^2 1.5EI/L^2)$, $\rho_{ce} = 0.67\rho_{ab}$. Stiffness matrix is

$$\begin{bmatrix} M_A \\ M_c \end{bmatrix} = (EI/L)$$
$$\times \begin{bmatrix} S_1(\rho) + 4 & S_2(\rho) \\ S_2(\rho) & 1.5S_1(0.67\rho) + S_1(\rho) + 4 \end{bmatrix}\begin{bmatrix} \theta_a \\ \theta_c \end{bmatrix}$$

Determinant = $\{S_1(\rho) + 4\}\{1.5 S_1(0.67\rho) + S_1(\rho) + 4 - S_2(\rho)^2$

ρ	$S_1(\rho)$	$S_2(\rho)$	0.67ρ	$S_1(0.67\rho)$	Determinant
2.2	−0.52	3.90	1.47	1.53	4.89
2.3	−0.89	4.13	1.53	1.39	−0.90

$P_{cr} = 2.2 + (2.3 - 2.2)4.89/(4.89 + 0.90) = 2.89$,
$P_{cr} = 2.29\pi^2 EI/L^2$

7. Figure E8.5 shows a rigid-jointed frame. The flexural rigidity of column CE is 1.5 times that

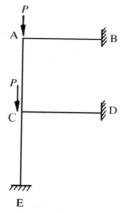

P
A ▮ ▮ B

P
C ▮ ▮ D

E

Figure E8.5

of other members. Determine the elastic critical load.

Answer: The axial force in AC is P and that in CE is 2P. $\rho_{ac} = P/(\pi^2 EI/L^2)$, $\rho_{ce} = 2P/(\pi^2 1.5EI/L^2) = 1.33\rho_{ab}$. Therefore the stiffness matrix is given by

$$\begin{bmatrix} M_A \\ M_c \end{bmatrix} = (EI/L)$$
$$\times \begin{bmatrix} S_1(\rho) + 4 & S_2(\rho) \\ S_2(\rho) & 1.5S_1(1.33\rho) + S_1(\rho) + 4 \end{bmatrix}\begin{bmatrix} \theta_a \\ \theta_c \end{bmatrix}$$

Determinant = $\{S_1(\rho) + 4\}\{1.5S_1(1.33\rho) + S_1(\rho) + 4\} - S_2(\rho)^2$

ρ	$S_1(\rho)$	$S_2(\rho)$	1.33ρ	$S_1(1.33\rho)$	Determinant
1.8	0.72	3.22	2.40	−1.30	7.31
1.9	0.44	3.37	2.53	−1.90	−4.30

$P_{cr} = 1.8 + (1.9 - 1.8)7.31/(7.31 + 4.30) = 1.86$,
$P_{cr} = 1.86\pi^2 EI/L^2$

Chapter nine

Structural dynamics

Previous chapters discussed the analysis of structures subjected to static loads. Engineers have to design structures subjected not only to static loads but also to dynamic loads. Typical dynamic loads are vehicles travelling on a bridge, loads due to gusting of wind, vibrating machinery, seismic disturbance, impact of vehicles against bridge or building, forces arising due to blasting, etc. It is therefore important that the engineer should have a clear understanding of the factors that affect the response of structures to dynamic loading. This chapter gives a brief introduction.

9.1 ● Dynamic equilibrium of a mass point

Let a set of N coplanar forces F_i, $i = 1, 2, \ldots, N$, with a resultant R act on a particle of mass M. A particle with a mass is often called a mass point. According to Newton's second law of motion if R is not equal to zero then the particle moves with an acceleration a directed along the line of action of the resultant force and related to R by

$$R = Ma$$

The above equation can be written as

$$R + (-Ma) = 0$$

In the case of static problems, $a = 0$ and therefore for equilibrium $R = 0$. In the case of the corresponding dynamic problem, the sum of the resultant of the applied force R and $(-Ma)$ must be equal to zero. Because $(-Ma)$ has the same units as force, it is called inertia force. Thus the requirement for 'dynamic equilibrium' of a mass point is that the sum of the resultant of applied force and inertia force must be zero. This way of considering 'dynamic equilibrium' as an equivalent static equilibrium problem by including the inertia force equal to $(-Ma)$ is known as D'Alembert's principle. As will be shown presently, in dynamic problems, forces other than applied forces and inertia forces need to be considered.

9.2 ● Vibration of a simple spring–mass–damper system

Practical structures such as buildings and bridges have mass and the load-resisting components such as the beams and columns act as springs resisting the deformation

Figure 9.1 ● A mass–spring–dashpot vibrating system

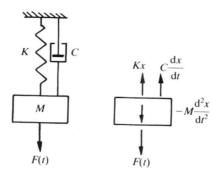

of the structure. Further, it is often the case that when a structure is disturbed from its position of rest it vibrates and eventually the vibration dies out. The damping of vibration is generally due to 'friction' caused by the movement of the structure relative to the surrounding medium such as air or other fluids like water. Damping is also caused by friction due to relative movement between the component parts at joints, cracks, etc. Therefore in dynamic problems we have to take into account

(a) the mass of the structure so that the inertia forces can be accounted for;
(b) elastic stiffness of the structure contributed by the structural elements;
(c) damping forces;
(d) the time-varying external forces acting on the structure.

As an example of a vibrating structure consider a simple spring–mass–damper system shown in Fig. 9.1. The damper (shown as a dashpot) represents the damping forces and should not be assumed in general to represent actual dampers in the real system. In considering the dynamic equilibrium of the mass at a time when the displacement of the mass is x downwards, the forces to be considered are

(i) spring force $= Kx$ acting upwards;
(ii) damping force acting upwards to resist motion;
(iii) inertia force $= -(M \times \text{acceleration}) = -M \, d^2x/dt^2$;
(iv) time-varying applied force $F(t)$ acting downwards.

The equation of dynamic equilibrium is thus given by

$$F(t) - Kx - \text{damping force} + (-M \, d^2x/dt^2) = 0$$

i.e. $M \, d^2x/dt^2 + \text{damping force} + Kx = F(t)$

For simplicity, damping force is assumed to be viscous in nature, i.e. proportional to velocity dx/dt. With this assumption, the damping force is equal to $C \, dx/dt$, where $C =$ damping constant. The equation of dynamic equilibrium becomes

$$M \, d^2x/dt^2 + C \, dx/dt + Kx = F(t) \tag{9.1}$$

This is a second-order non-homogeneous differential equation for which the solution is

$$x = C_1 e^{\alpha_1 t} + C_2 e^{\alpha_2 t} + \text{particular solution depending on } F(t)$$

where C_1 and C_2 are constants of integration, $\beta = C/(2M)$, $\omega^2 = K/M$, $\alpha_1 = -\beta + \sqrt{(\beta^2 - \omega^2)}$, $\alpha_2 = -\beta - \sqrt{(\beta^2 - \omega^2)}$.

In the following sections the solution to the above equation is studied in some detail.

9.2.1 Free vibration

If a structure is disturbed from its equilibrium position and let go, then it vibrates in the absence of any external force $F(t)$. This type of vibration is called free vibration. In addition, in practical structures, damping is fairly small so that $\omega^2 > \beta^2$. In this case $\alpha_1 = -\beta + i\,\omega_d$ and $\alpha_2 = -\beta - i\,\omega_d$, where $\omega_d = \sqrt{(\omega^2 - \beta^2)}$ and $i = \sqrt{(-1)}$. Because $\sin \omega_d t = (e^{i\omega_d t} - e^{-i\omega_d t})/(2i)$ and $\cos \omega_d t = (e^{i\omega_d t} + e^{-i\omega_d t})/2$, the solution becomes

$$x = e^{-\beta t}[C_1 \cos \omega_d t + C_2 \sin \omega_d t]$$

The constants of integration are determined from the boundary conditions. As an example, if the structure is displaced by x_0 and released, then at the time of release $t = 0$, $x = x_0$ and $dx/dt = 0$. Therefore $C_1 = x_0$ and $C_2 = \beta x_0/\omega_d$. The solution to x as a function of t is given by

$$x(t) = x_0 e^{-\beta t}[\cos \omega_d t + (\beta/\omega_d) \sin \omega_d t]$$

Figure 9.2 shows a plot of the variation of x with t. As can be seen, the vibration gets damped out with increasing values of t. As damping is increased, then β increases and $e^{-\beta t}$ decreases. This causes faster decay of the vibration amplitude.

The solution obtained above has an interesting property. For example, consider the displacement at time t and time $(t + T)$, where $T = 2\pi/\omega_d$.

$$x(t) = x_0 e^{-\beta t}[\cos \omega_d t + (\beta/\omega_d) \sin \omega_d t]$$

$$x(t + T) = x_0 e^{-\beta(t + T)}[\cos \omega_d(t + T) + (\beta/\omega_d) \sin \omega_d(t + T)]$$

Because

$$\cos \omega_d(t + T) = \cos \omega_d(t + 2\pi/\omega_d) = \cos(2\pi + \omega_d t) = \cos \omega_d t$$

and similarly $\sin \omega_d(t + T) = \sin \omega_d t$, then

$$x(t + T) = x_0 e^{-\beta(t + T)}[\cos \omega_d t + (\beta/\omega_d) \sin \omega_d t]$$

$$= x_0 e^{-\beta T} e^{-\beta t}[\cos \omega_d t + (\beta/\omega_d) \sin \omega_d t]$$

$$= e^{-\beta T} x(t)$$

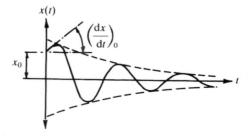

Figure 9.2 ● Response of damped free vibration

Therefore

$$x(t)/x(t + T) = e^{\beta T}$$

The displacement at time $(t + T)$ is $e^{-\beta T}$ times the displacement at time t. In other words, the motion has a period of T except that the displacement is damped. Obviously if damping is zero then $\beta = 0$ and $e^{-\beta T} = 1$ and $\omega_d = \omega$. In that case $x(t + T) = x(t) = x_0[\cos \omega t]$ and the motion is simple harmonic and repeats itself endlessly. These observations can be summarized as follows.

Undamped free vibration ($\beta = 0$). The motion is periodic with a period of $T = 2\pi/\omega$ where $\omega^2 = K/M$. The period T depends on both the mass M and the stiffness K of the system. Because damping is zero, the motion repeats itself endlessly. The circular frequency of vibration ω is expressed in radians per second and $f = \omega/(2\pi)$ is the frequency of vibration expressed in cycles per second or Hertz. $T = 1/f$ is the period or time for one cycle and is expressed in seconds.

Damped free vibration ($\beta < \omega$). The motion is damped periodic with a period of $T = 2\pi/\omega_d$, $\omega_d = \sqrt{(\omega^2 - \beta^2)}$. The ratio between successive displacements time T apart is $e^{\beta T}$. As damping increases, vibration decays exponentially.

9.2.2 Stiffness *K* for simple structures

In order to get an appreciation of the factors that affect K, consider some simple structures shown in Fig. 9.3.

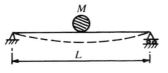
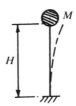
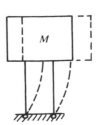

Figure 9.3 ● Some simple single degree freedom vibrating systems

Mass M at the centre of a simply supported beam (Fig. 9.3a). The beam obviously supplies the 'spring' support to the vibrating mass. As shown in Table 4.3, if a concentrated load W acts at the centre of a simply supported beam, the corresponding deflection Δ at the centre is given by

$$\Delta = WL^3/(48EI)$$

The stiffness K is defined as the load per unit deflection. Therefore $K = W/\Delta = 48EI/L^3$. The undamped free vibration period $T = 2\pi/\omega = 2\pi\sqrt{(M/K)} = 2\pi\sqrt{\{ML^3\}/(48EI)\}}$. It is easy to see from this equation why long span bridges tend to have fairly long periods of vibration.

Mass M at the end of a cantilever (Fig. 9.3b). Proceeding as in the previous case, in a cantilever loaded by W at the tip, the deflection Δ of the tip is given, from Table 4.1, by

$$\Delta = WH^3/(3EI) \text{ or } K = W/\Delta = 3EI/H^3$$

and

$$T = 2\pi\sqrt{\{MH^3/(3EI)\}}$$

As expected, the stiffness K of a cantilever is much less than that of the corresponding simply supported beam, with the result that it has a longer period of vibration than a simply supported beam. This formula can be used to calculate the period of vibration of a water tank supported on a single shaft. As can be inferred, tall buildings have long periods of vibration.

A building on stilts (Fig. 9.3c). Because of the very high stiffness of the building block compared to the columns, it can be assumed that the columns do not rotate at the top. Therefore each column acts as a cantilever fixed at the junction with the building block. Therefore $K = N(3EI/H^3)$, where N = number of column supports. $T = 2\pi\sqrt{(M/K)}$.

9.3 ● Critically damped system

It was pointed out in Section 9.2.1 that, as damping is increased, the vibrations decay exponentially. In fact, at a certain level of damping, the vibratory motion ceases altogether. This type of motion is called critically damped motion. The value of the damping constant for critically damped motion is called the critical damping constant C_{cr}. The vibratory motion results from the $\cos \omega_d t$ and $\sin \omega_d t$ terms and if $\omega_d = 0$, then the vibratory motion ceases.

Because $\omega_d = \sqrt{(\omega^2 - \beta^2)}$, $\omega_d = 0$ if $\omega = \beta$. But $\omega = \sqrt{(K/M)}$ and $\beta = \{C_{cr}/(2M)\}$. Therefore

$$\sqrt{(K/M)} = C_{cr}/(2M)$$

and

$$C_{cr} = 2\sqrt{(KM)}$$

When ω_d tends to 0, both $\cos \omega_d$ and $\sin \omega_d t/(\omega_d t)$ tend to 1. Therefore the solution

$$x(t) = x_0 e^{-\beta t}[\cos \omega_d t + (\beta/\omega_d) \sin \omega_d t]$$

becomes

$$x(t) = x_0 e^{-\beta t}[1 + \beta t], \beta = \omega = \sqrt{(K/M)}$$

Figure 9.4 shows the motion of a critically damped system.

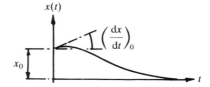

Figure 9.4 ● Response of a critically damped vibrating system

9.4 ● Logarithmic decrement

It was shown in Section 9.2.1 that, in the case of damped free systems, the ratio of displacements separated by time equal to period T apart is given by

$$x(t)/x(t + T) = e^{\beta T}$$

where $\beta = C/(2M)$ and $T = 2\pi/\omega_d$, $\omega_d = \surd\{\omega^2 - \beta^2\}$. This ratio can be used as a measure of the damping present in the structures. For convenience, instead of the ratio $x(t)/x(t + T)$ it is common to use $ln\,\{x(t)/x(t + T)\}$. This factor is known as the logarithmic decrement δ.

logarithmic decrement $\delta = ln\,\{x(t)/x(t + T)\} = \log\,\{e^{\beta T}\} = \beta T$

Therefore

$$\delta = \beta T = \beta(2\pi)/\omega_d = 2\pi\beta/\surd(\omega^2 - \beta^2)$$

$$= 2\pi(\beta/\omega)/\{\surd(1 - (\beta/\omega)^2\}$$

$\beta = C/(2M)$ and $\omega = \surd(K/M)$, therefore

$$\beta/\omega = C/\{2\surd(KM)\}$$

However $2\surd(KM) = C_{cr}$. Therefore

$$\beta/\omega = C/C_{cr}$$

Because the ratio of C/C_{cr} is very small in practical structures (generally much less than 0.10), logarithmic decrement δ can be expressed as

$$\delta = 2\pi(\beta/\omega)/\{\surd\{1 - (\beta/\omega)^2\} \simeq 2\pi\beta/\omega \simeq 2\pi(C/C_{cr})$$

Therefore, logarithmic decrement δ can be used directly to evaluate the damping present in the structure.

Note that $\omega_d = \surd(\omega^2 - \beta^2) = \omega\surd\{1 - (C/C_{cr})^2\}$.

9.5 ● Forced vibration

In the previous sections, free vibrations, i.e. the behaviour of systems vibrating in the absence of external forces, was considered. Because structures are subjected to vibratory forces, it is useful to study the response of spring–mass–damper systems to such forces. As an introduction to the study of forced vibration problems, consider the system shown in Fig. 9.1 where $F = F_0 \sin \Omega t$. In other words, the system is subjected to a sinusoidally varying force with an amplitude of F_0 and frequency of Ω radians per second. The differential equation to be solved is given by

$$M\,d^2x/dt^2 + C\,dx/dt + Kx = F_0 \sin \Omega t$$

The solution is given by

$$x(t) = \text{complementary function} + \text{particular integral}$$

where complementary function $= e^{-\beta t}[C_1 \cos \omega_d t + C_2 \sin \omega_d t]$, and the particular solution $= C_3 \sin \Omega t + C_4 \cos \Omega t$.

The constants C_3 and C_4 are determined so as to satisfy the differential equation. It can be shown that

$$C_3 = (F_0/M)(\omega^2 - \Omega^2)/D^2, \quad C_4 = -(F_0/M)2\beta\Omega/D^2,$$

$$D^2 = (\omega^2 - \Omega^2)^2 + (2\beta\Omega)^2$$

Therefore

Particular integral $= \{F_0/(MD^2)\}[(\omega^2 - \Omega^2) \sin \Omega t - 2\beta\Omega \cos \Omega t]$

The equation can be simplified further as follows.

Let $(\omega^2 - \Omega^2) = D \cos \alpha$ and $-2\beta\Omega = D \sin \alpha$, where $D^2 = (\omega^2 - \Omega^2)^2 + (2\beta\Omega)^2$, $\tan \alpha = (-2\beta\Omega)/(\omega^2 - \Omega^2)$. Therefore

Particular integral $= \{F_0/(MD)\}[\cos \alpha \sin \Omega t + \sin \alpha \cos \Omega t]$

$$= \{F_0/(MD)\} \sin (\Omega t + \alpha)$$

Therefore

$$x(t) = e^{-\beta t}[C_1 \cos \omega_d t + C_2 \sin \omega_d t] + \{F_0/(MD)\} \sin (\Omega t + \alpha)$$

The complementary part of the solution is known as the starting transient because the constants C_1 and C_2 depend on the initial boundary conditions. However, because of the effect of damping as represented by the $e^{-\beta t}$ term, the starting transient is very quickly damped out, with the result that the particular integral term dominates the response. It is for this reason that the particular integral is also known as the steady state solution. Considering the steady state solution,

$$x(t) = \{F_0/(MD)\} \sin (\Omega t + \alpha)$$

Because $\omega^2 = K/M$, substituting for $M = K/\omega^2$ gives

$$x(t) = \{F_0\omega^2/(KD)\} \sin (\Omega t + \alpha)$$

$$= (F_0/K)(\omega^2/D) \sin (\Omega t + \alpha)$$

Because $F_0 =$ amplitude of the force, $K =$ stiffness of the system, $F_0/K =$ maximum static displacement. Therefore the maximum dynamic displacement is given by

Maximum dynamic displacement $= (F_0/K)(\omega^2/D)$

$=$ maximum static displacement $\times$ dynamic magnification factor

where dynamic magnification factor $= \omega^2/D = \omega^2/\sqrt{[(\omega^2 - \Omega^2)^2 + (2\beta\Omega)^2]}$.

The dynamic magnification factor depends only on the frequency ratio Ω/ω and damping ratio $\beta = C/C_{cr}$. Figure 9.5 shows a plot of the dynamic magnification factor as a function of the frequency ratio Ω/ω. It can be seen that as the frequency ratio Ω/ω approaches unity, then the dynamic magnification factor becomes very large indeed, especially when damping is small, as happens in most structures. When the frequency of applied force and the natural frequency of the structure almost coincide to produce maximum dynamic displacement, the structure is said to resonate. When an undamped structure resonates, the displacements tend to infinity.

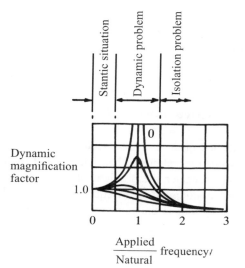

Figure 9.5 ● Variation of dynamic magnification factor with the ratio of applied/natural frequency and damping ratio. The curves are drawn for damping ratios of 1, 0.7, 0.5, 0.2 and 0

It is therefore important to ensure that as far as possible the applied force frequency Ω does not approach the natural frequency ω of the structure. For example, in the case of low-rise structures, the large structural stiffness precludes any possibility of resonance occurring under dynamic loads due to gusting of wind so that the dynamic magnification factor is practically unity. Thus, in the design of such structures, wind load can be treated as a quasi-static rather than as a dynamic load. As shown in Fig. 9.5, only over a small range of Ω/ω is it necessary to consider dynamic analysis in the design of structures. The majority of structures subjected to vibratory forces can be designed on the basis of quasi-static loading.

9.6 ● Suddenly applied load

Consider the case of a spring–mass–damper system to which a load F_0 is suddenly applied. The differential equation is given by

$$M \, d^2x/dt^2 + C \, dx/dt + Kx = F_0$$

The solution is given by

$$x(t) = e^{-\beta t}[C_1 \cos \omega_d t + C_2 \sin \omega_d t] + F_0/K$$

If the system starts from rest, then $x = dx/dt = 0$ at $t = 0$. Therefore $C_1 = -F_0/K$, $C_2 = (-F_0/K)\beta/\omega_d$. Thus

$$x(t) = (F_0/K)[1 - e^{-\beta t}\{\cos \omega_d t + (\beta/\omega_d) \sin \omega_d t\}]$$

If the system is undamped, then $\beta = 0$ and $\omega_d = \omega$. Therefore

$$x(t) = (F_0/K)[1 - \cos \omega t]$$

The maximum value of $x(t)$ is equal to $(F_0/K)2$, when $\cos \omega t = -1$. Because (F_0/K) is the maximum static displacement, the dynamic magnification factor in this case is 2. As the force in the spring is equal to $Kx(t)$, the maximum force in the spring is equal to twice the value of the force if the load F_0 is slowly applied. This is an important result. Obviously the presence of damping would reduce the dynamic magnification factor and hence the force in the spring.

9.7 ● Longitudinal vibrations of bars

The ideas developed in the previous sections can be extended to study the axial, flexural and torsional vibrations of bars, beams and shafts.

As an example consider the problem of a bar of distributed mass m per unit length vibrating longitudinally at a frequency of ω. For the sake of simplicity, it is assumed that the system is undamped. As shown in Section 4.1, the differential equation of equilibrium of a bar is given by

$$dF/dx + f = 0$$

where F = axial force and f = distributed axial force. As was explained in Section 9.1, in the case of vibration problems, using D'Alembert's principle

$$f = \text{(externally applied load + inertia force) per unit length}$$

If the bar is vibrating, then the inertia force per unit length is given by $(-m\partial^2 u/\partial t^2)$ where u is the axial displacement and $\partial^2 u/\partial t^2$ is the acceleration in the axial direction. Note the use of partial derivatives because the axial displacement u is a function not only of x but also of time t.

The axial tensile force $F = AE\ \partial u/\partial x$ where AE = axial rigidity and $\partial u/\partial x$ is the axial strain. The differential equation of motion is given by

$$\partial/\partial x\{AE\ \partial u/\partial x\} + [q(t) - m\partial^2 u/\partial t^2] = 0$$

where $q(t)$ = external axial force per unit length.

If we are dealing with the free vibration of a uniform bar, then AE = constant and $q = 0$. Further, because the system is undamped, the motion is simple harmonic, therefore (see Section 9.2.1)

$$u(x, t) = U(x) \cos \omega t$$

where $U(x)$ is the amplitude of motion and ω = circular frequency of vibration.

Substituting for u and simplifying, the differential equation of motion becomes

$$AE\ d^2U/dx^2 + m\omega^2 U = 0$$

This is the basic differential equation governing the undamped free vibration of a uniform bar of distributed mass of m per unit length.

9.7.1 Solution of the differential equation for the longitudinal vibration of a uniform bar

For the bar shown in Fig. 9.6, the equation of dynamic equilibrium is a second-order ordinary differential equation for which the solution is

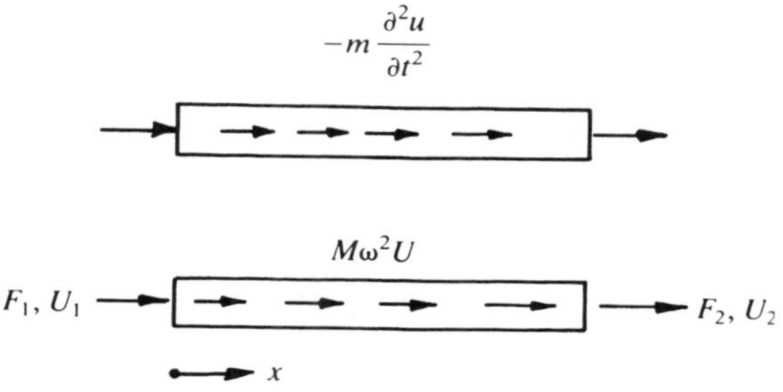

Figure 9.6 ● Free longitudinal vibration

$$U = C_1 \cos \beta x + C_2 \sin \beta x$$

where $\beta^2 = m\omega^2/(AE)$, C_1 and C_2 are integration constants. If the boundary conditions are $U = U_1$ at $x = 0$ and $U = U_2$ at $x = L$, then

$$C_1 = U_1, \; C_2 = \{U_2 - U_1 \cos \beta L\}/\sin \beta L$$

Therefore

$$U = \{\cos \beta x - \cos \beta L \sin \beta x/\sin \beta L\}U_1 + \{\sin \beta x/\sin \beta L\}U_2$$

The corresponding axial force F is given by

$$F = AE \, dU/dx$$

Substituting for U,

$$F = (AE/L)(\beta L)[-\{\sin \beta x + \cos \beta L \cos \beta x/\sin \beta L\}U_1 + \{\cos \beta x/\sin \beta L\}U_2]$$

The forces at the ends are $F = -F_1$ at $x = 0$ and $F = F_2$ at $x = L$. Therefore

$$F_1 = (AE/L)(\beta L)[\{\cos \beta L/\sin \beta L\}U_1 - \{1/\sin \beta L\}U_2]$$

$$F_2 = (AE/L)(\beta L)[-\{1/\sin \beta L\}U_1 + \{\cos \beta L/\sin \beta L\}U_2]$$

The above relationship can be represented as the following stiffness relationship

$$\begin{bmatrix} F_1 \\ F_2 \end{bmatrix} = (AE/L) \begin{bmatrix} A_{11} & -A_{12} \\ -A_{12} & A_{11} \end{bmatrix} \begin{bmatrix} U_1 \\ U_2 \end{bmatrix}$$

where $A_{11} = \beta L\{\cos \beta L/\sin \beta L\}$, $A_{12} = (\beta L)/\sin \beta L$.
Table 9.1 gives the numerical values of A_{11} and A_{12} as a function of βL.

9.7.2 Example of the longitudinal vibration of bars

As an example consider the vibration of a bar fixed at end 1 and free at end 2. In this case $U_1 = 0$ and $F_2 = 0$. Therefore from the stiffness matrix

$$A_{11}U_2 = 0$$

Because $U_2 \neq 0$, then

$$A_{11} = \beta L \cos \beta L/\sin \beta L = 0$$

Table 9.1 ●
Coefficients of the
longitudinal or torsional
vibration dynamic
stiffness matrix

$$\beta L = \sqrt{\left(\frac{m\omega^2}{AW}\right)}L \text{ or } \sqrt{\left(\frac{m\omega^2 I_p}{GJA}\right)}L$$

βL	A_{11}	A_{12}	βL	A_{11}	A_{12}
0	1.0	−1.0	3.6	7.295	8.135
0.1	0.997	−1.002	3.8	4.912	6.211
0.2	0.987	−1.007	4.0	3.455	5.285
0.3	0.970	−1.015	4.4	1.421	4.624
0.4	0.946	−1.027	4.8	−0.422	4.815
0.5	0.915	−1.043	5.2	−2.758	5.886
0.6	0.877	−1.063	5.6	−6.880	8.871
0.7	0.831	−1.087	6.0	−20.618	21.473
0.8	0.777	−1.115	6.4	54.538	−54.912
0.9	0.714	−1.149	6.8	11.965	−13.762
1.0	0.642	−1.188	7.2	5.519	−9.072
1.2	0.467	−1.288	7.6	1.973	−7.852
1.4	0.242	−1.421	8.0	−1.177	−8.086
1.6	−0.047	−1.601	8.4	−5.104	−9.829
1.8	−0.420	−1.848	8.8	−12.203	−15.045
2.0	−0.915	−2.200	9.2	−40.238	−41.276
2.2	−1.601	−2.721	9.6	54.226	55.069
2.4	−2.620	−3.553	10.0	15.424	18.382
2.6	−4.322	−5.044	10.4	7.048	12.563
2.8	−7.876	−8.359	10.8	2.140	11.010
3.0	−21.046	−21.259	11.2	−2.322	11.438
3.2	54.725	54.819	11.6	−8.012	9.545
3.4	12.863	13.305	12.0	−18.872	22.364

Therefore

$$\cos \beta L = 0$$

Therefore

$$\beta L = (2n - 1)\pi/2, \quad n = 1, 2, \ldots$$

Because

$$\beta^2 = m\omega^2/(AE), \quad \omega = (2n - 1)(\pi/2)\sqrt{\{AE/(mL^2)\}}$$

$AE/L =$ stiffness K and $mL =$ total mass M, therefore

$$\omega = (2n - 1)(\pi/2)\sqrt{\{K/M\}}$$

This expression is very similar to the undamped frequency of vibration of a simple spring–mass system.

The displacement amplitude U is given by

$$U = \{\sin \beta x/\sin \beta L\}U_2$$

Figure 9.7 shows the mode shape and the corresponding frequency of vibration. As in the solution of problems of elastic instability in Section 8.3.1, the value of U_2 remains undetermined.

Figure 9.7 ●
Frequencies and
corresponding modes
of free longitudinal
vibration of a bar fixed
at one end and free at
the other end. Arrows
show the direction of
displacement at any
particular time

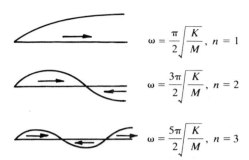

$$\omega = \frac{\pi}{2}\sqrt{\frac{K}{M}}, \ n = 1$$

$$\omega = \frac{3\pi}{2}\sqrt{\frac{K}{M}}, \ n = 2$$

$$\omega = \frac{5\pi}{2}\sqrt{\frac{K}{M}}, \ n = 3$$

9.8 ● Flexural vibration of beams

In Section 1.12, the differential equation governing the bending of beams was given by

$$d^2M/dx^2 = q(x)$$

Using D'Alembert's principle

q = (externally applied load + inertia force) per unit length

If a beam with a mass per unit length of m is vibrating, then the inertia force per unit length is given by $(-m\partial^2v/\partial t^2)$ where v is the lateral displacement and $\partial^2v/\partial t^2$ is the acceleration in the lateral direction. Note the use of partial derivatives because the displacement v is a function not only of x but also of time t.

The bending moment M is given by

$$M = EI\ \partial^2v/\partial x^2$$

where EI = flexural rigidity. Substituting for M and q in the differential equation of motion, we have

$$\partial^2v/\partial x^2\{EI\ \partial^2v/\partial x^2\} = [q(t) - m\partial^2v/\partial t^2]$$

where $q(t)$ = external lateral force per unit length. If we are dealing with free vibration, then $q(t) = 0$. If the beam is uniform, then EI = constant. In such a case the differential equation for free vibration becomes

$$EI\ \partial^4v/\partial x^4 + m\partial^2v/\partial t^2 = 0$$

Further if the system is undamped, then motion is simple harmonic. Therefore

$$v(x, t) = V(x)\cos \omega t$$

where $V(x)$ is the amplitude of motion and ω = circular frequency of vibration. Substituting for v and simplifying, the differential equation of motion becomes

$$EI\ d^4V/dx^4 - m\omega^2V = 0$$

This is the basic differential equation governing the undamped free vibration of uniform beams carrying a distributed mass m per unit length.

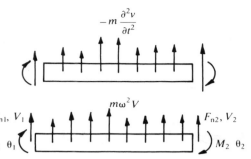

Figure 9.8 ● Free
flexural vibration of a
beam element

9.8.1 Solution of flexural vibration equation

For the beam shown in Fig. 9.8, the solution to the dynamic flexural differential equation is given by

$$V = C_1 \sin \lambda x + C_2 \cos \lambda x + C_3 \sinh \lambda x + C_4 \cosh \lambda x$$

where C_1 to C_4 = constants of integration, and $\lambda^4 = m\omega^2/(EI)$. Let the boundary conditions be

at $x = 0$, $V = V_{n1}$, $dV/dx = -\theta_1$

at $x = L$, $V = V_{n2}$, $dV/dx = -\theta_2$

Using the boundary conditions, the constants of integration are given by

$$C_1 = [-V_{n1}(Sc + Cs) + V_{n2}(S + s) + \theta_1\lambda^{-1}(Ss + Cc - 1) + \theta_2\lambda^{-1}(C - c)]/D$$

$$C_2 = [-V_{n1}(Ss + Cc - 1) - V_{n2}(C - c) - \theta_1\lambda^{-1}(Cs - Sc) - \theta_2\lambda^{-1}(S - s)]/D$$

$$C_3 = [V_{n1}(Sc + Cs) - V_{n2}(S + s) - \theta_1\lambda^{-1}(Ss - Cc + 1) - \theta_2\lambda^{-1}(C - c)]/D$$

$$C_4 = [-V_{n1}(Ss + Cc - 1) + V_{n2}(C - c) + \theta_1\lambda^{-1}(Cs - Sc) + \theta_2\lambda^{-1}(S - s)]/D$$

where $S = \sinh \lambda L$, $s = \sin \lambda L$, $C = \cosh \lambda L$, $c = \cos \lambda L$, $D = 2(1 - Cc)$.
The forces at the ends of the element are given by

At $x = 0$, $M_1 = EI\,d^2V/dx^2$, $F_{n1} = -EI\,d^3V/dx^3$

At $x = L$, $M_2 = -EI\,d^2V/dx^2$, $F_{n2} = EI\,d^3V/dx^3$

After carrying out the lengthy algebra while noting that $s^2 + c^2 = 1$ and $C^2 - S^2 = 1$, the following stiffness relationship can be established:

$$\begin{bmatrix} F_{n1} \\ M_1 \\ F_{n2} \\ M_2 \end{bmatrix} = (EI/L) \begin{bmatrix} T_1/L^2 & -Q_1/L & -T_2/L^2 & -Q_2/L \\ & S_1 & Q_2/L & S_2 \\ \text{symmetrical} & & T_1/L^2 & Q_1/L \\ & & & S_1 \end{bmatrix} \begin{bmatrix} V_{n1} \\ \theta_1 \\ V_{n2} \\ \theta_2 \end{bmatrix}$$

where $S_1 = B(Cs - Sc)$, $S_2 = B(S - s)$, $Q_1 = B\lambda LSs$, $Q_2 = B\lambda L(C - c)$, $T_1 = B(\lambda L)^2(Cs + Sc)$, $T_2 = B(\lambda L)^2(S + s)$, and $B = \lambda L/(1 - Cc)$.

Table 9.2 gives the numerical values of the coefficients of the dynamic stiffness matrix. Note that the magnitude of the shear force at the ends of the beam is not the same. This is because of the distributed inertial force on the beam.

Table 9.2 ●
Coefficients of dynamic
flexural vibration
stiffness matrix

$$\lambda L = \sqrt{\sqrt{\left(\frac{m\omega^2}{EI}\right)}}\, L$$

λL	S_1	S_2	Q_1	Q_2	T_1	T_2
0	4.00	2.00	6.00	6.00	12.00	12.00
1.0	3.99	2.01	5.95	6.03	11.63	12.13
1.5	3.95	2.04	5.73	6.16	10.11	12.66
1.6	3.94	2.05	5.65	6.21	9.55	12.86
1.7	3.92	2.06	5.56	6.26	8.87	13.10
1.8	3.90	2.08	5.44	6.33	8.06	13.39
1.9	3.87	2.10	5.30	6.42	7.10	13.73
2.0	3.84	2.12	5.14	6.51	5.96	14.14
2.2	3.77	2.18	4.73	6.77	3.09	15.20
2.4	3.67	2.26	4.17	7.11	−0.75	16.66
2.6	3.53	2.36	3.43	7.58	−5.81	18.64
2.8	3.35	2.51	2.45	8.21	−12.39	21.33
3.0	3.10	2.70	1.16	9.07	−20.92	25.01
3.2	2.78	2.97	−0.55	10.25	−31.97	30.10
3.4	2.33	3.34	−2.85	11.91	−46.41	37.30
3.6	1.71	3.87	−6.02	14.29	−65.61	47.78
3.8	0.81	4.67	−10.56	17.89	−92.06	63.80
4.0	−0.60	5.95	−17.53	23.73	−130.7	90.09
4.2	−3.08	8.28	−29.54	34.41	−193.9	138.6
4.4	−8.54	13.56	−55.49	58.77	−323.1	250.6
4.6	−30.66	35.47	−159.0	160.4	−813.9	721.1
4.8	73.22	−68.67	323.0	−323.9	1415	−1531
5.0	22.99	−18.74	88.72	−92.17	312.4	−456.6
5.2	15.37	−11.48	52.21	−58.80	127.5	−304.3
5.4	12.15	−8.69	36.01	−46.34	34.76	−249.9
5.6	10.26	−7.32	25.77	−40.59	−32.98	−227.5
5.8	8.90	−6.61	17.77	−38.05	−93.38	−221.2
6.0	7.79	−6.29	10.53	−37.51	−153.9	−225.8
6.2	6.74	−6.25	3.22	−38.57	−219.3	−240.1
6.4	5.67	−6.46	−4.82	−41.24	−293.9	−264.9
6.6	4.45	−6.96	−14.32	−45.86	−382.8	−303.7
6.8	2.94	−7.83	−26.35	−53.22	−494.4	−363.0
7.0	0.90	−9.30	−42.80	−65.06	−643.5	−456.6
7.2	−2.20	−11.85	−67.80	−85.35	−862.3	−615.8
7.4	−7.79	−16.90	−112.5	−125.1	−1239	−927.6
7.6	−21.76	−30.34	−223.4	−230.7	−2139	−1756
7.8	−138.7	−146.7	−1143	−1145	−9400	−8938
8.0	62.11	54.69	433.2	437.9	2956	3505
8.2	30.89	24.13	186.2	198.0	976.4	1624
8.4	22.20	16.16	116.0	135.8	382.4	1141
8.6	17.90	12.66	79.98	108.9	52.15	936.9
8.8	15.14	10.84	55.82	95.46	−190.0	840.0
9.0	13.07	9.87	36.63	88.90	−399.2	800.0
9.2	11.30	9.44	19.35	86.82	−600.5	798.7
9.4	9.63	9.40	2.19	88.39	−809.9	830.7
9.6	7.90	9.75	−16.31	93.59	−1041	898.3
9.8	5.94	10.53	−37.82	103.2	−1312	1011
10.0	3.52	11.92	−64.83	119.2	−1648	1192

9.8.2 Procedure for the calculation of natural frequencies

The procedure for calculating the natural frequency of structures is similar to the procedure used for the calculation of elastic buckling loads of structures. However, there is an important difference. In elastic stability calculations, the designer is interested mainly in the lowest load at which the structure will buckle. In natural frequency calculations the designer is interested in all those natural frequencies which are likely to coincide with the frequencies of applied forces (such as that due to wind, seismic forces, vibrating machinary, etc.) which have sufficient energy to excite one or more of the natural frequencies of the structure. The main concept used in determining the natural frequency is that if an applied vibratory force with a frequency equal to a natural frequency of the structure acts on the structure, then the structure resonates. In the absence of damping, when the structure is resonating the displacements tend to infinity. The criterion for displacements to tend to infinity is that the determinant of the dynamic structural stiffness matrix tends to zero. The criterion used is similar to that used in the determination of elastic critical loads of structures.

9.8.3 Examples

Example 1

As a simple example of the use of the dynamic structural stiffness matrix to calculate the natural frequencies of a flexural member, consider the simply supported uniform beam of span L and flexural rigidity EI carrying a mass of m per unit length.

Solution In this case V_{n1} and V_{n2} are zero. The dynamic stiffness matrix relationship is given by

$$\begin{bmatrix} M_1 \\ M_2 \end{bmatrix} = EI/L \begin{bmatrix} S_1 & S_2 \\ S_2 & S_1 \end{bmatrix} \begin{bmatrix} \theta_1 \\ \theta_2 \end{bmatrix}$$

The determinant of the stiffness matrix is equal to $(S_1^2 - S_2^2)$. Substituting for $S_1 = (Cs - Sc)/(1 - Cc)$ and $S_2 = (S - s)/(1 - Cc)$, the determinant is zero when

$$(Cs - Sc)^2 - (S - s)^2 = 0$$

i.e. $C^2s^2 + S^2c^2 - 2SCsc - S^2 - s^2 - 2Ss = 0$

$$s^2(C^2 - 1) + S^2(c^2 - 1) - 2Ss(SC + 1) = 0$$

Substituting $C^2 - 1 = S^2$ and $c^2 - 1 = -s^2$, the expression for the determinant becomes

$$s^2S^2 + S^2(-s^2) - 2Ss(SC + 1) = 0$$

Therefore

$$Ss(SC + 1) = 0$$

Because S is zero only when $\lambda L = 0$, and SC is always positive, the determinant is zero when $s = \sin \lambda L = 0$. Therefore

$$\lambda L = n\pi, \quad n = 1, 2, \ldots$$

Because $\lambda^4 = (m\omega^2)/(EI)$, the natural circular frequency ω is given by

$$\omega = \pi^2 \sqrt{\{EI/(mL^4)\}}$$

The above expression can be interpreted as follows. The total mass M on the beam is $M = mL$. As was shown in Chapter 4, the deflection Δ at the centre of a beam under any system of loading is given by

$$\Delta = CWL^3/(EI)$$

where C = coefficient depending on the distribution of the load, W = total load on the beam. The stiffness defined as load per unit deflection is

$$\text{Stiffness} = K = W/\Delta = EI/(CL^3)$$

This shows that the stiffness of a beam is proportional to EI/L^3. Therefore the expression for ω can be expressed as

$$\omega = \pi^2 \sqrt{\{EI/(mL^4)\}} = \sqrt{C}\pi^2 \sqrt{(K/M)}$$

where $M = mL$ total mass, $K = EI/L^3$ represents the stiffness of the beam. This shows that the formula derived for the undamped natural frequency of the simple spring–mass system of Fig. 9.1 is qualitatively the same as the formula for the frequency of simply supported beams under uniformly distributed mass.

Mode shape. The shape of deflection amplitude can be calculated as follows. Because the determinant of the structural stiffness matrix is zero, it is not possible to solve for the unknowns θ_1 and θ_2. All that we can do is to fix the value of say θ_1 and determine θ_2 in terms of θ_1. Taking the first row of the stiffness matrix

$$M_1 = (EI/L)[S_1\theta_1 + S_2\theta_2]$$

Because this is free vibration, M_1 = externally applied force = 0. Therefore

$$S_1\theta_1 + S_2\theta_2 = 0$$

When the beam is vibrating at its natural frequency, $s = 0$. Substituting $s = 0$ in the expressions for S_1 and S_2,

$$S_1 = -Sc/(1 - Cc) \text{ and } S_2 = S/(1 - Cc)$$

Therefore, $S_1\theta_1 + S_2\theta_2 = 0$ becomes

$$\{S/(1 - Cc)\}[-c\,\theta_1 + \theta_2] = 0$$

$$\therefore\ -c\,\theta_1 + \theta_2 = 0$$

Multiplying the above equation by c we have

$$-c^2\,\theta_1 + c\theta_2 = 0$$

But $c^2 = 1$ because $s = 0$. Therefore

$$-\theta_1 + c\,\theta_2 = 0$$

The deflection amplitude V is given by

$$V = C_1 \sin \lambda x + C_2 \cos \lambda x + C_3 \sinh \lambda x + C_4 \cosh \lambda x$$

Substituting $V_{n1} = V_{n2} = 0$ and $s = 0$ in the expressions for C_1 to C_4 given in Section 9.8.1, we have

$$C_1 = \lambda^{-1}[\theta_1(Cc - 1) + \theta_2(C - c)]/D$$
$$= \lambda^{-1}[C(c\,\theta_1 + \theta_2) - (\theta_1 + c\,\theta_2)]/D$$

Substituting $\theta_2 = c\theta_1$ and $c\theta_2 = \theta_1$ and simplifying, C_1 becomes

$$C_1 = [\lambda^{-1}\{2Cc - 1 - c^2\}/D]\theta_1 = [\lambda^{-1}\{2Cc - 2\}/D]\theta_1$$
$$= [-\lambda^{-1}2\{1 - Cc\}/D]\theta_1$$

Because $D = 2(1 - Cc)$, C_1 is given by

$$C_1 = -\lambda^{-1}\theta_1$$
$$C_2 = \lambda^{-1}[\theta_1 Sc - \theta_2 S]/D$$
$$= -\lambda^{-1}S[-c\,\theta_1 + \theta_2 = 0$$
$$C_3 = \lambda^{-1}[\theta_1(Cc - 1) - \theta_2(C - c)]/D$$
$$= \lambda^{-1}[-C(-c\,\theta_1 + \theta_2) - (\theta_1 - c\,\theta_2)] = 0$$
$$C_4 = \lambda^{-1}[\theta_1(-Sc) + \theta_2 S]/D$$
$$= \lambda^{-1}S[-c\,\theta_1 + \theta_2] = 0$$

Therefore

$$V = -\lambda^{-1}\,\theta_1 \sin \lambda x$$

Using $\lambda L = n\pi$, $V = -\{L\,\theta_1/(n\pi)\} \sin n\pi x/L$. The mode shape is a sine curve as shown in Fig. 9.9.

Figure 9.9 ● First three modes of vibration of a simply supported beam. $K = EI/(CL^3)$, $M = mL$

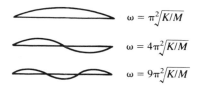

$$\omega = \pi^2\sqrt{K/M}$$

$$\omega = 4\pi^2\sqrt{K/M}$$

$$\omega = 9\pi^2\sqrt{K/M}$$

Example 2

Calculate the first (also called the fundamental) natural frequency of a uniform cantilever beam of uniformly distributed mass m per unit length.

Solution   Assuming that the beam is fixed at end 1, then $V_{n1} = \theta_1 = 0$. Therefore the dynamic stiffness relationship is given by

Table 9.3

λL	S_1	Q_1	T_1	$(T_1 S_1 - Q_1{}^2)$
1.8	3.90	5.44	8.06	1.840
1.9	3.87	5.30	7.10	−0.613

$$\begin{bmatrix} F_{n2} \\ M_2 \end{bmatrix} = EI/L \begin{bmatrix} T_1/L^2 & Q_1/L \\ Q_1/L & S_1 \end{bmatrix} \begin{bmatrix} V_{n2} \\ \theta_2 \end{bmatrix}$$

Determinant $= T_1 S_1 - Q_1{}^2$. For the determinant $= 0$,

$$T_1 S_1 - Q_1{}^2 = 0$$

It is simpler to use the numerical values in Table 9.2 rather than develop analytical expressions. The results of calculating the determinant at $\lambda L = 1.8$ and 1.9 are shown in Table 9.3. Because the determinant is +1.840 at $\lambda L = 1.8$ and −0.613 at $\lambda L = 1.9$, the value of λL at which the determinant is zero is given, using linear interpolation, by

$$\lambda L = 1.8 + (1.9 - 1.8)(1.840)/\{1.840 - (-0.613)\} = 1.875$$

Therefore

$$\omega = 1.875^2 \sqrt{\{K/M\}}, \quad M = mL, \quad K = EI/L^3.$$

9.8.4 Fundamental frequencies of beams with idealized boundary conditions

The procedure used in the previous examples can be used to calculate the natural frequencies of beams with idealized boundary conditions. Table 9.4 gives the natural frequencies of beams under typical idealized boundary conditions:

$$\omega = C\sqrt{\{K/M\}}, \quad M = mL, \quad K = EI/L^3$$

where $C =$ constant shown in Table 9.4.

Table 9.4 ●
Fundamental frequencies of beams

Boundary condition	C
Simply supported ends	$\pi^2 \approx 10.0$
Cantilever	3.516
Fixed ends	22.373
Propped cantilever	15.421
Simply supported and guided end*	2.468
Fixed end and guided end	5.593

*Guided end has zero rotation and zero shear force

9.9 ● Dynamic stiffness matrix for an element of a 2-D rigid-jointed structure

The dynamic stiffness matrix for a bar element undergoing longitudinal vibration was derived in Section 9.7. Similarly the dynamic stiffness matrix for a beam element undergoing flexural vibrations was derived in Section 9.8.1. The two matrices can be combined to derive the general dynamic stiffness matrix for a member in a 2-D rigid-jointed structure. The procedure to be followed was explained in detail in Sections 6.1 to 6.3. The final matrix is shown in equation (9.1). The dynamic stiffness relationship for an element in a 2-D rigid-jointed structure:

$$
\begin{bmatrix} F_{x1} \\ F_{y1} \\ M_1 \\ F_{x2} \\ F_{y2} \\ M_2 \end{bmatrix} = \left(\frac{EI}{L}\right)
$$

$$
\times \begin{bmatrix}
\dfrac{T_1 m^2 + \alpha_1 l^2}{L^2} & \dfrac{(-T_1 + \alpha_1)lm}{L^2} & \dfrac{Q_1 m}{L} & \dfrac{(-T_2 m^2 + \alpha_2 l^2)}{L^2} & \dfrac{(T_2 + \alpha_2)lm}{L^2} & \dfrac{Q_2 m}{L} \\
 & \dfrac{T_1 l^2 + \alpha_1 m^2}{L^2} & \dfrac{-Q_1 l}{L} & \dfrac{(T_2 + \alpha_2)lm}{L^2} & \dfrac{(-T_2 l^2 + \alpha_2)lm}{L^2} & \dfrac{-Q_2 l}{L} \\
 & & S_1 & \dfrac{-Q_2 m}{L} & \dfrac{Q_2 l}{L} & S_2 \\
 & & & \dfrac{T_1 m^2 + \alpha_1 l^2}{L^2} & \dfrac{(-T_1 + \alpha_1)lm}{L^2} & \dfrac{-Q_1 m}{L} \\
 & & & & \dfrac{T_1 l^2 + \alpha_1 m^2}{L^2} & \dfrac{Q_1 l}{L} \\
 & \text{symmetrical} & & & & S_1
\end{bmatrix}
$$

$$
\times \begin{bmatrix} u_1 \\ v_1 \\ \theta_1 \\ u_2 \\ v_2 \\ \theta_2 \end{bmatrix}
\tag{9.1}
$$

$$
\alpha_1 = \frac{AL^2}{I} A_{11}, \ \alpha_2 = \frac{AL^2}{I} A_{12}
$$

9.10 ● Examples of the calculation of fundamental natural frequencies of 2-D rigid-jointed structures

Example 1

Calculate the fundamental frequency of the uniform continuous beam shown in Fig. 9.10.

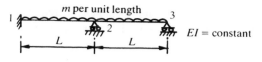

Figure 9.10 ● Free flexural vibration of a continuous beam

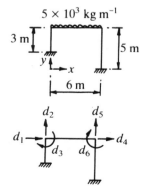

Solution   Let the rotation of the beam over supports 2 and 3 be d_1 and d_2 respectively. The dynamic stiffness matrix of the structure is given by

$$\begin{bmatrix} \Sigma F_1 \\ \Sigma F_2 \end{bmatrix} = (EI/L)\begin{bmatrix} S_1 + S_1 & S_2 \\ S_2 & S_1 \end{bmatrix}\begin{bmatrix} d_1 \\ d_2 \end{bmatrix}$$

The determinant is zero when $2S_1^2 - S_2^2 = 0$. Since the right-hand span is stiffer than a corresponding simply supported beam, the natural frequency of the whole struc-ture will also be such that $\lambda L > \pi$. Using the data in Table 9.5 and by linear inter-polation, the value of λL for zero determinant is given by

$$\lambda L = 3.2 + (3.4 - 3.2)(6.636)/\{6.636 - (-0.298)\} = 3.39$$

The corresponding value of ω is $\omega = 3.39^2\sqrt{\{K/M\}}$, $M = mL$ the total load on one beam, $K = EI/L^3$.

Table 9.5

λL	S_1	S_2	$2S_1^2 - S_2^2$
3.2	2.78	2.97	6.636
3.4	2.33	3.34	-0.298

Example 2

Determine the fundamental natural frequency of the structure shown in Fig. 9.11. Assume that the members have uniform properties given by $E = 210 \times 10^9$ N m^{-2}, $I = 1698.43 \times 10^{-6}$ m^4, $A = 21\ 630 \times 10^{-6}$ m^2. It is expected that the natural circular frequency ω is in the 50 radians per second range.

Figure 9.11 ● Free vibration of a portal frame

Solution The basic idea is to determine the smallest frequency at which the dynamic structural stiffness matrix has a zero determinant. An analytical approach to the solution is not feasible. Therefore the solution is sought numerically using an iterative technique. The dynamic structural stiffness matrix is formed at an estimated natural frequency and its determinant is evaluated. If it is positive, then the dynamic stiffness matrix is recalculated at a slightly higher frequency and if its corresponding determinant is still positive, then the process is repeated until a negative determinant is recorded. The natural frequency is then calculated by linear interpolation between two frequencies which are near to each other such that at the lower value of the frequency the determinant is positive and at the higher value of the frequency the determinant is negative. The steps to be followed are as follows.

(a) The structure has six degrees of freedom as shown in Fig. 9.11. The details required for forming the element and structural stiffness matrices are given in Tables 9.6 and 9.7.

Table 9.6

Node	u	v	θ
1	0	0	0
2	d_1	d_2	d_3
3	d_4	d_5	d_6
4	0	0	0

Table 9.7

Element	End 1	End 2	x_1	y_1	x_2	y_2	L	l	m	u_1	v_1	θ_1	u_2	v_2	θ_2
1–2	1	2	0	2	0	5	3	0	1	0	0	0	d_1	d_2	d_3
2–3	2	3	0	5	6	5	6	1	0	d_1	d_2	d_3	d_4	d_5	d_6
4–3	4	3	6	0	6	5	5	0	1	0	0	0	d_4	d_5	d_6

(b) Form element dynamic stiffness matrices.

Element 1–2. The element has no distributed mass. Therefore there are no distributed inertia forces and hence the dynamic stiffness matrix is the same as the static stiffness matrix as given in equation [6.2].

$$
\begin{bmatrix} F_{x1} = * \\ F_{y1} = * \\ M_1 = * \\ F_{x2} = F_1 \\ F_{y2} = F_2 \\ M_2 = F_3 \end{bmatrix} = 10^5 \begin{bmatrix} 1.59 & 0 & 2.39 & -1.59 & 0 & 2.38 \\ 0 & 15.14 & 0 & 0 & -15.14 & 0 \\ 2.39 & 0 & 4.76 & -2.38 & 0 & 2.38 \\ -1.59 & 0 & -2.38 & 1.59 & 0 & -2.38 \\ 0 & -15.14 & 0 & 0 & 15.14 & 0 \\ 2.38 & 0 & 2.38 & -2.38 & 0 & 4.76 \end{bmatrix}
$$

$$
\times \begin{bmatrix} u_1 = 0 \\ v_1 = 0 \\ \theta_1 = 0 \\ u_2 = d_1 \\ v_2 = d_2 \\ \theta_2 = d_3 \end{bmatrix}
$$

from which

$$
\begin{bmatrix} F_1 \\ F_2 \\ F_3 \end{bmatrix} = 10^5 \begin{bmatrix} 1.59 & 0 & -2.38 \\ 0 & 15.14 & 0 \\ -2.38 & 0 & 4.76 \end{bmatrix} \begin{bmatrix} d_1 \\ d_2 \\ d_3 \end{bmatrix}
$$

Element 2–3. This element has a mass of 5000 kg m^{-1}. Therefore inertial forces have to be considered. The dynamic stiffness matrix is a function of two parameters:

(i) $\beta L = L\sqrt{\{m\omega^2/(EA)\}}$ a factor from longitudinal vibration
(ii) $\lambda L = L\sqrt{\sqrt{\{m\omega^2/(EI)\}}}$ a factor from flexural vibration

Using consistent units of N, m, kg and seconds only we have $E = 210 \times 10^9$ N m^{-2}, $L = 6$ m , $I = 1698.43 \times 10^{-6}$ m^4, $A = 21\,630 \times 10^{-6}$ m^2, $m = 5000$ kg m^{-1}. The dynamic stiffness matrix will be formulated in the first instance at $\omega = 52$ radians/second. Therefore

$$\beta L = 6\sqrt{\{5000 \times 52^2/(210 \times 10^9 \times 21630 \times 10^{-6})\}} = 0.327$$

$$\lambda L = 6\sqrt{\sqrt{\{5000 \times 52^2/(210 \times 10^9 \times 1698.43 \times 10^{-6})\}}} = 2.648$$

Using these values and using the stiffness matrix in equation [9.1], the following stiffness matrix is evaluated:

$$
\begin{bmatrix} F_{x1} = F_1 \\ F_{y1} = F_2 \\ M_1 = F_3 \\ F_{x2} = F_4 \\ F_{y2} = F_5 \\ M_2 = F_6 \end{bmatrix} = 10^5 \begin{bmatrix} 7.30 & 0 & 0 & -7.71 & 0 & 0 \\ 0 & -0.12 & -0.32 & 0 & -0.32 & -0.76 \\ 0 & -0.32 & 2.09 & 0 & 0.76 & 1.44 \\ -7.71 & 0 & 0 & 7.30 & 0 & 0 \\ 0 & -0.32 & 0.76 & 0 & -0.12 & 0.32 \\ 0 & -0.76 & 1.44 & 0 & 0.32 & 2.09 \end{bmatrix}
$$

$$
\times \begin{bmatrix} u_1 = d_1 \\ v_1 = d_2 \\ \theta_1 = d_3 \\ u_2 = d_4 \\ v_2 = d_5 \\ \theta_2 = d_6 \end{bmatrix}
$$

from which

$$
\begin{bmatrix} F_1 \\ F_2 \\ F_3 \\ F_4 \\ F_5 \\ F_6 \end{bmatrix} = 10^5 \begin{bmatrix} 7.30 & 0 & 0 & -7.71 & 0 & 0 \\ 0 & -0.12 & -0.32 & 0 & -0.32 & -0.76 \\ 0 & -0.32 & 2.09 & 0 & 0.76 & 1.44 \\ -7.71 & 0 & 0 & 7.30 & 0 & 0 \\ 0 & -0.32 & 0.76 & 0 & -0.12 & 0.32 \\ 0 & -0.76 & 1.44 & 0 & 0.32 & 2.09 \end{bmatrix} \times \begin{bmatrix} d_1 \\ d_2 \\ d_3 \\ d_4 \\ d_5 \\ d_6 \end{bmatrix}
$$

Note: Not all the *diagonal* terms are positive as in the static stiffness matrix for a 2-D rigid-jointed structure element.

Element 4–3. The element has no distributed mass. Therefore there are no distributed inertia forces and hence the dynamic stiffness matrix is the same as the static stiffness matrix as given in equation [6.2].

$$
\begin{bmatrix} F_{x1} = * \\ F_{y1} = * \\ M_1 = * \\ F_{x2} = F_4 \\ F_{y2} = F_5 \\ M_2 = F_6 \end{bmatrix} = 10^5
\begin{bmatrix}
0.34 & 0 & 0.86 & -0.34 & 0 & 0.86 \\
0 & 9.09 & 0 & 0 & -9.09 & 0 \\
0.86 & 0 & 2.85 & -0.86 & 0 & 1.43 \\
-0.34 & 0 & -0.86 & 0.34 & 0 & -0.86 \\
0 & -9.09 & 0 & 0 & 9.09 & 0 \\
0.86 & 0 & 1.43 & -0.86 & 0 & 2.85
\end{bmatrix}
$$

$$
\times
\begin{bmatrix}
u_1 = 0 \\
v_1 = 0 \\
\theta_1 = 0 \\
u_2 = d_4 \\
v_2 = d_5 \\
\theta_2 = d_6
\end{bmatrix}
$$

from which

$$
\begin{bmatrix} F_4 \\ F_5 \\ F_6 \end{bmatrix} = 10^5
\begin{bmatrix}
0.34 & 0 & -0.86 \\
0 & 9.09 & 0 \\
-0.86 & 0 & 2.85
\end{bmatrix}
\begin{bmatrix} d_4 \\ d_5 \\ d_6 \end{bmatrix}
$$

(c) Form structural stiffness matrix at $\omega = 52$ radians/second.

$$
\begin{bmatrix}
\Sigma F_1 \\ \\ \\
\Sigma F_2 \\ \\ \\
\Sigma F_3 \\ \\ \\
\Sigma F_4 \\ \\ \\
\Sigma F_5 \\ \\ \\
\Sigma F_6
\end{bmatrix}
= 10^5
\begin{bmatrix}
\begin{matrix}7.30\\ \underline{1.59}\\ 8.89\end{matrix} & 0 & 0 & -7.71 & 0 & 0 \\
0 & \begin{matrix}-0.12\\ \underline{15.14}\\ 15.02\end{matrix} & -0.32 & 0 & -0.32 & -0.76 \\
0 & -0.32 & \begin{matrix}2.09\\ \underline{4.76}\\ 6.85\end{matrix} & 0 & -0.76 & 1.44 \\
-7.71 & 0 & 0 & \begin{matrix}7.30\\ \underline{0.34}\\ 7.64\end{matrix} & 0 & \begin{matrix}0\\ \underline{-0.86}\\ -0.86\end{matrix} \\
0 & -0.32 & 0.76 & 0 & \begin{matrix}-0.12\\ \underline{9.09}\\ 8.97\end{matrix} & 0.32 \\
0 & -0.76 & 1.44 & \begin{matrix}0\\ \underline{-0.86}\\ -0.86\end{matrix} & 0.32 & \begin{matrix}2.09\\ \underline{2.85}\\ 4.94\end{matrix}
\end{bmatrix}
\begin{bmatrix}
d_1 \\ \\ \\
d_2 \\ \\ \\
d_3 \\ \\ \\
d_4 \\ \\ \\
d_5 \\ \\ \\
d_6
\end{bmatrix}
$$

Note: In the structural stiffness matrix, additional entries appear:
- Row ΣF_1: -2.38 and -2.38 under the column for d_3.
- Row ΣF_3: -2.38 and -2.38 in the column for d_1.

(d) Forward reduction of the structural stiffness matrix. The matrix after reduction becomes

$$10^5 \begin{bmatrix} 8.89 & 0 & -2.38 & -7.71 & 0 & 0 \\ 0 & 15.02 & -0.32 & 0 & -0.32 & 0.76 \\ 0 & 0 & 6.20 & -2.06 & 0.76 & 1.42 \\ 0 & 0 & 0 & 0.27 & 0.25 & -0.38 \\ 0 & 0 & 0 & 0 & 8.63 & 0.50 \\ 0 & 0 & 0 & 0 & 0 & 3.99 \end{bmatrix}$$

The determinant of the structural stiffness matrix is equal to the continued product of the diagonal terms of the reduced matrix. Therefore

$$\text{Determinant} = (10^5)^6[8.89 \times 15.02 \times 6.20 \times 0.27 \times 8.63 \times 3.99]$$

$$= 7.59 \times 10^{33}$$

The determinant is positive. Therefore the natural frequency is greater than $\omega = 52$ radians/second.

(e) Increase ω to 64 and recalculate the stiffness matrices for elements and repeat all the previous steps. Fortunately only the stiffness matrix of element 2–3 which carries the uniformly distributed mass is affected.

Element 2–3. Stiffness matrix at $\omega = 64$.

$$\beta L = 6\sqrt{\{5000 \times 64^2/(210 \times 10^9 \times 21\,630 \times 10^{-6})\}}$$

$$= 0.403$$

$$\lambda L = 6\sqrt{\sqrt{\{5000 \times 64^2/(210 \times 10^9 \times 1698.43 \times 10^{-6})\}}}$$

$$= 2.94$$

Using these values and using the stiffness matrix in equation [9.1] the following stiffness matrix is evaluated:

$$\begin{bmatrix} F_1 \\ F_2 \\ F_3 \\ F_4 \\ F_5 \\ F_6 \end{bmatrix} = 10^5 \begin{bmatrix} 7.15 & 0 & 0 & -7.78 & 0 & 0 \\ 0 & -0.30 & -0.16 & 0 & -0.40 & -0.87 \\ 0 & -0.16 & 1.90 & 0 & 0.87 & 1.58 \\ -7.78 & 0 & 0 & 7.15 & 0 & 0 \\ 0 & -0.40 & 0.87 & 0 & -0.30 & 0.16 \\ 0 & -0.87 & 1.58 & 0 & 0.16 & 1.90 \end{bmatrix} \begin{bmatrix} d_1 \\ d_2 \\ d_3 \\ d_4 \\ d_5 \\ d_6 \end{bmatrix}$$

Form the structural stiffness matrix at $\omega = 64$

$$
\begin{bmatrix} \Sigma F_1 \\[2pt] \hline \Sigma F_2 \\[2pt] \hline \Sigma F_3 \\[2pt] \hline \Sigma F_4 \\[2pt] \hline \Sigma F_5 \\[2pt] \hline \Sigma F_6 \end{bmatrix}
= 10^5
\left[
\begin{array}{c|c|c|c|c|c}
\begin{smallmatrix}7.15\\ \underline{1.59}\\ 8.74\end{smallmatrix} & 0 & \begin{smallmatrix}0\\ -2.38\\ -2.38\end{smallmatrix} & -7.78 & 0 & 0 \\ \hline
0 & \begin{smallmatrix}-0.30\\ \underline{15.14}\\ 14.84\end{smallmatrix} & -0.16 & 0 & -0.40 & -0.87 \\ \hline
\begin{smallmatrix}0\\ -2.38\\ -2.38\end{smallmatrix} & -0.16 & \begin{smallmatrix}1.90\\ \underline{4.76}\\ 6.66\end{smallmatrix} & 0 & 0.87 & 1.58 \\ \hline
-7.78 & 0 & 0 & \begin{smallmatrix}7.15\\ \underline{0.34}\\ 7.49\end{smallmatrix} & 0 & \begin{smallmatrix}0\\ -0.86\\ -0.86\end{smallmatrix} \\ \hline
0 & -0.40 & 0.87 & 0 & \begin{smallmatrix}-0.30\\ \underline{9.09}\\ 8.79\end{smallmatrix} & 0.16 \\ \hline
0 & -0.87 & 1.58 & \begin{smallmatrix}0\\ -0.86\\ -0.86\end{smallmatrix} & 0.16 & \begin{smallmatrix}1.90\\ \underline{2.85}\\ 4.75\end{smallmatrix}
\end{array}
\right]
\begin{bmatrix} d_1 \\[2pt] d_2 \\[2pt] d_3 \\[2pt] d_4 \\[2pt] d_5 \\[2pt] d_6 \end{bmatrix}
$$

Forward reduction of the above structural stiffness matrix by Gaussian elimination leads to

$$
10^5
\begin{bmatrix}
8.74 & 0 & -2.38 & -7.78 & 0 & 0 \\
0 & 14.84 & -0.16 & 0 & -0.40 & -0.87 \\
0 & 0 & 6.01 & -2.12 & 0.87 & 1.57 \\
0 & 0 & 0 & -0.18 & 0.31 & -0.30 \\
0 & 0 & 0 & 0 & 9.17 & -0.62 \\
0 & 0 & 0 & 0 & 0 & 4.77
\end{bmatrix}
$$

The determinant $= (10^5)^6[8.74 \times 14.84 \times 6.01 \times (-0.18) \times 9.17 \times 4.77] = -6.06 \times 10^{33}$

(f) At $\omega = 52$, determinant $= 7.59 \times 10^{33}$ and at $\omega = 64$, determinant $= -6.06 \times 10^{33}$, so the zero determinant occurs when

$$\omega = 52.0 + (64 - 52) \times 7.59 / \{7.59 - (-6.06)\}$$

$$= 58.67 \text{ radians/second}$$

It can be shown that this is the smallest frequency at which the determinant becomes zero. Therefore this is the fundamental frequency of the structure. In fact, as shown in Fig. 9.12, the determinant becomes zero at an infinite number of frequencies. These are associated with the higher modes of vibration which will be discussed in the next section. In practice, the designer is interested in

Figure 9.12 ●
Variation of the
determinant of the
dynamic structural
stiffness matrix with
the frequency of the
applied force

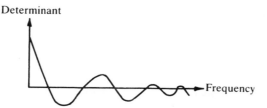

only those frequencies which might be near the frequencies of applied forces
which have sufficient power to excite resonance.

9.11 ● Concentrated mass

In discussing the longitudinal and flexural vibrations, it was assumed that the element carries a uniformly distributed mass. In many cases the structure has to support concentrated masses in addition to distributed mass. The procedure for handling this is best illustrated by an example.

Consider the rigid-jointed frame shown in Fig. 9.13. The beam 2–3 carries a uniformly distributed mass m and a concentrated mass M. When the structure vibrates, inertia forces are set up due to the acceleration of mass m and M. The inertia force is given by

Inertial force = −mass (acceleration)

Dealing with undamped free vibration, the motion is simple harmonic. Therefore if the beam is vibrating at a frequency of ω, then starting with zero velocity the displacements of the beam at any point in the x- and y-directions are respectively

$u(x, t) = U(x) \cos \omega t$ in the horizontal direction

$v(x, t) = V(x) \cos \omega t$ in the vertical direction

where $U(x)$ and $V(x)$ are respectively the amplitudes of displacement in the x- and y-directions respectively.

The accelerations are

$\partial^2 u/\partial t^2 = -\omega^2 U(x) \cos \omega t$ in the horizontal direction

$\partial^2 v/\partial t^2 = -\omega^2 V(x) \cos \omega t$ in the vertical direction

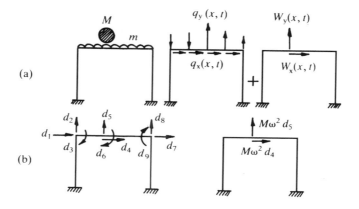

Figure 9.13 ● (a), (b)
Inertial forces due to
uniformly distributed
and concentrated
masses

The inertia forces set up by these accelerations are

(a) Due to uniformly distributed mass m:

$$q_x(x, t) = -m\partial^2 u/\partial t^2$$

$$= m\omega^2 U(x) \cos \omega t \text{ in the horizontal direction}$$

$$q_y(x, t) = -m\partial^2 v/\partial t^2$$

$$= m\omega^2 V(x) \cos \omega t \text{ in the vertical direction}$$

(b) Due to concentrated mass M:

$$W_x(t) = -M\partial^2 u/\partial t^2$$

$$= M\omega^2 U(x) \cos \omega t \text{ in the horizontal direction}$$

$$W_y(t) = -M\partial^2 v/\partial t^2$$

$$= M\omega^2 V(x) \cos \omega t \text{ in the vertical direction}$$

Because the motion is simple harmonic, we need to consider only the amplitudes because the value of a displacement at any time is the corresponding amplitude multiplied by $\cos \omega t$.

The amplitude of the distributed inertia forces $q_y(x, t)$ in the y-direction due to the uniformly distributed mass m is included in calculating the element stiffness coefficients S_1, S_2, Q_1, Q_2, T_1 and T_2 for flexural vibration. Similarly, the amplitude of the distributed inertial forces $q_x(x, t)$ in the x-direction due to the uniformly distributed mass m is included in calculating the element stiffness A_{11} and A_{12} for longitudinal vibration. Therefore we need not pay any further attention to these forces.

The amplitudes of the inertial forces associated with the concentrated mass, i.e. $W_x(t)$ and $W_y(t)$, have to be included in the analysis. These forces are linear functions of the amplitudes of displacement at the position where the concentrated mass acts. Therefore, when using the stiffness method, the simplest approach is to have additional 'joints' at the positions where the concentrated mass acts and treat the inertia forces associated with the concentrated mass as external loads. For the frame shown in Fig. 9.13(a), if the degrees of freedom at the joints, including the additional joint, are as shown in Fig. 9.13(b), then the dynamic structural stiffness relationship using the dynamic element stiffness matrix in equation (9.1) is

$$\begin{bmatrix} \Sigma F_1 \\ \Sigma F_2 \\ \Sigma F_3 \\ \Sigma F_4 \\ \Sigma F_5 \\ \Sigma F_6 \\ \Sigma F_7 \\ \Sigma F_8 \\ \Sigma F_9 \end{bmatrix} = \begin{bmatrix} S_{11} & S_{12} & S_{13} & S_{14} & S_{15} & S_{16} & S_{17} & S_{18} & S_{19} \\ & S_{22} & S_{23} & S_{24} & S_{25} & S_{26} & S_{27} & S_{28} & S_{29} \\ & & S_{33} & S_{34} & S_{35} & S_{36} & S_{37} & S_{38} & S_{39} \\ & & & S_{44} & S_{45} & S_{46} & S_{47} & S_{48} & S_{49} \\ & & & & S_{55} & S_{56} & S_{57} & S_{58} & S_{59} \\ & & & & & S_{66} & S_{67} & S_{68} & S_{69} \\ & & & & & & S_{77} & S_{78} & S_{79} \\ & & & & & & & S_{88} & S_{89} \\ \text{symmetrical} & & & & & & & & S_{99} \end{bmatrix} \begin{bmatrix} d_1 \\ d_2 \\ d_3 \\ d_4 \\ d_5 \\ d_6 \\ d_7 \\ d_8 \\ d_9 \end{bmatrix}$$

Because we are dealing with free vibration, there are no *external* forces acting on the structure. The distributed inertia forces due to uniformly distributed mass are already included in the structural stiffness matrix. All that we have to do is to include the concentrated inertia forces. In this particular example $\Sigma F_i = 0$ for all values of i except $\Sigma F_4 = M\omega^2 d_4$ and $\Sigma F_5 = M\omega^2 d_5$. Since these forces are functions of the displacements, the above set of equations of equilibrium can be written as

$$
\begin{bmatrix} 0 \\ 0 \\ 0 \\ 0 \\ 0 \\ 0 \\ 0 \\ 0 \\ 0 \end{bmatrix} =
\begin{bmatrix}
s_{11} & s_{12} & s_{13} & s_{14} & s_{15} & s_{16} & s_{17} & s_{18} & s_{19} \\
 & s_{22} & s_{23} & s_{24} & s_{25} & s_{26} & s_{27} & s_{28} & s_{29} \\
 & & s_{33} & s_{34} & s_{35} & s_{36} & s_{37} & s_{38} & s_{39} \\
 & & & (s_{44} - M\omega^2) & s_{45} & s_{46} & s_{47} & s_{48} & s_{49} \\
 & & & & (s_{55} - M\omega^2) & s_{56} & s_{57} & s_{58} & s_{59} \\
 & & & & & s_{66} & s_{67} & s_{68} & s_{69} \\
 & & & & & & s_{77} & s_{78} & s_{79} \\
 & & & & & & & s_{88} & s_{89} \\
\text{symmetrical} & & & & & & & & s_{99}
\end{bmatrix}
\times
\begin{bmatrix} d_1 \\ d_2 \\ d_3 \\ d_4 \\ d_5 \\ d_6 \\ d_7 \\ d_8 \\ d_9 \end{bmatrix}
$$

This shows clearly that if a concentrated mass is present, then the procedure to be followed is as follows:

(a) set up the dynamic structural stiffness matrix in the usual way;
(b) subtract from the diagonal terms corresponding to the translations of the concentrated mass M, the value $M\omega^2$;
(c) follow the procedure for determining the value of ω at which the determinant of the structural stiffness matrix is zero.

Example

Figure 9.14 shows a building frame. In this type of frame, the floor slabs acting integrally with the beam tend to make the beams 'rigid' both flexurally and axially. Because the beam is flexurally rigid, the joints of the frame do not rotate. Similarly, because the beam is axially rigid, the horizontal displacement of the beam is the same at both its ends. If, further, it is assumed that the columns are also axially rigid and hence there are no vertical translations, the displacement of such a structure can be described in terms of one sway displacement per storey. Such a frame is called a shear frame. The beams therefore have only one horizontal displacement. The mass on the beams can therefore be treated as a concentrated mass. The

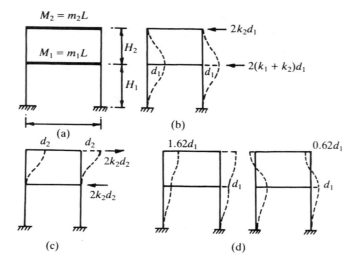

Figure 9.14 ● (a)–(d)
Free vibration of a shear
frame

structural stiffness matrix of the frame without any reference to the concentrated mass at floor levels can be set up. As was shown in Section 4.5, Example 4, Fig. 4.21, the translational stiffness for a beam element fixed against rotation at both ends is $12EI/L^3$. On the other hand, if it is fixed against rotation at one end only, then the corresponding translational stiffness is $3EI/L^3$.

For the present example let the translational stiffness of the first and second storey columns be k_1 and k_2 respectively. Then considering the displacement patterns shown in Fig. 9.14(b) and (c), the structural stiffness matrix is

$$\begin{bmatrix} \Sigma F_1 \\ \Sigma F_2 \end{bmatrix} = \begin{bmatrix} 2k_1 + 2k_2 & -2k_2 \\ -2k_2 & 2k_2 \end{bmatrix} \begin{bmatrix} d_1 \\ d_2 \end{bmatrix}$$

$\Sigma F_1 = M_1\omega^2 d_1$ and $\Sigma F_2 = M_2\omega^2 d_2$ due to the inertia forces associated with the concentrated masses, and so the above equation becomes

$$\begin{bmatrix} 0 \\ 0 \end{bmatrix} = \begin{bmatrix} (2k_1 + 2k_2 - M_1\omega^2) & -2k_2 \\ -2k_2 & (2k_2 - M_2\omega^2) \end{bmatrix} \begin{bmatrix} d_1 \\ d_2 \end{bmatrix}$$

For a non-trivial solution, the determinant is zero. Therefore

$$(2k_1 + 2k_2 - M_1\omega^2)(2k_2 - M_2\omega^2) - (-2k_2)^2 = 0$$

For simplicity if $k_2 = k_1 = k$, $M_1 = M_2 = M$, then

$$(4k - M\omega^2)(2k - M\omega^2) - 4k^2 = 0$$

$$M^2\omega^4 - 6kM\omega^2 + 4k^2 = 0$$

Let $k/M = \Omega^2$, then

$$\omega^4 - 6\omega^2\Omega^2 + 4\Omega^4 = 0$$

Solving the quadratic equation in ω^2, we have

$$\omega^2 = \Omega^2(0.76 \text{ or } 5.24)$$

Thus the lowest or fundamental frequency is $\omega^2 = 0.76\ \Omega^2$ and the next frequency for higher harmonic is $\omega^2 = 5.24\ \Omega^2$.

It is interesting to ascertain the mode shapes corresponding to these frequencies. It should be remembered that mode shape can only describe the shape and not the absolute value of the deformation of the structure.

(i) At $\omega^2 = 0.76\ \Omega^2 = 0.76\ k/M$, from the first row of the stiffness relationship, we have

$$(2k_1 + 2k_2 - M_1\omega^2)d_1 - 2k_2d_2 = 0$$

i.e.

$$(4k - M\omega^2)d_1 - 2kd_2 = 0$$

Substituting for ω^2 and simplifying

$$d_2 = 1.62d_1$$

(ii) At $\omega^2 = 5.24\ \Omega^2 = 5.24\ k/M$, from the first row of the stiffness relationship, we have

$$(2k_1 + 2k_2 - M_1\omega^2)d_1 - 2k_2d_2 = 0$$

i.e.

$$(4k - M\omega^2)d_1 - 2kd_2 = 0$$

Substituting for ω^2 and simplifying

$$d_2 = -0.62d_1$$

These two mode shapes are shown in Fig. 9.14(d).

This example shows that structures vibrate not just in the fundamental mode but also at higher harmonics. The particular frequency which is important in design depends very much on the frequency of the applied forces which can act on the structure.

9.12 ● Seismic forces

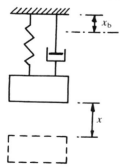

Figure 9.15 ●
A spring–mass–damper system subjected to support displacement

Consider the simple spring–mass–damper system shown in Fig. 9.15. Let the 'support' be subjected to a time dependent displacement $X_b(t)$ and the displacement of the mass be $X(t)$. Then

(i) the inertia force on the mass is $-M\ d^2X/dt^2$
(ii) the force in the spring is $K(X - X_b)$
(iii) the damping force in the dashpot is $C\ d/dt(X - X_b)$

Therefore the equilibrium equation for the mass is given by

$$M\ d^2X/dt^2 + C\ d/dt(X - X_b) + K(X - X_b) = 0$$

Adding and subtracting $M\ d^2X_b/dt^2$, the above equation can also be written as

$$M\ d^2X/dt^2 + M\ d^2X_b/dt^2 - M\ d^2X_b/dt^2 + C\ d/dt(X - X_b) + K(X - X_b) = 0$$

Substituting $X_r = X - X_b$, the displacement of the mass relative to the base, the above equation becomes

$$M \, \mathrm{d}^2 X_r/\mathrm{d}t^2 + C \, \mathrm{d}X_r/\mathrm{d}t + KX_r = M \, \mathrm{d}^2 X_b/\mathrm{d}t^2$$

The above equation can be written as

$$M \, \mathrm{d}^2 X_r/\mathrm{d}t^2 + C \, \mathrm{d}X_r/\mathrm{d}t + KX_r = F(t)$$

where $F(t) = M \, \mathrm{d}^2 X_b/\mathrm{d}t^2$ represents the external force. The above equation shows that the relative displacement of the mass is determined by considering the forced damped motion where the inertial force due to acceleration of the base acts as the external force.

In general, C is assumed to be zero. Therefore

$$M \, \mathrm{d}^2 X_r/\mathrm{d}t^2 + KX_r = M \, \mathrm{d}^2 X_b/\mathrm{d}t^2$$

Similar reasoning can be applied to the shear frame shown in Fig. 9.16. The dynamic equations become

$$\begin{bmatrix} M_1 & 0 \\ 0 & M_2 \end{bmatrix} \begin{bmatrix} \mathrm{d}^2 X_{1r}/\mathrm{d}t^2 \\ \mathrm{d}^2 X_{2r}/\mathrm{d}t^2 \end{bmatrix} + \begin{bmatrix} k_{11} & k_{12} \\ k_{21} & k_{22} \end{bmatrix} \begin{bmatrix} X_{1r} \\ X_{2r} \end{bmatrix} = \begin{bmatrix} M_1 \, \mathrm{d}^2 X_b/\mathrm{d}t^2 \\ M_2 \, \mathrm{d}^2 X_b/\mathrm{d}t^2 \end{bmatrix}$$

In general it is not possible to solve these equations analytically. In practice sophisticated numerical methods are used to obtain the solution of the above equations, considering if necessary, changes in the stiffness coefficients of the stiffness matrix to allow for 'damage' to the structural elements due to the earthquake forces.

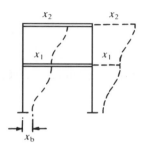

Figure 9.16 ● A shear frame subjected to support movement

9.13 ● Torsional vibration

It was shown in Section 4.16.2 that the equilibrium equation governing the torsional problem, assuming that Saint Venant's torsion is valid, is given by

$$GJ \, \mathrm{d}^2 \psi / \mathrm{d}x^2 + q(x) = 0$$

where q = distributed twisting moment per unit length. In the case of dynamic problems, the above equation becomes

$$GJ \, \partial^2 \psi / \partial x^2 + q(x, t) + \text{inertia torque} = 0$$

where twist $\psi = \psi(x, t)$.

The inertia torque can be obtained as follows. Consider an infinitesimal element of area $\mathrm{d}A$ in the cross-section as shown in Fig. 9.17. If r is the radial distance of the area $\mathrm{d}A$ from the axis of rotation, then the tangential displacement is $r\psi$. Therefore the tangential acceleration is $r\partial^2 \psi / \partial t^2$. If the mass per unit length is m, then the mass of the area $\mathrm{d}A$ is $m(\mathrm{d}A)/A$. Therefore the tangential inertia force on the area $\mathrm{d}A$ is

$$-\{m \, \mathrm{d}A/A\}\{r\partial^2 \psi / \partial t^2\} = -\{m/A\}\{\partial^2 \psi / \partial t^2\}r \, \mathrm{d}A$$

The torque produced by the tangential inertia force is r times the tangential force. Therefore the total inertia torque q_t is

$$q_t = -\int \{m/A\}\{\partial^2 \psi / \partial t^2\}r^2 \, \mathrm{d}A$$

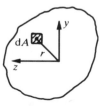

Figure 9.17 ● Cross-section of a shaft

Taking the constant terms outside the $\int$, we have

$$q_t = -\{m/A\}\{\partial^2\psi/\partial t^2\} \int r^2 \, dA$$

But $\int r^2 \, dA$ = polar second moment of area I_p about the x-axis

$$I_p = \int r^2 \, dA = \int (y^2 + z^2) \, dA = I_{zz} + I_{yy}$$

Therefore

$$q_t = -\{mI_p/A\}\{\partial^2\psi/\partial t^2\}$$

Therefore the differential equation is given by

$$GJ\partial^2\psi/\partial x^2 - \{mI_p/A\}\{\partial^2\psi/\partial t^2\} + q(x, t) = 0$$

If damping is absent and no external forces are present, then the motion is simple harmonic. Therefore

$$\psi(x, t) = \psi(x) \cos \omega t$$

Substituting for $\psi(x, t)$, the differential equation in terms of the amplitude $\psi(x)$ is given by

$$GJ \, d^2\psi/dx^2 + \{m \, \omega^2 I_p/A\}\psi(x) = 0$$

Dividing by GJ

$$d^2\psi/dx^2 + \beta^2\psi(x) = 0$$

where $\beta^2 = \{m\omega^2 I_p/(GJA)\}$. This differential equation is identical to the longitudinal vibration equation solved in Section 9.7.1 and therefore the stiffness relationship is given by

$$\begin{bmatrix} T_1 \\ T_2 \end{bmatrix} = (GJ/L) \begin{bmatrix} G_{11} & -G_{12} \\ -G_{12} & G_{11} \end{bmatrix} \begin{bmatrix} \psi_1 \\ \psi_2 \end{bmatrix}$$

where $G_{11} = \beta L\{\cos \beta L/\sin \beta L\}$, $G_{12} = \beta L/\sin \beta L$. Numerical values of G_{11} and G_{12} are tabulated in Table 9.1.

9.14 ● Dynamic stiffness matrix for an element of a plane grid structure

In Sections 6.2 to 6.4, the stiffness matrices for beam and shaft elements were combined to obtain the general plane grid member stiffness matrix. Following identical steps, the dynamic stiffness matrix for an element of a plane grid structure can be derived. This is shown in equation [9.2]: the dynamic stiffness relationship for an element in a plane grid structure.

$$
\begin{bmatrix} F_{y1} \\ M_{x1} \\ M_{z1} \\ F_{y2} \\ M_{x2} \\ M_{z2} \end{bmatrix} = \frac{EI}{L}
$$

$$
\times \begin{bmatrix} \dfrac{T_1}{L^2} & \dfrac{Q_1 n}{L} & \dfrac{-Q_1 l}{L} & \dfrac{-T_2}{L^2} & \dfrac{Q_2 n}{L} & \dfrac{-Q_2 l}{L} \\[2mm] & S_1 n^2 + \alpha_1 l^2 & -(S_1 - \alpha_1)ln & \dfrac{-Q_2 n}{L} & S_2 n^2 + \alpha_2 l^2 & -(S_2 - \alpha_2)ln \\[2mm] & & S_1 l^2 + \alpha_1 n^2 & \dfrac{Q_2 l}{L} & -(S_2 - \alpha_2)ln & S_2 l^2 + \alpha_2 m^2 \\[2mm] & & & \dfrac{T_1}{L^2} & \dfrac{-Q_1 n}{L} & \dfrac{Q_1 l}{L} \\[2mm] & & & & S_1 n^2 + \alpha_1 l^2 & -(S_1 - \alpha_1)ln \\[2mm] & & & & & S_1 l^2 + \alpha_1 n^2 \end{bmatrix}
$$

$$
\times \begin{bmatrix} v_1 \\ \theta_{x1} \\ \theta_{z1} \\ v_2 \\ \theta_{x2} \\ \theta_{z2} \end{bmatrix}
\tag{9.2}
$$

$$\alpha_1 = GJ\,G_{11}/(EI), \quad \alpha_2 = GJ\,G_{12}/(EI)$$

9.15 ● References

Bhatt P 1981 *Problems in Matrix Analysis of Skeletal Structures* Construction Press (Reprinted by A H Wheeler & Co., Allahabad, India, 1988.) (Contains many numerical examples and some additional theory.)

Bhatt P 1986 *Programming the Matrix Analysis of Skeletal Structures* Ellis Horwood (Contains two well documented programs in FORTRAN for the determination of natural frequencies of 2-D rigid-jointed frames and plane grids.)

Bolton A 1978 Natural Frequencies of Structures for Designers *Structural Engineer* **56A** (9): 245–53

Bolton A 1983, 1984 Design Against Wind-excited Vibration *Structural Engineer* **61A** (8): 237–45 Discussions May 1984: 159–61

Chopra A K n.d. *Dynamics of Structures: A Primer* Earthquake Engineering Research Institute, Berkeley, California, USA

Irvine M 1986 *Structural Dynamics for the Practising Engineer* Allen & Unwin (This uses only energy methods.)

Paz M 1985 *Structural Dynamics* (2nd edn), van Nostrand (A well written book. Contains additional theory.)

Wiesner K B 1986 Taming Lively Buildings *Civil Engineering* (American Society of Civil Engineers) 54–57 (Examples of dampers to control vibration of tall buildings.)

9.16 ● Problems

1. Calculate the fundamental natural frequency when two rigidly jointed members support a mass M as shown in Fig. E9.1. Assume EI constant.

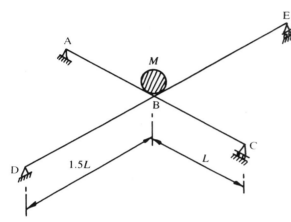

Figure E9.1

Answer: $K = 48EI\{1/(2L)^3 + 1/(3L)^3\} = 7.78EI/L^3$. Therefore $\omega = \sqrt{(K/M)} = 2.79\sqrt{\{EI/(ML^3)\}}$.

2. A reinforced concrete multiflue chimney was vibrated by a time varying force to determine the damping present. After excitation, the ratio of amplitudes 20 cycles apart had a value of 1.73. Determine logarithmic decrement and the percentage of critical damping present.

Answer: $x(t)/x(t + nT) = e^{n\beta T} = 1.73$ and $n = 20$. Log decrement $= \beta T = 0.027 = 2\pi(C/C_{cr})$. Therefore $C/C_{cr} = 0.44$ per cent.

3. Calculate the fundamental frequency of a propped cantilever carrying uniformly distributed mass m.

Answer: Dynamic stiffness matrix $= (EI/L)[S_1]\theta$. From Table 9.2, $S_1 = 0.81$, $\lambda L = 3.8$, $S_1 = -0.60$, $\lambda L = 4.0$. Therefore $S_1 = 0$ when

$$\lambda L = 3.8 + (4.0 - 3.8)0.81/\{0.81 - (-0.60)\}$$
$$= 3.915$$

$\lambda L = L\{\sqrt{\sqrt{\{m \ \omega^2(EI)\}}}$. Therefore $\omega = 15.33\sqrt{\{EI/(mL^4)\}}$. Exact value without linear interpolation gives $\omega = 15.42\sqrt{\{EI/(mL^4)\}}$.

4. Using the information in Table 9.3, for the structures shown in Fig. E9.2, calculate the first two natural frequencies.

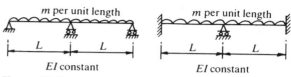

Figure E9.2

Answer:

(a) The fundamental mode is antisymmetric. The two spans behave as simply supported beams. Therefore

$$\omega = \pi^2\sqrt{\{EI/(mL^4)\}}.$$

(b) The fundamental mode is symmetric. The two spans behave as propped cantilevers. Therefore

$$\omega = 15.42\sqrt{\{EI/(ML^4)\}}.$$

5. For the shear frame shown in Fig. E9.3, determine the natural frequencies and corresponding mode shapes.

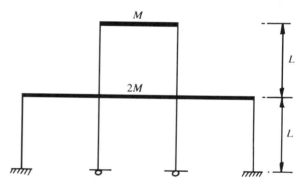

Assume flexurally rigid beams

Figure E9.3

Answer:

$$(EI/L^3)\begin{bmatrix} 2 \times 12 & -2 \times 12 \\ -2 \times 12 & 2 \times 12 + 2 \times 3 + 2 \times 2 \end{bmatrix}\begin{bmatrix} u_2 \\ u_1 \end{bmatrix}$$
$$= \begin{bmatrix} -M\,d^2u_2/dt^2 \\ -2M\,d^2u_1/dt^2 \end{bmatrix}$$

Because there is no damping, motion is simple harmonic. Therefore $u_i(t) = U_i \cos \omega t$ where U_i is the amplitude. Therefore $d^2u_i/dt^2 = -\omega^2 u_i$

$$(EI/L^3)\begin{bmatrix} 2 \times 12 - \omega^2 \ ML^3/(EI) & -2 \times 12 \\ -2 \times 12 & 2 \times 12 + 2 \times 12 + \\ & 2 \times 3 - \omega^2 \ 2ML^3/(EI) \end{bmatrix}$$
$$\times \begin{bmatrix} U_2 \\ U_1 \end{bmatrix} = \begin{bmatrix} 0 \\ 0 \end{bmatrix}$$

Let $\omega^2 ML^3/(EI) = \alpha$. For zero determinant,

$$(24 - \alpha)(54 - 2\alpha) - 24^2 = 0$$

$$\alpha^2 - 51\alpha + 360 = 0$$

Solving, $\alpha = 8.46$ and 42.54

$$\omega_1 = 2.91\sqrt{\{EI/(ML^3)\}}, \quad \omega_2 = 6.52\sqrt{\{EI/(ML^3)\}}$$

From the first equation, ratio of displacements U_2/U_1 = $24/(24 - \alpha)$. Therefore at $\omega = \omega_1$, $U_2/U_1 = 1.545$ and at $\omega = \omega_2$, $U_2/U_1 = -1.295$.

6. Determine the fundamental natural frequency of the continuous beam shown in Fig. E9.4. Ignore axial and shearing deformations.

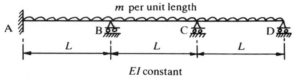

m per unit length

EI constant

Figure E9.4

Answer: Dynamic stiffness matrix is given by

$$(EI/L)\begin{bmatrix} 2S_1 & S_2 & 0 \\ S_2 & 2S_1 & S_2 \\ 0 & S_2 & S_1 \end{bmatrix}\begin{bmatrix} \theta_B \\ \theta_C \\ \theta_D \end{bmatrix}$$

The first natural frequency will be 'governed' by the span CD and will be slightly above the lowest frequency of a simply supported beam. Letting $mL = M$ and $EI/L^3 = K$, then for a simply supported beam, from Table 9.4, $\omega = \pi^2\sqrt{(K/M)}$. Since $\lambda L = \sqrt{\sqrt{(M\omega^2/K)}} = \pi$, start the iteration at $\lambda L = 3.2$, $S_1 = 2.78$, $S_2 = 2.97$

$$\begin{bmatrix} 5.56 & 2.97 & 0.0 \\ 2.97 & 5.56 & 2.97 \\ 0.0 & 2.97 & 2.78 \end{bmatrix}$$

After Gaussian elimination

$$\begin{bmatrix} 5.56 & 2.97 & 0.0 \\ 0 & 3.97 & 2.97 \\ 0 & 0 & 0.56 \end{bmatrix}$$

Determinant = $5.56 \times 3.97 \times 0.56 = 12.36$. The determinant is positive therefore $(\lambda L)_{cr} > 3.2$.

Continue the iteration at $\lambda L = 3.4$, $S_1 = 2.33$, $S_2 = 3.34$

$$\begin{bmatrix} 4.66 & 3.34 & 0.0 \\ 3.34 & 4.66 & 3.34 \\ 0.0 & 3.34 & 2.33 \end{bmatrix}$$

After Gauss elimination

$$\begin{bmatrix} 4.66 & 3.34 & 0.0 \\ 0 & 2.27 & 3.34 \\ 0 & 0 & -2.58 \end{bmatrix}$$

Determinant = $4.46 \times 2.27 \times (-2.58) = -27.29$. The determinant is negative therefore $(\lambda L)_{cr} < 3.4$. Thus

$$(\lambda L)_{cr} = 3.2 + (3.4 - 3.2) \times 12.36/\{12.36 - (-27.29)\} = 3.26$$

Therefore $\omega = 10.64\sqrt{(K/M)}$ which is only 8 per cent higher than the natural frequency of the corresponding simply supported beam.

7. Determine the fundamental natural frequency of the rigid-jointed structure shown in Fig. E9.5. Ignore axial and shearing deformations.

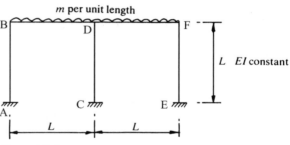

m per unit length

L EI constant

Figure E9.5

Answer: Note that the sway deformations will introduce a total horizontal inertia force of $-\omega^2$ total mass Δ. Total mass = $m(2L) = 2mL$. Horizontal inertia force = $-2mL\,\omega^2\Delta$ = $(EI/L^3)\{-2m\omega^2 L^4/(EI)\}\Delta$. Since $(\lambda L)^4 = m\omega^2 L^4/(EI)$. Therefore horizontal inertia force = $(EI/L^3)\{-2(\lambda L)^4\}\Delta$. Dynamic stiffness matrix:

$$\left(\frac{EI}{L}\right)\begin{bmatrix} S_1 + 4 & S_2 & 0 & -6 \\ S_2 & 2S_1 + 4 & S_2 & -6 \\ 0 & S_2 & S_1 + 4 & -6 \\ -6 & -6 & -6 & 3 \times 12 - 2(\lambda L)^4 \end{bmatrix}$$

$$\times \begin{bmatrix} \theta_B \\ \theta_D \\ \theta_F \\ \Delta/L \end{bmatrix}$$

If the beams are assumed to be flexurally 'rigid', then $K = 3$ columns $\times$ $12EI/L^3 = 36EI/L^3$. $M = m(2L) = 2mL$. Therefore $\omega = \sqrt{(K/M)} = \sqrt{\{18mL^4/(EI)\}}$ and $\lambda L = 2.06$.

Similarly, if the beams are assumed to be flexurally 'flexible', then $K = 3$ columns $\times$ $3EI/L^3$

= $9EI/L^3$. $M = m(2L) = 2mL$. Therefore $\omega = \sqrt{(K/M)}$ = $\sqrt{\{4.5mL^4/(EI)\}}$ and $\lambda L = 1.46$. Therefore

$$1.46 < (\lambda L)_{cr} < 2.06$$

Start iteration at $\lambda L = 1.8$, $S_1 = 3.90$, $S_2 = 2.08$

$$\begin{bmatrix} 7.90 & 2.08 & 0 & -6.0 \\ 2.08 & 11.80 & 2.08 & -6.0 \\ 0 & 2.08 & 7.90 & -6.0 \\ -6.0 & -6.0 & -6.0 & 15.01 \end{bmatrix}$$

After Gaussian elimination

$$\begin{bmatrix} 7.90 & 2.08 & 0 & -6.0 \\ 0 & 11.25 & 2.08 & -4.45 \\ 0 & 0 & 7.52 & -5.18 \\ 0 & 0 & 0 & 5.15 \end{bmatrix}$$

Determinant = $7.90 \times 11.25 \times 7.52 \times 5.15$ = 3435.28. The determinant is positive therefore $(\lambda L)_{cr} > 1.80$.

Continue iteration at $\lambda L = 1.9$, $S_1 = 3.87$, $S_2 = 2.10$

$$\begin{bmatrix} 7.87 & 2.10 & 0 & -6.0 \\ 2.10 & 11.74 & 2.10 & -6.0 \\ 0 & 2.10 & 7.87 & -6.0 \\ -6.0 & -6.0 & -6.0 & 9.94 \end{bmatrix}$$

After Gaussian elimination

$$\begin{bmatrix} 7.87 & 2.10 & 0 & -6.0 \\ 0 & 11.18 & 2.10 & -4.40 \\ 0 & 0 & 7.48 & -5.17 \\ 0 & 0 & 0 & 0.07 \end{bmatrix}$$

Determinant = $7.87 \times 11.18 \times 7.48 \times 0.07 = 44.10$. The determinant is positive, therefore $(\lambda L)_{cr} > 1.90$. From the decrease of the determinant from 3435.28 to 44.10 when λL is changed from 1.8 to 1.9, it is clear that $(\lambda L)_{cr}$ is very near to 1.9. Therefore interpolating for zero determinant we have

$$\{(\lambda L)_{cr} - 1.8\}/(3435.28 - 0)$$

$$= (1.9 - 1.8)/(3435.28 - 44.10)$$

Therefore

$$(\lambda L)_{cr} = 1.901$$

and

$$\omega = 3.61\sqrt{\{EI/(mL^4)\}}$$

Chapter ten

Virtual work and energy principles

In previous chapters, the equilibrium of the structure under a given set of forces was established using the equations of statics. Another procedure based on the principle of virtual work is frequently used to check equilibrium of structures. In addition, concepts of strain and complementary energies are also used in practice. These are important concepts which at one time were extensively used for the solution of simple problems by manual procedures. Nowadays they are used not so much for solving specific problems but for <u>deriving properties of an element for use in the general stiffness method of analysis and for deriving general properties of linear and non-linear structures</u>. Although in the majority of cases the differential equations occurring in structural analysis problems can be solved using either the exact or approximate solution techniques, many prefer to use virtual work and energy principles. In fact many books on finite element methods of structural analysis have been written using solely the virtual work and energy principles. It is therefore important that the student is aware of these principles. In this chapter the concepts will be explained at an elementary level and will be illustrated by simple examples. The examples are given mainly to clarify the principles being explained. *They are not intended to develop practical calculation procedures.* In practice computer programs based on the stiffness method presented in Chapter 6 are used.

10.1 ● Principle of virtual work applied to a single particle or mass point

Consider a single particle A as shown in Fig. 10.1, under the action of a set of concurrent forces F_i, i = 1, 2, 3, . . . having a resultant R.

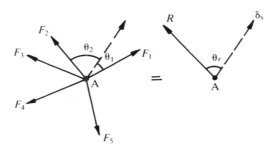

Figure 10.1 ● A set of concurrent forces in equilibrium

Imagine that the particle is given an arbitrarily small displacement δs. Then the work done by the forces is equal to δW where

$$\delta W = \sum F_i \delta s \cos \theta_i = R \cos \theta_r \delta s$$

where θ_i = angle between the displacement δs and the force F_i, θ_r is the angle between δs and the resultant force R.

If the particle is in equilibrium, then according to Newton's second law of motion $R = 0$, and if $R = 0$, then $\delta W = 0$. Therefore one way of testing the equilibrium of a particle under a system of forces is to *imagine* that the particle is given an arbitrarily small displacement δs. Because the displacement δs is imaginary, it can be assumed that it does not alter the line of action of the applied forces. If for the displacement δs the resulting work δW is equal to zero, then the particle is in equilibrium. Obviously equilibrium should be tested in any two orthogonal directions because virtual displacement in any arbitrary direction can be resolved into two orthogonal components. It should be noted that the displacement assumed need not actually occur. It is a fictitious displacement. Hence the name virtual (i.e. not real) displacement and the corresponding work is therefore called virtual work. In this chapter, arbitrarily small virtual quantities such as displacements, forces and work are denoted by δs, δF and δW respectively. It is again emphasized that the displacement considered is virtual and is restricted to small values so that neither the magnitude nor the direction of forces are altered during the virtual displacement.

10.2 ● Principle of virtual work applied to rigid bodies

Consider the two-dimensional rigid body in the x–y plane under the action of a set of external forces as shown in Fig. 10.2. Imagine that the rigid body is made up of n particles connected to each other.

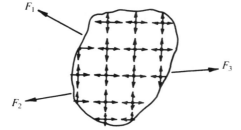

Figure 10.2 ● A rigid body in equilibrium under the action of external forces

During a virtual displacement, let the virtual work on the ith particle be δW_i. For it to be in equilibrium, $\delta W_i = 0$. For the rigid body to be in equilibrium, the virtual work done on each of the n particles must be zero. Therefore for equilibrium of the rigid body

$$\delta W = \sum \delta W_i = 0$$

It should be appreciated that as shown in Fig. 10.2, forces acting on a particle are either 'internal' or 'external'. Therefore the total virtual work done can be split into two parts as

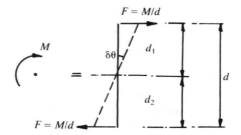

Figure 10.3 ●
Work done by a
couple = $M\delta\theta$

$$\delta W = \delta W_{\text{external}} + \delta W_{\text{internal}}$$

The internal forces occur in self-equilibrating pairs. Because the body is rigid, it is not strained. Therefore there is no *relative* displacement between the particles. Therefore the virtual work $\delta W_{\text{internal}}$ done by the internal forces cancel themselves out because the work done on a particle by an internal force is equal and opposite to the work done on a neighbouring particle by an internal force which is equal and opposite to the first one. The total virtual work δW done by the forces acting on all the particles constituting the rigid body therefore amounts to the virtual work $\delta W_{\text{external}}$ done by the external forces acting on the rigid body. Thus for equilibrium the total virtual work done by the external forces only acting on a rigid body during any virtual displacement must be zero. In the case of two-dimensional rigid bodies, it is important to test equilibrium with respect to translation in two orthogonal directions and rotation about an axis normal to the body. The reason for this is that any arbitrary displacement can be 'resolved' into more than three components.

It should be noted that the words force and displacement have been used in the general sense to include both (forces and couples) and (translations and rotations) respectively. The virtual work δW done by a force F going through a virtual translation δs in the *direction of the force* is given by $\delta W = F\delta s$. Similarly, if a couple M rotates by a virtual rotation $\delta\theta$, then the virtual work $\delta W = M\delta\theta$. This is easily proved as follows. Let the couple M be replaced by a set of equal and opposite forces F at distance d apart as shown in Fig. 10.3. The value of each force $F = M/d$. The rotation $\delta\theta$ causes a translation of $d_1\delta\theta$ and $d_2\delta\theta$ in the direction of the forces. Therefore $\delta W = (M/d)(d_1\delta\theta) + (-M/d)(-d_2\delta\theta) = (M/d)(d_1 + d_2)\delta\theta = M\delta\theta$, because $d = d_1 + d_2$.

10.2.1 Examples

In order to demonstrate the application of the virtual work equation to rigid bodies, some simple examples of application to a simply supported beam are given below. These examples are for demonstration purposes only.

Example 1

Figure 10.4 shows a simply supported beam carrying a concentrated load W at a distance of a from the left-hand support. Calculate the reactions.

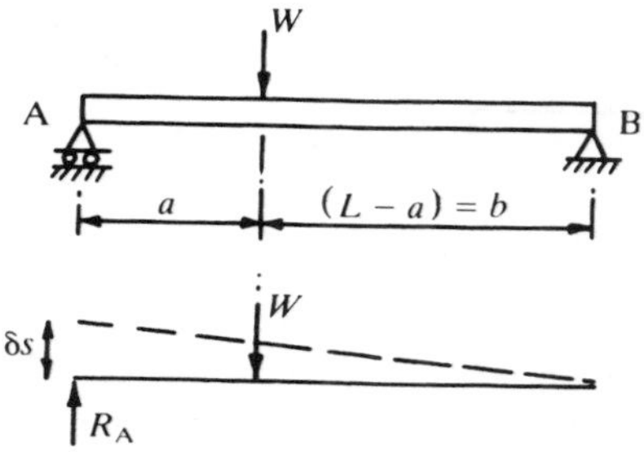

Figure 10.4 ● Virtual translation at a support of a simply supported beam

Solution   Imagine that the left support is replaced by an external force V_A. Provide a virtual displacement δs at the left support in the vertically upward direction. Because the beam is statically determinate it can be assumed to be rigid during the virtual displacement. The displacement of the load W is $(b/L)\delta s$, where $b = (L - a)$. The total virtual work done is given by

$$\delta W = V_A\delta s + W(-b/L)\delta s - V_B 0 = 0$$

Note that the virtual work done by the load W is negative because the virtual displacement of W is upwards but the force W acts downwards. Solving the above equation, $V_A = Wb/L$.

Example 2

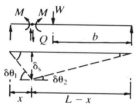

Figure 10.5 ● Virtual relative rotation at a section in the span of a simply supported beam

Calculate the bending moment and shear force at a section x from the left-hand support in the beam shown in Fig. 10.5.

1. Bending moment at x. As in Example 1, to determine the bending moment, assume that the beam is pinned at that section and apply the bending moment at that section as a pair of equal and opposite external couples M. Then giving the system a virtual displacement as shown in Fig. 10.5, the virtual work is done only by the external force W and the external couples M. The virtual work done is given by

$$\delta W = M(-\delta\theta_1) + M(-\delta\theta_2) + W\{b\delta s/(L - x)\} = 0$$

From geometry, $\delta\theta_1 = \delta s/x$ and $\delta\theta_2 = \delta s/(L - x)$, so M is given by

$$M\{1/(L - x) + 1/x)\} = Wb/(L - x) \text{ or}$$

$$M = \{Wb/L\}x = V_A x$$

Note that in the above calculation the shear force Q at the section which consists of a pair of equal and opposite forces does not do any work because both the forces undergo the same value of virtual displacement δs.

2. Shear force at x. To determine the shear force at the section, replace the shear force at the section by equal and opposite external force Q and provide a virtual displacement such that only the external load W and the shear force Q do work. It is important to ensure that the bending moment M at the section does not do any work. This can be done by ensuring that the displacement pattern adopted allows

Figure 10.6 ● Virtual relative translation at a section in the span of a simply supported beam

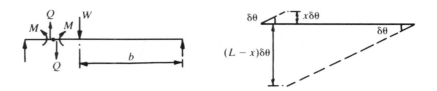

for continuity of rotation at the 'cut section'. Adopting the virtual displacement pattern shown in Fig. 10.6, it is clear that the bending moment M at x does not do net work because the left-hand and right-hand couples rotate by the same amount $\delta\theta$. Therefore δW is given by

$$\delta W = Q(x\delta\theta) + Q\{(L - x)\delta\theta\} + W(b\delta\theta) = 0$$

i.e. $Q = -Wb/L = -V_A$

It should be appreciated that the technique adopted is successful precisely because the structure considered is a statically determinate structure and therefore it can undergo 'support' displacements during virtual displacement without the structure straining. This cannot be done in the case of statically indeterminate structures.

10.3 ● Principle of virtual work applied to deformable bodies

When the virtual work equation was applied to rigid bodies, it was noticed that because the internal forces exist in self-equilibrating pairs and because there are no strains, meaning that there was no relative displacement between the particles, $\delta W_{internal}$ was zero leading to $\delta W = \delta W_{external}$. If the body is deformable, then there is relative displacement between particles and therefore the internal forces do work and this has to be included in the virtual work calculations. The procedure for doing this is best illustrated by examples of application to specific types of structures.

10.4 ● Principle of virtual work applied to the calculation of joint displacements of pin-jointed structures

Consider the statically determinate pin-jointed structure shown in Fig. 10.7. Let the forces in the members due to the external load W_2 and W_4 be F_1, F_2, F_3 and F_4 assumed to be tensile. Let the structure be given a virtual displacement which results in joint displacements in the x- and y-directions of $(\delta u_2, \delta v_2)$ at joint 2 and $(\delta u_3, \delta v_3)$ at joint 3. Considering joint 2, the virtual work done by the forces acting at the joint is given by

$$W_2(-\delta v_2) + F_1(-\delta u_2) + F_2(-\delta v_2) = 0$$

Similarly, considering joint 3

$$W_4(-\delta v_3) + F_4(-\delta u_3) + F_2(\delta v_3) + F_3\{-(\delta u_3 \cos \alpha - \delta v_3 \sin \alpha)\} = 0$$

where $(\delta u_3 \cos \alpha - \delta v_3 \sin \alpha)$ is the displacement of joint 3 in the direction of 1–3 which is the line of action of F_3.

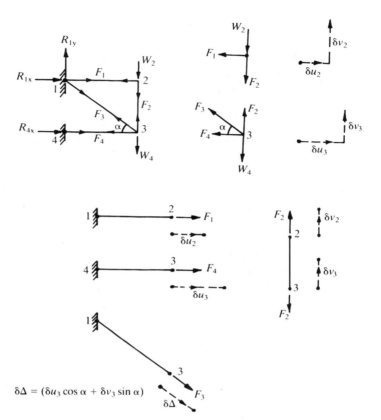

Figure 10.7 ● Forces in a pin-jointed truss and the associated virtual translations

$$\delta\Delta = (\delta u_3 \cos\alpha + \delta v_3 \sin\alpha)$$

Adding the above two equations and transferring the virtual work done by the member forces to the right-hand side, we have

$$W_2(-\delta v_2) + W_4(-\delta v_3)$$

$$= F_1\delta u_2 + F_2(\delta v_2 - \delta v_3) + F_3(\delta u_3 \cos\alpha - \delta v_3 \sin\alpha) + F_4\delta u_3$$

In the above expression, the left-hand terms are obviously the virtual work done by the external forces and the right-hand terms are the virtual work done by the forces in the members going through the corresponding virtual *extension* of the member associated with the assumed virtual displacement pattern. Therefore the principle of virtual work can be stated as follows when applied to pin-jointed structures:

Virtual work done by external loads

$$= \sum(\text{member force} \times \text{corresponding virtual extension})$$

Note: In the case of rigid bodies, virtual extension is zero. Therefore virtual work done by external loads is equal to zero.

10.4.1 Deflection of pin-jointed structures

The application of the equation of virtual work to determine the deflection of the joints of a pin-jointed truss is discussed for three separate scenarios.

Case 1: Single concentrated load.

The structure is subjected to a single concentrated load and the deflection of the loaded point is required. Let F be the force in a member due to the external load, and the corresponding extension be $FL/(AE)$, where L = length of member, A = area of cross-section of the member and E = Young's modulus. Thus the equation

Virtual work done by external loads

$$= \sum (\text{member force} = F) \times \{\text{corresponding virtual extension} = FL/(AE)\}$$

becomes

Virtual work by external loads = applied load × deflection under the load

$$= \sum F^2 L/(AE)$$

∴ Deflection under load = $\{\sum F^2 L/(AE)\}$/load at the joint

Note: The summation sign Σ applies to all members of the truss.

Note in the above for virtual extension and virtual displacement, real extension and displacement values have been used. This is perfectly acceptable as long as compatibility between extensions and deflections remains.

Example 1

For the pin-jointed truss shown in Fig. 5.3, determine the vertical deflection at joint 5. Take Young's modulus as 210 kN mm^{-2}.

Table 10.1

Member	Force, kN	Area A, mm^2	Length L, m	AE/L, kN/mm	$F^2L/(AE)$, kN mm
1–3	10.0	50.0	2.0	5.25	19.0476
1–4	10√2	66.67	2√2	4.95	40.4040
2–4	−20.0	100.0	2.0	10.50	38.0952
3–4	−10.0	50.0	2.0	5.25	19.0476
3–5	10√2	70.0	2√2	5.197	38.4837
4–5	−10.0	50.0	2.0	5.25	19.0476

$$\Sigma 174.1257$$

Solution Calculations are set out in Table 10.1.

Virtual work by external load = applied load × deflection under the load = 10 kN $\times v_5 = 174.1257$

$$v_5 = 17.4126 \text{ mm}$$

Example 2

For the pin-jointed truss shown in Fig. 5.4, determine the vertical deflection at joint 3. Take Young's modulus as 210 kN mm^{-2}.

Solution Calculations are set out in Table 10.2.

Table 10.2

Member	Force, kN	Area A, mm²	Length L, m	AE/L, kN/mm	F²L/(AE), kN mm
1–2	$-50/\sqrt{3}$	150	3	10.5	79.3671
1–3	$25/\sqrt{3}$	75	3	5.25	39.6825
2–3	$50/\sqrt{3}$	150	3	10.5	79.3651
2–4	$-50/\sqrt{3}$	150	3	10.5	79.3651
3–4	$50/\sqrt{3}$	150	3	10.5	79.3651
3–5	$25/\sqrt{3}$	75	3	5.25	39.6825
4–5	$-50/\sqrt{3}$	150	3	10.5	79.3651

$$\Sigma 476.1906$$

Virtual work by external load $= 50 \text{ kN} \times v_3 = 476.1906$

$v_3 = 9.5238$ mm

Case 2: Deflection at any joint.

In the above two examples for Case 1, because there was only one load acting on the truss and the deflection required was also that at the loaded joint, the virtual extension was simply made equal to the real extension. If the deflection is required at some joint other than the loaded joint or if several loads act on the structure, then the strategy adopted is slightly different. Let the extensions of the members under the action of external forces be $FL/(AE)$ and the desired deflection at a specific joint be Δ.

Consider another load system, where the truss is loaded by a *single external force of unity at the joint and in the direction in which the actual deflection Δ is required*. Let the force developed in the ith member be f_i. Then if the forces in the new system are subjected to deflections and extensions caused by the actual loads, then clearly

$$1 \times \Delta = \sum f_i\{FL/(AE)\}_i$$

where the summation Σ is over all members of the truss.

This procedure is some times known as the dummy load method.

Example 3

For the pin-jointed truss shown in Fig. 5.4, determine the vertical deflection at joint 4. Take Young's modulus as 210 kN mm⁻².

Solution The first step is to determine the force in members due to unit vertical load at joint 4. This is shown in Fig. 10.8. The rest of the calculations are set out in Table 10.3.

Virtual work by external load $= 1 \text{ kN} \times v_4 = 5.5506$

$v_4 = 5.5506$ mm (downwards)

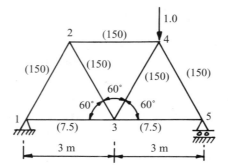

Figure 10.8 ●
A unit vertical load
acting at joint 4 of
a pin-jointed truss

Table 10.3

Member	f, kN	F, kN	Area A mm²	Length L, m	AE/L, kN/mm	f{FL/(AE)} kN mm
1–2	−0.2887	−50/√3	150	3	10.5	0.7937
1–3	0.1443	25/√3	75	3	5.25	0.3967
2–3	0.2887	50/√3	150	3	10.5	0.7937
2–4	−0.2887	−50/√3	150	3	10.5	0.7937
3–4	−0.2887	50/√3	150	3	10.5	−0.7937
3–5	0.4331	25/√3	75	3	5.25	1.1907
4–5	−0.8661	−50/√3	150	3	10.5	2.3811

$$\Sigma 5.5560$$

Example 4

For the pin-jointed truss shown in Fig. 5.4, determine the horizontal deflection at
joint 5. Take Young's modulus as 210 kN mm⁻².

Solution The first step is to determine the force in members due to unit horizontal load
at joint 5. This is shown in Fig. 10.9. The rest of the calculations are set out in
Table 10.4.

Virtual work by external load = 1 kN × u_5 = 5.4986

u_5 = 5.4986 mm (to left)

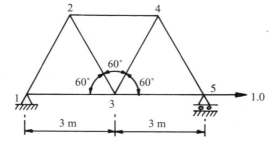

Figure 10.9 ● A unit
horizontal load acting at
joint 5 of a pin-jointed
truss

Table 10.4

Member	f, kN	F, kN	Area A mm²	Length L, m	AE/L, kN/mm	f{FL/(AE)} kN mm
1–2	0	−50/√3	150	3	10.5	0
1–3	1	25/√3	75	3	5.25	2.7493
2–3	0	50/√3	150	3	10.5	0
2–4	0	−50/√3	150	3	10.5	0
3–4	0	50/√3	150	3	10.5	0
3–5	1	25/√3	75	3	5.25	2.7493
4–5	0	−50/√3	150	3	10.5	0

$$\Sigma 5.4986$$

Case 3: Deflection at any joint due to thermal changes.

If the deflection is required due to thermal changes, then the strategy adopted is again slightly different. Let the extension of the ith member under the action of thermal changes, be $\alpha_t \, T_i \, L_i$ and the required deflection at a specific joint be Δ, where α_t = coefficient of thermal expansion, T_i = change in temperature in the ith member and L_i = length of the ith member.

Consider another load system, where the truss is loaded by a *single external force of unity at the joint and in the direction in which the actual deflection Δ is required*. Let the force developed in the ith member be f_i. Then if the forces in the new system are subjected to deflections and extensions caused by temperature changes, then clearly

$$1 \times \Delta = \sum f_i \, \alpha_t \, T_i \, L_i$$

where the summation Σ is over all members of the truss.

Example 5

For the pin-jointed truss shown in Fig. 5.4, determine the vertical deflection at joint 4 due to the change in the ambient temperature of member 2–4 by 30 °C. Take coefficient of thermal expansion $\alpha_t = 10 \times 10^{-6}$.

Solution The first step is to determine the force in members due to unit vertical load at joint 4. This is shown in Fig. 10.8. The rest of the calculations are set out in Table 10.5.

Table 10.5

Member	f, kN	T, °C	Length L, mm	$\alpha_t \, T \, L$, mm	f{$\alpha_t \, T \, L$} kN mm
1–2	−0.2887	0	3	0	0
1–3	0.1443	0	3	0	0
2–3	0.2887	0	3	0	0
2–4	−0.2887	30	3	0.9	−0.26
3–4	−0.2887	0	3	0	0
3–5	0.4331	0	3	0	0
4–5	−0.8661	0	3	0	0

$$\Sigma -0.26$$

Virtual work by external load = 1 kN × v_4 = −0.26

v_4 = −0.26 mm (upwards)

10.5 ● Principle of virtual work applied to the analysis of rigid-jointed structures

The ideas developed for the pin-jointed structures,

Virtual work done by external loads

$= \sum$ member axial force × virtual extension

is also valid for the analysis of rigid-jointed structures. The equation only needs interpretation in terms of the member force and the corresponding extension. In the case of flexural members, flexural stress varies not only in the cross-section but also along the length of the member. If the bending moment at a section is m, then assuming symmetrical bending, the flexural stress σ_x at a distance y from the neutral axis is given by

$\sigma_x = -my/I$

where the usual positive convention has been adopted.

As shown in Fig. 10.10, the axial force F over an element of area dA is

$F = \sigma_x \, dA = -my \, dA/I$

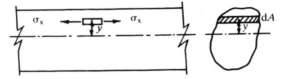

Figure 10.10 ●
Internal bending stress
in a beam

If the virtual displacement is caused by a virtual bending moment distribution of δM, then the virtual normal strain $\delta \varepsilon_x$ at the level y is

$\delta \varepsilon_x = (-\delta My/I)/E = -\delta My/(EI)$

The virtual extension δu of an element of length dx is

$\delta u = \delta \varepsilon_x \, dx = -\delta My \, dx/(EI)$

Member force × virtual extension = $\{-my \, dA/I\}\delta u$

$\{-my \, dA/I\}\{-\delta My \, dx/(EI)\} = [m\delta M \, dx/(EI^2)]y^2 \, dA$

For the whole member, member force × total virtual extension is obtained by

member force × virtual extension = $\displaystyle\int m\delta M \, dx/(EI^2)y^2 \, dA$

The integration is carried over the cross-section as well as over the length of the member. Therefore

$$\sum \text{member force} \times \text{virtual extension} = \int [m\delta M \, dx/(EI^2)] \left\{\int y^2 \, dA\right\}$$

$$= \int m\delta M \, dx/(EI)$$

because $\int y^2 \, dA = I$.

Therefore, in the case of rigid-jointed structures, the virtual work equation is given by

$$\text{virtual work done by external loads} = \int m\delta M \, dx/(EI)$$

Note that $\delta M/(EI)$ = virtual curvature.

The integration is done over the length of the whole structure. In the case of commonly occurring rigid-jointed structures, the above equation can be written as

$$\text{virtual work done by external loads} = \sum \int \text{bending moment} \times \text{virtual curvature}$$

where $\int$ refers to any one member and Σ is the summation for all the members of the structure.

10.5.1 Deflection of rigid-jointed structures

The application of the equation of virtual work to determine the deflection at specific points of a beam or a rigid-jointed structure is demonstrated by a few simple examples. As in the case of pin-jointed structures, three basic scenarios are considered. However, since the evaluation of

$$\int (\text{bending moment}) \times \{\text{corresponding virtual curvature}\} \, dx$$

involves integration, and in practice typical cases occur, it is useful to tabulate the results of integrals such as $\int f_1(x) \, f_2(x) \, dx$ for various values of $f_1(x)$ and $f_2(x)$ and integrated between the limits $x = 0$ and L. As examples,

(i) $f_1(x) = A$ and $f_2(x) = C$, then

$$\int f_1(x) \, f_2(x) \, dx = LAC$$

(ii) $f_1(x) = A$ and $f_2(x) = Cx/L$ or $f_2(x) = C(L - x)/L$ then

$$\int f_1(x) \, f_2(x) \, dx = LAC/2$$

(iii) $f_1(x) = A \, x/L$ and $f_2(x) = Cx/L$, then

$$\int f_1(x) \, f_2(x) \, dx = LAC/3$$

(iv) $f_1(x) = Ax/L$ and $f_2(x) = C(L - x)/L$ then

$$\int f_1(x) \, f_2(x) \, dx = LAC/6$$

Similarly integrals can be evaluated in the usual way if, say, $f_1(x) = (4C/L^2)\{x(L -x)\}$, arising from parabolic variation of bending moment as in a simply supported beam subjected to uniformly distributed load or $f_1(x) = (2C/L)\{x - 2\langle x - 0.5L\rangle\}$ arising from triangular variation of bending moment as in a simply supported beam subjected to a concentrated load at midspan interacting with different variations of $f_2(x)$. The values of integrals for very many typical cases are given in Fig. 10.11. These integrals are sometimes known as Mohr integrals.

f_j \ f_i	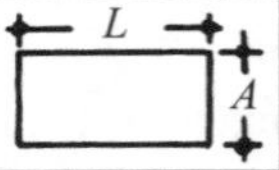 $\square$ (const) A	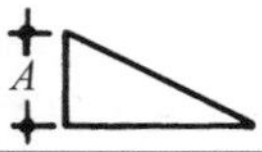 $\searbackslash$ A	 $\diagup$ A	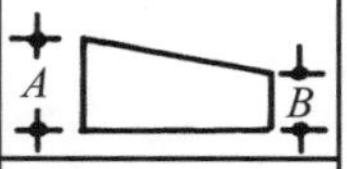 trapezoid A B
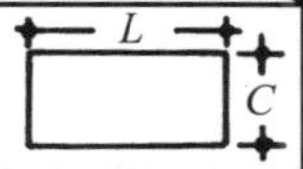 $\square$ C	LAC	$\frac{1}{2}LAC$	$\frac{1}{2}LAC$	$\frac{1}{2}L(A+B)C$
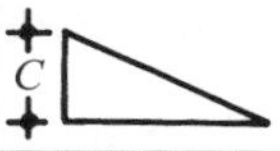 $\searbackslash$ C	$\frac{1}{2}LAC$	$\frac{1}{3}LAC$	$\frac{1}{6}LAC$	$\frac{1}{6}L(2A+B)C$
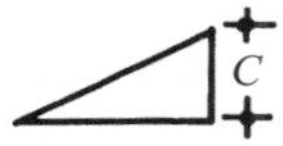 $\diagup$ C	$\frac{1}{2}LAC$	$\frac{1}{6}LAC$	$\frac{1}{3}LAC$	$\frac{1}{6}L(A+2B)C$
 $\frown$ C	$\frac{2}{3}LAC$	$\frac{1}{3}LAC$	$\frac{1}{3}LAC$	$\frac{1}{3}L(A+B)C$
 $\triangle$ C	$\frac{1}{2}LAC$	$\frac{1}{4}LAC$	$\frac{1}{4}LAC$	$\frac{1}{4}L(A+B)C$
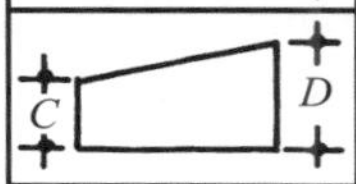 trapezoid C D	$\frac{1}{2}LA(C+D)$	$\frac{1}{6}LA(2C+D)$	$\frac{1}{6}LA(C+2D)$	$\frac{1}{6}\{A(2C+D)+B(C+2D)\}$
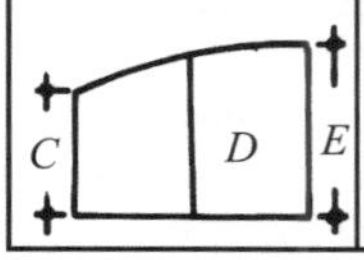 parabolic C D E	$\frac{1}{6}L(C+4D+E)$	$\frac{1}{6}LA(C+2D)$	$\frac{1}{6}LA(2D+E)$	$\frac{1}{6}\{A(C+2D)+B(D+3E)\}$

Figure 10.11 ●
Values of integrals

Case 1: Single concentrated load.

The application of the virtual work equation to determine the deflection at the loaded point of a rigid-jointed structure loaded by a single concentrated load is demonstrated by a few simple examples. Note that if M is the bending moment at a section in a member, then the corresponding curvature is $M/(EI)$, where I = relevant second moment of area of the cross-section of the member and, E = Young's modulus. Thus the equation

Virtual work done by external loads

$$= \sum \int (\text{bending moment} = M) \times \{\text{curresponding virtual curvature} = M/(EI)\}\,dx$$

becomes

$$\text{Virtual work by external loads} = \sum \int \{M^2/(EI)\}\,dx$$

where the summation sign Σ applies to all members of the frame.

Example 1

For the cantilever beam shown in Fig. 1.46, determine the vertical deflection at the tip.

Solution   As shown in Fig. 1.28,

$$M = Wx$$

$$\int \{M^2/(EI)\}\ \mathrm{d}x = \int \{(Wx)^2/(EI)\}\ \mathrm{d}x = W^2/(EI) \int x^2\ \mathrm{d}x = W^2L^3/(3EI)$$

where the integration has been carried over the entire length of the member from $x = 0$ to L.

If using the tabular values of integrals, then $A = C = WL$ for triangular variation of the same type, $\int M^2\ \mathrm{d}x = LAC/3 = W^2L^3/3$

$$\text{Virtual work by external loads} = W\ \Delta = \sum \int \{M^2/(EI)\}\mathrm{d}x = W^2L^3/(3EI)$$

where Δ = tip deflection

$$\Delta = WL^3/(3EI)$$

Example 2

For the rigid-jointed frame shown in Fig. 5.6, determine the horizontal deflection of the load at joint C.

Solution   As detailed in Section 5.3, the bending moments in various members are
Member BC: $M = Wy$, with the origin at C
Thus $A = Wa$, $C = Wa$, $L = a$, $EI = EI_c$

$$\int \{M^2/(EI)\}\ \mathrm{d}x = W^2a^3/(3EI_c)$$

Member AB: $M = Wy$, with the origin at A
Thus $A = Wa$, $C = Wa$, $L = a$, $EI = EI_c$

$$\int \{M^2/(EI)\}\ \mathrm{d}x = W^2a^3/(3EI_c)$$

Member BD: $M = (2Wa/L)x$, with the origin at D
Thus $A = 2Wa$, $C = 2Wa$, $L = L$, $EI = EI_b$

$$\int \{M^2/(EI)\}\ \mathrm{d}x = 4W^2La^2/(3EI_b)$$

$$\text{Virtual work by external loads} = W\ \Delta = \sum \int \{M^2/(EI)\}\mathrm{d}x$$

$$= 2W^2a^3/(3EI_c) + 4W^2La^2/(3EI_b)$$

where Δ = horizontal deflection at C

$$\Delta = 2Wa^3/(3EI_c)\{1 + 2(L/a)(I_c/I_b)\}$$

Case 2: Deflection at any joint.

In the above two examples for Case 1, there was only one load acting on the frame and the deflection required was also that at the loaded joint; the virtual curvature and deflection at the joint were simply made equal to the real values caused by loads. If the deflection is required at a joint other than the loaded joint or if several loads act on the structure, then the strategy adopted is similar to that for pin-jointed structures. Let the curvature of the member at any section under the action of external forces, be $M/(EI)$ and the desired deflection at a specific joint be Δ.

Consider another load system, where the structure is loaded by a *single external force of unity at the joint and in the direction in which the actual deflection Δ is required*. Let the bending moment developed in the ith member be m_i. Then if the forces and bending moments in the new system are subjected to deflections and curvatures caused by the actual loads, then clearly

$$1 \times \Delta = \sum \int m_i \, \{M/(EI)\}_i \, dx$$

where the summation Σ is over all members of the frame.

This procedure is sometimes known as the dummy load method.

Example 3

For the cantilever beam subjected to uniformly distributed load q shown in Fig. 1.47, determine the vertical deflection at the tip.

Solution   As shown in Fig. 1.47,

$$M = -qx^2/2$$

If a unit load is placed at the tip, then

$$m = -1 \,.\, x$$

$$\int \{mM/(EI)\} \, dx = \int \{qx^2 \, x/(2EI)\} \, dx = q/(2EI) \int x^3 \, dx = qL^4/(8EI)$$

where the integration has been carried over the entire length of the member from $x = 0$ to L.

Virtual work by external loads $= 1 \times \Delta = \sum \int \{mM/(EI)\} \, dx = qL^4/(8EI)$

where Δ = tip deflection

$$\Delta = qL^4/(8EI)$$

Example 4

For the rigid-jointed frame shown in Fig. 5.6, determine the horizontal deflection at support D.

Solution *Step 1*: Calculate the bending moment distribution due to unit horizontal load at joint D as shown in Fig. 10.12. The reactions can be calculated as the structure is statically determinate. Taking moments about A,

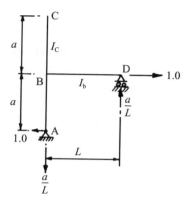

Figure 10.12 ● A unit horizontal load acting at joint D of a rigid-jointed frame

$$V_d L - 1 \cdot a = 0, \quad V_d = a/L$$

$$V_a = -V_d = -a/L$$

$$H_a = 1, \text{ from right to left}$$

The bending moments in various members can now be calculated.
Member BC: No bending moment. $m = 0$.
Member AB: $m = y$, origin at A and causing tension on the inside of the frame.
Member BD: $m = ax/L$, origin at D and causing tension on the under-side of the beam.

Step 2: Evaluate the integrals.
Member BC: $M = Wy$, with the origin at C
Thus $A = Wa$, $C = 0$, $L = a$, $EI = EI_c$

$$\int \{mM/(EI)\} \, dx = 0$$

Member AB: $M = Wy$, with the origin at A

$$m = y$$

Thus $A = Wa$, $C = a$, $L = a$, $EI = EI_c$

$$\int \{mM/(EI)\} \, dx = Wa^3/(3EI_c)$$

Member BD: $M = (2Wa/L)x$, with the origin at D

$$m = ax/L$$

Thus $A = 2Wa$, $C = a$, $L = L$, $EI = EI_b$

$$\int \{m \, M/(EI)\} \, dx = 2W \, La^2/(3EI_b)$$

Virtual work by external loads $= W \Delta = \sum \int \{M^m/(EI)\}dx$

$$= Wa^3/(3EI_c) + 2WLa^2/(3EI_b)$$

where Δ = horizontal deflection at D

$\Delta = Wa^3/(3EI_c)\{1 + 2(L/a)(I_c/I_b)\}$

Case 3: Deflection at any joint due to thermal changes.

If the deflection is required due to thermal changes, then the strategy adopted is similar to that used in the case of pin-jointed structures. As shown in Section 5.3.1, if a beam element is subjected to thermal changes of T_t and T_b at the top and bottom face respectively, then the corresponding curvature of the beam (the beam hogs up) is given by

curvature $= \alpha_t(T_t - T_b)/d$

where d = depth of the beam.

Consider another load system, where the frame is loaded by a *single external force of unity at the joint and in the direction in which the actual deflection* Δ *is required*. Let the bending moment developed in the ith member be m_i. Then if the forces in the new system are subjected to deflections and curvatures caused by the actual loads, then clearly

$1 \times \Delta = \sum \{m_i \, \alpha_t(T_t - T_b)/d\}dx$

where the summation Σ is over all members of the frame.

Example 5

For the rigid-jointed truss shown in Fig. 5.8, determine the horizontal deflection at roller support D due to the change in the ambient temperature of members AB and BC only by T °C on the outside of the frame.

Solution   Calculate the bending moment distribution due to unit horizontal load at joint D as shown in Fig. 10.13. The horizontal reaction due to unit load at A is also unity acting from right to left. All other reactions are zero.

Member AB: $T_t = T$, $T_b = 0$

Curvature $= \alpha_t \, T/d_c$ 'tension' outside

$m = 1 \times y$, tension on the inside, taking the origin at A. The moment at B is $1 \times H = H$

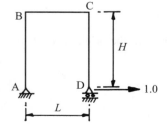

For evaluating the integral, $A = \alpha_t\, T/d_c$, $C = -H$, $L = H$

$$\int \{m \times \text{curvature}\}\, dx = -H^2 \alpha_t\, T/(2d_c)$$

Member BC: $T_t = T$, $T_b = 0$
Curvature $= \alpha_t\, T/d_b$ 'tension' outside
$m = 1 \times H$, tension on the inside
Thus $A = \alpha_t\, T/d_c$, $C = -H$, $L = L$

$$\int \{m \times \text{curvature}\}\, dx = -H\, L\alpha_t\, T/(d_b)$$

Member CD: Curvature $= 0$
$m = 1 \times y$, origin at Δ and causing tension on the inside of the frame.

$$\int \{m \times \text{curvature}\}\, dx = 0$$

Thus $1 \times \Delta = \sum \{m_i\, \alpha_t (T_t - T_b)/d\}dx$
$$= -H^2 \alpha_t\, T/(2d_c) - HL\alpha_t\, T/(d_b)$$

$$\Delta = -H^2 \alpha_t\, T/(2d_c)\{1 + 2(L/H)/(d_c/d_b)\}$$

The negative sign indicates that the deflection is to the left (i.e. opposite to the direction of the applied force).

10.6 ● Strain energy and complementary energy

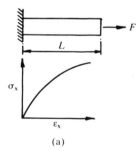

(a)

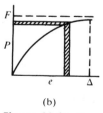

(b)

Figure 10.14 ●
(a), (b) Strain and complementary energies for a bar under tensile forces

In addition to the principle of virtual work, two additional concepts known as strain energy and complementary energy are also used extensively. These concepts are of considerable assistance in deriving mathematical expressions and general properties of structures in structural mechanics. In the following sections the concepts of strain and complementary energy will be explained in some detail.

As an example to illustrate the concepts, consider a prismatic bar subjected to an end load F as shown in Fig. 10.14(a). Under the influence of this load, the bar stretches by Δ. For generality let it be assumed that the material of the bar shows non-linear stress–strain behaviour as in Fig. 10.14(b). At any load stage P the corresponding extension is e, because the only stress present in the bar is an axial stress $\sigma_x = P/A$ and the strain $\varepsilon_x = e/L$, and the shape of the load–extension curve for the bar is identical to the stress–strain curve. The infinitesimal work dW done by P due to an infinitesimal change in extension de is equal to $dW = P\, de$. Therefore the total work W done when the extension is Δ is given by

$$W = \int_0^\Delta P\, de = \text{area 'under' the load–extension curve}$$

10.6.1 Strain energy

The work W is obviously external work done and because of the fact that energy must be conserved, in the absence of any loss of energy, the above work is stored as energy in the bar which is recoverable when the bar is unloaded. The energy stored in the bar is called strain energy. In this case therefore

$$\text{Strain energy (generally represented by } U) = W = \int P\, de$$

It is useful to derive an expression for the strain energy in terms of the strains themselves. Consider a bar of area dA and length dL. If the stress in the bar is σ_x, then the axial force is $\sigma_x\, dA$. If the strain is increased by $d\varepsilon_x$, then the extension is changed by $d\varepsilon_x\, dL$. The strain energy dU stored in the infinitesimal element volume of dV of the bar is

$$dU = (\sigma_x\, dA)(d\varepsilon_x\, dL) = \sigma_x\, d\varepsilon_x\, dV$$

where $dA\, dL = dV$, the volume of the infinitesimal element. Therefore the strain energy U stored in the bar is equal to

$$U = \int_0^{\Delta/L} \sigma_x\, d\varepsilon_x\, dV$$

In the special case of stress–strain law being linear, then for a prismatic bar of cross-sectional area A and length L, $\sigma_x = E\varepsilon_x$, $\varepsilon_x = e/L$, and because the state of strain is constant along the bar,

$$U = \int dV \int \sigma_x\, d\varepsilon_x = V \int E\varepsilon_x\, d\varepsilon_x$$

$$= EV \int_0^{\Delta/L} \varepsilon_x\, d\varepsilon_x = \frac{EV\,\Delta^2}{2L^2}$$

because $V = AL$ and $U = \{AE/(2L)\}\Delta^2$. It should be appreciated that the above expression for strain energy is applicable to *linearly elastic* materials only.

10.6.2 Complementary energy

A concept related to strain energy U is the complementary energy C. In terms of the 'external' force P and the corresponding displacement e, this is defined as (see Fig. 10.14(b))

$$C = \int_0^{F} e\, dP = \text{area 'above' the load–extension curve}$$

Note that strain energy $U = \int P\, de$.

In terms of the internal stresses and strains, this can be expressed as

$$C = \int \varepsilon_x\, d\sigma_x\, dV$$

In the case of a prismatic bar of linearly elastic material, $\varepsilon_x = \sigma_x/E = F/(AE)$. Therefore

$$C = \int dV \int \varepsilon_x \, d\sigma_x = V \int (\sigma_x/E) \, d\sigma_x = (V/E)\sigma_x^2/2$$

Substituting $V = AL$ and $\sigma_x = F/A$,

$$C = F^2L/(2AE)$$

It should be noted that irrespective of the stress–strain law

$$F\Delta = U + C$$

If the response is linearly elastic, then $U = C = 0.5F\Delta$. In all cases, irrespective of material laws, $U + C$ is equal to total loss of potential energy $F\Delta$ of the external loads.

10.7 ● Expressions for strain and complementary energies in terms of stress resultants and corresponding strains

In the previous section, expressions for strain energy U and complementary energy C stored in an infinitesimal volume dV were given in terms of normal stress σ and normal strain ε as

$$U = \int \sigma \, d\varepsilon \, dV, \quad C = \int \varepsilon \, d\sigma \, dV$$

In a similar way, if the stress and strain under consideration are shear stress τ and shear strain γ, then

$$U = \int \tau \, d\gamma \, dV, \quad C = \int \gamma \, d\tau \, dV$$

These are general expressions in terms of stresses and strains applicable to all cases. However, in the approximate analysis considered in this book, we deal with stress resultants such as axial force, shear force, bending moment and twisting moment rather than with the more general normal or shear stresses. It is therefore convenient to derive expressions for U and C in terms of stress resultants and corresponding 'strains'. It is assumed that the material is linearly elastic with a Young's modulus of E. Therefore $U = C$ in all cases except that U is expressed in terms of 'strains' and C is expressed in terms of 'stresses'.

1. Axial force F and axial extension Δ. This was dealt with in the previous section where it was shown that $U = 0.5AE \, \Delta^2/L$, $C = 0.5F^2L/(AE)$, AE = axial rigidity.
2. Bending moment M and curvature d^2v/dx^2. In this case

$$\sigma_x = -My/I$$

where I = second moment of area and y = distance from the neutral axis to point under consideration. Therefore

$$d\sigma_x = -dM(y/I)$$

$$\varepsilon_x = \sigma_x/E = -My/(EI)$$

where it is assumed that σ_x is the only stress present. Thus

$$C = \int_0^M -My/(EI)\{-dM(y/I)\}\,dV$$

Substituting $dV = dA\,dx$, where dA = element of area as shown in Fig. 10.10, the expression for C becomes:

$$C = \int_0^M M\,dM\,dx/(EI^2)\left[\int y^2\,dA\right]$$

$$= \int_0^M M\,dM\,dx/(EI) = \frac{M}{2EI}\,dx$$

where $\int y^2\,dA$ has been replaced by I.

The above expression for C can be expressed for a member as

$$C = \int \{M^2/(2EI)\}\,dx$$

$M = EI\,d^2v/dx^2$ and for linearly elastic structures $U = C$, therefore U can be expressed as

$$U = 0.5\int EI(d^2v/dx^2)^2\,dx$$

Therefore, for linearly elastic members,

$$U = 0.5\int EI(d^2v/dx^2)^2\,dx, \quad C = 0.5\int \{M^2/(EI)\}\,dx$$

3. Shear force Q and 'average' shear strain γ. The distribution of shear stress τ at a cross-section depends on the particular cross-section considered. However, by making some simplifying assumptions, it is possible to derive expressions for U and C in terms of γ and Q. Initially consider a rectangular section of breadth b and depth d with the shear force parallel to the depth as shown in Fig. 3.24. For this case, the shear stress τ is given by

$$\tau = 1.5(Q/A)\{1 - (2y/d)^2\}$$

where y is the distance to the section from the neutral axis. Because $\gamma = \tau/G$, where G = shear modulus, and $C = \int \gamma\,d\tau\,dV$, substituting for τ, C can be expressed as

$$C = \int_0^Q \{1.5/A\}^2/G\{Q\,dQ\}\{1 - (2y/d)^2\}^2\,dV$$

$dV = dA\,dx$ and $dA = b\,dy$, and so the above expression for C becomes

$$C = \int \{1.5/A\}^2/G\{Q\,dQ\}\{1 - (2y/d)^2\}^2 b\,dy\,dx$$

Integrating with respect to y between the limits $y = -0.5d$ to $+0.5d$,

$$C = (6/5)\int_0^Q Q\,dQ/(GA)\,dx = (6/5)Q^2/(2GA)\,dx$$

Setting $A_{shear} = 5A/6 = A/1.2$,

$$C = 0.5\int Q^2/(GA_{shear})\,dx \text{ for the element}$$

The ratio A_{shear}/A is called Area Shear factor

Although the above expression was derived for the case of a rectangular section, it is assumed to be applicable to all shapes of sections with A_{shear} being a suitable proportion of the cross-sectional area A. For example, in the case of I sections with shear force parallel to the web, because most of the shear force is carried by the web and the shear stress is parabolically distributed in the web, it is fair to say that $A_{shear} = A_{web}/1.2$. Similarly, if the shear force is parallel to the flanges, A_{shear} = area of two flanges/1.2.

Assuming that the 'average' shear strain $\gamma = \tau_{average}/G$ and $\tau_{average} = Q/A$ then

$$U = 0.5 \int GA\gamma^2 \, dx, \quad C = 0.5 \int Q^2/(GA) \, dx$$

where

A = area of cross-section resisting shear

4. Torque T and twist ψ. If the relationship between torque T and twist ψ is given by $T = GJ \, d\psi/dx$, where J = Saint Venant's torsional inertia, then it can be shown that

$$U = 0.5 \int GJ(d\psi/dx)^2 \, dx, \quad C = 0.5 \int T^2/(GJ) \, dx$$

10.8 ● Energy theorems

The concepts developed in the previous sections yield very powerful methods of structural analysis. Two theorems named after Castigliano and Engesser are particularly important. They are discussed in the next section.

10.8.1 Castigliano's theorem

Consider a simple 2-D pin-jointed truss subjected to a set of forces W_i, $i = 1$, $2 \ldots n$. Let the axial force in the members of the truss be F_i, $i = 1, 2 \ldots m$. Let the system be subjected to virtual displacements corresponding to forces W_i of $\delta \Delta_i$. Let the corresponding virtual extension of the ith member be δe_i. Then from the principle of virtual work

$$\sum W_i \, \delta \Delta_i = \sum F_i \, \delta e_i \tag{i}$$

Now consider another system of virtual displacement where, say, $\delta \Delta_1$ is increased to $\delta(\Delta_1 + d\Delta_1)$ but other displacements are unchanged from the original system. Let the corresponding extensions be $\delta(e_i + de_i)$. Again from the principle of virtual work

$$W_1 \, \delta d\Delta_1 + \sum W_i \, \delta \Delta_i = \sum F_i \, \delta(e_i + de_i) = \sum F_i \, \delta e_i + \sum F_i \, \delta de_i \tag{ii}$$

Subtracting the first equation from the second,

$$W_1 \, \delta \Delta_1 = \sum F_i \, \delta de_i$$

Obviously $\sum F_i \, \delta de_i$ = change in the strain energy of the system due to a change in $\delta \Delta_1$ by $\delta d\Delta_1$. Therefore if U is the strain energy of the system, then $\sum F_i \, \delta de_i = \{\partial U/\partial \Delta_1\} \, \delta d\Delta_1$. Thus

$$W_1 \, \delta \, d\Delta_1 = \{\partial U/\partial \Delta_1\} \, \delta \, d\Delta_1 \text{ or}$$

$$W_1 = \partial U/\partial \Delta_1$$

In general

$$W_i = \partial U/\partial \Delta_i$$

This result is known as Castigliano's first theorem. It is stated as, the partial derivative of the strain energy of the structure with respect to the displacement Δ_i gives the corresponding force. It should be noted that the above result was derived from the principle of virtual displacement and therefore is valid irrespective of whether the structure is linear or not. Although the above equation was derived for the case of a pin-jointed structure, as was shown in Section 10.7, because $\Sigma F_i e_i$ can be interpreted in terms of bending moments for members undergoing bending deformations, it is applicable to all types of structures and not simply to pin-jointed structures.

Application of Castigliano's theorem

The use of Castigliano's theorem for the derivation of the element stiffness matrices for some typical members is shown by examples.

Example 1

Derive the element stiffness matrix for an axial element.

Solution   As explained in Section 10.7, the elastic strain energy U is given by

$$U = (0.5AE/L) \, \Delta^2$$

As shown in Section 6.3, for a pin-jointed member, extension $\Delta = \Delta_{a2} - \Delta_{a1}$
where Δ_{a2} = axial displacement at end 2
$\quad\quad \Delta_{a1}$ = axial displacement at end 1
The axial displacements can be expressed in terms of displacements along the x- and y-axes as follows.

$$\Delta_{a2} = u_2 \, l + v_2 \, m$$

$$\Delta_{a1} = u_1 \, l + v_1 \, m$$

where l and m are the direction cosines of the member and u_i and v_i are the displacements along the x- and y-axis respectively at the ith end.
Thus

$$U = (0.5AE/L)\{(u_2 - u_1)l + (v_2 - v_1)m\}^2$$

It is worth pointing out that all the four translations are independent displacements. Using Castigliano's theorem

$$\partial U/\partial u_1 = F_{x1} = (0.5AE/L) \, 2\{(u_2 - u_1)l + (v_2 - v_1)m\}(-l)$$

$$F_{x1} = (AE/L)\{(u_1 - u_2)l^2 + (v_1 - v_2)lm\}$$

Similarly

$$\partial U/\partial v_1 = F_{y1} = (0.5AE/L)\,2\{(u_2 - u_1)l + (v_2 - v_1)m\}(-m)$$

$$F_{y1} = (AE/L)\{(u_1 - u_2)lm + (v_1 - v_2)m^2\}$$

$$\partial U/\partial u_2 = F_{x2} = (0.5AE/L)\,2\{(u_2 - u_1)l + (v_2 - v_1)m\}(l)$$

$$F_{x2} = (AE/L)\{(u_2 - u_1)l^2 + (v_2 - v_1)lm\}$$

$$\partial U/\partial v_2 = F_{y2} = (0.5AE/L)\,2\{(u_2 - u_1)l + (v_2 - v_1)m\}(m)$$

$$F_{y2} = (AE/L)\{(u_2 - u_1)lm + (v_2 - v_1)m^2\}$$

which clearly leads to the stiffness matrix shown in Section 6.4.
Note that

$$\partial F_{x1}/\partial u_1 = (AE/L)l^2, \; \partial F_{x1}/\partial v_1 = (AE/L)lm$$

which are clearly the stiffness coefficients.
Because

$$F_{x1} = \partial U/\partial u_1, \; \partial F_{x1}/\partial u_1 = \partial^2 U/\partial u_1^2 = (AE/L)l^2$$

then

$$\partial F_{x1}/\partial v_1 = \partial^2 U/\partial u_1\,\partial v_1 = (AE/L)lm$$

Thus in general, $\partial^2 U/\partial u_i\,\partial v_j$ = stiffness coefficient s_{ij} corresponding to the ith row and jth column of the stiffness matrix.

Example 2

Derive the element stiffness matrix for a bending element.

Solution   As explained in Section 10.7, the elastic strain energy U is given by

$$U = (0.5EI) \int (d^2v/dx^2)^2 \, dx$$

The deflection of a beam element is governed by the four 'displacements' (i.e. two translations and two rotations) at the ends of the beam. Thus the four boundary conditions govern the deflection of the beam. Therefore a general polynomial expression for v is given by

$$v = C_1 + C_2x + C_3x^2 + C_4x^3$$

where C_1, C_2, C_3 and C_4 are constants to be determined from boundary conditions.
At $x = 0$, $v = \Delta_{n1}$ and $dv/dx = -\theta_1$
At $x = L$, $v = \Delta_{n2}$ and $dv/dx = -\theta_2$
In the above, the sign convention adopted is translation along positive y-direction and clockwise rotation are positive.
Using the boundary conditions and solving for the constants,

$$C_1 = \Delta_{n1}$$

$$C_2 = -\theta_1$$

$$C_3 = (-3\Delta_{n1} + 3\Delta_{n2} + 2\theta_1 L + \theta_2 L)/L^2$$

$$C_4 = (2\Delta_{n1} - 2\Delta_{n2} - \theta_1 L - \theta_2 L)/L^3$$

Substituting for the constants,

$$v = \Delta_{n1} - \theta_1 x + x^2(-3\Delta_{n1} + 3\Delta_{n2} + 2\theta_1 L + \theta_2 L)/L^2$$
$$+ x^3(2\Delta_{n1} - 2\Delta_{n2} - \theta_1 L - \theta_2 L)/L^3$$

Rearranging,

$$v = \Delta_{n1}\{1 - 3(x/L)^2 + 2(x/L)^3\} - \theta_1 L\{x/L - 2(x/L)^2 + (x/L)^3\}$$
$$+ \Delta_{n2}\{3(x/L)^2 - 2(x/L)^3\} - \theta_2 L\{(x/L)^3 - (x/L)^2\}$$

$$d^2v/dx^2 = 2C_3 + 6C_4 x$$

Substituting for C_3 and C_4,

$$d^2v/dx^2 = (-1 + 2x/L) 6\Delta_{n1}/L^2 + (1 - 2x/L) 6\Delta_{n2}/L^2$$
$$+ (4 - 6x/L) \theta_1/L + (2 - 6x/L) \theta_2/L$$

As was shown in Section 6.3, the translations Δ_{n2} and Δ_{n1} which are normal to the beam axis can be expressed in terms of the displacements along the x- and y-axes as follows:

$$\Delta_{n2} = -u_2 l + v_2 m$$

$$\Delta_{n1} = -u_1 l + v_1 m$$

where l and m are the direction cosines of the member and u_i and v_i are the displacements along the x- and y-axis respectively at the ith end.

Thus

$$d^2v/dx^2 = (-u_1 l + v_1 m) f_1 + (-u_2 l + v_2 m) f_2 + f_3 \theta_1 + f_4\theta_2$$

where

$$f_1 = (-1 + 2x/L) 6/L^2$$

$$f_2 = (1 - 2x/L) 6/L^2$$

$$f_3 = (4 - 6x/L)/L$$

$$f_4 = (2 - 6x/L)/L$$

$$U = (0.5EI) \int (d^2v/dx^2)^2 \, dx$$

$$= (0.5EI) \int \{(-u_1 l + v_1 m) f_1 + (-u_2 l + v_2 m) f_2 + f_3 \theta_1 + f_4\theta_2\}^2 \, dx$$

Applying Castigliano's theorem,

$$\partial U/\partial u_1 = F_{x1}$$

$$= (0.5EI) 2 \int \{(-u_1 l + v_1 m) f_1 + (-u_2 l + v_2 m) f_2 + f_3 \theta_1 + f_4\theta_2\}(-l f_1) \, dx$$

$$F_{x1} = EI \int \{(u_1 \, l^2 - v_1 \, lm) \, f_1 + (u_2 \, l^2 - v_2 \, lm) \, f_2 - l \, f_3 \, \theta_1 - l \, f_4\theta_2\} \, f_1 \, dx$$

Similarly

$$\partial U/\partial v_1 = F_{y1}$$

$$= (0.5EI) \, 2 \int \{(-u_1 \, l + v_1 \, m) \, f_1 + (-u_2 \, l + v_2 \, m) \, f_2 + f_3 \, \theta_1 + f_4\theta_2\}(m \, f_1) \, dx$$

$$F_{y1} = EI \int \{(-u_1 \, l \, m + v_1 \, m^2) \, f_1 + (-u_2 \, l \, m + v_2 \, m^2) \, f_2 + m \, f_3 \, \theta_1 + m \, f_4\theta_2\} \, f_1 \, dx$$

$$\partial U/\partial \theta_1 = M_1$$

$$= (0.5EI) \, 2 \int \{(-u_1 \, l + v_1 \, m) \, f_1 + (-u_2 \, l + v_2 \, m) \, f_2 + f_3 \, \theta_1 + f_4\theta_2\} \, f_3 \, dx$$

$$M_1 = EI \int \{(-u_1 \, l + v_1 \, m) \, f_1 + (-u_2 \, l + v_2 \, m) \, f_2 + f_3 \, \theta_1 + f_4\theta_2\} \, f_3 \, dx$$

$$\partial U/\partial u_2 = F_{x2}$$

$$= (0.5EI) \, 2 \int \{(-u_1 \, l + v_1 \, m) \, f_1 + (-u_2 \, l + v_2 \, m) \, f_2 + f_3 \, \theta_1 + f_4\theta_2\}(-l \, f_2) \, dx$$

$$F_{x2} = EI \int \{(u_1 \, l^2 - v_1 \, lm) \, f_1 + (u_2 \, l^2 - v_2 \, lm) \, f_2 - l \, f_3 \, \theta_1 - l \, f_4\theta_2\} \, f_2 \, dx$$

$$\partial U/\partial v_2 = F_{y2}$$

$$= (0.5EI) \, 2 \int \{(-u_1 \, l + v_1 \, m) \, f_1 + (-u_2 \, l + v_2 \, m) \, f_2 + f_3 \, \theta_1 + f_4\theta_2\}(m \, f_2) \, dx$$

$$F_{y1} = EI \int \{(-u_1 \, lm + v_1 \, m^2) \, f_1 + (-u_2 \, lm + v_2 \, m^2) \, f_2 + m \, f_3 \, \theta_1 + m \, f_4\theta_2\} \, f_2 \, dx$$

$$\partial U/\partial \theta_2 = M_2$$

$$= (0.5EI) \, 2 \int \{(-u_1 \, l + v_1 \, m) \, f_1 + (-u_2 \, l + v_2 \, m) \, f_2 + f_3 \, \theta_1 + f_4\theta_2\} \, f_4 \, dx$$

$$M_2 = EI \int \{(-u_1 \, l + v_1 \, m) \, f_1 + (-u_2 \, l + v_2 \, m) \, f_2 + f_3 \, \theta_1 + f_4\theta_2\} \, f_4 \, dx$$

Before the evaluation of the stiffness matrix is completed, it is necessary to evaluate the following integrals. It can be easily shown that

$$\int f_1 f_1 \, dx = 12/L^3, \int f_1 f_2 \, dx = -12/L^3, \int f_1 f_3 \, dx = -6/L^2, \int f_1 f_4 \, dx = -6/L^2$$

$$\int f_2 f_2 \, dx = 12/L^3, \int f_2 f_3 \, dx = 6/L^2, \int f_2 f_4 \, dx = 6/L^2$$

$$\int f_3 f_3 \, dx = 4/L, \int f_3 f_4 \, dx = 2/L$$

$$\int f_4 f_4 \, dx = 4/L$$

Substituting the value of the integrals clearly leads to the stiffness matrix shown in Section 6.4. It should be appreciated that the above stiffness matrix includes contributions from the bending effects only. For the full matrix, the contribution from the axial deformation and shear deformation have to be added.

10.8.2 Engesser's theorem

Consider the same structure as used in deriving Castigliano's theorem in Section 10.8.1 so that using the principle of virtual work we have

$$\sum W_i \delta \Delta_i = \sum F_i \delta e_i \qquad \text{(i)}$$

Consider another system, where the load W_1 is increased to $W_1 + dW_1$. Let the bar force in this case be $F_i + dF_i$. Then applying the principle of virtual work

$$dW_1 \delta \Delta_1 + \sum W_i \delta \Delta_i = \sum (F_i + dF_i) \delta e_i \qquad \text{(ii)}$$

Taking the difference between the two equations (i) and (ii),

$$dW_1 \delta \Delta_i = \sum dF_i \delta e_i$$

But $\sum dF_i \delta e_i$ is equal to the change in the complementary energy of the system due to a change in W_1 by dW_1. Therefore

$$dW_1 \delta \Delta_1 = \{\partial C / \partial W_i\}\, dW_1$$

i.e.

$$\delta \Delta_1 = \partial C / \partial W_1$$

or in general

$$\delta \Delta_i = \partial C / \partial W_i$$

Engesser's theorem is generally stated as, if the complementary energy of the structure is expressed in terms of a set of independent forces, then the partial derivative of the complementary energy with respect to a force is equal to the corresponding deflection.

As in the case of Castigliano's theorem, Engesser's theorem is also quite general and applicable to all types of structures.

Application of Engesser's theorem

The use of Engesser's theorem for the derivation of the element stiffness matrices for some typical members is shown by examples.

Example 1

A cantilever beam is subjected at the tip to a clockwise couple P and a vertical force of F. Calculate the 'force–deflection' relationship. Include both bending and shear deformation.

Solution Taking the origin at the tip, the bending moment M and shear force Q at a section distance x from the tip are given by

$$M = P + Fx$$

$$Q = -F$$

The complementary energy is given by

$$C = \int M^2 \, dx/(2EI) + \int Q^2 \, dx/(2GA_{\text{shear}})$$

Substituting for M and Q,

$$C = \int (P + Fx)^2 \, dx/(2EI) + \int F^2 \, dx/(2EIGA_{\text{shear}})$$

$$\partial C/\partial P = \theta = (1/EI) \int \{P + Fx\} \, dx = 1/(EI) \{PL + FL^2/2\}$$

$$\partial C/\partial F = \Delta = (1/EI) \int \{P + Fx\} \, xdx + 1/(GA_{\text{shear}}) \int F \, dx$$

$$= 1/(EI) \{P \, L^2/2 + FL^3/3\} + 1/(GA_{\text{shear}}) \, F \, L$$

The above two equations can be expressed in matrix form as

$$\begin{bmatrix} \Delta \\ \theta \end{bmatrix} = \begin{bmatrix} L^3/(3EI) + L/(GA_{\text{shear}}) & L^2/(2EI) \\ L^2/(2EI) & L/(EI) \end{bmatrix} \begin{bmatrix} F \\ M \end{bmatrix}$$

This matrix relating given forces to their corresponding displacements is known as a flexibility matrix.

Example 2

A cantilever arch in the form of a quarter circle as shown in Fig. 10.15 is subjected at the tip to a couple and two forces in the horizontal and vertical directions. Derive the 'force–deflection' relationship. Include only bending deformation.

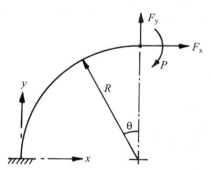

Figure 10.15 ●
A semicircular
arch element

Solution The bending moment M at a section θ from the vertical is given by

$$M = -P - F_x R \sin \theta + F_y R(1 - \cos \theta)$$

The complementary energy, if only bending deformations are included, is given by

$$C = \int M^2 \, ds/(2EI)$$

Note that integration has to be carried out over the curved length of the member. Therefore an element of length of the arch is taken as ds, where $ds = R \, d\theta$.

Substituting for M

$$C = \int \{-P - F_x R \sin \theta + F_y R(1 - \cos \theta)\}^2 R \, d\theta/(2EI)$$

where the limits for θ are 0 and $\pi/2$

$$\partial C/\partial P = \theta = (1/EI) \int \{-P - F_x R \sin \theta + F_y R(1 - \cos \theta)\}(-1) R \, d\theta$$

$$= 1/(EI)\{P R \pi/2 + F_x R^2 - F_y R^2(\pi/2 - 1)\}$$

Similarly

$$\partial C/\partial F_x = \Delta_x = (1/EI) \int \{-P - F_x R \sin \theta + F_y R(1 - \cos \theta)\}(-R \sin \theta) R \, d\theta$$

$$= 1/(EI)\{P R^2 + F_x R^3 \pi/4 - F_y R^3/2\}$$

$$\partial C/\partial F_y = \Delta_y$$

$$= (1/EI) \int \{-P - F_x R \sin \theta + F_y R(1 - \cos \theta)\} \theta)\}R(1 - \cos \theta) R \, d\theta$$

$$= 1/(EI)\{-P R^2 (\pi/2 - 1) - F_x R^3/2 + F_y R^2(3\pi/4 - 2)\}$$

The above three equations can be expressed in matrix form leading to the flexibility matrix relationship for the given structure:

$$\begin{bmatrix} \theta \\ \Delta_x \\ \Delta_y \end{bmatrix} = 1/(EI) \begin{bmatrix} R\pi/2 & R^2 & -R^2(\pi/2 - 1) \\ R^2 & R^3 \pi/4 & -R^3/2 \\ -R^2(\pi/2 - 1) & -R^3/2 & R^3(3\pi/4 - 2) \end{bmatrix} \begin{bmatrix} P \\ F_x \\ F_y \end{bmatrix}$$

Example 3

A cantilever bow girder in the form of a quarter circle as shown in Fig. 10.16 is subjected to two couples along the x- and z-axis and a force normal to the plane of the structure. Derive the 'force–deflection' relationship. Include only bending and twisting deformations.

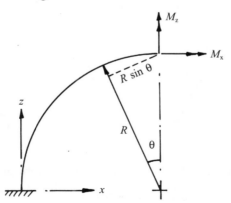

Figure 10.16 ●
A semicircular bow girder element

Solution The bending moment M and twisting moment T at a section θ from the vertical is given by

$$M = M_z \cos \theta - M_x \sin \theta - F_y R\sin \theta$$

$$T = M_z \sin \theta + M_x \cos \theta - F_y R(1 - \cos \theta)$$

Note that as shown in Fig. 10.16, the lever arms for the force F_y are $R(1 - \cos \theta)$ and $R \sin \theta$ for determining T and M respectively.

The complementary energy is given by

$$C = \int M^2 \, ds/(2EI) + \int T^2 \, ds/(2GJ)$$

Note that as in the case of the arch element, integration has to be carried out over the curved length of the member. Therefore an element of length of the bow girder is taken as ds, where $ds = R \, d\theta$.

Substituting for M and T

$$C = \int \{M_z \cos \theta - M_x \sin \theta - F_y R\sin \theta\}^2 \, R \, d\theta/(2EI)$$
$$+ \int \{M_z \sin \theta + M_x \cos \theta - F_y R(1 - \cos \theta)\}^2 \, R \, d\theta/(2GJ)$$

where the limits for θ are 0 and $\pi/2$.

Using Engesser's theorem,

$$\partial C/\partial M_z = \theta_z$$
$$= 1/(EI) \int \{M_z \cos \theta - M_x \sin \theta - F_y R \sin \theta\} \cos \theta \, R \, d\theta$$
$$+ 1/(GJ) \int \{M_z \sin \theta + M_x \cos \theta - F_y R(1 - \cos \theta)\} \sin \theta \, R \, d\theta$$
$$= R/(EI)\{\pi/4 \, M_z - M_x/2 - F_y R/2\} + R/(GJ)\{\pi/4 \, M_z + M_x/2 - F_y R/2\}$$

Similarly

$$\partial C/\partial M_x = \theta_x = 1/(EI) \int \{M_z \cos \theta - M_x \sin \theta - F_y R \sin \theta\}(-\sin \theta) \, R \, d\theta$$
$$+ 1/(GJ) \int \{M_z \sin \theta + M_x \cos \theta - F_y R(1 - \cos \theta)\}(\cos \theta) \, R \, d\theta$$
$$= R/(EI) \int \{-M_z/2 + \pi/4 \, M_x + \pi/4 \, F_y R\} + R/(GJ)\{M_z/2 + \pi/4 \, M_x$$
$$- F_y R(1 - \pi/4)\}$$

$$\partial C/\partial F_y = \Delta_y$$
$$= 1/(EI) \int \{M_z \cos \theta - M_x \sin \theta - F_y R\sin \theta\}(-R\sin \theta) \, R \, d\theta$$
$$+ 1/(GJ) \int \{M_z \sin \theta + M_x \cos \theta - F_y R(1 - \cos \theta)\}(-R(1 - \cos \theta)) \, R \, d\theta$$
$$= R^2/(EI)\{-M_z/2 + \pi/4 \, M_x + F_y R(\pi/4)\}$$
$$+ R^2/(GJ)\{-M_z/2 - (1 - \pi/4) M_x + (3\pi/4 - 2) F_y R\}$$

The above three equations can be expressed in matrix form leading to the flexibility matrix relationship for the given structure:

$$\begin{bmatrix} \theta_z \\ \theta_x \\ \Delta_y \end{bmatrix} = R/(4EI) \begin{bmatrix} \pi(1 + \alpha) & 2(\alpha - 1) & -2R(1 + \alpha) \\ 2(\alpha - 1) & \pi(1 + \alpha) & \{\pi(1 + \alpha) - 4\alpha\}R \\ -2R(1 + \alpha) & \{\pi(1 + \alpha) - 4\alpha\}R & \{\pi + \alpha(3\pi - 8)\}R^2 \end{bmatrix} \begin{bmatrix} M_z \\ M_x \\ F_y \end{bmatrix}$$

where $\alpha = EI/(GJ)$

10.9 ● Reciprocal theorem and its corollaries

It was pointed out at various places in Chapter 6 that, for linear elastic structures, both the element and structural stiffness matrices are symmetrical. It will be shown in this section that symmetry was not something that happened to be true only for the structures considered but is true for all linear elastic structures. Before this important result is proved, it is necessary to establish an important theorem called the Maxwell–Betti theorem.

10.9.1 Maxwell–Betti theorem

Consider a linear elastic structure supported so as to prevent rigid body motion. Let it be subjected to two independent *systems of loads* W_I and W_{II}. Consider two separate load cases as follows:

1. Load system W_I only. Let the structure deflect $\Delta_{I,I}$ at the point of application of loads W_I and in the direction of the W_I loads. Similarly, let the deflection at the point of application of W_{II} and in the direction of W_{II} loads be $\Delta_{I,II}$.
2. Load system W_{II} only. Let the structure deflect $\Delta_{II,II}$ at the point of application of loads W_{II} and in the direction of the W_{II} loads. Similarly, let the deflection at the point of application of W_I and in the direction of W_I loads be $\Delta_{II,I}$.

Because the structure is linearly elastic, the principle of superposition holds good. Therefore the final stress–strain state of the structure and hence its strain energy is dependent only on the final state of external loads on the structure and not on their order of application.

Consider the following two separate orders of load application.

1. Load system W_I first, followed by system W_{II}. In this case initially loads W_I displace by $\Delta_{I,I}$. The external work done is the area under the load–deflection diagram which is equal to $0.5W_I\Delta_{I,I}$. However, when W_{II} loads are applied, W_{II} loads displace by $\Delta_{II,II}$ doing an external work of $0.5W_{II}\Delta_{II,II}$. But as W_I loads are still on the structure W_I loads displace by $\Delta_{II,I}$ doing an additional external work of $W_I\Delta_{II,I}$. Note that the 0.5 term does not occur for work done by the W_I load when W_{II} loads are applied because only the displacement increases gradually but the loads are constant. This is clear from the load–displacement diagram in Fig. 10.17(a).

    Therefore the total external work done, which is equal to the strain energy U_1 stored in the structure is

    $$U_1 = 0.5W_I\Delta_{I,I} + 0.5W_{II}\Delta_{II,II} + W_I\Delta_{II,I}$$

2. Load W_{II} first, followed by W_I. Following the arguments for the first order of load application, it is easily shown that the strain energy U_2 stored in this case is (see Fig. 10.17(d))

    $$U_2 = 0.5W_{II}\Delta_{II,II} + 0.5W_I\Delta_{I,I} + W_{II}\Delta_{I,II}$$

Because in both systems of loading, the structure supports loads $W_I + W_{II}$, the final strain energies in the structure must be equal. Equating $U_1 = U_2$, it is seen that

$$W_I\Delta_{II,I} = W_{II}\Delta_{I,II}$$

This is the Maxwell–Betti theorem, which can be stated as follows.

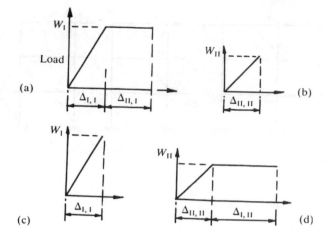

Figure 10.17 ●
(a)–(d) Force–
displacement
relationships

If a linearly elastic structure so supported as to prevent rigid body motion is subjected to two systems of loads W_I and W_{II}, then the work done by the W_I system of loads moving through the displacement caused by the W_{II} system of loads is equal to the work done by the W_{II} system of loads moving through the displacement caused by the W_I system of loads.

The following two corollaries can now be established.

10.9.2 Structural stiffness and flexibility matrices are symmetrical

(a) Let the stiffness relationship be given by

$$
\begin{bmatrix} \vdots \\ W_i \\ \vdots \\ W_j \\ \vdots \end{bmatrix} = \begin{bmatrix} \cdot & \cdot & \cdot & \cdot & \cdot & \cdot & \cdot \\ \cdot & \cdot & s_{ii} & \cdot & \cdot & s_{ij} & \cdot \\ \cdot & \cdot & \cdot & \cdot & \cdot & \cdot & \cdot \\ \cdot & \cdot & s_{ji} & \cdot & \cdot & s_{jj} & \cdot \\ \cdot & \cdot & \cdot & \cdot & \cdot & \cdot & \cdot \end{bmatrix} \begin{bmatrix} \vdots \\ \Delta_i \\ \vdots \\ \Delta_j \\ \vdots \end{bmatrix}
$$

$$\{W\} = [S]\{\Delta\}$$

where Δ and W are displacement and force vectors respectively and S is the structure stiffness matrix.

Let a set of forces $\{W_I\}$ be applied such that only the displacement Δ_i results. Typical forces W_j and W_i of the system are given by

$$W_j = s_{ji}\Delta_i, \quad W_i = s_{ii}\Delta_i$$

Similarly, let a set of forces $\{W_{II}\}$ be applied such that only the displacement Δ_j results. Typical forces W_i and W_j of the system are given by

$$W_i = s_{ij}\Delta_j, \quad W_j = s_{jj}\Delta_j$$

Using the Maxwell–Betti theorem

$$W_j\Delta_j + \sum W_i 0 = W_i\Delta_i + \sum W_j 0$$

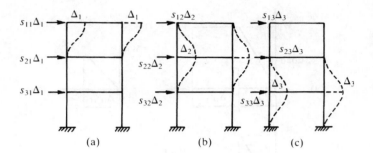

Figure 10.18 ● (a)–(c)
A 2-D rigid-jointed
structure under three
independent systems
of loading

Therefore

$$s_{ji}\Delta_i\Delta_j = s_{ij}\Delta_j\Delta_i$$

and

$$s_{ji} = s_{ij}$$

Therefore the stiffness matrix is symmetrical.

As an example consider the structure shown in Fig. 10.18. The stiffness relationship is given by

$$\begin{bmatrix} W_1 \\ W_2 \\ W_3 \end{bmatrix} = \begin{bmatrix} s_{11} & s_{12} & s_{13} \\ s_{21} & s_{22} & s_{23} \\ s_{31} & s_{32} & s_{33} \end{bmatrix} \begin{bmatrix} \Delta_1 \\ \Delta_2 \\ \Delta_3 \end{bmatrix}$$

If $\Delta_2 = \Delta_3 = 0$, then from the stiffness matrix $W_1 = s_{11}\Delta_1$, $W_2 = s_{21}\Delta_1$ and $W_3 = s_{31}\Delta_1$. Fig. 10.18(a) shows the forces and the associated displacement Δ_1.

Similarly, if $\Delta_1 = \Delta_3 = 0$, then $W_1 = s_{12}\Delta_2$, $W_2 = s_{22}\Delta_2$ and $W_3 = s_{32}\Delta_2$ and Fig. 10.18(b) shows the forces and the associated displacement Δ_2. Finally, if $\Delta_1 = \Delta_2 = 0$, then $W_1 = s_{13}\Delta_3$, $W_2 = s_{23}\Delta_2$ and $W_3 = s_{33}\Delta_3$. Fig. 10.18(c) shows the forces and the associated displacement Δ_3.

If the forces in Fig. 10.18(a) are considered as system I forces and forces in Fig. 10.18(b) are considered as system II forces, then

$$W_I\Delta_{II,I} = s_{11}\Delta_1 \times (0) + s_{21}\Delta_1 \times (\Delta_2) + s_{31}\Delta_1 \times (0) = s_{21}\Delta_1\Delta_2$$

$$W_{II}\Delta_{I,II} = s_{12} \times (\Delta_1) + s_{22}\Delta_2 \times (0) + s_{32}\Delta_2 \times (0) = s_{12}\Delta_2\Delta_1$$

Therefore, according to the reciprocal theorem, $s_{21}\Delta_1\Delta_2 = s_{12}\Delta_2\Delta_1$. Therefore

$$s_{12} = s_{21}$$

Similarly, if the forces in Fig. 10.18(a) are considered as system I forces and forces in Fig. 10.18(c) are considered as system II forces, then

$$W_I\Delta_{II,I} = s_{11}\Delta_1 \times (0) + s_{21}\Delta_1 \times (0) + s_{31}\Delta_1 \times (\Delta_3) = s_{31}\Delta_1\Delta_3$$

$$W_{II}\Delta_{I,II} = s_{13}\Delta_3 \times (\Delta_1) + s_{23}\Delta_3 \times (0) + s_{33}\Delta_3 \times (0) = s_{13}\Delta_3\Delta_1$$

Therefore, according to the reciprocal theorem, $s_{31}\Delta_1\Delta_3 = s_{13}\Delta_3\Delta_1$, and

$$s_{31} = s_{13}$$

Similarly, considering the forces and displacements shown in Fig. 10.18(b) and (c), it can be shown that $s_{23} = s_{32}$.

(b) The flexibility relationship for a linear elastic structure can be expressed as

$$
\begin{bmatrix} \cdot \\ \cdot \\ \cdot \\ \Delta_i \\ \cdot \\ \cdot \\ \Delta_j \\ \cdot \end{bmatrix} =
\begin{bmatrix}
\cdot & \cdot & \cdot & \cdot & \cdot & \cdot & \cdot & \cdot \\
\cdot & \cdot & \cdot & \cdot & \cdot & \cdot & \cdot & \cdot \\
\cdot & \cdot & f_{ii} & \cdot & \cdot & f_{ij} & \cdot & \\
\cdot & \cdot & & \cdot & \cdot & & \cdot & \\
\cdot & \cdot & & \cdot & \cdot & & \cdot & \\
\cdot & \cdot & f_{ji} & \cdot & \cdot & f_{jj} & \cdot & \\
\cdot & \cdot & \cdot & \cdot & \cdot & \cdot & \cdot & \cdot
\end{bmatrix}
\begin{bmatrix} \cdot \\ \cdot \\ W_i \\ \cdot \\ \cdot \\ W_j \\ \cdot \end{bmatrix}
$$

$$\{\Delta\} = [F]\{W\}$$

where Δ and W are displacement and force vectors respectively and F is the structure flexibility matrix.

If, for example, system I loads consist only of force W_i, then the displacement Δ_j in the direction of W_j is given by

$$\Delta_j = f_{ij} W_i$$

Similarly, if system II loads consists only of W_j, then the displacement Δ_i in the direction of W_i is given by

$$\Delta_i = f_{ij} W_j$$

Using the Maxwell–Betti theorem $W_i \Delta_i = W_j \Delta_j$. Substituting for Δ_i and Δ_j we have

$$W_i f_{ij} W_j = W_j f_{ji} W_i$$

or

$$f_{ij} = f_{ji}$$

Therefore the flexibility matrix is symmetrical.

10.9.3 Muller–Breslau principle

The reciprocal theorem leads to a very convenient procedure for visualizing the shape of an influence line. The ideas involved are best explained by means of simple examples.

1. Influence line for reaction. Consider a continuous beam as shown in Fig. 10.19(a). It is required to determine the shape of the influence line for reaction at C. Let a unit load act at X at a distance x from the left-hand end and the reaction at C for this load position be R_c. The external load W and the reactions at the supports form system I loading. Consider another system of loading where an external load is applied at C such that the deflection at C and X are Δ_c and Δ_x respectively as shown in Fig. 10.19(b). The applied load at C together with the reactions at the supports form system II loading. Applying the reciprocal theorem we have

$$W_I \Delta_{II,I} = W \Delta_x + R_c(-\Delta_c) + \text{(reaction from system I)(zero displacement)}$$

$$W_{II} \Delta_{I,II} = \text{applied load at C} \times 0 + \text{(reactions from system)(zero displacement)}$$

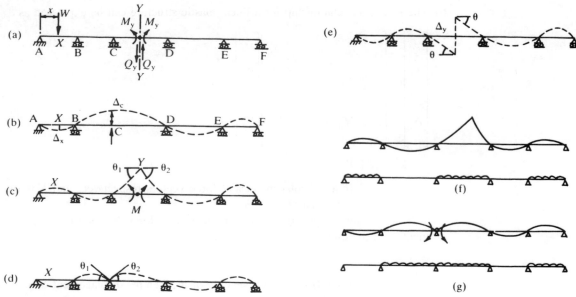

Figure 10.19 ● Influence lines for forces in a continuous beam: (a) Beam with a moving load at x from LHS, (b) influence line for a vertical reaction, (c) influence line BM in a span section, (d) influence line for BM at a support section, (e) influence line for SF at a span section. (f) Determination of load positions for maximum bending moment at a span section and (g) at a support section

Therefore

$$W\Delta_x + R_c(-\Delta_c) = 0$$

and

$$R_c = W(\Delta_x/\Delta_c)$$

If $W = 1$ and $\Delta_c = 1$, then $R_c = \Delta_x$. What the above equation implies is that if the influence line for the reaction R_c is required, then apply a force in the direction of R_c such that the deflection Δ_c in the direction of R_c is unity. The deflected shape of the structure gives the influence line for the reaction at C. This idea is extremely useful in practice because in the majority of cases we are not interested in the ordinates of the influence line. All we are interested in is determining which spans should be fully loaded with dead and live load to obtain the maximum reaction. It is clear from the influence line given by the deflected shape of the structure that the maximum value of the reaction is obtained by fully loading the spans adjacent to the reaction and every alternate span. The maximum reaction is not obtained by loading all the spans.

2. Influence line for bending moment. It is required to obtain the influence line for the bending moment at section Y. Imagine that a hinge has been inserted at Y and apply a pair of equal and opposite couples as shown in Fig. 10.19(c). Let the deflection at x be Δ_x and rotation of the beam on either side of the hinge be θ_1 and θ_2. Using the reciprocal theorem on the system in Fig. 10.19(a) as system I and the system in Fig. 10.19(c) as system II, we have

$$W_I \Delta_{II,I} = W(-\Delta_x) + M_y(\theta_1) + (-M_y)(-\theta_2) - (\text{reactions} \times \text{zero deflections})$$

$$W_{II} \Delta_{I,II} = M(-\theta_y) + M(\theta_y) - (\text{reactions} \times \text{zero deflections})$$

Note that because of continuity of displacement there is only one value of rotation θ_y at Y in system I. Therefore

$$W(-\Delta_x) + M_y(\theta_1 + \theta_2) = 0$$

or

$$M_y = W[\Delta_x/(\theta_1 + \theta_2)]$$

If $W = 1$, then the deflection Δ_x gives the influence line for the bending moment at Y–Y to some unspecified scale. But as was remarked before, if we are not interested in the absolute values of the influence ordinates then the deflected shape gives the shape of the influence line. The deflected shape will give the actual influence line if $(\theta_1 + \theta_2) = 1$.

The procedure for obtaining the shape of the influence line for the bending moment involves inserting a hinge at the section where the influence line for the bending moment is required and applying at the hinge equal and opposite couples. The resulting deflected shape of the structure is the desired influence line.

The shape of the influence line for the bending moment at a section in the span of the beam is shown in Fig. 10.19(c). It clearly indicates that, for maximum moment causing tension at the bottom face, it is necessary to fully load the span under consideration and every alternate span, as shown in Fig. 10.19(f).

Similarly, the shape of the influence line for the bending moment at a support section of the beam is shown in Fig. 10.19(d). It clearly indicates that, for maximum moment causing tension at the top face, it is necessary to fully load the spans adjacent to the support section under consideration and every alternate span, as shown in Fig. 10.19(f).

3. Influence line for shear force. The influence line for shear force at a section can also be obtained in a similar fashion. A 'sliding hinge' is introduced at the section where the influence line for shear force is required and equal and opposite forces and couples are applied such that the net relative displacement at the cut section is Δ_y but there is *no relative rotation* between the two ends of the cut section.

Proceeding as in (2) above, and using the two systems of loading shown in Fig. 10.19(a) and (e), it can be shown that

$$Q_y = W(\Delta_x/\Delta_y)$$

Therefore if $W = 1$ and $\Delta_y = 1$, the deflected shape gives the influence line for shear force.

10.9.4 Numerical evaluation of the influence line for statically determinate structures

As discussed in Chapter 2, influence lines for statically determinate structures consist of straight lines and are easily calculated using Muller–Breslau's principle, as shown by the following example.

Example 1

Figure 10.20(a) shows a balanced cantilever structure. Draw the influence lines for

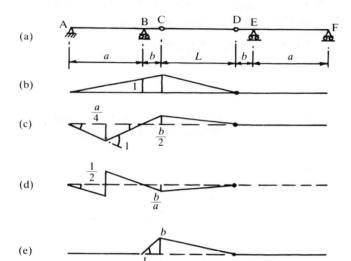

(a)

(b)

(c)

(d)

Figure 10.20 ● (a)–(e)
Influence lines for forces
in a balanced cantilever
structure

(e)

(a) reaction at support B;
(b) midspan bending moment in AB;
(c) shear force at midspan of AB;
(d) support bending moment at B.

This is the same structure as considered in Section 2.8 in Chapter 2.

Solution   *(a) Reaction at B:* Give a vertical displacement of unity at B. The structure displaces as shown in Fig. 10.20(b). This is the influence line for the reaction at B.

(b) Midspan bending moment in AB: Insert a hinge at the midspan of AB and create a total angle change at the hinge of unity as shown in Fig. 10.20(c). This gives the influence line for the bending moment at the midspan of AB.

(c) Shear force at midspan of AB: Cut the beam at midspan of AB and displace the cut ends so as to give a unit relative displacement between the ends but preserving continuity of rotation at the cut section. The displaced form of the structure is given in Fig. 10.20(d). Note that the rotation at the cut ends is $\theta = 0.5/a$ on both sides of the cut, thus preserving the continuity of rotation with respect to rotations.

(d) Support bending moment at B: Introduce a hinge at B and create a total angle change at the hinge of unity. The displaced form of the structure is shown in Fig. 10.20(e), which is the required influence line.

10.9.5 Numerical evaluation of the influence line for statically indeterminate structures

The influence lines for statically indeterminate structures invariably consist of curved lines. The procedure to be adopted for numerical evaluation of influence lines can be explained with respect to the example in Fig. 10.21.

Consider the calculation of the influence line for the vertical reaction at the middle. It is necessary to calculate the deflected shape due to the application of a vertical force at the position of the reaction so as to create a vertical displacement of unity. The operation can be accomplished in two stages as follows.

1. Clamp all the supports and apply a vertical force so as to create a vertical displacement at the middle support of unity as shown in Fig. 10.21(b). Calculate the resulting fixed end forces.

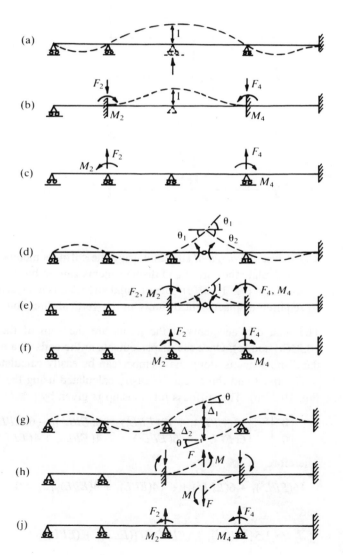

Figure 10.21 ●
(a)–(j) Calculation of influence lines for a continuous beam

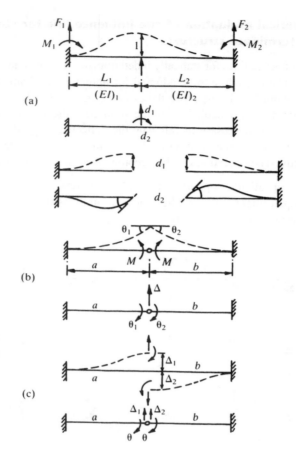

Figure 10.22 ● (a)–(c) Fixed end forces for the generation of influence lines for vertical reaction, bending moment at a span section and shear force at a span section

2. Analyse the *original* structure using the stiffness method explained in Chapter 6 to calculate the forces and displacements caused by the *removal* of restraints as shown in Fig. 10.21(c). The original structure is analysed so as not to disturb the required displacement of unity in the vertical direction.

The final displacements at the joints are the sum of the displacements in Figs 10.21(b) and (c). Once the displacements at the ends of a member are known, then the displacements along the member can be easily calculated.

The fixed end forces can be easily calculated using the 'substructure' shown in Fig. 10.22(a). The stiffness relationship is given by

$$\begin{bmatrix} W \\ 0 \end{bmatrix} = \begin{bmatrix} 12(EI/L^3)_1 + 12(EI/L^3)_2 & 6(EI/L^2)_1 - 6(EI/L^2)_2 \\ 6(EI/L^2)_1 - 6(EI/L^2)_2 & 4(EI/L)_1 + 4(EI/L)_2 \end{bmatrix} \times \begin{bmatrix} d_1 = 1 \\ d_2 \end{bmatrix}$$

Therefore

$$\{6(EI/L^2)_1 - 6(EI/L^2)_2\} + \{4(EI/L)_1 + 4(EI/L)_2\}d_2 = 0$$

and

$$d_2 = -1.5\{(EI/L^2)_1 - (EI/L^2)_2\}/[(EI/L)_1 + (EI/L)_2]$$

The fixed end forces are

$$F_1 = -12(EI/L^3)_1 - 6(EI/L^2)_1 \, d_2$$

$$M_1 = 6(EI/L^2)_1 + 2(EI/L)_1 \, d_2$$

$$F_2 = -12(EI/L^3)_2 + 6(EI/L^2)_2 \, d_2$$

$$M_2 = 6(EI/L^2)_2 - 2(EI/L)_2 \, d_2$$

Similar calculations can be done for the influence lines for the bending moment and shear force at a span section as shown in Fig. 10.21(d)–(f) and (g)–(j) respectively. The necessary fixed end forces can be calculated as follows.

(a) Fixed end forces for influence line calculation for the bending moment at a span section. Considering the 'substructure' shown in Fig. 10.22(b), the stiffness relationship is given by

$$EI \begin{bmatrix} 4/a & 0 & 6/a^2 \\ 0 & 4/b & -6/b^2 \\ 6/a^2 & -6/b^2 & 12/a^3 + 12/b^3 \end{bmatrix} \begin{bmatrix} \theta_1 \\ \theta_2 \\ \Delta \end{bmatrix} = \begin{bmatrix} -M \\ M \\ 0 \end{bmatrix}$$

where a and b are the distances to the 'hinge' from the left and right support respectively. Note the discontinuity in slope but not of translation at the 'hinge'.
Solving:

$$EI \, \theta_1 = -(Ma/4)\{1 + 3b^2/C\}$$

$$EI \, \theta_2 = (Mb/4)\{1 + 3a^2/C\}$$

$$EI \, \Delta = (M/2)\{a^2b^2/C\}$$

where $C = a^2 - ab + b^2$. For $-\theta_1 + \theta_2 = 1$, $M = 4EI \, C/(a - b)^3$

$$\theta_1 = -a[a^2 - ab + 4b^2]/(a + b)^3$$

$$\theta_2 = b[4a^2 - ab + b^2]/(a + b)^3$$

$$\Delta = 2a^2b^2/(a - b)^3$$

Then $M_1 = (0.5M/C)\{-a^2 + ab + 2b^2\}$, $M_2 = (0.5M/C)\{2a^2 + ab + b^2\}$,

$$F_1 = F, \; F_2 = -F, \qquad F = 1.5(M/C)(b - a)$$

(b) Fixed end forces for influence line calculation for shear force at a span section. Considering the 'substructure' shown in Fig. 10.22(c), the stiffness relationship is given by

$$EI \begin{bmatrix} 4/a + 4/b & 6/a^2 & -6/b^2 \\ 6/a^2 & 12/a^3 & 0 \\ -6/b^2 & 0 & 12/b^3 \end{bmatrix} \begin{bmatrix} \theta \\ \Delta_1 \\ \Delta_2 \end{bmatrix} = \begin{bmatrix} 0 \\ F \\ -F \end{bmatrix}$$

Note the discontinuity of translation but not of rotation at the 'hinge'.
Solving:

$$EI \, \Delta_1 = (Fa^3/12)\{1 + 3b/a\}$$

$$EI \, \Delta_2 = -(Fb^3/12)\{1 + 3a/b\}$$

$$EI \, \theta = -(F/2)\{ab\}$$

where Δ_1 and Δ_2 are considered positive upwards.

For $\Delta_1 - \Delta_2 = 1$, $F = 12EI/(a + b)^3$

$$\Delta_1 = a^3[1 + 3b/a]/(a + b)^3$$

$$\Delta_2 = -b^3[1 + 3a/b]/(a + b)^3$$

$$\theta = -6ab/(a + b)^3$$

Then $M_1 = 0.5F(a + b)$, $M_2 = -0.5F(a + b)$, $F_1 = -F$, $F_2 = +F$. It should be noted that although a continuous beam was used as an example, the above procedure is applicable to all statically indeterminate structures.

Examples of numerical evaluation of influence line coefficients

Example 1

As an example of the use of the above equations, consider a three-span continuous beam. The details are as follows:

Span AB = 4 m, $EI = 4.93 \times 10^3$ kNm^2

Span BC = 6 m, $EI = 11.64 \times 10^3$ kNm^2

Span CD = 5 m, $EI = 7.15 \times 10^3$ kNm^2

The ends A and D are simply supported. Assuming that shearing and axial deformations are neglected, the stiffness matrix for the beam is given by:

$$10^3 \begin{bmatrix} 4 \times 4.93/4 & 2 \times 4.93/4 & 0 & 0 \\ 2 \times 4.93/4 & 4 \times 4.93/4 + 4 \times 11.64/6 & 2 \times 11.64/6 & 0 \\ 0 & 2 \times 11.64/6 & 4 \times 11.64/6 + 4 \times 7.15/5 & 2 \times 7.15/5 \\ 0 & 0 & 2 \times 7.15/5 & 4 \times 7.15/5 \end{bmatrix}$$

which simplifies to

$$10^3 \begin{bmatrix} 4.93 & 2.465 & 0 & 0 \\ 2.465 & 12.69 & 3.88 & 0 \\ 0 & 3.88 & 13.48 & 2.86 \\ 0 & 0 & 2.86 & 5.72 \end{bmatrix}$$

The displacement and load vectors are respectively

$$\begin{bmatrix} \theta_A \\ \theta_B \\ \theta_C \\ \theta_D \end{bmatrix} \quad \text{and} \quad \begin{bmatrix} \Sigma M_A \\ \Sigma M_B \\ \Sigma M_B \\ \Sigma M_D \end{bmatrix}$$

(a) Influence line for vertical reaction at B

L_1 = Span AB = 4 m, $(EI)_1 = 4.93 \times 10^3$

$(EI/L^3)_1 = 77.03$, $(EI/L^2)_1 = 3.0813 \times 10^2$, $(EI/L)_1 = 1.2325 \times 10^3$

L_2 = Span BC = 6 m, $(EI)_2 = 11.64 \times 10^3$

$(EI/L^3)_2 = 53.89$, $(EI/L^2)_2 = 3.2333 \times 10^2$, $(EI/L)_2 = 1.94 \times 10^3$

Step 1. Calculation of fixed forces due to unit displacement at the support. Substituting in the formulae,

$$d_2 = -1.5\{3.0813 \times 10^2 - 3.2333 \times 10^2\}/\{1.2325 \times 10^3 + 1.94 \times 10^3]$$

$$= 7.1868 \times 10^{-3}$$

The fixed end forces are:

$$F_1 = -12 \times 77.03 - 6 \times 3.0813 \times 10^2 \times 7.1868 \times 10^{-3} = -937.6468 \text{ kN}$$

$$M_1 = 6 \times 3.0813 \times 10^2 + 2 \times 1.2325 \times 10^3 \times 7.1868 \times 10^{-3}$$

$$= 1866.481 \text{ kN m (clockwise positive)}$$

$$F_2 = -12 \times 53.89 + 6 \times 3.2333 \times 10^2 \times 7.1868 \times 10^{-3} = -632.71 \text{ kN}$$

$$M_2 = 6 \times 3.2333 \times 10^2 - 2 \times 1.94 \times 10^3 \times 7.1868 \times 10^{-3}$$

$$= 1912.095 \text{ kN m (anticlockwise positive)}$$

Step 2. Remembering that in matrix analysis, the 'forces' at the joints are the opposite of the sum of fixed forces at the joint, and noting that M_1 is clockwise and M_2 is anticlockwise, the joint forces for matrix analysis are:

$$\sum M_A = 0, \ \sum M_B = -M_1 = -1866.481, \ \sum M_C = M_2 = 1912.095, \ \sum M_D = 0$$

Carrying out Gaussian elimination, the stiffness matrix and the load vector become:

$$10^3 \begin{bmatrix} 4.93 & 2.465 & 0 & 0 \\ 0 & 11.458 & 3.88 & 0 \\ 0 & 0 & 12.166 & 2.86 \\ 0 & 0 & 0 & 5.048 \end{bmatrix}$$

and

$$10^3 \begin{bmatrix} -1.8665 \\ 0.9333 \\ 1.5961 \\ -0.3752 \end{bmatrix}$$

Solving, the corresponding displacements become:

$$\{\theta_A, \theta_B, \theta_C, \theta_D\} = 10^{-3} \{-394.153, 31.095, 148.662, -74.331\}$$

Step 3. Calculation of final 'displacements' at the ends of members. Thus the final translations and rotations at the ends of members are:

Span AB:

End A: $\theta_1 = \theta_A = -0.3942, \ \Delta_{n1} = 0$

End B: $\theta_2 = \theta_B + d_2 = (31.1095 + 7.1868) \times 10^{-3} = 0.0383, \ \Delta_{n2} = 1$

Span BC:

End B: $\theta_1 = \theta_B + d_2 = (31.1095 + 7.1868) \times 10^{-3} = 0.0383, \ \Delta_{n1} = 1$

End C: $\theta_2 = \theta_C = 0.1487, \ \Delta_{n2} = 0$

Span CD:

End C: $\theta_1 = \theta_C = 0.1487$, $\Delta_{n1} = 0$

End D: $\theta_2 = \theta_D = -0.0743$, $\Delta_{n2} = 0$

Step 4. Calculation of influence line ordinates. As was shown in the section describing the application of Engesser's theorem, the deformation of a beam element in terms of the translations and rotations at the ends of the element is given by:

$$v = \Delta_{n1}\{1 - 3(x/L)^2 + 2(x/L)^3\} - \theta_1 L\{x/L - 2(x/L)^2 + (x/L)^3\}$$
$$+ \Delta_{n2}\{3(x/L)^2 - 2(x/L)^3\} - \theta_2 L\{(x/L)^3 - (x/L)^2\}$$

Thus the deflected shape of all the elements can be determined. The deformed shape of the continuous beam is the influence line for the vertical reaction at B. Table 10.6 gives the influence line ordinates in each span at 20 points along the span.

Table 10.6 ● Influence ordinates for reaction at B

Span AB		Span BC		Span CD	
x	Ordinate	x	Ordinate	x	Ordinate
0	0	4	1	10	0
0.2	0.078759	4.3	0.984498	10.25	−0.03442
0.4	0.157085	4.6	0.961412	10.5	−0.06355
0.6	0.234546	4.9	0.931402	10.75	−0.08766
0.8	0.310709	5.2	0.895125	11	−0.10704
1	0.385143	5.5	0.853242	11.25	−0.12195
1.2	0.457415	5.8	0.80641	11.5	−0.13268
1.4	0.527091	6.1	0.755288	11.75	−0.13951
1.6	0.593741	6.4	0.700534	12	−0.14272
1.8	0.656931	6.7	0.642808	12.25	−0.14258
2	0.716229	7	0.582768	12.5	−0.13937
2.2	0.771202	7.3	0.521073	12.75	−0.13338
2.4	0.821418	7.6	0.458381	13	−0.12488
2.6	0.866445	7.9	0.39535	13.25	−0.11414
2.8	0.905849	8.2	0.332641	13.5	−0.10146
3	0.9392	8.5	0.27091	13.75	−0.08711
3.2	0.966063	8.8	0.210818	14	−0.07136
3.4	0.986008	9.1	0.153022	14.25	−0.05449
3.6	0.9986	9.4	0.098181	14.5	−0.03679
3.8	1.003408	9.7	0.046954	14.75	−0.01854
4	1	10	0	15	0

(b) Influence line for midspan bending moment in span BC

Step 1. Calculation of fixed end forces. Because the influence line is required at midspan of span BC,

$EI = 11.64 \times 10^3$ kN m^2, $a = b = 6/2 = 3$ m

Using the formulae previously derived,

$$C = a^2 - ab + b^2 = 9$$

$$M = 4EIC/(a + b)^3 = 4 \times 11.63 \times 10^3 \times 9/(6^3) = 1.9383 \times 10^3 \text{ kN m}$$

$$M_1 = \{0.5M/C\}[-a^2 + ab + 2b^2] = 1.9383 \times 10^3 \text{ (clockwise positive)}$$

$$M_2 = \{0.5M/C\}[2a^2 + ab - b^2] = 1.9383 \times 10^3 \text{ (anticlockwise positive)}$$

$$F_1 = -F_2 = 1.5(M/C)\{b - a\} = 0$$

$$\theta_1 = -a[a^2 - ab + 4b^2]/(a + b)^3 = -0.5$$

$$\theta_2 = b[4a^2 - ab + b^2]/(a + b)^3 = 0.5$$

$$\Delta = 2a^2b^2/(a + b)^3 = 0.75$$

Step 2. The joint forces for matrix analysis are:

$$\sum M_A = 0, \sum M_B = -M_1 = -1.9383 \times 10^3, \sum M_C = M_2 = 1.9383 \times 10^3, \sum M_D = 0$$

Carrying out Gaussian elimination, the load vector becomes:

$$10^3 \begin{bmatrix} 0 \\ -1.9383 \\ 2.5946 \\ -0.6100 \end{bmatrix}$$

Solving, the corresponding displacements become:

$$\{\theta_A, \theta_B, \theta_C, \theta_D\} = \{0.1255, -0.2509, 0.2414, -0.1209\}$$

Step 3. Calculation of final 'displacements' at the ends of members. Thus the final translations and rotations at the ends of members are:

Span AB:

End A: $\theta_1 = \theta_A = 0.1255$, $\Delta_{n1} = 0$

End B: $\theta_2 = \theta_B = -0.2509$, $\Delta_{n2} = 0$

Span BC:

End B: $\theta_1 = \theta_B = -0.2509$, $\Delta_{n1} = 0$

End C: $\theta_2 = \theta_C = 0.2414$, $\Delta_{n2} = 0$

In addition to the above displacements caused by the overall deformation of the structure, we have to superpose local displacements caused in the span BC only when a hinge was inserted at the midspan and equal and opposite moments were applied at the hinge.

From B to midspan of BC:

End B: $\theta_B = 0$, $\Delta_{n1} = 0$

Midspan end of BC: $\theta = \theta_1 = -0.5$, $\Delta_{n2} = 0.75$

From midspan of BC to C:

Midspan end of BC: $\theta = \theta_2 = 0.5$, $\Delta_{n1} = 0.75$

End C: $\theta = 0$, $\Delta_{n2} = 0$

Span CD:

End C: $\theta_1 = \theta_C = 0.2414$, $\Delta_{n1} = 0$

End D: $\theta_2 = \theta_D = -0.1209$, $\Delta_{n2} = 0$

The influence ordinates are shown in Table 10.7.

Table 10.7

Span AB		Span BC				Span CD	
x	Coefficient	x	Overall deformation	Local deformation	Coefficient	x	Coefficient
0.00	0.0000	4.00	0.0000	0.0000	0.0000	10.00	0.0000
0.20	−0.0250	4.30	0.0714	0.0098	0.0811	10.25	−0.0559
0.40	−0.0497	4.60	0.1350	0.0380	0.1730	10.50	−0.1032
0.60	−0.0736	4.90	0.1908	0.0833	0.2741	10.75	−0.1424
0.80	−0.0964	5.20	0.2390	0.1440	0.3830	11.00	−0.1738
1.00	−0.1176	5.50	0.2796	0.2188	0.4983	11.25	−0.1981
1.20	−0.1370	5.80	0.3125	0.3060	0.6185	11.50	−0.2155
1.40	−0.1541	6.10	0.3379	0.4043	0.7422	11.75	−0.2266
1.60	−0.1686	6.40	0.3558	0.5120	0.8678	12.00	−0.2318
1.80	−0.1801	6.70	0.3662	0.6278	0.9940	12.25	−0.2316
2.00	−0.1882	7.00	0.3692	0.7500	1.1192	12.50	−0.2264
2.20	−0.1925	7.30	0.3648	0.6278	0.9926	12.75	−0.2167
2.40	−0.1927	7.60	0.3531	0.5120	0.8651	13.00	−0.2029
2.60	−0.1884	7.90	0.3340	0.4043	0.7383	13.25	−0.1855
2.80	−0.1792	8.20	0.3078	0.3060	0.6138	13.50	−0.1649
3.00	−0.1647	8.50	0.2742	0.2188	0.4930	13.75	−0.1416
3.20	−0.1445	8.80	0.2336	0.1440	0.3776	14.00	−0.1160
3.40	−0.1184	9.10	0.1858	0.0833	0.2690	14.25	−0.0886
3.60	−0.0858	9.40	0.1309	0.0380	0.1689	14.50	−0.0598
3.80	−0.0465	9.70	0.0689	0.0098	0.0787	14.75	−0.0301
4.00	0.0000	10.00	0.0000	0.0000	0.0000	15.00	0.0000

(c) Influence line for shear force at midspan in span BC

Step 1. Calculation of fixed end forces. Because the influence line is required at midspan of span BC,

$EI = 11.64 \times 10^3$ kN m^2, $a = b = 6/2 = 3$ m

Using the formulae previously derived,

$$F = 12EI/(a + b)^3 = 12 \times 11.63 \times 10^3/(6^3) = 0.6461 \times 10^3 \text{ kN}$$

$$M_1 = 0.5F[a + b] = 1.9383 \times 10^3 \text{ kN m (clockwise positive)}$$

$$M_2 = -M_1 = -1.9383 \times 10^3 \text{ kN m (anticlockwise positive)}$$
$$F_1 = -F = -0.6461 \times 10^3, \; F_2 = F = 0.6461 \times 10^3 \text{ kN}$$
$$\Delta_1 = a^3[1 + 3b/a]/(a + b)^3 = 0.5$$
$$\Delta_2 = -b^3[1 + 3a/b]/(a + b)^3 = -0.5$$
$$\theta = -6ab/(a + b)^3 = -0.25$$

Step 2. The joint forces for matrix analysis are:

$$\sum M_A = 0, \sum M_B = -M_1 = -1.9383 \times 10^3,$$

$$\sum M_C = M_2 = 1.9383 \times 10^3, \sum M_D = 0.$$

Carrying out Gaussian elimination, the load vector becomes:

$$10^3 \begin{bmatrix} 0 \\ -1.9383 \\ -1.2819 \\ 0.3014 \end{bmatrix}$$

Solving, the corresponding displacements become:

$$\{\theta_A, \theta_B, \theta_C, \theta_D\} = \{0.0644, -0.1287, -0.11944, 0.0597\}$$

Step 3. Calculation of final 'displacements' at the ends of members. Thus the final translations and rotations at the ends of members are:

Span AB:

End A: $\theta_1 = \theta_A = 0.1255, \Delta_{n1} = 0$

End B: $\theta_2 = \theta_B = -0.2509, \Delta_{n2} = 0$

Span BC:

End B: $\theta_1 = \theta_B = -0.2509, \Delta_{n1} = 0$

End C: $\theta_2 = \theta_C = 0.2414, \Delta_{n2} = 0$

In addition to the above displacements caused by the overall deformation of the structure, we have to superpose local displacements caused in the span BC only when a hinge was inserted at the midspan and equal and opposite forces were applied at the hinge.

From B to midspan of BC:

End B: $\theta_B = 0, \Delta_{n1} = 0$

Midspan end of BC: $\theta = -0.25, \Delta_{n2} = 0.5$

From midspan of BC to C:

Midspan end of BC: $\theta = -0.25, \Delta_{n1} = -0.5$

End C: $\theta = 0, \Delta_{n2} = 0$

Table 10.8

Span AB		Span BC				Span CD	
x	Coefficient	x	Overall deformation	Local deformation	Coefficient	x	Coefficient
0.00	0.0000	4.00	0.0000	0.0000	0.0000	10.00	0.0000
0.20	−0.0128	4.30	0.0331	0.0084	0.0415	10.25	0.0276
0.40	−0.0255	4.60	0.0561	0.0320	0.0881	10.50	0.0510
0.60	−0.0378	4.90	0.0700	0.0686	0.1386	10.75	0.0704
0.80	−0.0494	5.20	0.0759	0.1160	0.1919	11.00	0.0860
1.00	−0.0604	5.50	0.0750	0.1719	0.2469	11.25	0.0979
1.20	−0.0703	5.80	0.0684	0.2340	0.3024	11.50	0.1066
1.40	−0.0791	6.10	0.0571	0.3001	0.3573	11.75	0.1120
1.60	−0.0865	6.40	0.0424	0.3680	0.4104	12.00	0.1146
1.80	−0.0924	6.70	0.0253	0.4354	0.4607	12.25	0.1145
2.00	−0.0966	7.00	0.0070	0.5000	0.5070	12.50	0.1119
2.00	−0.0966	7.00	0.0070	−0.5000	−0.4930	12.50	0.1119
2.20	−0.0988	7.30	−0.0115	−0.4354	−0.4469	12.75	0.1071
2.40	−0.0989	7.60	−0.0290	−0.3680	−0.3970	13.00	0.1003
2.60	−0.0966	7.90	0.0445	−0.3001	−0.3446	13.25	0.0917
2.80	−0.0919	8.20	−0.0567	−0.2340	−0.2907	13.50	0.0815
3.00	−0.0845	8.50	−0.0645	−0.1719	−0.2364	13.75	0.0700
3.20	−0.0741	8.80	−0.0670	−0.1160	−0.1830	14.00	0.0573
3.40	−0.0607	9.10	−0.0629	−0.0686	−0.1315	14.25	0.0438
3.60	−0.0440	9.40	−0.0511	−0.0320	−0.0831	14.50	0.0296
3.80	−0.0238	9.70	−0.0305	−0.0084	−0.0389	14.75	0.0149
4.00	0.0000	10.00	0.0000	0.0000	0.0000	15.00	0.0000

Span CD:

End C: $\theta_1 = \theta_C = 0.2414$, $\Delta_{n1} = 0$

End D: $\theta_2 = \theta_D = -0.1209$, $\Delta_{n2} = 0$

The influence ordinates are shown in Table 10.8.

10.10 ● Principle of contragredience

The concept of elastic strain energy can be used to establish a very useful principle known as the principle of contragredience.

Consider a structure for which the stiffness relationship is given by

$$\{W_I\} = [S_I]\{\Delta_I\}$$

where $\{W_I\}$, $[S_I]$ and $\{\Delta_I\}$ are the load vector, stiffness matrix and displacement vector respectively.

The strain energy U stored in the structure is given by the work done by the external loads which is given by

$$U = 0.5\{W_I\}^T\{\Delta_I\}$$

where $\{W_I\}^T$ is the transform of the vector $\{W_I\}$. This simply means that $\{W_I\}^T$ has the same elements as the vector $\{W_I\}$ except that it is a row instead of a column vector. This has been done to ensure that the rules of matrix multiplication are not violated.

Consider the same structure with the stiffness relationship expressed in a different system as

$$\{W_{II}\} = [S_{II}]\{\Delta_{II}\}$$

where $\{W_{II}\}$, $[S_{II}]$ and $\{\Delta_{II}\}$ are the load vector, stiffness matrix and displacement vector respectively.

The strain energy U stored in the structure is given by the work done by the external loads which is given by

$$U = 0.5\{W_{II}\}^T\{\Delta_{II}\}$$

where $\{W_{II}\}^T$ is the transform of the vector $\{W_{II}\}$.

From equilibrium considerations, let the two systems of loads $\{W_I\}$ and $\{W_{II}\}$ be related by

$$\{W_{II}\} = [R]\{W_I\}$$

In this case the strain energy stored in the structure under the two load systems must be identical. Therefore

$$U = 0.5\{W_{II}\}^T\{\Delta_{II}\} = 0.5\{W_I\}^T\{\Delta_I\}$$

Substituting for W_{II} in terms of W_I, we have

$$([R]\{W_I\})^T\{\Delta_{II}\} = \{W_I\}^T\{\Delta_I\}$$

Note that in matrix algebra, $([R]\{W_I\})^T = \{W_I\}^T[R]^T$. Therefore

$$\{W_I\}^T[R]^T\{\Delta_{II}\} = \{W_I\}^T\{\Delta_I\}$$

and

$$[R]^T\{\Delta_{II}\} = \{\Delta_I\}$$

The principle of contragredience is mathematically expressed as

If $\{W_{II}\} = [R]\{W_I\}$, then $\{\Delta_I\} = [R]^T\{\Delta_{II}\}$

Because $\{W_I\} = [S_I]\{\Delta_I\}$, substituting for $\{\Delta_I\}$ in terms of $\{\Delta_{II}\}$ as

$$\{\Delta_I\} = [R]^T\{\Delta_{II}\}$$

we have

$$\{W_I\} = [S_I][R]^T\{\Delta_{II}\}$$

Premultiplying both sides of the above equation by $[R]$, we have

$$[R]\{W_I\} = [R][S_I][R]^T\{\Delta_{II}\}$$

But

$$[R]\{W_I\} = \{W_{II}\}$$

Therefore

$$\{W_{II}\} = [R][S_I][R]^T\{\Delta_{II}\}$$

Because

$$\{W_{II}\} = [S_{II}]\{\Delta_{II}\}$$

then

$$[S_{II}] = [R][S_I][R]^T$$

Summarizing, if $\{W_I\} = [S_I]\{\Delta_I\}$ and $\{W_{II}\} = [R]\{W_I\}$ then

$$\{\Delta_I\} = [R]^T\{\Delta_{II}\} \text{ and } \{W_{II}\} = [S_{II}]\{\Delta_{II}\}$$

where $[S_{II}] = [R][S_I][R]^T$.

10.10.1 Examples of the use of the contragredience principle

Example 1

The stiffness relationship for a pin-jointed member is given in Section 4.1.2 by

$$\begin{bmatrix} F_{a1} \\ F_{a2} \end{bmatrix} = (AE/L)\begin{bmatrix} 1 & -1 \\ -1 & 1 \end{bmatrix}\begin{bmatrix} \Delta_{a1} \\ \Delta_{a2} \end{bmatrix}$$

This is the stiffness relationship in 'member axes', i.e.

$$\{W_I\} = [S_I]\{\Delta_I\}$$

In order to obtain the more general relationship with forces and displacements along the orthogonal x–y system, as shown in Section 6.3, we have

From equilibrium considerations as shown in Fig. 6.3, $F_{xi} = F_{ai}l$, $F_{yi} = F_{ai}m$, where $i = 1, 2$. Therefore

$$\begin{bmatrix} F_{x1} \\ F_{y1} \\ F_{x2} \\ F_{y2} \end{bmatrix} = \begin{bmatrix} l & 0 \\ m & 0 \\ 0 & l \\ 0 & m \end{bmatrix}\begin{bmatrix} F_{a1} \\ F_{a2} \end{bmatrix}$$

i.e. $\{W_{II}\} = \{R\}\{W_I\}$. Therefore

$$\{R\} = \begin{bmatrix} l & 0 \\ m & 0 \\ 0 & l \\ 0 & m \end{bmatrix}$$

Similarly from geometrical considerations, as shown in Fig. 6.3,

$$\begin{bmatrix} \Delta_{a1} \\ \Delta_{a2} \end{bmatrix} = \begin{bmatrix} l & m & 0 & 0 \\ 0 & 0 & l & m \end{bmatrix}\begin{bmatrix} u_1 \\ v_1 \\ u_2 \\ v_2 \end{bmatrix}$$

i.e. $\{\Delta_I\} = \{R\}^T\{\Delta_{II}\}$. Note that

$$\begin{bmatrix} l & m & 0 & 0 \\ 0 & 0 & l & m \end{bmatrix} = \{R\}^T$$

We have shown in this case that if $\{W_{II}\} = [R]\{W_I\}$ from considerations of equilibrium, then from purely geometrical considerations, $\{\Delta_I\} = [R]^T\{\Delta_{II}\}$. The general stiffness matrix can be obtained by carrying out the matrix multiplication

$$[S_{II}] = \begin{bmatrix} l & 0 \\ m & 0 \\ 0 & l \\ 0 & m \end{bmatrix} (AE/L) \times \begin{bmatrix} 1 & -1 \\ -1 & 1 \end{bmatrix} \begin{bmatrix} l & m & 0 & 0 \\ 0 & 0 & l & m \end{bmatrix}$$

This will yield the stiffness matrix shown in equation [6.1].

Example 2

As shown in Fig. 6.4, for an element of a 2-D rigid-jointed structure, the relationship between the forces in the normal and axial directions and their corresponding components in the orthogonal x–y system is given by

$$\begin{bmatrix} F_{x1} \\ F_{y1} \\ M_1 \\ F_{x2} \\ F_{y2} \\ M_2 \end{bmatrix} = \begin{bmatrix} l & -m & 0 & 0 & 0 & 0 \\ m & l & 0 & 0 & 0 & 0 \\ 0 & 0 & 1 & 0 & 0 & 0 \\ 0 & 0 & 0 & l & -m & 0 \\ 0 & 0 & 0 & m & l & 0 \\ 0 & 0 & 0 & 0 & 0 & 1 \end{bmatrix} \begin{bmatrix} F_{a1} \\ F_{n1} \\ M_1 \\ F_{a2} \\ F_{n2} \\ M_2 \end{bmatrix}$$

i.e. $\{W_{II}\} = [R]\{W_I\}$

Note that the relationship $M_1 = M_1$ and $M_2 = M_2$ in the third and sixth rows has been added for completeness. The couples M_1 and M_2 are vectors normal to the x–y plane and therefore remain unaffected by rotation of axes in the x–y plane.

The relationship between the displacements in the x–y system and their corresponding components in the axial and normal directions are given from geometrical considerations by

$$\begin{bmatrix} \Delta_{a1} \\ \Delta_{n1} \\ \theta_1 \\ \Delta_{a2} \\ \Delta_{n2} \\ \theta_2 \end{bmatrix} = \begin{bmatrix} l & m & 0 & 0 & 0 & 0 \\ -m & l & 0 & 0 & 0 & 0 \\ 0 & 0 & 1 & 0 & 0 & 0 \\ 0 & 0 & 0 & l & m & 0 \\ 0 & 0 & 0 & -m & l & 0 \\ 0 & 0 & 0 & 0 & 0 & 1 \end{bmatrix} \begin{bmatrix} u_1 \\ v_1 \\ \theta_1 \\ u_2 \\ v_2 \\ \theta_2 \end{bmatrix}$$

i.e. $\{\Delta_I\} = [R]^T\{\Delta_{II}\}$

Note that the matrices relating the two sets of forces and two sets of displacements are transpose of each other.

Once again we have shown in this case that if $\{W_{II}\} = [R]\{W_I\}$ from considerations of equilibrium, then from purely geometrical considerations, $\{\Delta_I\} = [R]^T\{\Delta_{II}\}$.

The general stiffness matrix shown in equation [6.2] can be obtained by carrying out the matrix multiplication $[S_{II}] = \{R\}[S_I]\{R\}^T$, where $[S_I]$ is the matrix given in Chapter 6, Section 6.2.

Example 3

The beam shown in Fig. 10.23 rests on an inclined support on the right-hand side. Establish the stiffness relationship for the beam and determine the displacements at the supports and the forces in the beam. Assume $E = 210 \times 10^6$ kN m^{-2}, $G = 80.77 \times 10^6$ kN m^{-2}, $I = 55.44 \times 10^{-6}$ m^4, $A = 4740 \times 10^{-6}$ m^2, $ASF = 4.116$, $L = 3$ m, $\beta = 12EI/(GA_s) = 0.167$.

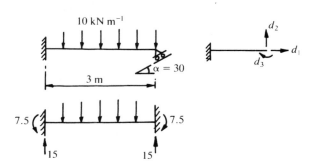

Figure 10.23 ●
Analysis of a beam on
an inclined support

Solution (a) Let the displacements at end 2 be d_1, d_2 and d_3 as shown in Fig. 10.23. The element stiffness matrix is given by

$$\begin{bmatrix} F_{x1} = * \\ F_{y1} = * \\ M_1 = * \\ F_{x2} = F_1 \\ F_{y2} = F_2 \\ M_2 = F_3 \end{bmatrix} = 10^3 \begin{bmatrix} 331.80 & 0 & 0 & -331.80 & 0 & 0 \\ 0 & 4.43 & -6.65 & 0 & -4.43 & -6.65 \\ 0 & -6.65 & 13.86 & 0 & 6.65 & 6.10 \\ -331.80 & 0 & 0 & 331.80 & 0 & 0 \\ 0 & -4.43 & 6.65 & 0 & 4.43 & 6.65 \\ 0 & -6.65 & 6.10 & 0 & 6.65 & 13.86 \end{bmatrix}$$

$$\times \begin{bmatrix} u_1 = 0 \\ v_1 = 0 \\ \theta_1 = 0 \\ u_2 = d_1 \\ v_2 = d_2 \\ \theta_2 = d_3 \end{bmatrix}$$

from which

$$\begin{bmatrix} F_1 \\ F_2 \\ F_3 \end{bmatrix} = 10^3 \begin{bmatrix} 331.80 & 0 & 0 \\ 0 & 4.43 & 6.65 \\ 0 & 6.65 & 13.86 \end{bmatrix} \begin{bmatrix} d_1 \\ d_2 \\ d_3 \end{bmatrix}$$

This gives the stiffness relationship $\{W_I\} = [S_I]\{\Delta_I\}$

(b) The displacements (d_1, d_2 and d_3) have been treated as independent displacements. However, $d_2/d_1 = \tan \alpha$. Therefore $d_2 = d_1 \tan \alpha$, where $\alpha = 30°$. Therefore the relationship between the independent and dependent displacements is given by

$$\begin{bmatrix} d_1 \\ d_2 \\ d_3 \end{bmatrix} = \begin{bmatrix} 1 & 0 \\ \tan \alpha & 0 \\ 0 & 1 \end{bmatrix} \begin{bmatrix} d_1 \\ d_3 \end{bmatrix}$$

This establishes the relationship $\{\Delta_I\} = [R]^T\{\Delta_{II}\}$. Therefore

$$\begin{bmatrix} 1 & 0 \\ \tan \alpha & 0 \\ 0 & 1 \end{bmatrix} = \{R\}^T$$

Therefore $[R]$ is given by

$$\begin{bmatrix} 1 & \tan \alpha & 0 \\ 0 & 0 & 1 \end{bmatrix}$$

(c) The fixed end forces can be easily calculated and as shown in Fig. 10.23 are given by $\{F_{x1} = 0(*),\ F_{y1} = 15(*),\ M_1 = -7.5(*),\ F_{x2} = 0(d_1),\ F_{y2} = 15(d_2),\ M_2 = 7.5(d_3)\}$.

Note displacements inside ( ) indicate the direction of the displacement in which the force acts. (*) indicates it is the direction of a reaction. The force vector causing displacement is given by

$$\begin{bmatrix} F_1 \\ F_2 \\ F_3 \end{bmatrix} = \begin{bmatrix} 0 \\ -15.0 \\ -7.5 \end{bmatrix}$$

This gives the $\{W_I\}$ force vector.

(d) The $\{W_{II}\}$ force vector is given by $[R]\{W_I\}$. Therefore

$$\{W_{II}\} = \begin{bmatrix} 1 & \tan \alpha & 0 \\ 0 & 0 & 1 \end{bmatrix} \begin{bmatrix} 0 \\ -15.0 \\ -7.5 \end{bmatrix} = \begin{bmatrix} -8.66 \\ -7.50 \end{bmatrix}$$

$\tan \alpha = \tan 30 = 1/\sqrt{3}$.

(e) The stiffness matrix $[S_{II}]$ is given by $[S_{II}] = \{R\}[S_I]\{R^T\}$

$$S_{II} = \begin{bmatrix} 1 & \tan \alpha & 0 \\ 0 & 0 & 1 \end{bmatrix} \times 10^3 \begin{bmatrix} 331.80 & 0 & 0 \\ 0 & 4.43 & 6.65 \\ 0 & 6.65 & 13.86 \end{bmatrix} \begin{bmatrix} 1 & 0 \\ \tan \alpha & 0 \\ 0 & 1 \end{bmatrix}$$

where $\tan \alpha = \tan 30 = 1/\sqrt{3}$. Therefore

$$[S_{II}] = 10^3 \begin{bmatrix} 338.28 & 3.839 \\ 3.389 & 13.86 \end{bmatrix}$$

(f) Solve the set of equations given by

$$10^3 \begin{bmatrix} 338.28 & 3.839 \\ 3.389 & 13.86 \end{bmatrix} \begin{bmatrix} d_1 \\ d_3 \end{bmatrix} = \begin{bmatrix} -8.66 \\ -7.50 \end{bmatrix}$$

The displacements are given by $d_1 = -0.0198 \times 10^{-3}$, $d_3 = -0.5386 \times 10^{-3}$. Calculate $\{\Delta_I\}$ displacements

$$\begin{bmatrix} d_1 \\ d_2 \\ d_3 \end{bmatrix} = \begin{bmatrix} 1 & 0 \\ \tan\alpha & 0 \\ 0 & 1 \end{bmatrix} \begin{bmatrix} d_1 \\ d_3 \end{bmatrix} = \begin{bmatrix} -0.0198 \times 10^{-3} \\ -0.0114 \times 10^{-3} \\ -0.5356 \times 10^{-3} \end{bmatrix}$$

In the usual way calculate the forces due to displacements as explained in Chapter 6. This gives

Axial force $= (AE/L)(d_1) = -6.56$ kN

$M_1 = \{EI/L\,(1+\beta)\}[(2-\beta)d_3 + 6d_2/L] = -3.35$

$M_2 = \{EI/L\,(1+\beta)\}[(4-\beta)d_3 + 6d_2/L] = -7.50$

$F_{n1} = -F_{n2} = 3.62$ kN

(g) Calculate the sum of fixed end and displacement forces.

Axial force $= -6.56$ kN

$M_1 = -7.50 - 3.35 = -10.85$ kN m

$M_2 = 7.50 - 7.50 = 0$

$F_{n1} = 15.0 + 3.62 = 18.62$ kN

$F_{n2} = 15.0 - 3.62 = 11.38$ kN

As a check calculate the force along the inclined support. This is equal to

$-6.56\cos\alpha + 11.38\sin\alpha = 0$

This is an example using the principle of contragredience to incorporate constraints on displacements. This approach is commonly used in practice in the analysis of large-scale structures.

10.11 ● References

Bhat P 1981 *Solution of Problems in the Matrix Analysis of Skeletal Structures* Construction Press (Reprinted by A W Wheeler & Co., Allahabad, India, 1988.) (Contains many examples on the derivation of stiffness matrices using energy theorems and use of the principle of contragredience. Also contains material on approximate solution of differential equations using Galerkin's weighted residual method.)

Davies Glyn A O 1982 *Virtual Work in Structural Analysis* Wiley (A useful book. Contains advanced material.)

Hall A, Kabaila A P 1977 *Basic Concepts of Structural Analysis* Pitman (Somewhat old-fashioned book.)

Harker R J 1986 *Elastic Energy Methods of Design Analysis* Elsevier Applied Science (An old-fashioned book.)

Kong F K, Prentis J M, Charlton T M 1983 Principle of virtual work for a general deformable body – a simple proof *Structural Engineer* **61A** Discussions, **62A** (2) 1984: 67–70.

Neal B G 1964 *Structural Theorems and Applications* Pergamon (A handy book.)

Shaw F S 1972 *Virtual Displacements and Analysis of Structures* Prentice Hall (Somewhat old-fashioned book.)

Tauchert T R 1974 *Energy Principles in Structural Mechanics* McGraw-Hill (Uses energy theorems for the solution of static, dynamic and stability problems.)

Thomson F, Haywood G B 1986 *Structural Analysis Using Virtual Work* Chapman and Hall 1982 (An old-fashioned book.)

10.12 ● Problems

1. Derive the flexibility matrix for a beam element taking the moments M_1 and M_2 as the independent variables. Include both flexural and shear complementary energies,

Answer: If M_1 and M_2 are the clockwise moments at ends 1 and 2 respectively, then the reactions are

$$-V_1 = V_2 = (M_1 + M_2)/L$$

The bending moment and shear forces at a section x from end 1 are

$$M = M_1 - (M_1 + M_2)x/L = M_1(1 - x/L) - M_2(x/L)$$

$$Q = (M_1 + M_2)/L$$

Complementary energy C is given by

$$C = \int M^2/(2EI)\, dx + \int Q^2/(2GA_s)\, dx$$

Using Engesser's theorem, the rotations θ_1 and θ_2 at the ends are given by

$$\theta_1 = \frac{\partial C}{\partial M_1} = \int \frac{M}{EI}\frac{\partial M}{\partial M_1}\, dx + \int \frac{Q}{GA_s}\frac{\partial Q}{\partial M_1}\, dx$$

$$\theta_2 = \frac{\partial C}{\partial M_2} = \int \frac{M}{EI}\frac{\partial M}{\partial M_2}\, dx + \int \frac{Q}{GA_s}\frac{\partial Q}{\partial M_2}\, dx$$

Carrying out the differentiation,

$$\frac{\partial M}{\partial M_1} = 1 - \frac{x}{L}, \frac{\partial M}{\partial M_2} = -\frac{x}{L}, \frac{\partial Q}{\partial M_1} = \frac{1}{L}, \frac{\partial Q}{\partial M_2} = \frac{1}{L}$$

The integrals are given by

$$\int_0^L \left(1 - \frac{x}{L}\right)\left(1 - \frac{x}{L}\right) dx = \frac{L}{3}, \int_0^L \left(1 - \frac{x}{L}\right)\left(\frac{x}{L}\right) dx = \frac{L}{6},$$

$$\int_0^L \left(\frac{x}{L}\right)\left(\frac{x}{L}\right) dx = \frac{L}{3}$$

Substituting for the integrals and simplifying,

$$\theta_1 = \frac{(2M_1 - M_2)L}{6EI} + \frac{(M_1 + M_2)}{GA_s L},$$

$$\theta_2 = \frac{(-M_1 + 2M_2)L}{6EI} + \frac{(M_1 + M_2)}{GA_s L}$$

Setting $\beta = \dfrac{12EI}{GA_s L^2}$

$$\theta_1 = \left\{\frac{(4 + \beta)M_1 - (2 - \beta)M_2)}{12EI}\right\}L,$$

$$\theta_2 = \left\{\frac{-(2 - \beta)M_1 + (4 + \beta)M_2)}{12EI}\right\}L$$

The flexibility matrix is given by

$$\begin{bmatrix} \theta_1 \\ \theta_2 \end{bmatrix} = L/(12EI)\begin{bmatrix} (4 + \beta) & -(2 - \beta) \\ -(2 - \beta) & (4 + \beta) \end{bmatrix}\begin{bmatrix} M_1 \\ M_2 \end{bmatrix}$$

2. Calculate the fixed end moments in a clamped beam carrying a concentrated load at distances a and b respectively from left and right supports. Include flexural and shear complementary energies.

Answer: If M_1 is the the anticlockwise moment and M_2 the clockwise moment at ends 1 and 2 respectively, then the reaction at support 1 is

$$V_1 = Wb/L + (M_1 - M_2)/L$$

The bending moment and shear forces at a section x from end 1 are

$$M = -M_1 + V_1 x - W\langle x - a\rangle$$

$$Q = V_1 - W\langle x - a\rangle^0$$

Substituting for V_1,

$$Q = (M_1 - M_2)/L + Wb/L - W\langle x - a\rangle^0$$

$$M = -M_1(1 - x/L) - M_2(x/L) + Wbx/L - W\langle x - a\rangle$$

Complementary energy C is given by

$$C = \int M^2/(2EI)\, dx + \int Q^2/(2GA_s)\, dx$$

Using Engesser's theorem, the rotations θ_1 and θ_2 at the ends are given by

$$\theta_1 = \frac{\partial C}{\partial M_1} = \int \frac{M}{EI}\frac{\partial M}{\partial M_1}\, dx + \int \frac{Q}{GA_s}\frac{\partial Q}{\partial M_1}\, dx$$

$$\theta_2 = \frac{\partial C}{\partial M_2} = \int \frac{M}{EI}\frac{\partial M}{\partial M_2}\, dx + \int \frac{Q}{GA_s}\frac{\partial Q}{\partial M_2}\, dx$$

Carrying out the differentiation, we have

$$\frac{\partial M}{\partial M_1} = -\left(1 - \frac{x}{L}\right), \frac{\partial M}{\partial M_2} = -\frac{x}{L}, \frac{\partial Q}{\partial M_1} = \frac{1}{L}, \frac{\partial Q}{\partial M_2} = -\frac{1}{L}$$

Noting that, as in the previous example,

$$\int_0^L \left(1 - \frac{x}{L}\right)\left(1 - \frac{x}{L}\right) dx = \frac{L}{3}, \int_0^L \left(1 - \frac{x}{L}\right)\left(\frac{x}{L}\right) dx = \frac{L}{6},$$

$$\int_0^L \left(\frac{x}{L}\right)\left(\frac{x}{L}\right) dx = \frac{L}{3}$$

and

$$\int_0^L \langle x - a\rangle\frac{x}{L}\, dx = \int_a^L (x - a)\frac{x}{L}\, dx = \frac{(2L^3 - 3aL^2 + a^3)}{6L}$$

$$\int_0^L \langle x - a\rangle\left(1 - \frac{x}{L}\right) dx = \int_a^L (x - a)\left(1 - \frac{x}{L}\right) dx$$

$$= \frac{(L^3 - 3aL^2 + 3a^2L - a^3)}{6L}$$

$$\int_0^L <x - a>^0 \, dx = \int_a^L dx = (L - a)$$

Note that $\int_0^L \left\{ \dfrac{b}{L} - <x - a>^0 \right\} dx = 0$

Substituting for the integrals, simplifying and setting the rotations at the ends to zero as the supports are fixed, we have

$$2M_1 + M_2 + \frac{\beta(M_1 - M_2)}{2}$$

$$= W \left\{ b - \left(L - 3a + 3\frac{a^2}{L} - \frac{a^3}{L^2} \right) \right\}$$

$$= Wa \left\{ 2 - 3\frac{a}{L} + \frac{a^2}{L^2} \right\}$$

$$M_1 + 2M_2 + \frac{\beta(-M_1 + M_2)}{2}$$

$$= W \left\{ 2b - \left(2L - 3a + \frac{a^3}{L^2} \right) \right\}$$

$$= Wa \left\{ 1 - \frac{a^2}{L^2} \right\}$$

Solving for the moments,

$$M_1 = \frac{Wab^2}{L^2} \left[\frac{\left\{ 1 + 0.5\beta \dfrac{L}{b} \right\}}{(1 + \beta)} \right],$$

$$M_2 = \frac{Wa^2b}{L^2} \left[\frac{\left\{ 1 + 0.5\beta \dfrac{L}{a} \right\}}{(1 + \beta)} \right]$$

3. For the continuous beam ABCD with a hinge at C as shown in Fig. E2.5, draw the influence line for the following forces. Use Muller–Breslau's theorem.
   (i) Bending moment at midspan of AB
   (ii) Bending moment at support B
   (iii) Reaction at support at A
   (iv) Reaction at support at B
   (v) Reaction at support at D
   (vi) Shear force at midspan of AB

   Answer: The influence lines are shown in Fig. E2.5.

4. Taking the origin at the fixed support, the expressions for deflection in a cantilever subjected to a couple M_s or a force W at the tip are given respectively by (see Section 4.3)

$$v = -M_s \frac{x^2}{2EI}, \quad v = W \frac{(3Lx^2 - x^3)}{6EI}$$

Use the Maxwell–Betti theorem to derive the slope and deflection at the tip of a cantilever subjected to a load W at $x = L/2$.

Answer: $\theta = -W \dfrac{L^2}{8EI}, \ v = -W \dfrac{5L^3}{48EI}$

5. Sketch qualitative influence lines for the following 'forces' for the statically indeterminate structure shown in Fig. E10.1.

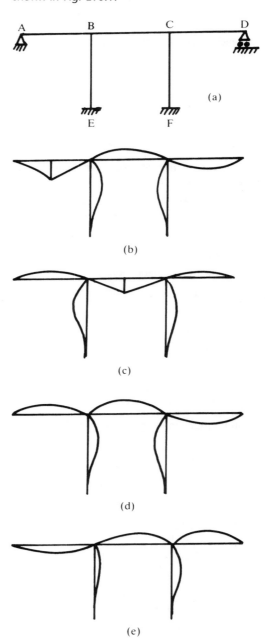

(a)

(b)

(c)

(d)

(e)

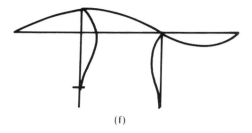

(f)

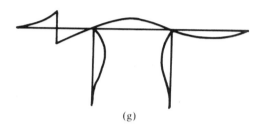

(g)

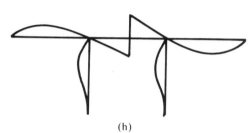

(h)

Figure E10.1(a)–(h)

(i) Midspan moments in AB and BC
(ii) Support moment at B in AB and C in BC
(iii) Column moment at B
(iv) Vertical reactions at E
(v) Shear force at midspan of BC

Answer: Influence lines are shown in Fig. E10.1.

6. Use the virtual work method to determine the vertical and horizontal deflections at joint 4 of the pin-jointed structure shown in Fig. 5.3(a).

Answer: 5.95 and 1.91 mm.

7. For the rigid-jointed frame shown in Fig. 5.5(a), determine the horizontal deflections at joints B and D. Take $H = 4$ m, $L = 6$ m, $q = 20$ kN m and $W = 100$ kN.

Answer:

$$\Delta_B = \frac{1}{EI_c} \int_0^4 \left\{80y - 20\frac{y^2}{2}\right\} y \, dx + \frac{1}{EI_b} \int_0^6 \{160$$

$$+ \ 23.33x - 100{<}x - 3{>}\}(4 - 0.67x) \, dx$$

$$= \frac{1066.67}{EI_c} + \frac{2180}{EI_b}$$

$$\Delta_D = \frac{1}{EI_c} \int_0^4 \left\{80y - 20\frac{y^2}{2}\right\} y \, dx + \frac{1}{EI_b} \int_0^6 \{160$$

$$+ \ 23.33x - 100{<}x - 3{>}\}(4) \, dx$$

$$= \frac{1066.67}{EI_c} + \frac{3720}{EI_b}$$

Chapter eleven

Limit analysis of plane frames

The elastic behaviour of structures was analysed in the previous chapters. It was based on the assumption that the material is linearly elastic and the stresses are proportional to the corresponding strains. One important assumption made was that the structures were subjected to loads such that stresses were less than that to cause the yielding of the material. In the case of linearly elastic structures, the stresses at a point are directly proportional to the loads acting on the structure and the principle of superposition holds true. As the loads on the structure are increased, then, at a certain load level called the yield load, the stresses at the most highly stressed point satisfy the criterion for the yielding of the material. Once this happens the stresses at the point cannot increase but the strains can increase (see Fig. 3.14). Therefore the linear relationship between stress and strain becomes invalid. In this chapter the behaviour of structures where the stresses due to bending dominate and where the structure is loaded beyond yield load is examined.

11.1 ● Plastic bending

Consider a rectangular beam of breadth b and depth d subjected to a bending moment M. Let the stress–strain relationship for the material of the beam be elastic perfectly plastic as shown in Fig. 11.1(a). Initially the bending stress is linearly distributed and because the material is elastic the corresponding strain is also linearly distributed, as shown in Fig. 11.1(b). However, as the bending moment is gradually increased, the stress in the extreme fibres reaches the yield stress σ_0 and the strain in the top fibre level is equal to ε_0, the yield strain. At this stage, the bending moment is given by $M_0 = \sigma_0 I/(0.5d)$. $I = bd^3/12$, therefore $M_0 = bd^2\sigma_0/6$.

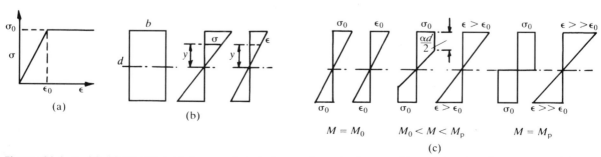

Figure 11.1 ● (a)–(c) Variation of stress and strain in a rectangular beam subjected to a bending moment M

The moment M_0 is called the yield moment of the cross-section because at this moment the maximum stress reaches the yield stress σ_0. When the bending moment is increased beyond M_0, the stress in the extreme fibres cannot increase beyond σ_0 but the strain can increase beyond ε_0. Assuming that plane sections remain plane, then the increased strain causes the yielding to spread towards the neutral axis of the beam. If the depth of plastification is assumed to be $\alpha d/2$, then using the stress distribution shown in Fig. 11.1(c), the bending moment can be calculated as

$$M = bd^2\sigma_0\{(1 - \alpha)^2/6 \text{ due to the linear stress part}$$
$$+ \alpha(2 - \alpha)/4\} \text{ due to the constant stress part}$$

$$= M_0\{(1 - \alpha)^2 + 1.5\alpha(2 - \alpha)\}, \text{ where } M_0 = bd^2\sigma_0/6$$

Note that when $\alpha = 0$, then $M = M_0$.

As M is increased still further, the whole section (except for a very small area near the neutral axis) reaches the yield stress. The value of moment at this level is obtained by substituting $\alpha = 1$ and is given by $M = bd^2\sigma_0/4$. Because the whole section (practically) has yielded at this stage, the moment is called the plastic moment capacity M_p of the section:

$$M_p = \frac{bd^2}{4}\sigma_0$$

The bending moment M_p represents the maximum moment that the section can sustain. Summarizing,

$M_0 \geqslant M \geqslant 0$, section is elastic

$M_p \geqslant M \geqslant M_0$, section is partly elastic and partly plastic

$M = M_p$, the whole section is plastic and the section reaches its maximum bending capacity. The ratio M_p/M_0 is called the shape factor and for a rectangular section this is equal to 1.5. This means that the section can carry 50 per cent more moment than its yield moment M_0 before it reaches its ultimate moment capacity M_p.

In the case of a symmetrical I section as shown in Fig. 11.2, the second moment of area is given by

$$I = bd^3/12 - (b - t_w)(d - 2t_f)^3/12$$

$$\therefore I = bd^3/12\{1 - (1 - t_w/b)(1 - 2t_f/d)^3\}$$

Therefore $M_0 = \sigma_0 I/(0.5d) = \{bd^3\sigma_0/6\}[1 - (1 - t_w/b)(1 - 2t_f/d)^3]$.
The plastic moment capacity is given by

$$M_p = M_p \text{ of rectangle } b \times d - M_p \text{ of section } (b - t_w) \times (d - 2t_f)$$

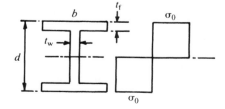

Figure 11.2 ●
Bending stress distribution in an I beam when the section is fully plastic

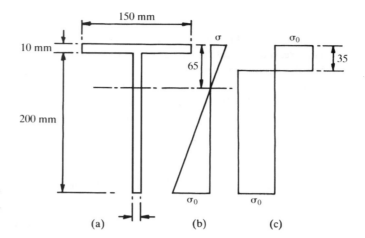

Figure 11.3 ● (a)–(c)
Stress distribution in a *T*
section at yield and fully
plastic moment

Therefore

$$M_p = bd^2\sigma_0/4 - (b - t_w)(d - 2t_f)^2\sigma_0/4$$

$$= bd^2\sigma_0/4\{1 - (1 - t_w/b)(1 - 2t_f/d)^2\}$$

Shape factor $= M_p/M_0$

$$= \frac{1.5\{1 - (1 - t_w/b)(1 - 2t_f/d)^2\}}{\{1 - (1 - t_w/b)(1 - 2t_f/d)^3\}}$$

If, say, $t_w = 0.1b$ and $t_f = 0.05d$, then the shape factor $= 1.18$. In general, for rolled steel *I* and channel sections, shape factor is approximately equal to 1.15. The small shape factor indicates that the difference between the yield moment M_0 and the plastic moment M_p is small.

When a section is bent about an axis of symmetry, the position of neutral axis remains the same. This is not the case if the axis about which the beam is bent is not an axis of symmetry. The reason for this is that, in the elastic stage, the neutral axis coincides with the centroidal axis. However, when the section is fully plastic, because the stress in the entire section is σ_0 tension or compression, and because for equilibrium in the axial direction total compressive force must be equal to the total tensile force, the neutral axis divides the section into two equal areas. In other words, the neutral axis in the fully plastic state coincides with the equal area division axis.

As an example consider the *T* section shown in Fig. 11.3 bent about an axis parallel to the flange. The position of the neutral axis in the fully elastic and fully plastic stages can be calculated as follows:

1. Fully elastic stage. Taking moments about the top face of the flange,

First moment of area $= 150 \times 10 \times (10/2) + 200 \times 10 \times (200/2 + 10)$

$$= 227\ 500$$

Area $= 150 \times 10 + 200 \times 10 = 3500$

Therefore the centroid is at a distance of $227\ 500/3500 = 65$ mm from top of the flange.

2. Fully plastic section. Let the equal area axis be at a distance y from the top face of the flange and lie in the web. For equal area axis, tension area = compression area. Therefore

$$150 \times 10 + (y - 10)10 = \{200 - (y - 10)\}10$$

thus $150 + 7 - 10 = 200 - y + 10$, or $y = 35$ mm from the top face of the flange. Therefore, as the plasticity spreads in the section, the neutral axis shifts upwards towards the flange.

The yield moment and plastic moment capacities are calculated as follows.

3. Calculation of yield moment M_0. The second moment of area of the section about the relevant centroidal axis is given by

$$I = 150 \times 10^3/12 + 150 \times 10 \times (65 - 10/2)^2 \text{ for the flange}$$
$$+ 10 \times 200^3/12 + 200 \times 10 \times (200/2 + 10 - 65)^2 \text{ for the web}$$

$$= 16.13 \times 10^6 \text{ mm}^4$$

$$M_0 = \sigma_0 I/y_{max}$$

$y_{max} = 145$ mm to the bottom of the web

Therefore

$$M_0 = 16.13 \times 10^6 \sigma_0/145 = 1.11 \times 10^5 \sigma_0$$

The section starts yielding at the bottom of the web.

4. Calculation of plastic moment capacity M_p : M_p is equal to the moment of forces in the section about the plastic neutral axis. Because the stress in all areas is equal to $\pm \sigma_0$, M_p is given by

$$M_p = \sigma_0[150 \times 10 \times (35 - 10/2) \text{ for flange}$$
$$+ 25 \times 20 \times 25/2 \text{ for web above neutral axis}$$
$$+ 175 \times 10 \times 75/10 \text{ for web below neutral axis}]$$

Therefore $M_p = 2.01 \times 10^5 \sigma_0$, and shape factor $= M_p/M_0 = 2.01/1.11 = 1.81$, a value much higher than for a rectangular section.

11.2 ● Elastic–plastic moment–curvature relationship

It was shown in Section 4.2.1 that for a beam whose neutral axis is bent into an arc of circle of radius R as shown in Fig. 11.4, the strain ε_x at distance y from the neutral axis is given by

$$\varepsilon_x = y/R \text{ or curvature } 1/R = \varepsilon_x/y$$

1. Elastic stage. In the elastic stage, $\varepsilon_x = \sigma/E$, where E = Young's modulus and $\sigma = My/I$. Therefore

$$1/R = \varepsilon_x/y = (\sigma/E)\{M/(I\sigma)\} = M/(EI)$$

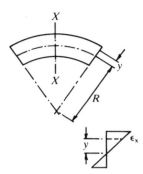

Figure 11.4 ●
Relationship between
strain and curvature

This is a linear relationship between M and $1/R$. This is valid up to $M = M_0$, when the stress in the extreme fibres is σ_0. When $M = M_0 = bd^2\sigma_0/6$ and $I = bd^3/12$ for a rectangular section, $1/R$ is equal to $2\sigma_0/(Ed)$.

2. Elastic–plastic stage. As shown in Fig. 11.1(c), at a distance $y = (1 - \alpha)0.5d$ from the neutral axis, stress $\sigma = \sigma_0$ and the material has just reached the yield stress. In the elastic portion of the beam, $\varepsilon_x = \sigma_0/E$. Therefore

$$1/R = \varepsilon_x/y = (\sigma_0/E)1/\{(1 - \alpha)0.5d\} = 2\sigma_0/(Ed)\{1/(1 - \alpha)\}$$

The corresponding moment acting on the section is $M = M_0\{(1 - \alpha)^2 + 1.5\alpha(2 - \alpha)\}$.
     Thus fixing the value of α at values in the range $0 \leqslant \alpha \leqslant 1$, the value of $1/R$ and the corresponding M can be calculated. The values are shown in Table 11.1 and are also plotted as curve OCB in Fig. 11.5. The beam shows elastic–plastic behaviour from $M_y = 0.67M_p$ to M_p. When $M_p = 1.15M_0$, the range of elastic–plastic behaviour is much smaller than for a rectangular section. Thus the moment–curvature

Table 11.1 ●
Elastic–plastic moment–
curvature relationship

α	M/M_0	$Ed/(2\sigma_0 R)$
0	0	0
0	1.0	1.0
0.1	1.095	1.11
0.2	1.180	1.25
0.3	1.255	1.43
0.4	1.320	1.67
0.5	1.375	2.00
0.6	1.420	2.50
0.7	1.455	3.33
0.8	1.480	5.00
0.9	1.495	10.00
0.95	1.499	20.00
1.00	1.50	∞

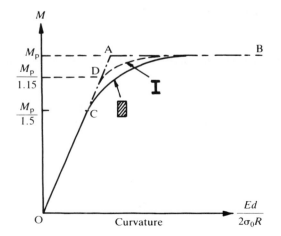

Figure 11.5 ●
Moment–curvature
relationship for
rectangular and
I section beams

relationship remains linear up to a much higher value of moment, i.e. $M = M_p/1.15 = 0.87M_p$. The curve for I sections is shown as ODB. In practice rolled steel sections rather than rectangular sections are commonly used. For convenience in calculations, it is commonly assumed that the shape factor is unity and hence $M_p = M_0$. The corresponding moment–curvature relationship becomes OAB. Therefore the section exhibits elastic–perfectly plastic behaviour as shown in Fig. 11.5. The moment–curvature relationship has the same shape as the idealized stress–strain relationship for the material. For the material presented in the rest of this chapter, it is assumed that the shape factor is unity.

11.3 ● Behaviour of a simply supported beam

Consider the centrally-loaded simply supported beam shown in Fig. 11.6. Assume that the shape factor for the section is unity. The maximum moment of $WL/4$ occurs at a single cross-section at midspan. If the load-point deflection Δ is recorded while the load W increases from zero, the initial behaviour will be elastic to give the straight line OA. When the maximum moment $WL/4$ reaches the fully plastic moment M_p, i.e. $WL/4 = M_p$ or $W = 4M_p/L$, the central deflection at this stage is $\Delta = WL^3/(48EI) = M_pL^2/(12EI)$. This is shown in Fig. 11.6(c). For any attempt to increase the load beyond $4M_p/L$, the bending moment at the centre cannot increase because the moment there is equal to M_p. Therefore equilibrium cannot be maintained for $W > 4M_p/L$. Thus the maximum load, also called the ultimate load, W_u that the beam can carry is given by

$$W_u = 4M_p/L$$

During the deformation in the range A–B, the bending moments everywhere in the beam are, except at midspan, less than M_p. Therefore, except the midspan section,

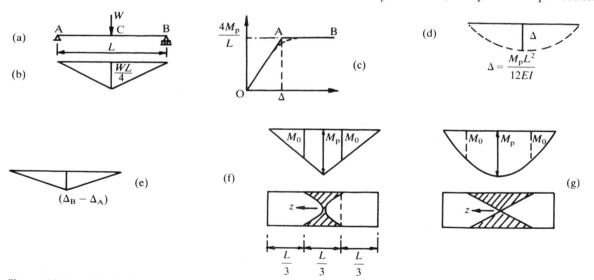

Figure 11.6 ● (a)–(g) Plastic zones in a simply supported beam subjected to midspan concentrated load and uniformly distributed load

the rest of the structure remains elastic. Figure 11.6(d) shows the deflected profile of the beam when the load has just reached $4M_p/L$. The *additional* deflection profile of the beam in the region A–B, when the beam is deforming under constant load, is shown in Fig. 11.6(e). During the displacement in the range A–B, large curvature is concentrated at midspan cross-section. There is a hinge action and the beam develops a clearly defined kink, i.e. a plastic hinge at midspan. This is clearly seen in the *additional* deflection profile shown in Fig. 11.6(e). The beam behaves as if there was a real hinge at the centre except that the moment at the hinge is equal to M_p. Once a hinge forms at the centre, the beam is converted into two hinged bars hinged at A and C and resting on a roller support at B. Therefore when the load is equal to $4M_p/L$, the deflection can increase with no change in load from A to B as in Fig. 11.6(c).

The load–deflection relationship is shown with a sharp kink at A. This is due to the assumption that the shape factor is unity, i.e. the yield moment and plastic moment are equal. This has the effect that as soon as the maximum fibre stress reaches the yield stress, the whole section becomes plastic. In practice yield does spread gradually through the section under increasing moment and this leads to rounding off of the kinks in the load–deflection relationship as shown by dotted lines in Fig. 11.6(c).

It is useful to think of a plastic hinge as a 'rusty hinge'. *Relative* rotation between the two sides of the hinge is zero if the moment at the section is below M_p. However, free *relative* rotation is possible once the moment at the hinge is equal to the plastic moment M_p of the section.

Some idea of the approximation involved in the concept of an ideal plastic hinge can be gained from studying the spread of plastic zones in the case of a simply supported rectangular beam. If the loading on the beam is a central concentrated load, then the bending moment distribution at collapse is triangular and the bending moment at centre is equal to M_p. Because, in the case of a rectangular beam, yield moment $M_0 = 0.67M_p$ is reached at $L/3$ from the supports, the beam will be partially plastic for a distance of $L/6$ on both sides of the load. In the partially plastic region, $M = M_p(1 - 2z/L)$, where the origin for z is at the midspan. The corresponding moment is given, from Section 11.1 with $M_0 = 0.67M_p$, by

$$M = 0.67M_p\{(1 - \alpha)^2 + 1.5\alpha(2 - \alpha)\}$$

Therefore the relationship between the depth $0.5\alpha d$ of the plastic zone and z is parabolic, as shown in Fig. 11.6(f). Thus in this case the beam is elastic–plastic over a third of its span.

Similarly, if the beam is loaded by a uniformly distributed load q, then the beam collapses when $q_u L^2/8 = M_p$. The bending moment distribution at collapse is parabolic with the central moment being M_p. The bending moment is equal to yield moment $M_0 = 0.67M_p$ at $0.21L$ from the supports. Thus, in the case of uniformly distributed loading, a much larger portion of the beam yields than in the case of a central concentrated load. Because in this case M is a second degree function of both z and α, the relationship between z and α is linear and the yielded zones are shown in Fig. 11.6(g).

In the case of I sections, because $M_p = 1.15M_0$, under central concentrated load, only a small section of the beam on either side of the load yields partially. Thus the plastic hinge concept is much closer to the truth in the case of *I* sections than for rectangular sections.

Although a simply supported beam was used as an example, the behaviour noted is common to all statically determinate structures. The structure reaches its ultimate load as soon as the maximum moment at one section reaches the plastic moment capacity of the section.

11.4 ● Behaviour of a statically indeterminate structure

Consider the propped cantilever beam of Fig. 11.7(a). Elastic methods of analysis show that the bending moments M_A and M_B at A and B respectively are

$M_A = (3/16)WL$ causing tension at the top

$M_B = (5/32)WL$ causing tension at the bottom

giving the bending moment diagram in Fig. 11.7(b).

The structure remains linearly elastic until the moment at A which is the maximum moment anywhere in the beam reaches the plastic moment M_p, i.e.

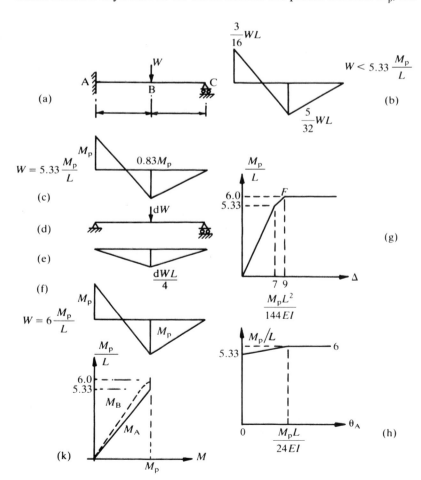

Figure 11.7 ● (a)–(k) Behaviour of a propped cantilever

$$(3/16)WL = M_p \text{ or } W = 5.33M_p/L$$

At this stage the moment at B is

$$M_B = (5/32)WL = (5/32)5.33M_p/L = 0.833M_p$$

The load $W = 5.33M_p/L$ signifies the end of elastic behaviour. At the end of elastic behaviour, the deflection at the load point is given by

$$\Delta = WL^3/(48EI) - M_AL^2/(16EI)$$

Substituting for $W = 5.33M_p/L$ and $M_A = M_p$, Δ is given by

$$\Delta = 7M_pL^2/(144EI)$$

The rotation at A is zero, because the beam behaves as a propped cantilever.

Because the moment at A is equal to M_p, a plastic hinge forms at A which will not resist any rotation for loads greater than $5.33M_p/L$. The moment at A cannot increase further but the curvature can increase indefinitely, so that additional load dW can cause no change in M_A, and because a plastic hinge offers no resistance to rotation, support A behaves as a pinned support except that the moment is held constant at M_p. This additional load must be carried as if the beam were simply supported (Fig. 11.7(d)); the additional moments for equilibrium are as in Fig. 11.7(e). Load dW can now be increased until the total moment at B of $0.833M_p$ due to $W = 5.33M_p/L$ acting on the propped cantilever and $dWL/4$ due to dW acting on the simply supported beam equals the plastic moment M_p, and during this stage unrestricted plastic rotation occurs at A. Equating the total moment at B to M_p,

$$0.833M_p + 0.25dWL = M_p$$

Therefore

$$dW = 0.67M_p/L$$

$$\text{Total load } W_u = W + dW = 5.33M_p/L + 0.67M_p/L = 6M_p/L$$

At this stage the additional deflection at the load point and the rotation at A due to the beam behaving as a simply supported beam are given by

$$d\Delta = dWL^3/(48EI) = M_pL^2/(72EI)$$

$$d\theta_A = dWL^2/(16EI) = M_pL/(24EI)$$

Thus when the total load is equal to $6M_p/L$, the central deflection is

$$\Delta + d\Delta = (7/144 + 1/72)M_pL^2/(EI) = 9M_pL^2/(144EI)$$

$$\text{Rotation at A} = 0 + d\theta_A = M_pL/(24EI)$$

Attempts to add further load fail, because neither M_A nor M_B can increase so that equilibrium is not possible and indefinite rotation will occur at A and B. The beam is then a two-bar mechanism hinged at A and B and resting on a roller at C and will collapse by excessive rotation at A and B. The physical behaviour can be demonstrated by recording the load point deflection as in Fig. 11.7(g). At F the plastic moment is reached at the load point and the structure can collapse at constant load.

The final bending moment diagram at collapse is recorded in Fig. 11.7(f). Figure 11.7(h) and (k) shows respectively the variation of θ_A, M_A and M_B with load.

As remarked in Section 11.3, the actual load–deflection relationship will not have sharp kinks which are due to the assumption of unit load factor, i.e. the yield moment is equal to the plastic moment. It should be noted that once the plastic hinges form, the moment at the section remains constant while curvature is increasing. This phenomenon relies upon the ductility of the material. For example, in a simple tension test it is found that the yield stress for mild steel is about 250 N mm^{-2} and the yield strain ε_0 is about 1200×10^{-6}. At the end of the plastic plateau the strain is approximately $10\varepsilon_0$ and the ultimate strain is about $200\varepsilon_0$. It is because of the 'large ductility' present that the sections which yield early in the loading history can continue to deform while sustaining the moment at M_p and allow redistribution of stresses to take place. Materials other than mild steel, such as aluminium and reinforced concrete, also exhibit ductile behaviour but the amount of ductility present is much less than in mild steel. In these cases unless care is taken, because of limited ductility, sections which yield early in the loading history might fail *before* the calculated collapse load is reached. Designers should bear these facts in mind when calculating the collapse load of structures.

It is useful to note that if the location of plastic hinges is known the collapse load can be calculated directly from equilibrium considerations without any reference to the intermediate stages of behaviour. For the present example, the vertical reaction V_C at C is

$$V_C = W_u/2 - M_A/L$$

The moment at the centre is

$$M_B = V_C L/2 = W_u L/4 - M_A/2$$

Therefore

$$2M_B + M_A = 0.5 W_u L$$

This is an equation of equilibrium. If $M_A = M_B = M_p$, then $W_u = 6M_p/L$.

This example highlights some very important features of the elastic–plastic behaviour of statically indeterminate structures. They are:

1. Unlike statically determinate structures, collapse does not occur at the formation of the first plastic hinge at the most highly stressed section.
2. Additional plastic hinges can form until the structure becomes a mechanism. During this process the structure deforms with increasing loss of stiffness, as can be observed from Fig. 11.7(g). The formation of each plastic hinge leads to further loss of stiffness of the structure.
3. The load–deflection behaviour is 'piecewise linear'. With reference to the propped cantilever, this can be explained as
   (a) between the stages of no plastic hinge and one plastic hinge at A, the structure behaves as a linear elastic propped cantilever;
   (b) between the stages of a plastic hinge at A and the next plastic hinge at B, the beam behaves as a linear elastic simply supported beam.
4. The structure experiences 'stress redistribution'. What this means is that during the loading process 'stresses', i.e. bending moments at some sections, cease to

increase while at others they might increase or even decrease from the previous stage. This can be observed from Fig. 11.7(k).

5. Once a plastic hinge forms at a section, the moment at that section remains constant at the plastic moment capacity at that section as shown in Fig. 11.7(k). It should be remembered that this assumes 'large ductility' is available.

6. Once a plastic hinge forms at a section, that section offers no additional resistance to rotation. Thus a section with a plastic hinge behaves as a section with a real hinge except that the bending moment at the hinge is M_p.

7. It was noticed that the ultimate load could be calculated from equilibrium considerations without any reference to stiffness characteristics of the structure. This implies that the ultimate load of a structure is not influenced by settlement of supports, thermal stresses, residual stresses, etc.

The fact that the behaviour is 'piecewise linear' means that the principle of superposition cannot be used. This means that the concept of influence lines cannot be used for the design of these structures. The designer has to ensure that if a structure is subjected to alternative load cases, then each load case has to be examined independently to ensure that the collapse load for each case is more than the corresponding design ultimate load.

Note: During loading, plastic hinges form at various locations. It is possible that at some sections where a plastic hinge has formed at some stage in loading, the moments at those sections decrease at a later stage in the loading. Plastic hinge action persists as long as the curvature is increasing. But as soon as the curvature starts to decrease, plastic hinge action disappears and elastic behaviour is restored at that section.

11.5 ● Definition of collapse mechanism

The insertion of a hinge into a statically indeterminate rigid-jointed frame reduces the number of statically indeterminate moments by one, so that, if the degree of statical indeterminacy is n, then the addition of n hinges produces a statically determinate structure (see Sections 1.15 and 1.21, Chapter 1). The addition of one more hinge will convert the structure to a mechanism, i.e. the structure can deform without any loads acting on it. Thus a structure becomes a mechanism when the number of plastic hinges is equal to $(n + 1)$. For example, a simply supported beam is a statically determinate structure, i.e. $n = 0$. Therefore formation of one plastic hinge causes the structure to collapse. Similarly the propped cantilever shown in Fig. 11.7(a) is one times statically indeterminate, i.e. $n = 1$. Therefore the formation of two plastic hinges (at A and B) signifies collapse. When a structure collapses by the formation of $(n + 1)$ plastic hinges it is said to form a complete collapse mechanism. A well-designed structure should experience total collapse. However, a poorly designed structure can collapse because of an 'under-designed' member in the structure. For example, in a continuous beam, if any span is under-designed, then that span alone can collapse with the formation of plastic hinges at its ends and in the span. Although this is collapse of a single span, as far as the user is concerned it is the collapse of the structure. When a collapse takes place by the

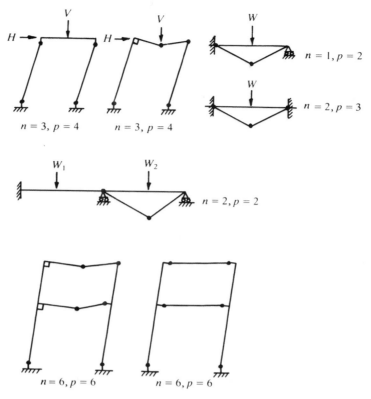

Figure 11.8 ●
Examples of complete
and partial collapse
mechanisms

p = number of plastic hinges, n = degree of statical indeterminacy

formation of less than $(n + 1)$ plastic hinges, it is called a partial collapse. Because of the fact that many plastic hinges could form simultaneously but be positioned so as not to convert the structure into a mechanism, at collapse it is possible to have more than $(n + 1)$ plastic hinges. Such a collapse is called 'over complete' collapse. Typical partial and complete collapse mechanisms are shown in Fig. 11.8.

11.6 ● Prediction of collapse loads of structures

In the design of structures, the designer would like to ensure that under working loads the structure behaves in an elastic manner so that large and permanent deformations are avoided. Elastic analysis of structures can be used to calculate the maximum load at which the most highly stressed section reaches the plastic moment capacity of the section. The designer would also like to ensure that the structure can resist the design ultimate loads that are likely to act during the lifetime of the structure. This can be ensured by making the ultimate or collapse load of the structure at least as great as the design ultimate loads. It is because of these design requirements that interest is centred on the calculation of yield and ultimate loads rather than in tracing the elastic–plastic behaviour of structures. It was noted in Section 11.4 that if the exact positions of plastic hinges at collapse are known, then it is possible to calculate the ultimate or collapse load of the structure from purely equilibrium considerations. The rest of this chapter will concentrate on the calculation of collapse loads of rigid-jointed structures.

Example 1

Calculate the collapse load of the fixed beam shown in Fig. 11.9. Assume that the plastic moment capacity is a constant value of M_p.

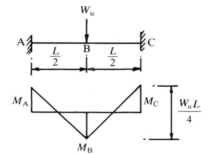

Figure 11.9 ● A fixed-ended beam with a midspan concentrated load

In this structure the plastic hinges will form at the supports and midspan. Therefore, from statics, the bending moment at midspan is equal to

$$M_B = W_u L/4 - (M_A + M_C)/2$$

If $M_A = M_B = M_C = M_p$, then $W_u = 8M_p/L$.

Example 2

Calculate the collapse load of the fixed beam shown in Fig. 11.10. Assume that the plastic moment capacity is a constant value of M_p. In this structure the plastic hinges will form at the supports and midspan. Therefore from statics, the bending moment at midspan is equal to

$$M_B = q_u L^2/8 - (M_A + M_C)/2$$

If $M_A = M_B = M_C = M_p$, then $q_u = 16M_p/L^2$.

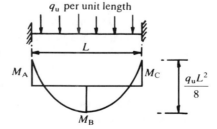

Figure 11.10 ● A fixed-ended beam under uniformly distributed load

Example 3

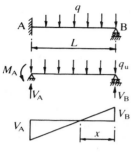

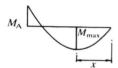

Figure 11.11 ●
A propped cantilever
under uniformly
distributed load

Calculate the collapse load of the propped cantilever shown in Fig. 11.11. Assume that the plastic moment capacity is a constant value of M_p.

In this structure the plastic hinges will form at the support A and in the span. However, the position of the plastic hinge in the span is not known but it will form at the maximum moment position. From statics, the vertical reactions at A and B are given by

$$V_B = 0.5q_uL - M_A/L, \quad V_A = 0.5q_uL + M_A/L$$

The maximum moment M_{max} occurs at x from the right-hand support where the shear force is zero. Therefore $q_ux = V_B$, and $x = 0.5L - M_A/(q_uL)$.

$$M_{max} = V_Bx - 0.5q_ux^2$$

Substituting for x and equating $M_{max} = M_A = M_p$

$$M_p = \{0.5q_uL - M_p/L\}\{0.5L - M_p/(q_uL)\} - 0.5q_u\{0.5L - M_p/(q_uL)\}^2$$
$$= 0.5q_u\{0.5L - M_p/(q_uL)\}^2$$

Setting $\alpha = q_uL^2$ and simplifying the above equation, we have

$$0.25\alpha^2 - 3M_p\alpha + M_p^2 = 0$$

Therefore

$$\alpha = (6 \pm 4\sqrt{2})M_p$$

choosing $\alpha = (6 + 4\sqrt{2})M_p$, $q_u = 11.65M_p/L^2$ and $x = 0.4142L$. (*Note:* Choosing $\alpha = (6 - 4\sqrt{2})M_p$ leads to a negative value of x which is inadmissible.)

It is useful to explore the consequence of *assuming* that the plastic hinge forms at the centre. In this case $x = 0.5L$ and M_{centre} is given by

$$M_{centre} = V_BL/2 - 0.5q_u(0.5L)^2$$

Substituting for V_B and equating $M_{centre} = M_A = M_p$

$$M_p = \{0.5q_uL - M_p/L\}\{0.5L\} - 0.5q_u\{0.5L\}^2$$
$$= 0.25q_uL^2 - 0.5M_p - 0.125q_uL^2$$

Solving

$$q_u = 12M_p/L^2$$

This load is higher than the one calculated from assuming that the maximum moment occurs off the midspan and towards the right-hand end. If now the true maximum moment is calculated using $q_u = 12M_p/L^2$, then the maximum moment occurs when $q_ux = V_B$. Therefore $x = 0.5L - M_A/(q_uL) = 0.4167L$

$$M_{max} = V_Bx - 0.5q_ux^2 = (0.5q_uL - M_p/L)(0.4167L) - 0.5q_u(0.4167L)^2$$
$$= 1.042M_p > M_p \text{ which is inadmissible.}$$

This example has some important features which should be emphasized. In the first case, plastic hinges were assumed to form only at the left-hand support and in the span such that the bending moment anywhere in the beam did not exceed the plastic moment capacity of the section. The collapse load was determined on the basis of equilibrium. In the second case the plastic hinge in the span was *assumed* to occur at midspan and the collapse load was determined on the basis of equilibrium. However, because the maximum moment in the span was greater than M_p, which is clearly inadmissible, a wrong value of the collapse load which was higher than the actual collapse load was obtained.

The lesson to be learnt from this example is that by assuming *any* collapse mechanism a value can be calculated for the collapse load. If it is a wrong mechanism, it can be recognized from the fact that at that collapse load the numerical value of the moment some where in the structure will be $> M_p$ of the corresponding section.

11.7 ● Use of virtual work to calculate the collapse load of a mechanism

Because the collapse load is calculated by assuming plastic hinges at possible hinge positions and using equations of equilibrium, it is possible to use virtual work to establish the equations of equilibrium more conveniently. This is illustrated by a few simple examples.

Example 1

For the propped cantilever shown in Fig. 11.12(a), the equation of equilibrium was shown to be

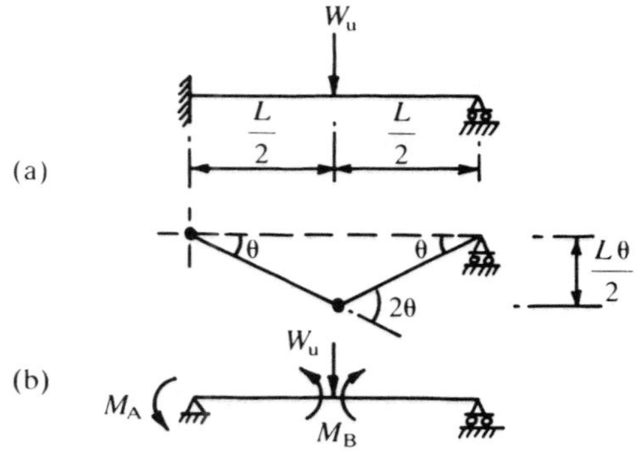

Figure 11.12 ● (a), (b) Collapse mechanism for a propped cantilever under midspan concentrated load

$$M_B = W_u L/4 - M_A/2$$

i.e. $M_A + 2M_B = W_u L/2$

This equation is established more conveniently as follows. At collapse plastic hinges form at A and B. Except at A and B, the bending moment everywhere is assumed to be less than M_p. Therefore if a virtual displacement as shown in Fig. 11.12(b) is given to the structure when it is carrying the ultimate load, virtual work is done by moments at A and B and the load W. Moments at other parts of the beam do no work because there are no curvature changes at those locations as the structure deforms as a straight line between plastic hinges. Deflection of the load is $0.5L\ \theta$. Therefore external work done by the load is $0.5L\ \theta W_u$. The internal work is done only by moments at A and B. The work done by moment at A is $(-M_A\theta)$ and the work done by moment at B is $(-M_B2\theta)$. Therefore for equilibrium virtual work done must be zero. Thus,

$$-M_A\ \theta - M_B2\theta + W_u(0.5L\ \theta) = 0$$

Cancelling θ, $M_A + 2M_B = 0.5W_uL$, an equation obtained from equilibrium considerations before in Section 11.4. If $M_A = M_B = M_p$, then

$$3M_p = 0.5W_uL$$

$$W_u = 6M_p/L$$

Example 2

Consider the fixed beam shown in Fig. 11.13. Assuming plastic hinges at A, B and C, the virtual work equation is established as

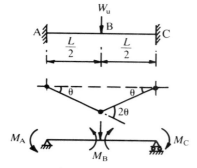

Figure 11.13 ●
Collapse mechanism for a clamped beam under midspan concentrated load

$$M_A\ \theta + M_B2\theta + M_C\ \theta = W_u(0.5L\ \theta)$$

If $M_A = M_B = M_C = M_p$, then

$$4M_p = 0.5W_uL$$

$$W_u = 8M_p/L$$

Example 3

Consider the fixed beam shown in Fig. 11.14. Assuming plastic hinges at A, B and C, the virtual work equation is established as

$$M_A\ \theta + M_B2\theta + M_C\ \theta = \text{External work done by the loads}$$

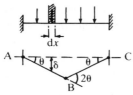

Figure 11.14 ●
Collapse mechanism for
a clamped beam under
uniformly distributed load

On an element of length dx of the beam, the load acting is q dx and if the corresponding deflection is δ, then the external work done is $= \int q_u \, dx\delta = q_u \int \delta \, dx = q_u$ [area of deflected shape under the load]. The area of deflected shape is a triangle with base L and height $0.5L\,\theta$. Therefore

$$M_A\,\theta + M_B 2\theta + M_C\,\theta = q_u 0.5L(0.5L\,\theta) = 0.25 q_u L^2$$

If $M_A = M_B = M_C = M_p$, then

$$4M_p = 0.25 q_u L^2$$

$$q_u = 16 M_p / L^2$$

Example 4

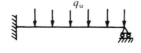

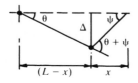

Note: $\Delta = (L - x)\theta = x\psi$

Figure 11.15 ●
Collapse mechanism for
a propped cantilever
under uniformly
distributed load

Consider the propped cantilever shown in Fig. 11.15. Assuming that the plastic hinge in the span is at a distance x from the prop, if the rotation at A is θ, then the deflection at B is $(L - x)\theta$ and the hinge rotation at C is $\psi = (L - x)\theta/x$. The total angle change at B is $\{\theta + (L - x)\theta/x\}$. The virtual work equation is given by

$$M_A\,\theta + M_B\{\theta + (L - x)\theta/x\} = q_u 0.5L(L - x)\theta$$

Setting $M_A = M_B = M_p$, then

$$q_u = 2M_p(L + x)/\{Lx(L - x)\}$$

For a minimum q_u, dq_u/d$x = 0$

$$x(L - x) - (L + x)(L - 2x) = 0$$

i.e. $x^2 + 2Lx - L^2 = 0$

Solving, $x = L(\sqrt{2} - 1) = 0.414L$. Substituting this value for x, q_u is given by

$$q_u = 11.66 M_p / L^2$$

It is worth remembering that although when writing the virtual work equation the sign convention for M and θ is important, because the work done at the plastic hinges will always be positive, no attention need to be paid to the sign of the product of M_p and the corresponding plastic rotation.

11.8 ● General collapse conditions

The discussion in Section 11.6 on the calculation of the collapse load of a propped cantilever under uniformly distributed load showed that although assuming an arbitrary collapse mechanism, a corresponding collapse load can be calculated, a *true collapse mechanism* satisfies the following three necessary and sufficient conditions.

1. Equilibrium condition. The bending moment distribution must be in equilibrium with the external loads.

2. Yield condition. The bending moments may nowhere exceed the plastic moment values of the members.

3. Mechanism condition. There must be sufficient plastic hinges to form a mechanism.

If a system of bending moments can be found which satisfies these three conditions then that system defines the true collapse load of the structure. Formal proof of the above theorem is available. In practice two corollaries of the above theorem are of considerable use. They are

(a) Upper bound load corollary. If a collapse mechanism is assumed and the corresponding collapse load is determined from equating the internal work at the plastic hinges to the corresponding external work, then the calculated collapse load will be equal to or greater than the true collapse load because to maintain equilibrium under the calculated collapse load, the bending moment at certain sections might be greater than the plastic moment capacity of the section.

(b) Lower bound load corollary. If a bending moment distribution is assumed such that the bending moment at any section does not exceed the plastic moment capacity at that section and the corresponding load is calculated from equilibrium considerations, then that 'collapse' load is less than or equal to the true collapse load because sufficient plastic hinges might not be present to form a collapse mechanism.

The use of the above corollaries for the calculation of the true collapse load of rigid-jointed structures will be demonstrated in the next section.

11.9 ● Calculation of collapse load – examples

The general procedure for the calculation of the true collapse load is as follows.

(a) A possible collapse mechanism is assumed and the corresponding collapse load is calculated.
(b) Step (a) is repeated for all other possible collapse mechanisms.
(c) Tentatively the collapse mechanism yielding the lowest collapse load is assumed to be correct.
(d) Using equilibrium considerations, the bending moment distribution in the whole structure is determined. If the bending moment distribution indicates that the plastic moment capacity is not exceeded, then the true collapse load has been determined.

The procedure is illustrated by some examples.

Example 1

Figure 11.16(a) shows a uniform two-span beam fixed at one end and simply supported at the other end. Figure 11.16(b)–(f) shows some of the possible collapse mechanisms. Note that the structure is two times statically indeterminate and therefore requires three plastic hinges for a complete collapse. The collapse loads for each mechanism can be determined as follows.

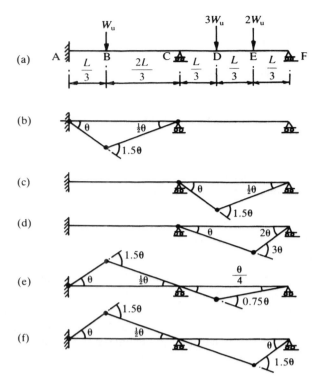

Figure 11.16 ● (a)–(f)
Collapse mechanisms for
a continuous beam

(a) Collapse in span ABC (Fig. 11.6(b)). Assuming plastic hinges at A, B and C, the collapse load can be calculated as follows. Let the virtual displacement be defined by a rotation at A of θ. The deflection at B is $(L/3)\theta$. For compatibility, the rotation at C is $\theta/2$. Using virtual work,

$$M_A \theta + M_B(\theta + 0.5\theta) + M_C(0.5\theta) = W_u(L/3)\theta$$

$M_A = M_B = M_C = M_p$, therefore

$$W_u = 9M_p/L$$

This is a case of complete collapse.

(b) Collapse in span CDEF (Fig. 11.6(c)). Assuming plastic hinges at C, D and a real hinge at F, the collapse load can be calculated as follows. Let the virtual displacement be defined by a rotation at C of θ. The deflection at D is $(L/3)\theta$. For compatibility, the rotation at F is $\theta/2$. The displacement at E is $(L/3)0.5\theta = (L/6)\theta$. Using virtual work,

$$M_C \theta + M_D(\theta + 0.5\theta) + M_F(0.5\theta)$$

$$= 3W_u(L/3)\theta \text{ at D} + 2W_u(L/6)\theta \text{ at E} = 1.33W_uL\,\theta$$

$M_C = M_D = M_p$, and $M_F = 0$, therefore

$$W_u = 1.875M_p/L$$

This is a case of partial collapse.

(c) Collapse in span CDEF (Fig. 11.6(d)). Assuming plastic hinges at C, E and a real hinge at F, the collapse load can be calculated as follows. Let the virtual displacement be defined by a rotation at C of θ. The deflection at E is $(2L/3)\theta$ and the deflection at D is $(L/3)\theta$. For compatibility, the rotation at F is 2θ. Using virtual work,

$$M_C\,\theta + M_E(\theta + 2\theta) + M_F(2\theta)$$

$$= 3W_u(L/3)\theta \text{ at D} + 2W_u(2L/3)\theta \text{ at E} = 2.33W_uL\,\theta$$

$M_C = M_E = M_p$, and $M_F = 0$, therefore

$$W_u = 1.714M_p/L$$

This is a case of partial collapse.

(d) Collapse of the whole beam (Fig. 11.6(e)). Assuming plastic hinges at A, B, D and a real hinge at F we have, assuming the plastic rotation at A is θ, deflection at B is $(L/3)\theta$ upwards, section BCD rotates about C by 0.5θ and the deflection at D is $(L/3)0.5\theta$ downwards, F rotates 0.25θ and E deflects $(L/3)0.25\theta$. Using virtual work

$$M_A\,\theta + M_B(\theta + 0.5\theta) + M_D(0.5\theta + 0.25\theta) + M_F(0.25\theta)$$

$$= -W_u(L/3)\theta \text{ at B} + 3W_u(L/3)0.5\theta \text{ at D} + 2W_u(L/3)0.25\theta \text{ at E}$$

Note that at B the load is acting downwards but the deflection is upwards. Therefore the load at B does negative external work. Assuming $M_A = M_B = M_D = M_p$, $M_F = 0$

$$W_u = 9.75M_p/L$$

This is a case of complete collapse.

(e) Collapse of the whole beam (Fig. 11.6(f)). Assuming plastic hinges at A, B, E and a real hinge at F we have, assuming the plastic rotation at A is θ, deflection at B is $(L/3)\theta$ upwards, section BCE rotates 0.5θ and the deflection at D and E are respectively $(L/3)0.5\theta$ and $(2L/3)0.5\theta$ both downwards. At F rotation is θ. Using virtual work

$$M_A\,\theta + M_B(\theta + 0.5\theta) + M_E(0.5\theta + \theta) + M_F\,\theta$$

$$= -W_u(L/3)\theta \text{ at B} + 3W_u(L/3)0.5\theta \text{ at D} + 2W_u(2L/3)0.5\theta \text{ at E}$$

Note that at B the load is acting downwards but the deflection is upwards. Therefore the load at B does *negative* external work. Assuming $M_A = M_B = M_E = M_p$, $M_F = 0$

$$W_u = 4.8M_p/L$$

This is another case of complete collapse.

Obviously the mechanism in Fig. 11.6(d) gives the lowest collapse load. However, before concluding that it is the true collapse load, it is necessary to show that this collapse load can be sustained without the moment at any section exceeding the plastic moment capacity of the section. This is best done by using the equations of equilibrium used in deriving the collapse load. Thus for the mechanism in Fig. 11.6(c), if $W_u = 1.714M_p/L$, then

$$M_C\,\theta + M_D(\theta + 0.5\theta) + M_F(0.5\theta) = 1.33W_uL\,\theta$$

where $M_F = 0$ and $M_C = M_p$ at collapse. Therefore

$$M_D = 0.86M_p$$

Alternatively, for the mechanism in Fig. 11.6(d), the actual moment value at D can be calculated as follows. Because the moment at E is M_p, the reaction V_F at F must be $3M_p/L$. As the total load on the span CDEF is $5W_u = 8.58M_p/L$, the reaction V_C at C for the span CDEF is $5.58M_p/L$. Therefore the moment at D is $(V_CL/3 - M_p$ at C$) = 0.86M_p$. Similarly for the mechanism in Fig. 11.6(b), equilibrium requires that

$$M_A\,\theta + M_B(\theta + 0.5\theta) + M_C(0.5\theta) = W_u(L/3)\theta$$

Thus for equilibrium,

$$M_A + 1.5M_B + 0.5M_C = W_uL/3$$

If $W_u = 1.714M_p/L$ and $M_C = M_p$ at collapse, then

$$M_A + 1.5M_B = -0.429M_p$$

For example, if $M_A = 0$, then $M_B = -0.28M_p$.

Once again it is clear that the above equilibrium equation can be satisfied without M_A and M_B exceeding M_p.

Similar calculations can be made for mechanisms in Fig. 11.6(e) and (f). It is worth emphasizing that it is not necessary to calculate the exact distribution of moments. All that is necessary is to show that a possible distribution of moments exists without violating the yield condition.

It is useful to note that the continuous beam is statically indeterminate to the second degree, i.e. $n = 2$. For a complete collapse $n + 1 = 3$ hinges must form. However, because one span of the structure has collapsed with the formation of only two plastic hinges, it is a case of partial collapse. In this example it would have been theoretically possible to use a 'weaker' beam for the span ABC. However, practical considerations such as connecting two different beam sections might make it more economical to use a stronger than required beam for the span ABC. The reader should not draw any general conclusion from the fact that in this particular case a partial rather than a complete collapse mechanism gave the smallest collapse load. It is purely accidental.

Example 2

Figure 11.17(a) shows a single-storey, single-bay portal frame with constant plastic moment capacity M_p. Calculate the true collapse load.

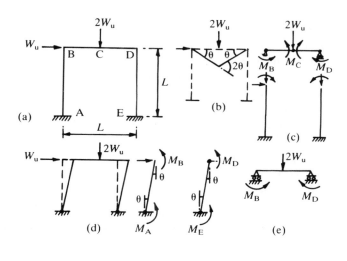

Figure 11.17 ● (a)–(g) Collapse mechanisms for a single-bay portal frame

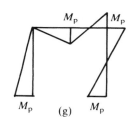

Solution Consider the two easy-to-visualize collapse mechanisms. They are called the beam mechanism because only the beam is involved in the collapse and the sway mechanism because collapse takes place by the horizontal sway failure of the whole frame. The corresponding collapse loads are calculated as follows.

(a) Beam mechanism. The virtual work equation is

$$M_B \theta + M_C 2\theta + M_D \theta = (2W_u)(0.5L\ \theta) = W_u L\ \theta$$

If $M_B = M_C = M_D = M_p$, then

$$W_u = 4M_p/L$$

Note that in this mechanism the horizontal load W_u is not involved in the external work. The sign assumed that M_B and M_D cause tension on the top face of the beam so that M_B is anticlockwise and M_D is clockwise.

(b) Sway mechanism. The virtual work equation is

$$M_A \theta + M_B \theta + M_D \theta + M_E \theta = W_u(L\ \theta) = W_u L\ \theta$$

If $M_A = M_B = M_D = M_E = M_p$, then

$$W_u = 4M_p/L$$

Note that in this mechanism the vertical load is not involved in the external work. This is because if a column of height H rotates by a small angle θ, then the horizontal displacement of the column is $H \sin \theta$ and the vertical displacement is $H(1 - \cos \theta)$. If θ is small, then the horizontal displacement is $h \, \theta$ and vertical displacement is 0.

Because both the mechanisms yield the same collapse load, it may be tempting to assume that the correct collapse load has been calculated. However, unless a check has been made to show that a bending moment distribution without violating the yield condition can be found, the lowest calculated value of collapse load cannot be accepted as correct because it is possible a mistake has been made or other possible collapse mechanisms have been missed out.

Check: Assuming $W_u = 4M_p/L$ as correct and using the beam mechanism with $M_B = M_p$ acting anticlockwise and $M_D = M_p$ acting clockwise then vertical equilibrium can be preserved. If now the sway mechanism is examined for horizontal equilibrium with $W_u = 4M_p/L$ and using the moments at B and D *from the beam mechanism*, the horizontal equilibrium equation becomes

$$M_A \, \theta + M_B \, \theta + M_D \, \theta + M_E \, \theta = W_u(L \, \theta) = W_u L \, \theta$$

where all the moments are assumed to be anticlockwise as acting on the columns. If M_B = anticlockwise when acting on the beam and therefore clockwise acting on the column = $-M_p$. M_D = clockwise when acting on the beam and anticlockwise when acting on the column = M_p. M_A and M_E are unknown moments. Therefore

$$M_A \, \theta - M_p \, \theta + M_p \, \theta + M_E \, \theta = W_u L \, \theta = 4M_p \, \theta$$

and

$$M_A + M_E = 4M_p$$

It is quite clear that there is no way of satisfying the above equation with both M_A and M_E being numerically less than M_p. In other words, $W_u = 4M_p/L$ is an upper bound to the true collapse load.

Alternative check: We can come to the same conclusion by starting with the sway mechanism and examining the vertical equilibrium using the beam mechanism.

From the sway mechanism, $M_B = M_D = M_p$ acting anticlockwise on the top of the columns. Therefore M_B and M_D act on the beam in a clockwise direction. Thus when using the beam mechanism equilibrium equation, $M_B = -M_p$, $M_D = M_p$ and $W_u = 4M_p/L$. Therefore

$$M_B \, \theta + M_C \, 2\theta + M_D \, \theta = W_u L \, \theta$$

$$-M_p \, \theta + M_C \, 2\theta + M_p \, \theta = 4M_p \, \theta$$

Therefore

$$M_C = 2M_p$$

which violates the yield condition.

Therefore there must be other possible mechanisms which result in a smaller collapse load. In fact, as shown in Fig. 11.17(f), a composite mechanism involving both beam and sway mechanisms is possible. It can be seen that in the beam mechanism beam segment BC rotates clockwise by θ and in the sway mechanism the column AB also rotates about A in the clockwise direction by θ. Thus when the two mechanisms are superimposed, it is possible to create a new mechanism where segment ABC rotates as a rigid body about A in the clockwise direction, column ED rotates about E in the clockwise direction and beam segment CD rotates about D in the anticlockwise direction. Thus the equilibrium equation becomes

$$M_A\,\theta + M_C(\theta + \theta) + M_D(\theta + \theta) + M_E\,\theta = 2W_u(L/2)\theta + W_uL\,\theta$$

Setting $M_A = M_C = M_D = M_E = M_p$,

$$W_u = 3Mp/L$$

This is a smaller collapse load than that for the beam or sway mechanisms. However, it is necessary to show that this load can be sustained without the moment exceeding the plastic moment capacity of the section.

Thus using the beam mechanism

$$M_B\,\theta + M_C2\theta + M_D\,\theta = W_uL\,\theta = 3M_p\,\theta$$

$M_C = M_D = M_p$, and so

$$M_B = 0$$

Similarly, using the sway mechanism

$$M_A\,\theta + M_B\,\theta + M_D\,\theta + M_E\,\theta = W_u(L\,\theta) = 3M_p\,\theta$$

If $M_A = M_D = M_E = M_p$, then $M_B = 0$. Thus, at $W_u = 3M_p/L$, it is possible to obtain a bending moment distribution without violating the yield condition. Therefore the true collapse load is $W_u = 3M_p/L$. The corresponding bending moment diagram is shown in Fig. 11.17(g).

It is useful to note that the fixed base portal frame is three times statically indeterminate, i.e. $n = 3$. In this example it is the complete collapse mechanism with the formation of four plastic hinges which gave the smallest collapse load. This is exactly the opposite of what happened in the previous example of a continuous beam structure. Thus whether a collapse mechanism is complete or partial is no guide to the resulting collapse load giving the smallest collapse load.

11.10 ● Elementary mechanisms

In the above example of a portal frame, it was noted that it was easy to visualize the beam and sway mechanisms. Having obtained these two mechanisms it was possible to think of the composite or combined mechanism. The beam mechanism represents the vertical equilibrium and the sway mechanism the horizontal equilibrium relationships connecting the bending moments at plastic hinges to the external loads on the structure. These are called 'elementary' mechanisms. When analysing a complex structure, it is best to start off with elementary mechanisms because they 'lead on to' more complex composite mechanisms. It is therefore useful to develop a simple procedure for calculating the number of elementary mechanisms.

It is worth remembering that each elementary mechanism represents an equilibrium condition relating some of the bending moments to the external load acting on the structure.

Considering a structure loaded by concentrated loads so that the bending moment diagram consists of straight lines. If there are m peak values of bending moments, then these m values completely define the moment profile and each peak is a possible plastic hinge location. If the structure is n times statically indeterminate, then the moments at n sections would, for an elastic structure, be determined from n compatibility equations. The remaining $(m - n)$ values must be obtained from $(m - n)$ independent equations of equilibrium. Each independent elementary mechanism with one degree of freedom defines an equilibrium equation. Therefore the number of elementary mechanisms $= (m - n)$.

Example 1

The bending moment diagram for the uniform rectangular plane frame ABCDE shown in Fig. 11.17(a) is defined by the five values at A, B, C, D and E and the frame has three redundants. Therefore $m =$ number of possible hinge locations $= 5$, $n =$ degree of statical indeterminacy $= 3$, and $(m - n) =$ number of elementary mechanisms $= (5 - 3) = 2$. The two elementary mechanisms are beam and sway mechanisms.

Example 2

For the two-bay portal shown in Fig. 11.18(a), using the formula developed in Chapter 1, Section 1.21, $n = 6$. The value of $m = 10$. Note that at the central joint it is necessary to specify three moments at the ends of the beams and the column meeting there. The number of elementary mechanisms $= 10 - 6 = 4$. They are the two beam mechanisms defining the vertical equilibrium of the individual beams, one sway mechanism defining the horizontal equilibrium of the structure and one 'joint mechanism' representing the equation of equilibrium that the sum of the moments at the ends of three members meeting at the central joint must be zero.

$m = 10, n = 6, m - n = 4$

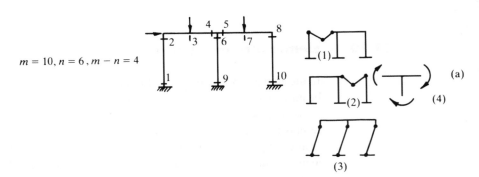

(a)

Figure 11.18 ● (a)
Elementary mechanisms

Example 3

Figure 11.18(b) shows a single-bay, two-storey frame. Therefore using the formula developed in Chapter 1, Section 1.21, $n = 6$. From Fig. 11.18(b), $m = 12$. Therefore

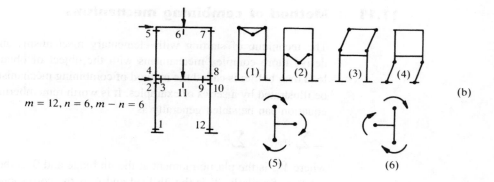

$$m = 12, n = 6, m - n = 6$$

(b)

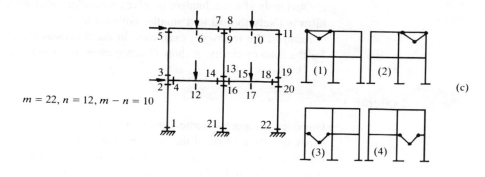

$$m = 22, n = 12, m - n = 10$$

(c)

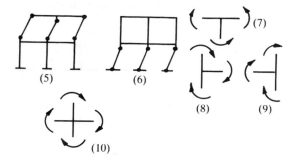

Figure 11.18 ● (b)–(c)
Elementary mechanisms

the number of elementary mechanisms is equal to $(m - n) = 6$. They consist of two beam mechanisms, two sway mechanisms and two joint mechanisms as shown in Fig. 11.18(b).

Example 4

Figure 11.18(c) shows a two-bay, two-storey frame. Using the formula developed in Chapter 1, Section 1.21, $n = 12$. From Fig. 11.18(c), $m = 22$. Therefore $m - n = 10$. The ten elementary mechanisms consist of four beam mechanisms, two sway mechanisms, and four joint mechanisms as shown in Fig. 11.18(c).

11.11 ● Method of combining mechanisms

The technique of starting with elementary mechanisms and combining them to derive more complex mechanisms with the object of obtaining a lower collapse load than before is called the method of combining mechanisms. The technique will be illustrated by a series of examples. It is worth remembering that the virtual work equation can be stated generally as

$$\sum M_{pi}\theta_i = \sum W_j \Delta_j$$

where M_{pi} is the plastic moment at the ith hinge and θ_i is the corresponding plastic rotation. Similarly W_j is the jth load and Δ_j is the corresponding displacement.

Obviously if a mechanism involves a smaller value of $\sum M_{pi}\theta_i$, then that mechanism is likely to lead to a smaller collapse load.

Note: In the following examples, in the diagrams the plastic hinges are shown, for the sake of clarity, a short distance away from the joint.

Example 1

Consider the portal frame shown in Fig. 11.19(a). The column plastic moment capacity is twice that of the beam plastic moment capacity. Therefore at the junction between the beam and the column the plastic hinge forms in the beam rather than in the column. In this example $m = 5$ and $n = 3$. Therefore the number of elementary mechanisms $= (5 - 3) = 2$. The virtual work equations are

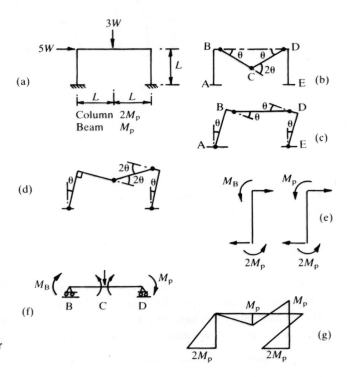

Figure 11.19 ● (a)–(g)
Collapse mechanisms for
a portal frame

(a) Beam mechanism. From Fig. 11.19(b)

$$M_p \theta + M_p(\theta + \theta) + M_p \theta = 3W_uL \theta$$

$$4M_p \theta = 3W_uL \theta$$

Therefore

$$W_u = 1.33M_p/L$$

(b) Sway mechanism. From Fig. 11.19(c)

$$2M_p \theta \text{ at } A + M_p \theta \text{ at } B + M_p \theta \text{ at } D + 2M_p \theta \text{ at } E = 5W_uL \theta$$

$$6M_p \theta = 5W_u \theta$$

Therefore

$$W_u = 1.2M_p/L$$

(c) Combined mechanism. Fig. 11.19(d)

The combined mechanism is obtained by combining the beam and sway mechanisms but removing the plastic hinge at B. The virtual work equation is given by

$$4M_p \theta = 3W_uL \theta \text{ from the beam mechanism}$$

$$6M_p \theta = 5W_uL \theta \text{ from the sway mechanism}$$

$$- M_p \theta - M_p \theta = 0 \text{ removing the plastic hinge at B.}$$

Note that when removing the work done at a hinge, $M_p \theta$ is removed twice because when adding the beam and sway mechanisms we have included the work done at the plastic hinge at B twice – once in the beam mechanism and once for the sway mechanism.

Therefore the virtual work equation for the combined mechanism is

$$(4 + 6 - 2)M_p \theta = 8W_uL \theta$$

$$8M_p \theta = 8W_uL \theta$$

Therefore

$$W_u = M_p/L$$

Thus the smallest collapse load is $W_u = M_p/L$

The actual bending moment at the various sections can be determined from statics because we have a complete collapse.

The bending moment at plastic hinges at A, D and E of the columns acts in the anticlockwise direction and assuming that the bending moment in the column at B is also in the anticlockwise direction, as shown in Fig. 11.19(e), then for horizontal equilibrium

$$(2M_p \text{ at } A + M_B + M_p \text{ at } D + 2M_p \text{ at } E)/L = 5W_u = 5M_p/L$$

Therefore

$$M_B = 0$$

As a check, considering the beam in Fig. 11.19(f),

$$M_C = M_p = (3W_u2L)/4 \text{ due to simply supported beam action}$$
$$- (M_p \text{ at D} - M_B)/2 \text{ due to end moments}$$

where M_B is assumed to act clockwise on the beam. Therefore

$$M_p = 1.5M_p - 0.5M_p + 0.5M_B$$
$$M_B = 0$$

The bending moment diagram at collapse is shown in Fig. 11.19(g) from which it is clear that the bending moment at no section exceeds the plastic moment capacity of the section.

Example 2

Figure 11.20(a) shows a two-bay portal frame. The plastic moment capacities are shown against the members. As discussed in Section 11.10, there are four elementary mechanisms.

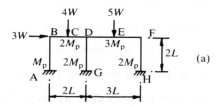

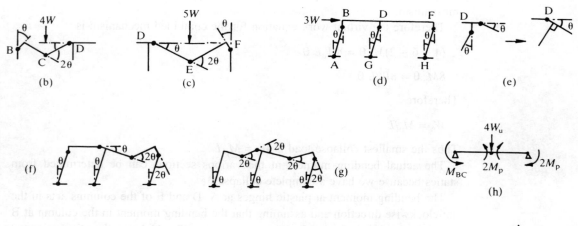

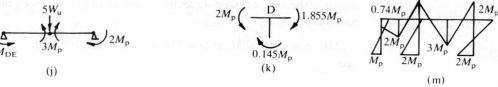

Figure 11.20 ● (a)–(m) Collapse mechanisms for a portal frame

They are: two beam mechanisms, one sway mechanism and one joint mechanism. The virtual work equation for the beam and sway mechanisms can be easily written as follows. It is worth remembering that when only two members meet at a joint, the plastic hinge always forms in the weaker member. When more than two members meet, it is best to assume that the plastic hinge forms at the end of individual members.

Mechanism 1. Consider the beam mechanism in member BCD shown in Fig. 11.20(b). At B, the plastic hinge forms in the column while at D it forms in the beam. Therefore $M_p \theta$ at B + $2M_p(\theta + \theta)$ at C + $2M_p \theta$ at D = $4W_uL \theta$. Thus

$$7M_p \theta = 4W_uL \theta \tag{1}$$

$$W_u = 1.75M_p/L$$

Mechanism 2. Consider the beam mechanism in member DEF shown in Fig. 11.20(c). At F, the plastic hinge forms in the column while at D it forms in the beam. Therefore $3M_p \theta$ at D + $3M_p(\theta + \theta)$ at E + $2M_p \theta$ at F = $7.5W_uL \theta$. Thus

$$11M_p \theta = 7.5W_uL \theta \tag{2}$$

$$W_u = 1.47M_p/L$$

Mechanism 3. Sway mechanism in which the plastic hinges form at the top and bottom of the columns shown in Fig. 11.20(d). Thus $M_p \theta$ at A + $M_p \theta$ at B + $2M_p \theta$ at G + $2M_p \theta$ at D + $2M_p \theta$ at H + $2M_p \theta$ at F = $3W_u(2L \theta)$

$$10M_p \theta = 6W_uL \theta \tag{3}$$

$$W_u = 1.67M_p/L$$

Having obtained the virtual work expressions for these three mechanisms, it is clear that the smallest collapse load is from mechanism 2. In combining mechanisms, mechanisms with low values of collapse load are combined provided there is a possibility of reducing the value of $\Sigma M_{pi} \theta_i$. In this case mechanism 2 and 3 can be combined because there is a possibility that the number of plastic hinges can be reduced at D. It is possible for mechanisms 1 and 3 to be combined to remove the hinge at B. However, because the collapse load from mechanism 1 is higher than from 2, it is advisable to start by combining 2 and 3. Because, when the two mechanisms are superimposed, DG and DE are inclined in the clockwise direction by θ, like the combined mechanism in a single-bay portal, for compatibility of displacements it is not necessary to have plastic hinges at D in beam DEF and column DG. Because BCD is horizontal, it is necessary, for compatibility of displacements, to have a plastic hinge at D in beam BCD. This is shown in Fig. 11.20(e). Thus the virtual work equation for the new mechanism is the sum of virtual work equations for mechanisms (2) and (3) from which the work at plastic hinges at D in DEF and DG are removed and the plastic work at the new hinge at D in beam BCD is added. Thus for the mechanism in Fig. 11.20(f)

$11M_p \theta = 7.5W_uL \theta$ from (2)

$10M_p \theta = 6.0W_uL \theta$ from (3)

$-2M_p \theta$ Remove hinge at D in DG

$-3M_p \theta$ Remove hinge at D in beam DEF

$+2M_p \theta$ Add the new hinge at D in beam BCD
___

$$18M_p \theta = 13.5W_uL \theta \tag{4}$$

Therefore

$$W_u = 1.33M_p/L$$

This value of collapse load is smaller than for mechanisms (1) and (3). This can be still further improved by noting that if mechanisms (4) and (1) are combined, then as in the combined mechanism for a single-bay portal frame, the hinge at B can be removed without adding any new hinge. Thus

$$18M_p \theta = 13.5W_uL \theta \tag{4}$$

$$7M_p \theta = 4.0W_uL \theta \tag{1}$$

$-M_p \theta - M_p \theta$ Remove plastic work at hinge at B
                               in mechanisms (1) and (4)
___

$$23M_p \theta = 17.5W_uL \theta \tag{5}$$

Therefore

$$W_u = 1.314M_p/L$$

The new mechanism is shown in Fig. 11.20(g).

Because we have exhausted all possible mechanisms, it may be tentatively assumed that this is the true collapse load to be confirmed by calculating the BM distribution at collapse.

Bending moment calculations:

Beam BCD (Fig. 11.20(h)).
$M_C = 2M_p = 4W_u(2L/4) - 0.5(M_{BC} + 2M_p$ at D)
Because $W_u = 1.314M_p/L$, $M_{BC} = -0.744M_p$ causing tension at the bottom face at B.

Beam DEF (Fig. 11.20(j)).
$M_E = 3M_p = 5W_u(3L/4) - 0.5(M_{DE} + 3M_p$ at F)
Because $W_u = 1.314M_p/L$, $M_{DE} = 1.855M_p$ causing tension at the top face at D.

Joint equilibrium at D (Fig. 11.20(k)). Moments acting at the joint are $M_{DC} = 2M_p$ causing tension at the top, $M_{DE} = 1.855M_p$ causing tension at the top and M_{DG}. For equilibrium $M_{DG} = 2M_p - 1.855M_p = 0.145M_p$ causing tension to the right in column DG.

Thus all the moments are determined. The horizontal equilibrium of the structure can be used as a check. As the moments at the top and bottom of the columns are all anticlockwise, then for horizontal equilibrium

$$(M_{AB} + M_{BA} + M_{GD} + M_{DG} + M_{HF} + M_{FH})/(2L) = 3W_u$$

$$(1.0 + 0.744 + 2.0 + 0.145 + 2.0 + 2.0)M_p/(2L) = 3.945M_p/L = 3W_u$$

Therefore

$$W_u = 1.315M_p/L$$

This structure, which was six times statically indeterminate, had seven plastic hinges at collapse. Therefore it is a case of complete collapse. The bending moment diagram at collapse is shown in Fig. 11.20(m).

Example 3

Figure 11.21(a) shows a single-bay, four-storey structure. The plastic moment capacities of the columns is M_p, that of the upper two beams is $1.5M_p$ and that of the two lower beams is $1.6M_p$. The structure is 12 times statically indeterminate and the number of sections at which moment values need to be known is 26. There are therefore $(26 - 12) = 14$ elementary mechanisms. These consist of four beam mechanisms, four sway mechanisms and six joint mechanisms at the six joints C,

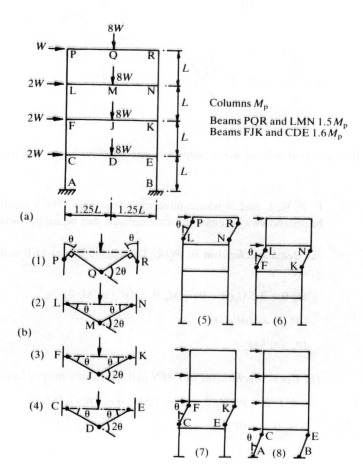

Figure 11.21 ●
(a), (b) Elementary
collapse mechanisms for
a multi-storey frame

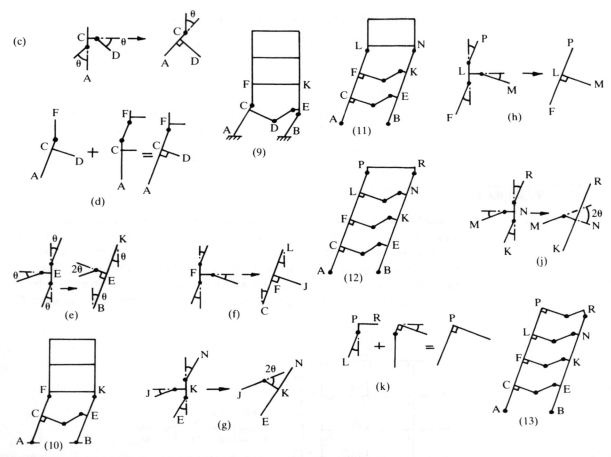

Figure 11.21 ● (c)–(k) Combined collapse mechanisms and associated joint mechanisms

E, F, K, L and N where three members meet. The virtual work equations for the beam and sway elementary mechanisms can be easily worked out as follows.

(a) Beam mechanism in PQR. The plastic hinges at P and R form in the column. Therefore

$$M_p \theta + 1.5M_p(\theta + \theta) + M_p \theta = 8W_u(1.25L \theta)$$

$$5M_p \theta = 10W_uL \theta \tag{1}$$

$$W_u = 0.5M_p/L$$

(b) Beam mechanism in LMN. All the plastic hinges form in the beam. Therefore

$$1.5M_p \theta + 1.5M_p(\theta + \theta) + 1.5M_p \theta = 8W_u(1.25L \theta)$$

$$6M_p \theta = 10W_uL \theta \tag{2}$$

$$W_u = 0.6M_pL$$

(c) Beam mechanism in FJK. All the plastic hinges form in the beam. Therefore

$$1.6M_p \, \theta + 1.6M_p \, (\theta + \theta) + 1.6M_p \, \theta = 8W_u(1.25L \, \theta)$$

$$6.4M_p \, \theta = 10W_uL \, \theta \tag{3}$$

$$W_u = 0.64M_p/L$$

(d) Beam mechanism in CDE. All the plastic hinges form in the beam. Therefore

$$1.6M_p \, \theta + 1.6M_p(\theta + \theta) + 1.6M_p \, \theta = 8W_u(1.25L \, \theta)$$

$$6.4M_p \, \theta = 10W_uL \, \theta \tag{4}$$

$$W_u = 0.64M_p/L$$

(e) Sway mechanism of storey LNPR. All the plastic hinges form in the columns. Therefore

$$M_p \, \theta + M_p \, \theta + M_p \, \theta + M_p \, \theta = W_uL \, \theta$$

$$4M_p \, \theta = W_uL \, \theta \tag{5}$$

$$W_u = 4M_p/L$$

(f) Sway mechanism of storey FLNK. All the plastic hinges form in the columns and external work is done by loads W at P and $2W$ at L both moving through $L \, \theta$. Therefore

$$M_p \, \theta + M_p \, \theta + M_p \, \theta + M_p \, \theta = W_uL \, \theta + 2W_uL \, \theta$$

$$4M_p \, \theta = 3W_uL \, \theta \tag{6}$$

$$W_u = 1.33M_p/L$$

(g) Sway mechanism of storey CFKE. Plastic hinges form in the columns and the external work is done by loads W at P, $2W$ each at L and F all moving through $L \, \theta$. Therefore

$$M_p \, \theta + M_p \, \theta + M_p \, \theta + M_p \, \theta = W_uL \, \theta + 2W_uL \, \theta + 2W_uL \, \theta$$

$$4M_p \, \theta = 5W_uL \, \theta \tag{7}$$

$$W_u = 0.8M_p/L$$

(h) Sway mechanism of storey ACBE. Plastic hinges form in the columns and the external work is done by loads W at P, $2W$ each at L, F and C all moving through $L \, \theta$. Therefore

$$M_p \, \theta + M_p \, \theta + M_p \, \theta + M_p \, \theta = W_uL \, \theta + 2W_uL \, \theta + 2W_uL \, \theta + 2W_uL \, \theta$$

$$4M_p \, \theta = 7W_uL \, \theta \tag{8}$$

$$W_u = 0.571M_p/L$$

Combine mechanisms. Having obtained the virtual work equations and the collapse loads for the elementary mechanisms, the next stage is to combine them judiciously to obtain an improved (i.e. lower) collapse load.

Step 1. Sway mechanism (8) can be combined with beam mechanism (4) with the object of removing the plastic hinges at C in CD and CA. However, for compatibility a new plastic hinge at C in CF will be introduced. This is shown in Fig. 11.21(c). Thus

$$4.0M_p \, \theta = \qquad 7W_uL \, \theta \qquad\qquad (8)$$

$$6.4M_p \, \theta = \qquad 10W_uL \, \theta \qquad\qquad (4)$$

$$-1.6M_p \, \theta \qquad \text{Remove the plastic hinge at C in CD}$$

$$-M_p \, \theta \qquad \text{Remove the plastic hinge at C in CA}$$

$$+M_p \, \theta \qquad \text{For the new hinge at C in CF}$$

Therefore

$$8.8M_p \, \theta = 17W_uL \, \theta \qquad\qquad (9)$$

$$W_u = 0.518M_p/L$$

This is still larger than $W_u = 0.5M_p/L$ from (1). The new mechanism (9) is shown in Fig. 11.21.

Step 2. Combine mechanism (9) with sway mechanism (7) because that will cancel out the hinge at C in CD (Fig. 11.21(d)). In addition, hinges at E in EB and EK can be cancelled but *altering* the rotation at the already existing plastic hinge at E in the beam CDE from θ to 2θ (Fig. 11.21(e)). Therefore

$$8.8M_p \, \theta = 17W_uL \, \theta \qquad\qquad (9)$$

$$4.0M_p \, \theta = 5W_uL \, \theta \qquad\qquad (7)$$

$$-M_p \, \theta \text{ Cancel the hinge at C in CF in mechanism} \qquad\qquad (9)$$

$$-M_p \, \theta \text{ Cancel the hinge at C in CF appearing in mechanism} \qquad\qquad (7)$$

$$-M_p \, \theta \text{ Cancel the hinge at E in EB in mechanism} \qquad\qquad (9)$$

$$-M_p \, \theta \text{ Cancel the hinge at E in EK appearing in mechanism} \qquad\qquad (7)$$
$$+ 1.6M_p \, \theta \text{ add the } \textit{additional} \text{ plastic work done at the hinge at E in ED}$$

Therefore

$$10.4M_p \, \theta = 22W_uL \, \theta \qquad\qquad (10)$$

$$W_u = 0.473M_p/L$$

This is the lowest value obtained so far. The new mechanism (10) is shown in Fig. 11.21.

Step 3. Combine (10) with (6) and (3) so that the procedure used at level CDE can be repeated at the level FJK (see Fig. 11.21(f, g)).

$$10.4M_p\,\theta = 22W_uL\,\theta \tag{10}$$

$$4.0M_p\,\theta = 3W_uL\,\theta \tag{6}$$

$$6.4M_p\,\theta = 10W_uL\,\theta \tag{3}$$

$-M_p\,\theta$ Cancel the hinge at F in FC in mechanism (10)

$-M_p\,\theta$ Cancel the hinge at F in FL in mechanism (6)

$-1.6M_p\,\theta$ Cancel the hinge at F in FJ in mechanism (3)

$-M_p\,\theta$ Cancel the hinge at K in KE in mechanism (10)

$-M_p\,\theta$ Cancel the hinge at K in KN in mechanism (6)

$+1.6M_p\,\theta$ Add the *increased* plastic work at the hinge at K in KJ in mechanism (3) because, for compatibility, the rotation at K in KJ changes from θ to 2θ

Therefore

$$16.8M_p\,\theta = 35W_uL\,\theta \tag{11}$$

$$W_u = 0.48M_p/L$$

Note that this is slightly higher than for mechanism (10). The new mechanism (11) is shown in Fig. 11.21.

Step 4. Combine (11) with (5) and (2) so that the procedure used at level FJK can be repeated at level LMN (Fig. 11.21(h, j)).

$$16.8M_p\,\theta = 35W_uL\,\theta \tag{11}$$

$$4.0M_p\,\theta = 1W_uL\,\theta \tag{5}$$

$$6.0M_p\,\theta = 10W_uL\,\theta \tag{2}$$

$-M_p\,\theta$ Cancel the hinge at L in LF in mechanism (11)

$-M_p\,\theta$ Cancel the hinge at L in LP in mechanism (5)

$-1.5M_p\,\theta$ Cancel the hinge at L in LM in mechanism (2)

$-M_p\,\theta$ Cancel the hinge at N in NK in mechanism (11)

$-M_p\,\theta$ Cancel the hinge at N in NR in mechanism (5)

$+1.5M_p\,\theta$ Add the *increased* plastic work at the hinge at N in NM in mechanism (2) because, for compatibility, the rotation at N in NM changes from θ to 2θ

Therefore

$$22.8M_p\,\theta = 46W_uL\,\theta \tag{12}$$

$$W_u = 0.496M_p/L$$

Note that this is slightly higher than for mechanism (11). It appears as though we are moving in the wrong direction because the calculated collapse load is increasing rather than decreasing. It may be that adding the beam mechanism (1) to (12) might improve the situation because all the hinges at P can be eliminated. This is worth trying. It is worth noting that we are progressing towards the multistorey equivalent of a combined mechanism for a single-bay, single-storey portal frame.

Step 5. Combine (12) with (1) so that the hinge at P can be eliminated (see Fig. 11.21(k)).

$$22.8M_p \, \theta = 46W_uL \, \theta \tag{12}$$

$$5.0M_p \, \theta = 10W_uL \, \theta \tag{1}$$

$-M_p \, \theta$ Cancel the hinge at P in mechanism (12)

$-M_p \, \theta$ Cancel the hinge at P in mechanism (1)

Therefore

$$25.8M_p \, \theta \ = 56W_uL \, \theta \tag{13}$$

$$W_u = 0.461M_p/L$$

This is the smallest collapse load so far, and anyway we have run out of mechanisms to combine. Therefore it is safe to tentatively assume that the lowest collapse load has been obtained and proceed to construct the bending moment diagram at collapse. The structure is 12 times statically indeterminate but there are only 10 plastic hinges at collapse. This is therefore a case of partial collapse.

From the point of view of verifying that the true collapse load has been obtained, all that is necessary is to show that a bending moment distribution that can maintain equilibrium and not exceed the plastic moment capacity of the section can be obtained. It should be noted that the collapse mechanism is partial and requires three more plastic hinges to form a complete collapse. Therefore at collapse, the structure is three times statically indeterminate and the moments at all the sections cannot be determined from statics.

Bending moment calculations (Fig. 11.22)

(a) Beams (Fig. 11.22(a)). The bending moment distribution in each beam is easily calculated because there is a plastic hinge in the middle and at the right-hand end. Assuming that the moment at the centre causes tension at the bottom and the support moments cause tension at the top, the moment at the left-hand end is calculated from equilibrium as

$$M_{centre} = 8W_u(2L)/4 - (M_{left} + M_{right})/2$$

For the top beam $M_{centre} = 1.5M_p$ and $M_{right} = M_p$. For the others $M_{centre} = M_{right} = 1.5M_p$ for the second beam from the top and $1.6M_p$ for the bottom two beams. $W_u = 0.461M_p/L$. From the above equation, the following bending moments can be calculated:

$$M_{PQ} = 0.607M_p \quad M_{LM} = 0.107M_p$$

$$M_{FJ} = M_{CO} = -0.193M_p$$

(b) Columns. Assuming that the bending moments in the columns are anticlockwise, then for horizontal equilibrium of the storey

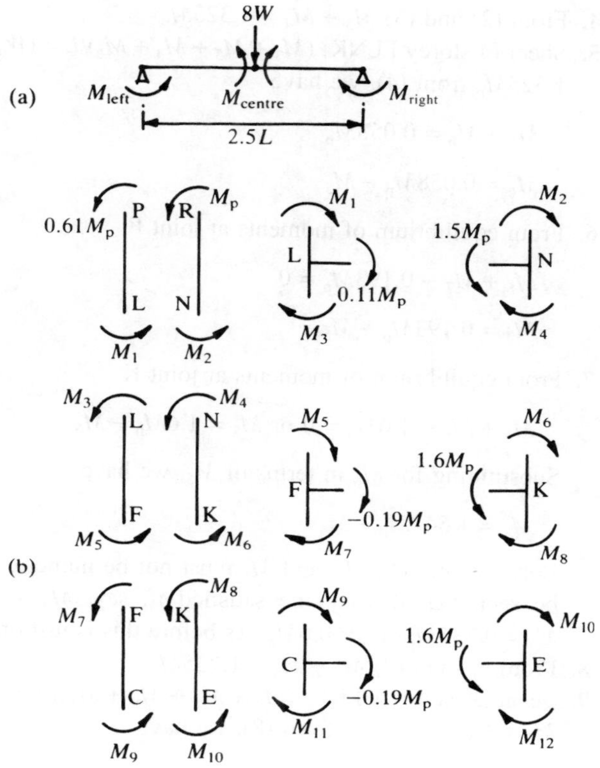

Figure 11.22 ● (a), (b) Bending moment distribution in a multistorey frame

$\{(M_{top} + M_{bottom})$ of left column
$+ \{(M_{top} + M_{bottom})$ of right column$\}/L =$ storey shear

As shown in Fig. 11.22(b), the following equations can be derived.

1. Shear of storey LPNR: $(-0.607M_p$ from $M_{PQ} + M_1 + M_p$ from $M_{RQ} + M_2)/L = W_u$ therefore

$$M_1 + M_2 = 0.068M_p$$

$$M_2 = 0.068M_p - M_1 \tag{1}$$

2. From equilibrium of moments at joint L

$$M_1 + M_3 + 0.107M_p = 0$$

$$M_3 = -0.107M_p - M_1 \tag{2}$$

3. From equilibrium of moments at joint N

$$M_2 + M_4 - 1.5M_p = 0 \text{ or } M_4 = 1.5M_p - M_2$$

Substituting for M_2 in terms of M_1, we have

$$M_4 = 1.432M_p + M_1 \tag{3}$$

Because M_1, M_2, M_3 and M_4 must not be numerically greater than M_p, it can be seen that if, say, $M_1 = -0.432M_p$, then $M_4 = M_p$, $M_2 = 0.364M_p$ and $M_3 = 0.325M_p$. This is just one possible set of moment distributions.

4. From (2) and (3) $M_3 + M_4 = 1.325M_p$ (4)

5. Shear of storey FLNK: $(M_3 + M_5 + M_4 + M_6)/L = (W_u + 2W_u)$. Using $M_3 + M_4 = 1.325M_p$ from (4), we have

$$M_5 + M_6 = 0.058M_p$$

$$M_6 = 0.058M_p - M_5$$ (5)

6. From equilibrium of moments at joint F

$$M_5 + M_7 - 0.193M_p = 0$$

$$M_7 = 0.193M_p - M_5$$ (6)

7. From equilibrium of moments at joint K

$$M_6 + M_8 - 1.6M_p = 0 \text{ or } M_8 = 1.6M_p - M_6$$

Substituting for M_6 in terms of M_5, we have

$$M_8 = 1.542M_p + M_5$$ (7)

Because M_5, M_6, M_7, and M_8 must not be numerically greater than M_p, it can be seen that this can be satisfied if, say, $M_5 = -0.542M_p$, then $M_8 = M_p$, $M_7 = 0.735M_p$, $M_6 = 0.6M_p$. As before this is just one possible set of moments.

8. From (6) and (7) $M_7 + M_8 = 1.735M_p$ (8)

9. Shear of storey CFKE: $(M_7 + M_8 + M_9 + M_{10})/L = (W_u + 2W_u + 2W_u)$. Using $M_7 + M_8 = 1.735M_p$ from (8), we have

$$M_9 + M_{10} = 0.57M_p$$

$$M_{10} = 0.57M_p - M_9$$ (9)

10. From equilibrium of moments at joint C

$$M_9 + M_{11} - 0.193M_p = 0$$

$$M_{11} = 0.193M_p - M_9$$ (10)

11. From equilibrium of moments at joint E

$$M_{10} + M_{12} - 1.6M_p = 0 \text{ or } M_{12} = 1.6M_p - M_{10}$$

Substituting for M_{10} in terms of M_9, we have

$$M_{12} = 1.03M_p + M_9$$ (11)

Because M_9, M_{10}, M_{11} and M_{12} must not be numerically greater than M_p, it can be seen that this can be satisfied if $M_3 = -0.03M_p$, then $M_{12} = M_p$, $M_{11} = 0.223M_p$ and $M_{10} = 0.6M_p$.

It is thus shown that a distribution of bending moments exists which can maintain equilibrium without exceeding the plastic moment capacity of the sections. Because, in addition, sufficient plastic hinges exist to convert the structure into a mechanism, it is established that the collapse load of the structure is $W_u = 0.461M_p/L$.

11.12 ● Pitched-roof frames

Pitched-roof portal frames are commonly used in practice. Because of the fact that the distance between eaves can change, this type of frame has certain collapse mechanisms which are not present in the rectangular portal frame. Consider the frame shown in Fig. 11.23(a) subjected to a vertical load V and a horizontal load H. It is assumed that the frame is of uniform section with the plastic moment value of M_p. The frame is three times statically indeterminate and there are five sections where the moment values need to be known. Therefore there are $(5 - 3) = 2$ elementary mechanisms. Consider the various collapse mechanisms shown in Fig. 11.23(b) to (d). Note that in this frame there is no beam mechanism similar to the beam mechanism in a rectangular portal. This is because plastic hinges at B, C and D do not constitute a collapse mechanism as the vertical load can be transferred to the columns by simple truss action.

(a) Sway mechanism (Fig. 11.23(b)). This is the same as in rectangular frames. The columns AB and ED rotate about A and E respectively. The virtual work equation is given by

$$M_p(\theta + \theta + \theta + \theta) = Hh\ \theta$$

$$M_p 4\theta = Hh\ \theta \tag{1}$$

(b) Panel mechanism (Fig. 11.23(c)). This mechanism is peculiar to pitched-roof frames. A concept called 'instantaneous centre' is useful when calculating the displacements of load points and the rotation at the plastic hinges. This is explained below.

Member BC is fixed in position at B and during collapse has to rotate about B. Similarly, member DE is fixed in position at E and has to rotate about E. Thus, for small displacements, it may be imagined that the mechanism rotates about a point O where BC and ED when produced meet. The whole mechanism can therefore be

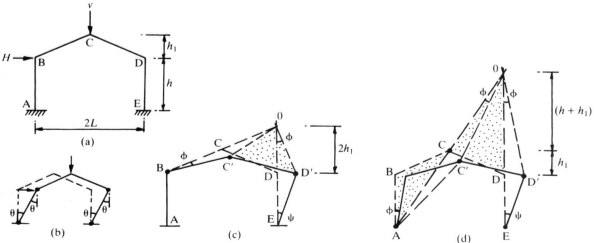

Figure 11.23 ● (a)–(d) Collapse mechanisms for a pitched-roof portal frame

assumed to consist of two rigid bars BC and DE and a rigid lamina OCD. Thus if the plastic hinge rotation at E is Ψ, then the horizontal displacement of D is $h\Psi$. The vertical distance OD is $2h_1$. Therefore, for compatibility, the rotation about O of OCD is Φ in the anticlockwise direction such that the horizontal displacement at D is

$$h\Psi = (2h_1)\Phi$$

The vertical displacement of C is downwards equal to $L\,\Phi$. For compatibility of displacement at C, the plastic hinge rotation at B is also Φ clockwise.

Because BC rotates clockwise by Φ and CD rotates anticlockwise by Φ, the plastic hinge rotation at C is $(\Phi + \Phi)$. The plastic hinge rotation at D is $(\Phi + \Psi)$ because CD rotates in an anticlockwise direction by Φ and DE rotates in the clockwise direction by Ψ. Thus the virtual work equation is given by

$$M_p\{\Phi \text{ at B} + (\Phi + \Phi) \text{ at C } + (\Phi + \Psi) \text{ at D} + \Psi \text{ at E}\} = VL\,\Phi$$

$$M_p\{4\Phi + 2\Psi) = VL\,\Phi$$

where $h\Psi = (2h_1)\Phi$. Therefore

$$M_p\{4 + 4h_1/h\}\Phi = VL\,\Phi \tag{2}$$

The above two mechanisms can be considered as the two elementary mechanisms.

(c) Combined mechanism (Fig. 11.23(d)). This can be obtained by combining mechanisms (1) and (2). However, it is sometimes preferable to work from first principles using the concept of instantaneous centre.

(i) Using the instantaneous centre approach. Column DE rotates about E. The triangular lamina ABC rotates about A. In fact the lamina ABC can be replaced by a rigid bar AC. The instantaneous centre O must lie where AC and ED produced meet. The mechanism consists of rigid bars AC, DE and the lamina OCD. Thus if the rotation at E is Ψ clockwise, then the horizontal displacement at D is $h\Psi$. For compatibility of displacement at D, the lamina OCD rotates in an anticlockwise direction about O by Φ such that

$$\Psi h = (h + 2h_1)\Phi$$

Because AC = OC, the rotation at A is Φ clockwise and the vertical displacement at C is $L\Phi$ and the horizontal displacement at B is $h\Phi$. The virtual work equation becomes

$$M_p\{\Phi \text{ at A} + (\Phi + \Phi) \text{ at C} + (\Phi + \Psi) \text{ at D} + \Psi \text{ at E}\} = Hh\,\Phi + VL\,\Phi$$

$$M_p(4\Phi + 2\Psi) = Hh\,\Phi + VL\,\Phi$$

Because $\Psi h = (h + 2h_1)\Phi$, the above equation can be expressed as

$$M_p\{6 + 4h_1/h)\Phi = Hh\,\Phi + VL\,\Phi \tag{3a}$$

(ii) Combined mechanism approach. Note that if θ in mechanism (1) is made equal to Φ in mechanism (2), then the plastic hinge at B can be cancelled and the virtual work equations for the two mechanisms are

$$M_p 4\Phi = Hh\ \Phi \tag{1}$$

where $\theta = \Phi$ has been substituted.

$$M_p\{4 + 4h_1/h\} = VL\ \Phi \tag{2}$$

$-M_p\ \Phi$ for cancelling the hinge at B in mechanism (1) $-M_p\ \Phi$ for cancelling the hinge at B in mechanism (2)

$$M_p(6 + 4h_1/h)\Phi = Hh\ \Phi + VL\ \Phi \tag{3b}$$

which is the same as equation (3a).

Example 1

Figure 11.24(a) shows a pitched-roof frame with individual loads on the two rafters. For this frame the number of sections where moment values are required is 7 and the structure is three times statically indeterminate. Thus the number of elementary mechanisms = 7 − 3 = 4.

These consist of two beam mechanisms, BCD and DEF, one sway mechanism and one panel mechanism as shown in Fig. 11.24(b) to (e) respectively.

The virtual work equations can be written as:

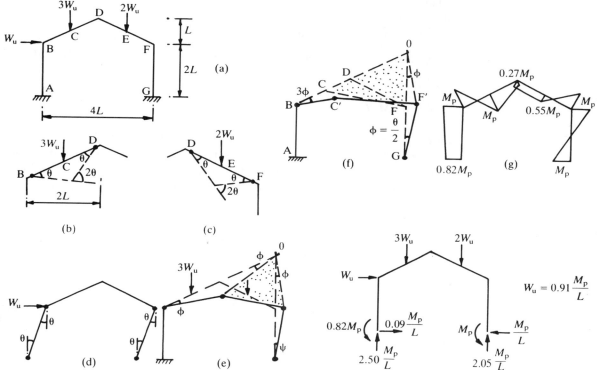

Figure 11.24 ● (a)–(g) Collapse mechanisms for a pitched-roof portal frame

(a) Beam mechanism BCD. The virtual work equation is

$$M_p\{\theta + (\theta + \theta) + \theta\} = 3W_uL\ \theta$$

$$4M_p\ \theta = 3W_uL\ \theta \tag{1}$$

$$W_u = 1.33M_p/L$$

(b) Beam mechanism DEF. The virtual work equation is

$$M_p\{\theta + (\theta + \theta) + \theta\} = 2W_uL\ \theta$$

$$4M_p\ \theta = 2W_uL\ \theta \tag{2}$$

$$W_u = 2.0M_p/L$$

(c) Sway mechanism. The virtual work equation is

$$M_p\{\theta + \theta + \theta + \theta\} = W_u2L\ \theta$$

$$4M_p\ \theta = 2W_uL\ \theta \tag{3}$$

$$W_u = 2M_p/L$$

(d) Panel mechanism. Using the instantaneous centre at O, because in this example, with reference to Fig. 11.23, height to the eaves is $h = 2L$ and the height from eaves to the ridge is $h_1 = L$, then $h = 2h_1$ and $\Phi = \Psi$. The plastic hinge rotations at B and G are Φ and at D and F are $(\phi + \phi)$. The vertical displacements at C and E are $L\Phi$.

The virtual work equation is

$$M_p\{\Phi \text{ at B} + (\Phi + \Phi) \text{ at D} + (\phi + \phi) \text{ at F} + \Phi \text{ at G}\} = 3W_uL\ \Phi + 2W_uL\ \Phi$$

$$6M_p\ \Phi = 5W_uL\ \Phi \tag{4}$$

$$W_u = 1.2M_p/L$$

(e) Combined mechanism. If mechanism (1), which has a plastic hinge at D with a rotation of θ producing tension at the top face, is combined with mechanism (4), which also has a plastic hinge at D with a rotation of 2Φ and producing tension at the bottom face, the plastic hinge at D can be eliminated if $\theta = 2\Phi$. Thus

$$6M_p\ \Phi = 5W_uL\ \Phi \tag{2}$$

Modifying the above equation by substituting $\theta = 2\Phi$, we have

$$3M_p\ \theta = 2.5W_uL\ \theta \tag{2a}$$

$$4M_p\ \theta = 3.0W_uL\ \theta \tag{1}$$

$-M_p\ \theta$ for cancelling the hinge at D in mechanism (1) $-M_p\ \theta$ for cancelling the hinge at D in mechanism (4)

$$\overline{}$$

$$5M_p\ \theta = 5.5W_uL\ \theta \tag{5}$$

$$W_u = 0.91M_p/L$$

It is interesting to note that this is another panel mechanism, except that the plastic hinge has shifted from D to C. The instantaneous centre of rotation is still at O and the mechanism consists of rigid bars BC, GF and the lamina OCF (Fig. 11.24(f)). For compatibility of horizontal displacement at F, if the rotation at G is $\Phi = 0.5\theta$ clockwise, then the rotation about the instantaneous centre is also $\Phi = 0.5\theta$ anticlockwise. Therefore for compatibility of displacement at B, the rotation at B must be $3\Phi = 1.5\theta$ clockwise. The rotation at C is $(3\phi + \phi) = 2\theta$ and at F it is $(\phi + \phi) = \theta$. Mechanisms (2) and (4) can be combined to remove the plastic hinge at D, but it will not produce a smaller collapse load than (5) because the collapse load from (2) is greater than from (1). Similarly, mechanisms (3) and (4) can be combined to remove the plastic hinge at B but again, because the collapse load from (3) is greater than that from (1), it will not produce a better result. So it can be concluded tentatively that the minimum collapse load has been obtained. This is to be confirmed by drawing the bending moment diagram at collapse.

(f) Bending moment calculation. The structure has four hinges at collapse which is one more than the degree of statical indeterminacy. The collapse is therefore complete. Using the equation for horizontal equilibrium, and assuming that the moments at the ends of the columns act in the anticlockwise direction, then

$$\{M_{AB} + M_{BA} + M_{GF} + M_{FG}\}/(2L) = W_u = 0.91M_p/L$$

However, $M_{GF} = M_{FG} = M_p$, $M_{BA} = -M_p$ as plastic hinges form at these locations. Therefore $M_{AB} = 0.82M_p$. Because the moments at the base of columns are known, vertical and horizontal reactions can be determined and the bending moment diagram constructed. Figure 11.24(g) shows the final bending moment diagram. The vertical and horizontal reactions at the supports are $V_A = 2.50M_p/L$, $V_G = 2.05M_p/L$, $H_A = 0.09M_p/L$ to the right and $H_G = M_p/L$ to the left.

11.13 ● Factors affecting the plastic moment capacity

In the discussion in the previous sections it was assumed that when the maximum moment capacity M_p of the section is reached, then the stress distribution in the section consists everywhere of σ_0 tension or compression. This stress distribution is acceptable provided that only a bending moment acts at the section. In reality at a section in a member of a rigid-jointed frame, not only a bending moment but also axial and shear forces act. In the presence of these forces, the maximum capacity of the section will be less than M_p. In this section formulae are derived for calculating the reduced bending capacity of the section in the presence of axial and shear forces.

11.13.1 Effect of axial force on the ultimate bending capacity

Consider the rectangular section in Fig. 11.25(a). Let the section be subjected to an axial compressive force F and a bending moment M so as to cause the yielding of

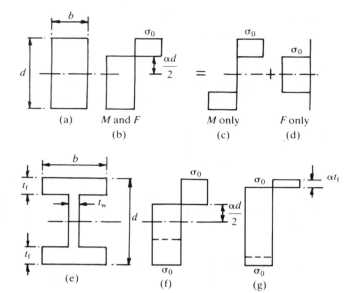

Figure 11.25 ●
(a)–(g) Stress
distribution under
bending moment and
axial force

the section. Due to the presence of the axial load F, the zero stress axis is assumed to be $0.5\alpha d$ from the centroidal axis. The stress distribution shown in Fig. 11.25(b) results in the stress everywhere in the cross-section being tension or compression σ_0. The resulting stress distribution can be split into two types. One type gives rise to a stress resultant F and the other one a stress resultant M. From the stress distribution, the stress resultants can be calculated. They are

$$M = (0.5 - 0.5\alpha)db\sigma_0(0.5 + 0.5\alpha)d = \{bd^2\sigma_0/4\}(1 - \alpha^2)$$

Because $bd^2\sigma_0/4 = M_p$, M can be expressed as

$$M = M_p(1 - \alpha^2)$$

$$F = b(\alpha d)\sigma_0 = F_p(\alpha)$$

$F_p = bd\sigma_0$ represents the 'squash' load, i.e. the maximum axial force that can be applied to a short length of the column resulting in the yielding of the whole cross-section. It should be noted that when the axial load F_p is applied, the column is assumed not to buckle.

Eliminating the parameter α from the equations for M and F, the 'interaction equation' between M and F is given by

$$M/M_p = 1 - (F/F_p)^2$$

In the above expression, because the term F/F_p appears squared, the formula is valid whether F is tensile or compressive. The above equation shows that if $F = 0$, then the maximum value of M that can be applied to the section is M_p. Similarly, if $F = F_p$, then the maximum M that the section can sustain is zero. In practice, because buckling failure has to be prevented, the ratio F/F_p will be reasonably small. The reduction in the ultimate bending capacity even in the presence of axial load is generally not serious.

In the case of an I section (Fig. 11.25(e)), a similar approach can be adopted except that two possible neutral axis positions need to be considered.

1. Neutral axis in the web (Fig. 11.25(f)). In this case the moment capacity is lost over portion $\alpha\, d$ of the web depth d. Therefore

$M = M_p$ − ultimate moment capacity of portion of the web of depth $\alpha\, d$ and breadth t_w

$$M = M_p - 0.25 t_w (\alpha\, d)^2 \sigma_0$$

$$F = (\alpha\, d) t_w \sigma_0$$

Eliminating α from the above two expressions, the relationship between M and F can be established.

2. Neutral axis in the flange (Fig. 11.25(g)). In this case if the neutral axis is at a distance $\alpha\, t_f$ from the top flange, then

$$M = b(\alpha\, t_f)\sigma_0 \{d - \alpha\, t_f\}$$

$$F = \{A - 2\alpha\, t_f\}\sigma_0 = F_p - 2\alpha\, t_f\, \sigma_0$$

where A = area of cross-section of the I beam and $F_p = A\, \sigma_0$.

Once again, eliminating the parameter α, the interaction between M and F can be established.

11.13.2 Effect of shear force on the ultimate bending capacity

In the case of interaction between axial load and bending moment, the stresses at any point in the cross-section due to the axial and bending stresses, both of which act normal to the cross-section, could be algebraically added. In the case of interaction between bending moment and shear force, we have to consider the interaction between normal stress σ due to the bending moment and shear force stress τ due to shear to cause yielding at a point. The problem is complex. It has been shown experimentally that in the case of two-dimensional problems, if at a point in the material the principal normal stresses are σ_1 and σ_2, then yielding occurs when

$$\sigma_1^2 + \sigma_2^2 - \sigma_1\sigma_2 = \sigma_0^2$$

This is known as von Mises criterion.

It can be seen that if $\sigma_1 = \sigma$ and $\sigma_2 = 0$, a case of uniaxial state of stress, yielding occurs when $\sigma = \sigma_0$. In the case of pure shear state of stress, as was shown in Chapter 3, Section 3.8.2, $\sigma_1 = \tau$ and $\sigma_2 = -\tau$. Therefore, according to von Mises criterion, yielding under pure shear stress occurs when

$$\tau^2 + \tau^2 - \tau(-\tau) = \sigma_0^2$$

Therefore

$$\tau = \sigma_0/\sqrt{3} = 0.577\sigma_0 = \tau_y$$

Therefore τ_y, the yield stress in shear, is 0.577 times the yield stress in uniaxial tension.

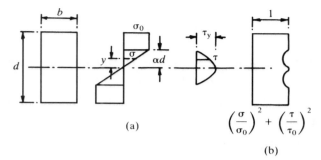

Figure 11.26 ●
(a), (b) Stress
distribution under
combined bending
moment and shear force

In the case of interaction between bending moment and shear force, at each point in the cross-section there exists a normal stress σ and a shear stress τ. The principal stresses are given by

$$\sigma_1 = 0.5\sigma + \sqrt{\{0.25\sigma^2 + \tau^2\}} \text{ and } \sigma_2 = 0.5\sigma - \sqrt{\{0.25\sigma^2 + \tau^2\}}$$

Substituting in von Mises criterion, and simplifying

$$\sigma^2 + 3\tau^2 = \sigma_0^2$$

Using $\tau_y = \sigma_0/\sqrt{3}$, the above expression becomes

$$(\sigma/\sigma_0)^2 + (\tau/\tau_y)^2 = 1$$

Therefore, for every point in the cross-section to yield, it is necessary to satisfy the above equation. This is not easily done analytically although the problem can be solved numerically. An acceptably approximate procedure is used here.

Let the stress distribution due to bending moment M and shear force Q be as shown in Fig. 11.26(a). Note that the parabolic shear stress distribution and linear bending stress distribution are confined to an area near the neutral axis while constant bending stress equal to σ_0 and zero shear stress are assumed in the region away from the neutral axis. The stress distribution in the symmetrical half of the cross-section is therefore given by

$$\sigma/\sigma_0 = y/(\alpha d), \quad \tau/\tau_y = 1 - \{y/(\alpha d)\}^2, \quad \alpha d \geqslant y \geqslant 0$$

$$\sigma = \sigma_0, \quad \tau = 0, \quad 0.5d \geqslant y \geqslant \alpha d$$

When the above values are substituted in von Mises criterion, and simplified

$$(\sigma/\sigma_0)^2 + (\tau/\tau_y)^2 = 1, \quad 0.5d \geqslant y \geqslant \alpha d$$

$$= 1 - (y/\alpha d)^2 + (y/\alpha d)^4, \quad \alpha d \leqslant y \leqslant 0$$

As shown in Fig. 11.26(b) and in Table 11.2, only over a small region of the cross-section is complete yielding of the cross-section not achieved. In other words, the section can resist more stress than shown in Fig. 11.26(a). From the stress distribution, the stress resultants M and Q can be calculated. They are given by

$$M = b(2\alpha d)^2 \sigma_0/6 \text{ due to linear stress distribution}$$
$$+ b(0.5d - \alpha d)\sigma_0\{0.5d + \alpha d\} \text{ due to constant stress distribution}$$

$$= 0.25bd^2\sigma_0(1 - 4\alpha^2/3) = M_p(1 - 4\alpha^2/3)$$

$$Q = (2/3)\,(2\alpha\,db\tau_y) = (4/3)\alpha Q_p$$

Table 11.2 ●

$y/(\alpha d)$	$(\sigma/\sigma_0)^2 + (\tau/\tau_y)^2$
0	1.0
0.1	0.99
0.2	0.96
0.3	0.92
0.4	0.87
0.5	0.81
0.7	0.75
0.8	0.77
0.9	0.85
1.0	1.00

where Q_p = plastic shear force = $bd\tau_y$. Eliminating the parameter from M and Q, we have

$$M = M_p\{1 - 0.75(Q/Q_p)^2\}$$

Note that if $Q = 0$, then $M = M_p$. However, if $Q = Q_p$, then $M = 0.25M_p$. The reason for this is that the interaction equation is only an approximate one as the whole section has not yielded.

As remarked when discussing the effect of axial force on the ultimate bending capacity, in the case of rolled steel sections, most of the bending capacity comes from the flanges where the shear stress will be small, and so in practice the effect of shear force on the bending capacity will be small.

11.14 ● Effect of plastic hinges on the buckling strength of structures

The procedure for the elastoplastic analysis of rigid-jointed structures and the determination of collapse load discussed in the previous sections assumes that the structure does not fail by buckling. In other words, the procedure discussed so far is similar to the linear elastic analysis of structures, except that when the plastic hinges form the behaviour is only 'piecewise' linear. It was noted that the formation of plastic hinges reduces the stiffness of the structure. As discussed in Chapter 8 on elastic stability analysis, the presence of axial compressive load also reduces the flexural stiffness of members. These two effects – the formation of plastic hinges and the stiffness reduction due to axial compressive forces – interact and can dramatically reduce the buckling load of structures.

In order to illustrate the phenomenon, consider the propped cantilever shown in Fig. 11.27(a). It is subjected to an axial force P and a midspan lateral load W. When the structure is elastic, the buckling load of the member, which is fixed at one end and simply supported at the other end, is given by

$$P_{cr} = 2.04P_{Euler}, \quad \text{where} \quad P_{Euler} = \pi^2 EI/L^2$$

As discussed in Section 11.4, a plastic hinge forms at the 'fixed' support A when $W = 5.33M_p/L$. Due to the presence of the axial load P, the ultimate bending

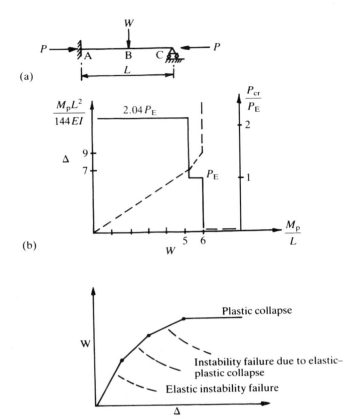

Figure 11.27 ●
(a)–(c) Interaction
of plastic hinges with
buckling load

capacity is slightly less than M_p but this has been ignored in the above calculation. After the formation of a plastic hinge at A, for further loading the structure behaves as if a real hinge was present at A. Because of this effect, for determining the buckling load, the beam is treated as a pin-ended column and the elastic buckling load is given by

$$P_{cr} = P_{Euler}, \quad \text{where} \quad P_{Euler} = \pi^2 EI/L^2$$

Note that the elastic buckling formula is still valid because the beam is elastic from support to support except for a small yielded zone at A. Thus the formation of the plastic hinge has the effect of decreasing the buckling load by a factor of more than 2.

When the lateral load is increased still further the second plastic hinge forms at the load point when $W = 6M_p/L$. In the absence of axial load, the structure can sustain this load although deflection can increase without further increase in load. When three hinges form the structure is a mechanism for which the buckling load is zero. Figure 11.27(b) shows the relationship between the lateral load W and load point deflection Δ as a discontinuous line. It also shows the relationship between W and the buckling load as a full line.

This simple example shows that the formation of plastic hinges dramatically reduces the buckling load of rigid-jointed structures. Therefore there is a possibility that the collapse load calculated by ignoring the possibility of buckling failure might not be reached in some instances. This is schematically shown in Fig. 11.27(c). In the absence of detailed elastic–plastic analysis, including a check for instability effects, various empirical formulae based on using the value of plastic collapse load calculated ignoring instability effects and the elastic buckling load calculated ignoring plasticity effects are used to estimate the true ultimate load which is affected by the formation of plastic hinges and instability effects. Interested readers should study the classic papers by Wood (1958), Horne (1979, 1985) and Anderson and Lok (1983).

11.15 ● Stiffness analysis of piecewise linear structures

The ideas developed in Chapter 6 for the analysis of linear elastic structures can be easily extended to the analysis of piecewise linear structures. The main difference is that because of the formation of new plastic hinges the degrees of freedom at any stage are dependent on the load level. In essence the analysis is iterative, with a new structure being analysed with the formation of a new set of plastic hinges.

With reference to the propped cantilever in Fig. 11.27(a), the structure is initially analysed as an elastic structure with three degrees of freedom, i.e. a translation Δ_B, rotation θ_B at B and a rotation θ_C at C. The 3×3 structural stiffness matrix is assembled and the analysis done as usual. The moment at A is given by

$$M_A = EI/(0.5L)\{2\theta_B + 6\Delta_B/(0.5L)\}$$

The bending moments at A and B are examined and when the moment at A reaches the plastic moment capacity at a load of $5.33M_p/L$, then a plastic hinge is introduced at A which increases the degrees of freedom by one. For the next loading level, the moment at A is held constant at M_p and the load on the structure is increased. Therefore for further loading

$$M_A = EI/(0.5L)\{4\theta_{pA} + 2\theta_B + 6\Delta_B/(0.5L)\}$$

where θ_{pA} is the plastic hinge rotation at A. The additional condition is $M_A = M_p$. Thus an additional degree of freedom is introduced and an additional equation is also obtained.

The new structure is analysed and the moments in the structure are examined. When the next plastic hinge forms at B, additional rotational degrees of freedom are introduced as

$$\theta_{BA} = \theta_B \quad \text{and} \quad \theta_{BC} = \theta_B + \theta_{Bp}$$

where θ_{Bp} is the *relative* plastic rotation at B. The additional condition is that

$$M_{BC} = M_p$$

and the process is continued if necessary until the next plastic hinge forms. The interested reader should study Majid (1972) and Bhatt (1981) for further information.

11.16 ● References

Anderson D, Lok T S 1983 Design studies on unbraced multistorey steel frames *Structural Engineer* **61B** (2): 29–34 (Explores the limitations of the empirical 'Rankine–Merchant' formula for estimating the true collapse load of structures.)

Baker J, Heyman J 1969 *Plastic Design of Frames* Vol. 1 Cambridge University Press (Contains a large number of exercises for practice.)

Bhatt P 1981 *Problems in Structural Analysis by Matrix Methods* Construction Press (Reprinted by A H Wheeler, Allahabad, India, 1988) (Contains a numerical example on the matrix analysis of piecewise linear elastic–plastic analysis of a rigid-jointed frame.)

Horne M R 1979 *Plastic Theory of Structures* (2nd Edn) Pergamon (A good text.)

Horne M R 1985 Frame instability and plastic design of rigid frames. In Narayanan R (ed) *Steel Framed Structures: Stability and Strength* Elsevier Applied Science: Ch 1 (An excellent state of the art summary.)

Majid K I 1972 *Nonlinear Structures* Butterworths (Details of elastic–plastic analysis using the stiffness method.)

Morris L J, Randall A L 1983 *Plastic Design* Constrado (A short design-orientated booklet.)

Moy S S J 1981 *Plastic Methods for Steel and Concrete Structures* Macmillan (A usable text.)

Neal B G 1965 *The Plastic Methods of Structural Analysis* Chapman & Hall (An early classic.)

Wood R H 1958 The stability of tall buildings *Proceedings of Institution of Civil Engineers* pp. 69–102 (An early classic.)

11.17 ● Problems

1. Calculate the yield moment and ultimate moment capacity of the following sections. Assume that the yield stress of the material is 400 N mm^{-2}.

(a) *T* section: Flange = 193 mm, flange thickness = 20 mm, web thickness = 12 mm, overall depth = 234 mm.

Answer:
Area = 6.43×10^3 mm^2, $\bar{y}$ from below = 177 mm, I_{zz} = 31.04 × 10^6 mm^4, Z_{zz} Minimum elastic modulus = 1.754 × 10^5 mm^3, M_{yield} = 70.16 kN m, $\bar{Y}$ plastic from below = 217.3 mm, Plastic modulus = 3.11 × 10^5 mm^3, $M_{plastic}$ = 124.4 kN m, Shape factor = 1.78

(b) A symmetrical *I* section: Flanges = 420 mm, flange thickness = 37 mm, web thickness = 22 mm, overall depth = 920 mm.

Answer:
Area = 49.69 × 10^3 mm^2, $\bar{y}$ from below = 460 mm, I_{zz} = 7.172 × 10^9 mm^4, Z_{zz} Minimum elastic modulus = 15.59 × 10^6 mm^3, M_{yield} = 6236.4 kN m, $\bar{Y}$ plastic from below = 460 mm, plastic modulus = 17.66 × 10^6 mm^3, $M_{plastic}$ = 7063.3 kN m, Shape factor = 1.13

(c) Unsymmetrical *I* section: Top flange = 400 mm, flange thickness = 40 mm, bottom flange = 200 mm, flange thickness = 25 mm, web thickness = 12 mm, overall depth = 800 mm.

Answer:
Area = 29.82 × 10^3 mm^2, $\bar{y}$ from below = 537 mm, I_{zz} = 2.90 × 10^9 mm^4, Z_{zz} Minimum elastic modulus = 5.41 × 10^6 mm^3, M_{yield} = 2164.3 kN m, $\bar{Y}$ plastic from below = 762.7 mm, plastic modulus = 7.30 × 10^6 mm^3, $M_{plastic}$ = 2918.4 kN m, Shape factor = 1.35.

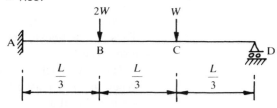

Figure E11.1

2. Determine the plastic collapse load for the propped cantilever shown in Fig. E11.1. Assume constant cross-section properties.

Answer:
(a) Plastic hinges at A and B only with rotation at A of θ:

$$M_p\, \theta + M_p(\theta + 0.5\theta) = 2W\theta\, L/3 + W0.5\theta\, L/3$$

therefore

$$W = 3M_p/L$$

(b) Plastic hinges at A and C only with rotation at A of θ:

$$M_p \theta + M_p(\theta + 2\theta) = 2W\theta\, L/3 + W2\theta\, L/3$$

therefore

$$W = 3M_p/L$$

At collapse, plastic hinges form at A, B and C. This is a case of overcomplete collapse. Reactions are $V_A = 6M_p/L$, $V_D = 3M_p/L$

3. Determine the plastic collapse load of the continuous beam shown in Fig. E11.2.

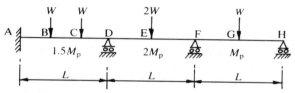

Figure E11.2

Answer:

(a) Plastic hinges at A, B and D with plastic moment = $1.5M_p$ at all the plastic hinges and rotation at A of θ:

$$1.5M_p\,\theta + 1.5M_p(\theta + 0.5\theta) + 1.5M_p(0.5\theta)$$
$$= W\theta\, L/3 + W0.5\theta\, L/3$$

therefore

$$W = 9M_p/L$$

(b) Plastic hinges at A, C and D with plastic moment = $1.5M_p$ at all the plastic hinges and rotation at A of θ:

$$1.5M_p\,\theta + 1.5M_p(\theta + 2\theta) + 1.5M_p(2\theta)$$
$$= W\theta\, L/3 + W2\theta\, L/3$$

therefore

$$W = 9M_p/L$$

(c) Plastic hinges at D, E and F with plastic moment equal to $1.5M_p$, $2M_p$ and M_p at plastic hinges at D, E and F respectively and rotation at D of θ:

$$1.5M_p\,\theta + 2M_p(\theta + \theta) + M_p(\theta) = 2W\theta\, L/2$$

therefore

$$W = 6.5M_p/L$$

(d) Plastic hinges at F and G and a real hinge at H with plastic moment of M_p at all the plastic hinges and rotation at F of θ:

$$M_p\,\theta + M_p(\theta + \theta) = W\theta\, L/2,\ \text{therefore}$$
$$W = 6M_p/L$$

Lowest plastic collapse load is $W = 6M_p/L$. This is a case of partial collapse.

Note if the plastic moment in span ABCD is made equal to M_p, then at the collapse load of $W = 6M_p/L$, plastic hinges form at A, B, C, D, F and G leading to overcomplete collapse.

4. Determine the plastic collapse load for the 2-D rigid-jointed structure shown in Fig. E11.3.

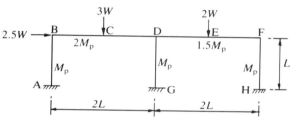

Figure E11.3

Answer:

(a) Beam collapse BCD: $M_p\,\theta + 2M_p(\theta + \theta) + 2M_p\,\theta$
= $3WL\,\theta$, therefore $7M_p\,\theta = 3WL\,\theta$,
$W = 2.33M_p/L$

(b) Beam collapse DEF:

$$1.5M_p\,\theta + 1.5M_p(\theta + \theta) + M_p\,\theta = 2WL\,\theta,$$
therefore $5.5M_p\,\theta = 2WL\,\theta$, $W = 2.75M_p/L$

(c) Sway collapse: $M_p\,\theta + M_p\,\theta + M_p\,\theta + M_p\,\theta + M_p\,\theta$
$+ M_p\,\theta = 2.5WL\,\theta$, therefore $6M_p\,\theta = 2.5WL\,\theta$,
$W = 2.40M_p/L$

(d) Combine (a) + (c) to remove the plastic hinge at B:

$$7M_p\,\theta = 3.0WL\,\theta$$

$$6M_p\,\theta = 2.5WL\,\theta$$

$$\underline{-M_p\,\theta - M_p\,\theta}$$

$$11M_p\,\theta = 5.5WL\,\theta$$

Therefore

$$W = 2M_p/L$$

Mechanism (b) cannot profitably be combined with (d) to reduce plastic work at the hinges.

There are seven plastic hinges in a structure which is six times statically indeterminate. This is a case of complete collapse. From statics $M_{BA} = M_{BC} = 0$. $M_{DE} = M_p$ causing tension at the top and at E the moment is M_p causing tension at the bottom face.

4. Determine the plastic collapse load for the 2-D rigid-jointed structure shown in Fig. E11.4.

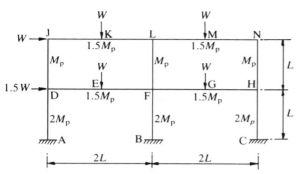

Figure E11.4

Answer:
(a) Beam collapse JKL: $M_p \theta + 1.5M_p(\theta + \theta) + 1.5M_p \theta = WL \theta$, therefore $5.5M_p \theta = WL \theta$, $W = 5.50M_p/L$
(b) Beam collapse LMN:

$$1.5M_p \theta + 1.5M_p(\theta + \theta) + M_p \theta = WL \theta,$$

therefore $5.5M_p \theta = WL \theta$, $W = 5.50M_p/L$
(c) Beam collapse DEF: $1.5M_p \theta + 1.5M_p(\theta + \theta) + 1.5M_p \theta = WL \theta$, therefore $6.0M_p \theta = WL \theta$, $W = 6.0M_p/L$
(d) Beam collapse FGH: $1.5M_p \theta + 1.5M_p(\theta + \theta) + 1.5M_p \theta = WL \theta$, therefore $6.0M_p \theta = WL \theta$, $W = 6.0M_p/L$
(e) Sway collapse – top storey:
$M_p(\theta + \theta + \theta + \theta + \theta + \theta) = WL \theta$,
therefore $6M_p \theta = WL \theta$, $W = 6.0M_p/L$
(f) Sway collapse – bottom storey:
$2M_p(\theta + \theta + \theta + \theta + \theta + \theta) = 1.5WL \theta + WL \theta$,
therefore $12M_p \theta = 2.5WL \theta$, $W = 4.8M_p/L$
Combine mechanisms:

(g) = (f) + (c) and reduce hinges at D by eliminating the plastic hinges in DA and DEF at D but introducing a new hinge in DJ at D.

$$12M_p \theta = 2.5WL \theta$$

$$6M_p \theta = 1.0WL \theta$$

$$-2M_p \theta - 1.5M_p \theta + M_p \theta$$
_____

$$15.5M_p \theta = 3.5WL \theta$$

Therefore

$$W = 4.43M_p/L$$

(h) Combine (g) + (e) and reduce hinges at D and H as there is no need for a hinge at D because the column ADJ can remain straight and at H by eliminating the plastic hinges in HC and HN at H but introducing a new hinge in HGF at H.

$$15.5M_p \theta = 3.5WL \theta$$

$$6.0M_p \theta = 1.0WL \theta$$

$$-M_p \theta - M_p \theta \qquad\qquad \text{at D}$$

$$-2M_p - M_p + 1.5M_p \qquad\qquad \text{at H}$$
_____

$$18.0M_p \theta = 4.5WL \theta$$

Therefore

$$W = 3.40M_p/L$$

(j) Combine (h) + (d) and reduce hinges at F as there is no need for a hinge in column BFL because it can remain straight and also there is no need for a hinge at F in beam FGH but an additional rotation is required at the hinge at F in beam DEF.

$$18.0M_p \theta = 4.5WL \theta$$

$$6.0M_p \theta = 1.0WL \theta$$

$$-M_p \theta - 2M_p \theta - 1.5M_p \theta + 1.5M_p \theta \text{ at F}$$
_____

$$21.0M_p \theta = 5.5WL \theta \text{ therefore } W = 3.82M_p/L$$

(k) Combine (j) + (a) and delete hinge at J

$$21.0M_p \theta = 5.5WL \theta$$

$$5.5M_p \theta = 1.0WL \theta$$

$$-M_p \theta - M_p \theta \text{ at J}$$
_____

$$24.5M_p \theta = 6.5WL \theta$$

Therefore

$$W = 3.77M_p/L$$

(l) Combine (k) + (b) and reduce hinges at L by eliminating the hinges at L in LF and LMN but introducing an additional rotation at L in beam JKL.

$$24.5M_p \theta = 6.5WL \theta$$

$$6M_p \theta = 1.0WL \theta$$

$$-M_p \theta - 1.5M_p \theta + 1.5M_p \theta \text{ at L}$$
_____

$$29.5M_p \theta = 7.5WL \theta$$

Therefore

$$W = 3.93M_p/L$$

Smallest collapse load = $3.77M_p/L$ for mechanism (k) with plastic hinges at A, B, C, E, F in FE, G, H in

HG, K, L in JL and N in NH. Only eleven plastic hinges in a structure twelve times statically indeterminate. This is a case of partial collapse.

11.18 ● Problems for practice

1. Calculate the ultimate load factor for the loads acting on a single-storey, two-bay frame *similar* to that shown in Fig. E11.3. It is given that:
Dimensions:
Spans: BD = 4 m, DF = 8 m, Height: AB = GD = HF = 4 m
Loads:
Vertical loads: 8 kN at C and 16 kN at E
Horizontal load = 8 kN at B
Plastic moment capacities:
Beam BCD = 22 kN m, Beam DEF = 30 kN m, Columns = 10 kN m.

Answer: Load factor = 1.48. Collapse mechanism is a complete collapse by the formation of plastic hinges at A, B, G, H and F in columns and in beams at D in BCD and at E in beam DEF.

2. Calculate the ultimate load factor for the loads acting on a two-storey, two-bay frame *similar* to that shown in Fig. E11.4. It is given that:
Dimensions:
Spans: JL = DF = 12 m, LN = FH = 10 m,
Columns = 5 m
Loads:
Vertical loads: 160 kN at E, 100 kN at G, 60 kN at K and 40 kN at M.
Horizontal load = 30 kN at D and J
Plastic moment capacities:
Beams: DF = 500, JL = 200, FH = 400 and LN = 150 kN m

Columns: AD = 130, DJ = 50, BF = 200, FL = 65, CH = 130 and HN = 50 kN m.

Answer: Load factor = 1.674. Collapse mechanism is a partial collapse by the formation of plastic hinges at D in DA, E, F in FED and FL, K, L in LKJ, H and N.

3. Calculate the ultimate load factor for the loads acting on a two-storey, single-bay frame shown in Fig. E11.5. It is given that:
Plastic moment capacities:
Columns: AC and BE = 165 kN m, all other members = 135 kN m.

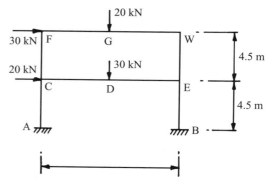

Figure E11.5

Answer: Load factor = 2.11. Collapse mechanism is a partial collapse by the formation of plastic hinges at A, B, F, H in columns and in D and E in beam CDE.

Appendix 1

Theory of elasticity and engineers' theory of stress analysis

In many sections in this book reference has been made to 'exact' solutions and to the approximate nature of the stress and strain analysis of axial and bending stress analysis based on the fundamental assumption that plane sections remain plane before and after stressing. The object of this appendix is to derive the 'exact' equations which need to be solved.

Theory of elasticity

The 'exact' method is called the theory of elasticity. For 2-D problems, the equations of the theory of elasticity can be summarized as follows.

1. Strain displacement relationship. As shown in Section 3.7.1, these are given, using the usual notation, by

$$\varepsilon_x = \partial u/\partial x, \ \varepsilon_y = \partial v/\partial y, \ \gamma_{xy} = \partial u/\partial y + \partial v/\partial x$$

   where $u = u(x, y)$ and $v = v(x, y)$ are the translations of a point in the x- and y-directions respectively.

2. Stress components. Following standard convention, the stress components are σ_x, σ_y and τ_{xy}.

3. Material stress–strain relationship
   (a) Isotropic materials:

   $$\varepsilon_x = \{\sigma_x - \nu\sigma_y\}/E + \alpha T,$$

   $$\varepsilon_y = \{\sigma_y - \nu\sigma_x\}/E + \alpha T, \ \gamma_{xy} = \tau_{xy}/G$$

   where $G = E/\{2(1 + \nu)\}$, T = temperature change at the point, and α = coefficient of linear thermal expansion. Therefore

   $$\sigma_x = \{E/(1 - \nu^2)\}(\varepsilon_x + \nu\varepsilon_y) - E\alpha T/(1 - \nu)$$

   $$\sigma_y = \{E/(1 - \nu^2)\}(\varepsilon_y + \nu\varepsilon_x) - E\alpha T/(1 - \nu)$$

   $$\tau_{xy} = G\gamma_{xy}$$

   Substituting for strains in terms of displacements, we have

   $$\sigma_x = \{E/(1 - \nu^2)\}\{\partial u/\partial x + \nu\partial v/\partial y) - E\alpha T/(1 - \nu)$$

   $$\sigma_y = \{E/(1 - \nu^2)\}\{\partial v/\partial y + \nu\partial u/\partial x) - E\alpha T/(1 - \nu)$$

   $$\tau_{xy} = G(\partial u/\partial y + \partial v/\partial x)$$

(b) Anisotropic materials:

$$\varepsilon_x = \sigma_x/E_x - v_x\sigma_y/E_y + \alpha T,$$

$$\varepsilon_y = \sigma_y/E_y - v_y\sigma_x/E_x + \alpha T, \; \gamma_{xy} = \tau_{xy}/G$$

Therefore

$$\sigma_x = \{E_x/(1 - v_xv_y)\}\{\varepsilon_x + v_x\varepsilon_y\} - E_x\alpha T\{(1 + v_x)/(1 - v_xv_y)\}$$

$$\sigma_y = \{E_y/(1 - v_xv_y)\}\{\varepsilon_y + v_y\varepsilon_x\} - E_y\alpha T\{(1 + v_y)/(1 - v_xv_y)\}$$

$$\tau_{xy} = G\gamma_{xy}$$

where $v_xE_x = v_yE_y$, and E and v denote Young's modulus and Poisson's ratio respectively, with the subscripts, if any, denoting the x- or y-direction.

The stresses can be expressed in terms of displacements using the strain–displacement relationship as was done for isotropic materials.

4. The fundamental conditions to be satisfied for the 'exact' solution are
   (a) Equilibrium. The internal stresses must be in equilibrium with the external loads on the structure.
   (b) Displacements and their first derivatives must be continuous.
   (c) The stress–strain law must be obeyed.
   (d) The prescribed stress and displacement conditions must be satisfied.

5. Equations of equilibrium: The state of stress varies from point to point. Considering an infinitesimal element in a general state of 2-D stress as shown in Fig. A1.1,

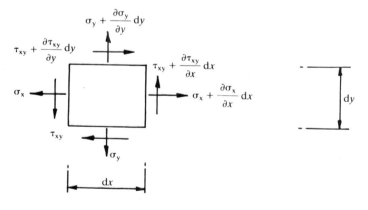

Figure A1.1

for equilibrium in the x-direction

$$[-\sigma_x + \{\sigma_x + (\partial\sigma_x/\partial x)\,dx\}]\,dy + [-\tau_{yx} + \{\tau_{yx} + (\partial\tau_{yx}/\partial y)\,dy\}]\,dx$$
$$+ X\,dx\,dy = 0$$

i.e. $\partial\sigma_x/\partial x + \partial\tau_{yx}/\partial y + X = 0$

$$[-\sigma_y + \{\sigma_y + (\partial\sigma_y/\partial y)\,dy\}]\,dx + [-\tau_{xy} + \{\tau_{xy} + (\partial\tau_{xy}/\partial x)\,dx\}]\,dy$$
$$+ Y\,dx\,dy = 0$$

i.e. $\partial\sigma_y/\partial y + \partial\tau_{xy}/\partial x + Y = 0$ where X and Y are force per unit volume in the x- and y-directions respectively. X and Y are known as body forces because they act on every particle in the body and not just on the surface. Typical body forces are gravitational, inertial and magnetic forces. Another typical body force is pore pressure.

Using the equations expressing the stress components in terms of the derivatives of displacement (for isotropic materials), the equations of equilibrium become

$$\{E/(1 - v^2)\}\{\partial^2 u/\partial x^2 + v\partial^2 v/\partial y\partial x\} - E\alpha\{\partial T/\partial x\}/(1 - v)$$
$$+ G\{\partial^2 u/\partial y^2 + \partial^2 v/\partial x\partial y\} + X = 0$$

$$\{E/(1 - v^2)\}\{\partial^2 v/\partial y^2 + v\partial^2 u/\partial y\partial x\} - E\alpha\{\partial T/\partial y\}/(1 - v)$$
$$+ G\{\partial^2 v/\partial x^2 + \partial^2 u/\partial x\partial y\} + Y = 0$$

If the material is anisotropic, then similar equations can be derived.

The above two equations of equilibrium in terms of the displacement derivatives which satisfy equilibrium and material laws (in this case the isotropic material law) are the fundamental equations of the theory of elasticity. The solution to these equations satisfying the prescribed boundary conditions yields the exact stress and strain distribution. In practice these differential equations are solved by numerical methods such as the finite element method.

The engineers' theory for the determination of axial and bending stress makes the fundamental assumption that plane sections remain plane. This immediately defines the form of displacement and hence strain normal to the plane. The next approximation that the theory makes is to assume either that only a stress normal to the plane exists or that the strain due to a normal stress perpendicular to the stress under consideration can be ignored. Having made these two important assumptions, the theory establishes the value of stress from the overall equilibrium with respect to the stress resultants corresponding to the stress normal to the plane. The reader should go back and study Sections 3.9.1 and 3.9.2 to appreciate how the above assumptions lead to equations such as $\sigma = P/A$ or $\sigma = My/I$. Thus the theory does not consider the general state of stress and strain in the material and makes no attempt to establish equilibrium in terms of the stress components. Naturally whenever the exact state of deformation and stress in a body is close enough to satisfy the assumptions in the engineers' theory, then the exact and approximate stress distributions agree to within acceptable error. This generally happens when the state of stress is reasonably uniform such as uniform axial or bending stress. However, as soon as the presence of concentrated forces or abrupt changes in cross-section, etc. disturb the 'smooth' state of stress, then the divergence between the exact and approximate values increases. The designer should be keenly aware of the situations which make the approximate solutions suspect so that he can obtain the 'exact' state of stress using the finite element method. Some common situations where this can occur were discussed in Section 3.17.

● References

Timoshenko S P, Goodier J N 1970 *Theory of Elasticity* (3rd edn) McGraw-Hill

Appendix 2

Approximate analysis of rigid-jointed frames

In the case of statically indeterminate structures, for a given structural arrangement (topology) and loading, the force at a section depends on the relative stiffness of the members as determined by their cross-section and material properties. Design of these structures is an iterative process. The designer estimates the required section sizes and then analyses the structure to determine the stress distribution. Depending on the resulting stress distribution, the designer may alter the section sizes which in turn will affect the stress distribution and so on. It is useful therefore to have a simple method which allows the prediction to reasonable accuracy of the forces in the members from the analysis of an equivalent statically determinate structure. This appendix describes simple methods for the approximate analysis of statically indeterminate rigid-jointed structures.

Continuous beams

It was shown in Section 4.10 that points of contraflexure in a member carrying a uniformly distributed load or a central point load and continuous at its ends are at approximately span/10 from each end. Once the positions of contraflexure points are fixed, then the support and span bending moments can be determined. Knowing the support moments, the bending moment distribution throughout the structure can be determined. Figure A2.1 shows the value of fixed end moments when only one end or both ends are continuous. The span and support moments are given by

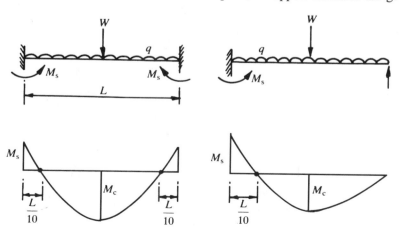

Figure A2.1

Both ends continuous: $M_{\text{support}} = (9/200)qL^2 + WL/20$

$$M_{\text{span}} = qL^2/8 + WL/4 - M_{\text{support}}$$

One end continuous: $M_{\text{support}} = qL^2/20 + WL/18$

$$M_{\text{span}} = qL^2/8 + WL/4 - 0.5M_{\text{support}}$$

In analysing continuous beams, for simplicity it is assumed that the bending moments are zero due to the load on the beam beyond one span at either end from the loaded beam. Figure A2.2 shows the distribution of moments due to load on an interior span and end span. Using the principle of superposition, the bending moment and shear force distribution throughout the structure can be determined.

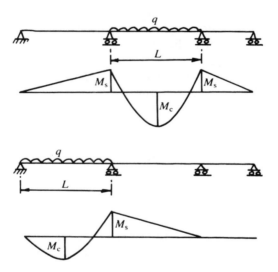

Figure A2.2

Rigid-jointed frames under gravity loading

The bending moment distribution in a single-bay rigid-jointed frame under uniformly distributed gravity loading is shown in Fig. A2.3. From exact analysis, the span and support moments are given by

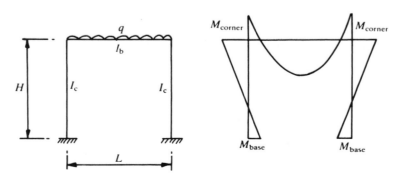

Figure A2.3

$$M_{corner} = \alpha_1 qL^2/12, \quad M_{base} = 0.5M_{corner}, \quad M_{span} = qL^2/8 - M_{corner}$$

where $\alpha = I_{beam}H/(I_{column}L)$, $\alpha_1 = 2/(2 + \alpha)$. Assuming that the point of contraflexure in the beams is at $0.1L$ from the ends and in the column the point of contraflexure is at $0.2H$ from the base, the bending moment distribution in the frame can be determined. The moments are given by

$$M_{corner} = (9/200)qL^2, \quad M_{base} = 0.25M_{corner}$$

In the case of multibay multistorey frames, the same approach can be adopted storey by storey and bay by bay, ignoring, for simplicity, any contribution from other storeys or bays as shown in Fig. A2.4.

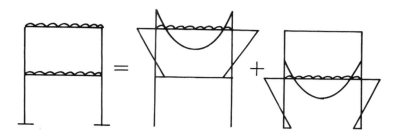

Figure A2.4

Rigid-jointed frames under 'wind' loading

The bending moment distribution in a single-bay rigid-jointed frame under concentrated 'wind' is shown in Fig. A2.5. In the column, the point of contraflexure is approximately at mid-height for fixed-base portals. The contraflexure point in the beam is always at midspan. Therefore, assuming for simplicity that the point of contraflexure is at mid-height of the columns, and the middle of the beam, then the bending moments at the 'joints' are $0.25WH$. Extending these ideas to single-bay multistorey frames, the bending moments in the whole structure can be determined from statics. These ideas can be extended to multistorey, multibay buildings provided some additional assumptions are made about the distribution among the columns in a storey of the total horizontal force from all the storeys above it. Such

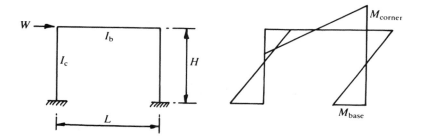

Figure A2.5

approximate methods – called portal and cantilever – are commonly used for the initial sizing of members in a multistorey building.

Example of portal method. In this method it is assumed that points of contraflexure are at mid-height of the columns and at the midspan of the beams. It is assumed that the shear force in the end columns is half that in the interior columns. This method is suitable for frames where the mode of deformation is essentially one of 'shear', as in moderately tall buildings, rather than 'flexural', as in very tall buildings. The two modes of deformation are shown in Fig. A2.6.

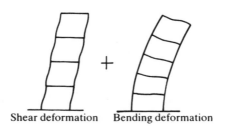

Figure A2.6

Shear deformation   Bending deformation

Consider the simple two-bay two-storey frame shown in Fig. A2.7.

Shear forces in columns. In the top storey, the total shear is W. This is distributed as $0.25W$ to the end columns and $0.50W$ to the central column. Similarly, in the middle storey the total shear is $(W + 2W) = 3W$. This is distributed as $0.75W$ to the end columns and $1.50W$ to the central column. Finally, in the bottom storey, the total shear is $(W + 2W + 2W) = 5W$. This is distributed as $1.25W$ to the end columns and $2.50W$ to the central column.

Calculation of bending moment in beams and columns. Because the points of contraflexure are at mid-height of the columns and beams, the bending moments are calculated as follows.

In the columns, bending moments at the ends of columns are equal to shear in the column × storey height/2. The bending moments in the beams can now be easily obtained so as to maintain equilibrium. Because the points of contraflexure in the beams are assumed at midspan, the bending at the two ends of a beam are equal and opposite.

Axial force in the columns. Knowing the bending moments in the beams, shear force in the beams can be calculated. Using free-body diagrams, axial force in the columns can now be calculated.

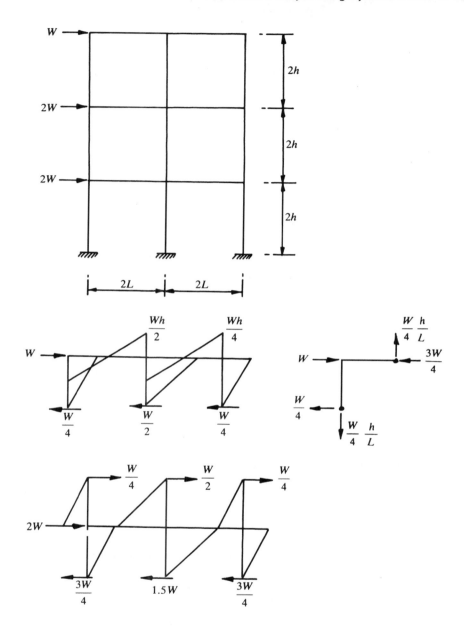

Figure A2.7

● **References**

Behr R A, Goodspeed C H, Henry R M 1989 Potential errors in approximate methods of structural analysis. *Journal of Structural Engineering (ASCE)* **115** (4): 1002–5

Fraser D J 1981 *Conceptual Design and Preliminary Analysis of Structures* Pitman Ch 10

Lin T Y, Stotesbury S D 1981 *Structural Concepts and Systems for Architects and Engineers* Wiley pp. 223–31

Appendix 3

Mohr's circle

In Section 3.3, the basic equations were set up for calculating the normal stress σ_n and shear stress τ_{nt} on a plane whose normal is inclined at an angle θ to the x-axis. In these equations, θ is measured in the anticlockwise direction from the x-axis as shown in Fig. 3.6. The equations are

$$\sigma_n = \frac{\sigma_x + \sigma_y}{2} + \frac{\sigma_x - \sigma_y}{2}\cos 2\theta + \tau_{xy} \sin 2\theta$$

$$\tau_{nt} = -\frac{\sigma_x - \sigma_y}{2}\sin 2\theta + \tau_{xy} \cos 2\theta$$

These equations can be transformed as follows:

Let $\dfrac{\sigma_x - \sigma_y}{2} = R \cos 2\alpha, \ \tau_{xy} = R \sin 2\alpha$

where

$$R = \sqrt{\left\{\left(\frac{\sigma_x - \sigma_y}{2}\right)^2 + \tau_{xy}^2\right\}}$$

$$\tan 2\alpha = \frac{2\tau_{xy}}{\sigma_x - \sigma_y}$$

Substituting in the expressions for σ_n and τ_{nt}, we have

$$\sigma_n = \frac{\sigma_x + \sigma_y}{2} + R \cos 2\alpha \cos 2\theta + R \sin 2\alpha \sin 2\theta$$

$$= \frac{\sigma_x + \sigma_y}{2} + R \cos (2\alpha - 2\theta)$$

Clearly the maximum and minimum values of σ_n are when $\cos (2\alpha - 2\theta)$ is equal to 1 and -1 respectively. $\cos (2\alpha - 2\theta) = 1$ when $(2\alpha - 2\theta) = 0$ or $\theta = \alpha$. Thus the principal normal stresses are

$$\sigma_1 = \frac{\sigma_x + \sigma_y}{2} + R$$

$$\sigma_2 = \frac{\sigma_x + \sigma_y}{2} - R$$

Substituting for R,

$$\sigma_1, \sigma_2 = \frac{\sigma_x + \sigma_y}{2} \pm \sqrt{\left\{\left(\frac{\sigma_x - \sigma_y}{2}\right)^2 + \tau_{yt}^2\right\}}$$

and the angle of the normal on which the principal stresses act is defined by

$$\tan 2\alpha = \frac{2\tau_{xy}}{\sigma_x - \sigma_y}$$

Similarly,

$$\tau_{nt} = - R \cos 2\alpha \sin 2\theta + R \sin 2\alpha \cos 2\theta = R \sin (2\alpha - 2\theta)$$

The maximum value of τ_{nt} is clearly when $\sin (2\alpha - 2\theta) = 1$ or $(2\alpha - 2\theta) = 90°$. Thus

$$\tau_{max} = R = \sqrt{\left\{\left(\frac{\sigma_x - \sigma_y}{2}\right)^2 + \tau_{xy}^2\right\}}$$

The above relationships can be given a graphical interpretation, which is frequently used in the field of soil mechanics. This is known as Mohr's circle and is shown in Fig. A3.1.

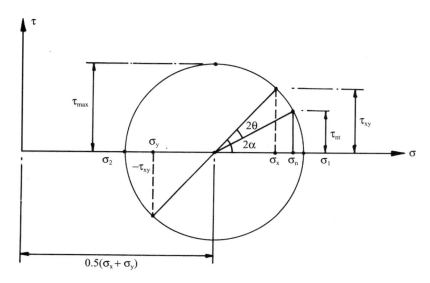

Figure A3.1

Let the normal stress σ be plotted on the horizontal axis and the shear stress τ on the vertical axis. Draw a circle of radius R given by

$$R = \sqrt{\left\{\left(\frac{\sigma_x - \sigma_y}{2}\right)^2 + \tau_{xy}^2\right\}}$$

The centre of the circle is located on the horizontal axis at $0.5(\sigma_x + \sigma_y)$. A point on the circle gives the normal and shear stresses on a plane whose normal is inclined at an angle θ to the x-axis. The points where the circle cuts the horizontal axis clearly give the values of the principal stress, and it is clear that the shear stress on these planes is zero. Similarly the planes on which the maximum shear stress acts also has a normal stress equal to $0.5(\sigma_x + \sigma_y)$.

Macaulay brackets

It was noticed in the examples on cantilever beams in Section 1.11.1, that when the load on the beam is discontinuous, it was necessary to write expressions for BM and SF which were valid only for specific sections. This can be avoided by using a convention called *Macaulay brackets*. The convention is based on the use of Step function in the Laplace Transform method in mathematics. The convention is as follows.

$$<x - a>^n = 0 \text{ if } x \le a$$
$$= (x - a)^n \text{ if } x > a$$

The brackets $< \ >$ are known as *Macaulay brackets*.
Using SF for shear force and BM for bending moment, if for example:

(a) A cantilever is subjected to a concentrated load W at $x = a$, then the SF and BM due to W can be expressed as follows.
SF $= 0$ for $x \le a$ but SF $= W$ for $x > a$, we can write a single expression as
SF $= <x - a>^0$
Similarly if
BM $= 0$ for $x \le a$ but BM $= W(x - a)$ for $x > a$, we can write a single expression as
BM $= W<x - a>^1$

(b) A cantilever is subjected to a uniformly distributed load q from $x = a$ onwards, then the SF and BM due to q can be expressed as follows.
SF $= 0$ for $x \le a$ but SF $= q(x - a)$ for $x > a$, we can write a single expression as
SF $= q<x - a>^1$
Similarly if
BM $= 0$ for $x \le a$ but BM $= q(x - 2)^2/2$ for $x > a$, we can write a single expression as
BM $= q<x - a>^2/2$

(c) If a uniformly distributed load q acts from $x = a$ to $x = b$ only, then due to q,
SF $= 0$ for $x \le a$
SF $= a(x - a)$ for $a < x \le b$
SF $= q(b - a)$ for $x > b$
we can write
SF $= q<x - a>^1 - q<x - b>^1$

Clearly both expressions are zero for $x \leq a$, only the second expression is zero for $a < x \leq b$ and finally both expressions are valid for $x > b$.

Clearly $q(x - a) - q(x - b) = q(b - a)$.

Similarly if

$$BM = 0 \text{ for } x \leq a$$
$$BM = q(x - a)^2/2 \quad \text{for} \quad a < x \leq b$$
$$BM = q(b - a)[x - 0.5(a + b)] \quad \text{for } x > b$$

we can write

$$BM = q<x - a>^2/2 - q<x - b>^2/2$$

Note that

$$q(x - a)^2/2 - q(x - b)^2/2 = 0.5(x^2 - 2ax + a^2 - x^2 + 2bx - b^2)$$
$$= x(b - a) + 0.5(a^2 - b^2).$$

Since $(a^2 - b^2) = (a + b)(a - b)$

$$q(x - a)^2/2 - q(x - b)^2/2 = q(b - a)[x - 0.5(a + b)]$$

The above manipulation also shows that when integrating, the expression inside the Macaulay brackets is treated as a single entity. Thus

$$\int <x - a>^n \, \mathrm{d}x = 0 \quad \text{if } x \leq a$$
$$= \{(x - a)^{n+1}\}/(n + 1) \quad \text{if } x > a$$

The use of this convention assists greatly the calculation of SF, BM and deformation in beams.

Index